Lecture Notes in Control and Information Sciences 440

For further volumes:
http://www.springer.com/series/642

Lecture Notes in Control and Information Sciences 440

Editors

Professor Dr.-Ing. Manfred Thoma
Institut fuer Regelungstechnik, Universität Hannover, Appelstr. 11, 30167 Hannover, Germany
E-mail: thoma@irt.uni-hannover.de

Professor Dr. Frank Allgöwer
Institute for Systems Theory and Automatic Control, University of Stuttgart, Pfaffenwaldring 9, 70550 Stuttgart, Germany
E-mail: allgower@ist.uni-stuttgart.de

Professor Dr. Manfred Morari
ETH/ETL I 29, Physikstr. 3, 8092 Zürich, Switzerland
E-mail: morari@aut.ee.ethz.ch

For further volumes:
http://www.springer.com/series/642

Bijnan Bandyopadhyay, S. Janardhanan,
and Sarah K. Spurgeon (Eds.)

Advances in Sliding Mode Control

Concept, Theory and Implementation

Springer

Editors
Prof. Bijnan Bandyopadhyay
Interdisciplinary Programme in Systems and Control Engineering
Indian Institute of technology Bombay
Powai
Mumbai
Maharashtra
India

Prof. Sarah K. Spurgeon
Department of Electronics
University of Kent
Kent
United Kingdom

Dr. S. Janardhanan
Control and Automation Group
Dept. of Electrical Engineering
Indian Institute of Technology Delhi
Hauz Khas
New Delhi
India

ISSN 0170-8643 ISSN 1610-7411 (electronic)
ISBN 978-3-642-36985-8 ISBN 978-3-642-36986-5 (eBook)
DOI 10.1007/978-3-642-36986-5
Springer Heidelberg New York Dordrecht London

Library of Congress Control Number: 2013933024

Printed on acid-free paper

Springer is part of Springer Science+Business Media (www.springer.com)

यदा ते मोहकलिलम् बुद्धिर्व्यतितरिष्यति।
तदा गन्तासि निर्वेदम् श्रोतव्यस्य श्रुतस्य च

Bhagavad Gita (2:52)

Transliteration : *yadA te moha-kalilam buddhir-vyatitariShyati taDa gantAsi nirvedam shrotavyasya shrutasya ca*

Translation : *When your spiritual intelligence overcomes this myriad of delusion at that time you will become indifferent towards all that has been heard and all that is to be heard.*

..in other words, a intelligently designed *(sliding mode)* control system, with the design being capable of rejecting uncertainties, would be able to become invariant to perturbations both known and unknown.

Preface

It has been well established that the variable structure and sliding mode control design methodologies are appropriate for robust control. This control design framework has many attractive features including the ability to counteract the effect of uncertainties and disturbances which are inevitable in most practical systems. Sliding mode control is a particular class of variable structure control which was first introduced by Emel'yanov and his co-workers. The sliding mode control paradigm has now become a mature technique for the design of robust controllers for a wide class of systems including nonlinear, uncertain and time-delayed systems.

This book is a collection of plenary and invited talks delivered at the *12th IEEE International Workshop on Variable Structure System* held at the Indian Institute of Technology, Mumbai, India in January 2012. The workshop organisers, together with the IEEE CSS Technical Committee on Variable Structure and Sliding Mode Control, invited leading international researchers to present plenary and invited talks at VSS 2012 in order to articulate the current state of the art both in terms of theory and practice in the discipline. After the workshop, these researchers were invited to develop book chapters for this edited collection in order to reflect the latest results and open research questions in the area.

The contributed chapters have been organised by the editors to reflect the various themes of sliding mode control which are the current areas of theorerical research and applications focus; namely articulation of the fundamental underpinning theory of the sliding mode design paradigm, sliding modes for decentralised system representations, control of time-delay systems,the higher order sliding mode concept, results applicable to nonlinear and underactuated systems, sliding mode observers, discrete sliding mode control together with cutting edge reseach contributions in the application of the sliding mode concept to real world problems.

The structure of the book is as follows:

The first chapter of the book has been authored by B. Draženović, Č. Milosavljević and B. Veselić. This chapter considers the design of reduced and integral sliding mode dynamics for state space systems. The resulting sliding mode dynamics have either a desired spectrum or optimal behavior in the linear quadratic regulator sense. The sliding subspaces obtained facilitate a fully decentralized solution of the reachability problem to determine the control which will ensure the sliding mode is attained. To facilitate analysis of the sliding mode performance, a new way to determine the dynamics, based on singular value decomposition, is also provided in the chapter.

Chapter 2, authored by V. Utkin and A. Poznyak, addresses the problem of chattering in sliding mode control using an adaptive sliding mode control concept. It is shown that adaptation based on the application of the equivalent control enables the control action magnitude to be reduced to a minimum possible value whilst keeping the property of finite-time convergence.

Chapter 3, dealing with the variable structure control of a class of multiple time varying delay interconnected systems with nonlinear disturbances, has been authored by Xing-Gang Yan and Sarah K. Spurgeon. A decentralised static output feedback variable structure control is synthesised, which is independent of the time delays, to stabilise the system globally uniformly asymptotically. A case study relating to a river pollution control problem is presented to illustrate the proposed approach.

Chapter 4, written by Y. Orlov, A. Pisano and E. Usai, addresses the Lyapunov-based design of distributed and boundary second order sliding mode controllers in the domain of distributed parameters systems. The proposed robust synthesis of the distributed control input is formed by linear PI-type feedback design and the "Super-Twisting" second-order sliding-mode control algorithm, suitably combined and re-worked in the infinite-dimensional setting.

The concept of practical relative degree and its relevance in sliding mode control is introduced in *Chapter 5*, by A. Levant. In this chapter, the notion of practical relative degree is proposed, which generalizes the standard relative-degree notion for the cases of uncertain systems for which an appropriate mathematical model is unavailable and thus for which the classical relative degree cannot be computed. Practical output regulation is ensured. Computer simulation and practical results confirm the efficacy of the theoretical approach.

The problem of accurate tracking of unmatched perturbed outputs using higher order sliding modes is tackled in *Chapter 6*, authored by L. Fridman, A. Estrada and A. Ferreira. Classical sliding mode theory cannot easily handle unmatched perturbations. In this chapter, a hierarchial design approach using integral HOSM is presented which addresses the above problem. Appropriate examples validate the proposed approach.

Chapter 7 has been authored by Prasiddh Trivedi and B. Bandyopadhyay. It focuses on a higher order sliding mode control strategy which uses an appropriately designed 2-sliding constraint which achieves finite time convergence for an integrator chain. The twisting controller is used for achieving a finite time convergent 2-sliding mode to the switching manifold. The fractional powers in the switching function are carefully designed to prevent the unboundedness, or singularity, arising because of the switching constraint being kept at zero.

In *Chapter 8*, Michael Basin and Pablo Rodriguez-Ramirez address the mean-square and mean-module filtering problems for stochastic polynomial systems subject to Gaussian white noise. The obtained solution contains a sliding mode term, signum of the innovations process. The designed sliding mode mean-module filter generates the mean-module estimate, which yields a better value of the mean-module criterion in comparison to the conventional polynomial mean-square filter. The theoretical results are complemented with illustrative examples verifying the performance of the designed filters.

Chapter 9, written by Luis T. Aguilar, Igor Boiko, Leonid Fridman, and Rafael Iriarte reviews a tool for the design of a periodic motion in an underactuated system via generating a self-excited oscillation of a desired amplitude and frequency driven by a variable structure control. Focussing on systems with one degree of underactuation, the chapter overviews the capabilities of the two-relay controller to induce oscillations in dynamical systems. Three methods to set the frequency and amplitude of oscillation are reviewed: the describing function method, Locus of the perturbed relay system design (LPRS), and Poincaré map based design. Theoretical and practical open problems are also discussed.

Chapter 10, written by D. Fulwani and B. Bandyopadhyay, presents a method for the design of the sliding surface with the constraint of actuator saturation. Here it is shown that the control amplitude can be regulated by a single parameter ε. Representative simulation results are included to illustrate the procedure.

Chapter 11 authored by Yan Yan and Xinghuo Yu contains a study of the quantization behaviour of the equivalent control based second order single input sliding mode control using quantized state feedback. The authors show that the class of sliding mode control (SMC) system with both uniform and logarithmic quantizers can make the system states converge into a band which can be expressed in terms of the quantization parameters. Several simulation results are presented to support the study.

Jaime Moreno proposes a unified method to design a class of discontinuous observers for second order systems in *Chapter 12*. This method generalizes and improves several other known approaches such as the high-gain observer, the super twisting observer and the uniform differentiator. Here the method is restricted to second order systems due to the fact that the proofs are Lyapunov function based.

A new method for the design of a functional observer authored by S. Janardhanan and N. Satyanarayana is presented in *Chapter 13*. The authors show that with the proposed method the number of output samples required is less than the observability index, which is in turn the minimum value for computation of the full state. The design approach is thus attractive as it potentially reduces the number of output samples required.

In *Chapter 14*, Elisabetta Punta proposes a full order observer for a class of nonlinear nonaffine systems by using only discrete measurements of the output. Convergence to the unique solution is established. The proposed methodology introduces integrators in the input channel and combines a sliding mode and Luenberger-like observer.

A. Levant and M. Livne show in *Chapter 15* that, unlike the case of homogeneous sliding mode based differentiators which provide high accuracy and robust finite-time exact estimation of derivatives, their discrete-time implementation loses the homogeneity and features decreased accuracy with respect to the sampling time.

The chapter also includes an analysis of the asymptotic accuracy of the differentiator for cases of both constant and variable sampling.

Chapter 16, authored by H. Imine and L. Fridman, presents a method for the prediction of the rollover risk for heavy vehicles. The results are validated experimentally. The method is based on the Load Transfer Ratio which depends on the vertical force estimated by a higher order sliding mode observer. The validation tests described were carried out on an instrumented truck rolling on the road at various speeds and undertaking lane change maneuvers.

In *Chapter 17*, C. Edwards, Halim Alwi and P. P. Menon address the application of a second order sliding mode observer scheme to the Advanced Fault Diagnosis for Sustainable Flight Guidance and Control (ADDSAFE) benchmark problem and a satellite formation flying problem. Two different fault detection and diagnosis (FDD) problems have been considered. Firstly, the detection and isolation problem associated with an actuator jam/runaway and secondly, an oscillatory failure case (OFC) scenario associated with aileron actuators. Simulation results based on the full nonlinear model of the ADDSAFE aircraft are presented.

The final chapter, authored by P.Ignaciuk and A. Bartoszewicz, deals with the application of sliding mode control concepts to the field of logistics. A discrete sliding mode (DSM) control policy for periodic-review inventory systems with fixed order quantity and uncertain demand is designed. The underlying algorithm employs only the fundamental arithmetical and logical operations to compute the moment at which the ordering decision shiould be taken.

We hope that this book provides the reader with a clear and complete picture of the current trends in Variable Structure Systems and Sliding Mode Control Theory.

We would like to take this opportunity to thank all the authors for their contributions and suggestions which helped in the successful compilation of this book. Our thanks also go to the 'Sliding Mode Control' research group of SYSCON, IIT Bombay, for their valuable help in the proof-reading process.

Mumbai, New Delhi, Kent
January, 2013

Bijnan Bandyopadhyay
S. Janardhanan
Sarah K. Spurgeon

Contents

List of Contributors

L. Aguilar
Instituto Politécnico Nacional, Ave. del parque 1310 Mesa de Otay,
Tijuana Baja California 22510
e-mail: laguilar@citedi.mx

Halim Alwi
Control and Instrumentation Group, Department of Engineering,
University of Leicester
e-mail: ha18@le.ac.uk

Bijnan Bandyopadhyay
Indian Institute of Technology Bombay, India
e-mail: bijnan@ee.iitb.ac.in

Andrzej Bartoszewicz
Institute of Automatic Control, Lodz University of Technology,
18/22 Stefanowskiego St., 90-924 Łódź
e-mail: andrzej.bartoszewicz@p.lodz.pl

Michael Basin
Department of Physical and Mathematical Sciences,
Autonomous University of Nuevo Leon, San Nicolas de los Garza,
Nuevo Leon, Mexico
e-mail: mbasin@fcfm.uanl.mx, mbasin2007@gmail.com

I. Boiko
Electrical Engineering Department, The Petroleum Institute, P.O. BOX 2533,
Abu Dhabi, U.A.E.
e-mail: i.boiko@ieee.org

Branislava Draženović
University of Sarajevo, Faculty of Electrical Engineering, Sarajevo,
Bosnia and Herzegovina
e-mail: brana_p@hotmail.com

Christopher Edwards
Center for Systems, Dynamics and Control, CEMPS, University of Exeter
e-mail: c.edwards@exeter.ac.uk

Antonio Estrada
LUNAM Universite, Ecole Centrale de Nantes,
IRCCyN UMR CNRS 6597, Nantes, France
e-mail: xheper@yahoo.com

Alejandra Ferreira de Loza
University of Bordeaux, IMS-Lab, Automatic Control Group,
351 Cours de la Libération, 33405 Talence, France
e-mail: da_ferreira@yahoo.com

Leonid Fridman
Department of Control, Division of Electrical Engineering, Engineering Faculty,
National Autonomous University of Mexico, UNAM, Ciudad Universitaria,
04510, Mexico, D.F.
e-mail: lfridman@unam.mx

Deepak Fulwani
Indian Institute of Technology Jodhpur, India
e-mail: fulwani@gmail.com

Przemyslaw Ignaciuk
Institute of Information Technology, Lodz University of Technology,
215 Wólczańska St., 90-924 Łódź
e-mail: przemyslaw.ignaciuk@p.lodz.pl

Hocine Imine
LEPSIS-IFSTTAR, Laboratory for road operation, perception, simulators and simulations, The French institute of science and technology for transport, development and networks, Paris, 75732, France
e-mail: hocine.imine@ifsttar.fr

R. Iriarte
Instituto Politécnico Nacional, Ave. del parque 1310 Mesa de Otay,
Tijuana Baja California 22510
e-mail: ririarte@citedi.mx

S. Janardhanan
Department of Electrical Engineering, Indian Institute of Technology Delhi, India
e-mail: janas@ee.iitd.ac.in

Arie Levant
Tel-Aviv University, 69978 Tel-Aviv, Israel
e-mail: levant@post.tau.ac.il

Miki Livne
Tel-Aviv University, Ramat-Aviv, 69978 Tel-Aviv, Israel
e-mail: miki.livne@gmail.com

Prathyush P Menon
Center for Systems, Dynamics and Control, Mathematical Research Institute, University of Exeter
e-mail: P.M.Prathyush@exeter.ac.uk

Čedomir Milosavljević
University of Istočno Sarajevo, Faculty of Electrical Engineering, Istočno Sarajevo, Bosnia and Herzegovina
e-mail: cedomir.milosavljevic@elfak.ni.ac.rs

Jaime A. Moreno
Eléctrica y Computación, Instituto de Ingeniería, Universidad Nacional Autónoma de México, 04510 Mexico D.F., Mexico
e-mail: JMorenoP@ii.unam.mx

Y. Orlov
CICESE Research Center, Ensenada, Mexico
e-mail: yorlov@cicese.mx

A. Pisano
Universitá degli studi di Cagliari, Cagliari, Italy, Dipartimento di Ingegneria Elettrica ed Elettronica (DIEE)
e-mail: pisano@diee.unica.it

Elisabetta Punta
National Research Council of Italy, Institute of Intelligent Systems for Automation (CNR-ISSIA), Via De Marini, 6 - 16149 Genoa, Italy
e-mail: elisabetta.punta@cnr.it

Alexander S. Poznyak
Dept. Control Automatico, CINVESTAV-IPN, Mexico
e-mail: apoznyak@ctrl.cinvestav.mx

Pablo Rodriguez-Ramirez
Department of Physical and Mathematical Sciences, Autonomous University of Nuevo Leon, San Nicolas de los Garza, Nuevo Leon, Mexico
e-mail: pablo.rodriguezrm@uanl.edu.mx

Neeli Satyanarayana
Department of Electrical Engineering, Indian Institute of Technology Delhi, India
e-mail: neeli.satya@gmail.com

Sarah K. Spurgeon
Instrumentation, Control and Embedded Systems Research Group,
School of Engineering and Digital Arts, University of Kent,
Canterbury, Kent CT2 7NT, United Kingdom
e-mail: S.K.Spurgeon@kent.ac.uk

Prasiddh Trivedi
IDP in Systems and Control Engineering, IIT Bombay, Mumbai, India
e-mail: prasiddh@sc.iitb.ac.in

E. Usai
Universitá degli studi di Cagliari, Cagliari, Italy,
Dipartimento di Ingegneria Elettrica ed Elettronica (DIEE)
e-mail: eusai@diee.unica.it

Vadim I. Utkin
Department of Electrical Engineering, The Ohio State University, USA
e-mail: utkin@ece.osu.edu

Boban Veselić
University of Niš, Faculty of Electronic Engineering, Niš, Serbia
e-mail: boban.veselic@elfak.ni.ac.rs

Xing-Gang Yan
Instrumentation, Control and Embedded Systems Research Group,
School of Engineering and Digital Arts, University of Kent, Canterbury,
Kent CT2 7NT, United Kingdom
e-mail: X.Yan@kent.ac.uk

Yan Yan
School of Information Science and Technology, Dalian Maritime University,
116026, Dalian, Liaoning, China
e-mail: yy_spring@163.com

Xinghuo Yu
School of Electrical and Computer Engineering, RMIT University,
Melbourne, VIC 3000, Australia, and School of Automation,
Southeast University, Nanjing, China
e-mail: x.yu@rmit.edu.au

Chapter 1
Comprehensive Approach to Sliding Mode Design and Analysis in Linear Systems

Branislava Draženović, Čedomir Milosavljević, and Boban Veselić

Abstract. This chapter considers the design of reduced and integral sliding mode (SM) dynamics for state space systems. The prescribed sliding mode dynamics are selected to have either a desired spectrum or optimal behavior in the linear quadratic regulator (LQR) sense. Due to the operator representation of the system equations, separate treatment of the discrete time (DT) and the continuous time (CT) cases is not needed. Fully decentralized design of the control used to satisfy the reachability problem is possible using the obtained sliding subspaces. For the sake of straightforward analysis of the SM dynamics, a new way to obtain the SM equation, based on singular value decomposition (SVD), is also provided. Algorithms are implemented in MATLAB. Simulations illustrating the usefulness of the developed design method conclude the chapter.

1.1 Introduction

In this section we first review various methods used for the design of linear sliding subspaces for LTI MIMO systems. The next subsection explains the main motivation for this chapter: to enable straightforward synthesis of the sliding subspace, coupled with a rapid analysis of the SM motion, by constructing both new algorithms and

Branislava Draženović
University of Sarajevo, Faculty of Electrical Engineering, Sarajevo,
Bosnia and Herzegovina
e-mail: brana_p@hotmail.com

Čedomir Milosavljević
University of Istočno Sarajevo, Faculty of Electrical Engineering, Istočno Sarajevo,
Bosnia and Herzegovina
e-mail: cedomir.milosavljevic@elfak.ni.ac.rs

Boban Veselić
University of Niš, Faculty of Electronic Engineering, Niš, Serbia
e-mail: boban.veselic@elfak.ni.ac.rs

B. Bandyopadhyay et al. (Eds.): *Advances in Sliding Mode Control*, LNCIS 440, pp. 1–19.
DOI: 10.1007/978-3-642-36986-5_1

simplified forms of existing algorithms that are valid for both discrete and continuous LTI systems. The ease of MATLAB implementation is a key requirement.

1.1.1 Previous Approaches

The basic idea of variable structure control with a SM, applied first to CT systems, is well established. Briefly, the control first drives the state to a sliding subspace. Once the state is in the subspace, the control enforces the state to stay in the subspace. Such motion is defined as the reduced order SM. An appropriate design provides beneficial features to the SM motion such as desired dynamics, suppression of disturbances, optimal behavior, robust stability etc. The advance of digital systems motivated the extension of this approach to DT systems, where samples of the state remain in the given sliding subspace.

CT and DT integral, or full order, SM systems were introduced later. In these systems a set of integrators is connected to the controlled system, and the sliding subspace is defined in terms of both the system states and the integrator outputs. The incentive to introduce integral systems was to remove the reaching phase by adjusting the initial values of the output of the integrators. The four basic types of SM considered here are: reduced order CT, reduced order DT, CT integral and DT integral SM.

The two main issues in the design of a control for systems with a SM are: how to determine an appropriate subspace to achieve the desired motion in the SM and how to design the control that guarantees the subspace is first attained and maintained, to ensure a lasting SM motion.

The topic of this chapter is design of the sliding subspace for SM control in LTI MIMO systems where the full state is available. Many papers have been devoted to this issue. They differ in the design aims, the type of SM, and, of course, in the approach to the design. A brief review of these papers follows. Consider first CT controllable systems with a reduced order SM represented by

$$\dot{x}(t) = Ax(t) + Bu(t), x \in \Re^n, u \in \Re^m, A \in \Re^{n \times n}, B \in \Re^{n \times m}, \text{rank}B = m, \quad (1.1)$$

where the design aim is to ensure a given spectrum or optimal behaviour in the SM.

First design of a SM motion in the sliding subspace, $g(t) = Cx(t) = 0$, of order $(n-m)$ was proposed in [1]. In that paper a nonsingular transformation of states $\hat{x} = Mx$, $\hat{x} \in \Re^n$ is applied first to obtain the so-called regular form, where nonzero elements of the transformed matrix B are in the last m rows only. The structure of the subspace matrix C in the transformed system ensures (CB) is full rank. For sliding subspace design, the last m equations of the transformed model are dropped, and the last m states denoted as $\hat{x}_2(t)$ are expressed in terms of the first $(n-m)$ states $\hat{x}_1(t)$. This procedure creates the following pair of equation in the SM

$$\begin{aligned} \dot{\hat{x}}_1(t) &= A_{11}\hat{x}_1(t) + A_{12}\hat{x}_2(t) \\ \hat{x}_2(t) &= -C_1\hat{x}_1(t) \end{aligned} \quad (1.2)$$

where $\hat{x}_2(t)$ plays the role of a virtual control vector and the matrix C_1 is a feedback matrix. The proper choice of this matrix may fulfill two design goals. The first goal is a given spectrum in the SM and the second aim is to have optimal behavior in the SM in the LQR sense. The first aim is achieved by using a pole placement method to find C_1. In the second aim the importance of the states $\hat{x}_1(t)$ and $\hat{x}_2(t)$ for the system is defined by matrices Q and R. The required value of C_1 is calculated by a classical LQR approach. This regular form approach provides a valid design, but requires many steps to obtain the result.

In the next group of papers, the right eigenvectors are used to obtain a sliding subspace. In [[2]] and [[3]] assignment of right eigenvectors was based on projector matrices. In [[4]] eigenvector construction using Kautsky's algorithm [[5]] was employed to obtain a robust pole assignment with feedback K. Then the matrix C should satisfy two matrix equations: $CB = Z$ and $CV = 0$, where Z is a chosen nonsingular matrix, and the matrix V has as columns $(n-m)$ selected right eigenvectors of the matrix $(A-BK)$.

The main point of interest in papers [[6]], [[7]] and [[8]] is how to obtain the feedback matrix K. Matrix $(A-BK)$ in this approach should have $(n-m)$ given desired eigenvalues, and m arbitrary stable real eigenvalues. The sliding subspace is spanned by the desired right eigenvectors. In [[6]] the arbitrary eigenvalues are all different, while in [[7]] they are all equal to a real value resulting in a closed form expression of C. This idea was exploited in [[8]] for the DT system:

$$x(k+1) = Ax(k) + Bu(k). \tag{1.3}$$

Again state feedback $u = -Kx$ should provide a given spectrum which has $(n-m)$ desired eigenvalues, while the other m eigenvalues are all equal to real number λ which may not be an eigenvalue of A. Then, the sliding subspace matrix is obtained in closed form as $C = K(A - \lambda I_n)^{-1}$.

The paper by [[9]] extends the application of Ackermann's formula, proposed first for SISO systems in [[10]], to MIMO systems.

The integral SM was introduced first for SISO systems, in order to eliminate the reaching phase. An integrator is added to the system, and the switching function encompasses both the state variables and integrator output. The integrator output is adjusted so that motion starts within the sliding subspace. This idea may be applied to MIMO systems, where the number of added integrators equals the number of controls. An explicit formula for CT linear integral MIMO systems is not available, although in [[11]] a recommendation how to obtain this has been given. In [[12]], DT full order SM was considered. The main result consists of a set of formulas defining the switching functions g_k, integrator outputs z_k, and matrix E defining the discrete integrator's outputs for a matrix D satisfying $\mathrm{rank}(DB) = m$

$$\begin{aligned} g_k &= Dx_k - Dx(0) + z_k \\ z_k &= z_{k-1} + Ex_{k-1} \\ E &= -D(A - I + BK) \end{aligned} \tag{1.4}$$

Here K provides a given spectrum to the matrix $(A-BK)$.

The second subsection explains the motivation in more detail and describes the chapter content.

1.1.2 Motivation

As may be seen above, the sliding subspace design for four types of SM were treated completely separately. For a control engineer wishing to implement SM control, identifying a suitable method, understanding the theory, encoding the design, and then checking the design by simulation may not appear straightforward. The main objective of this chapter is to both make application of the SM method more popular in the wider control community by offering simple solutions to effect sliding subspace design in all four cases, and also to enable rapid analysis of the SM system using many of the control system software design tools available for LTI systems. To achieve this goal some innovations are introduced, and some new algorithms are proposed.

First, the design formulas derived for CT and DT SMs show a striking similarity. The operator notation used in [5] is adopted for the first time in this chapter. The application of this notation results in design formulas valid for both CT and DT systems. Last, but not the least it eliminates the need to use brackets and subscripts to represent state variables.

This chapter exploits the similarity of the SM equations to the equations of first integrals. The proposed designs have two stages. In the first stage a linear state feedback is found, such that there exists a first integral of the closed loop system with desired properties. Then the motion of the system is restricted to a subspace by applying a SM control. The subspace equation at the same time defines the first integral of closed loop system. This approach permits the required calculations to be achieved by a few programming statements.

Although the design of the reaching control is not in the scope of this chapter, a sliding subspace is determined so that a fully decentralized reaching control is possible. Each control component effectively annihilates in finite time one switching function by providing a proper sign of the switching function increment, and thus there is no need to verify stability of the reaching and sliding phase.

The last phase of each design is its verification. One of the very important issues in each system is the sensitivity to disturbances. If there are unmatched disturbances in the system, they will affect the motion in the SM and their impact must be assessed. There are many tools to handle this problem, but the model must be in state space form. A novel coordinate transform is introduced, based on SVD, which needs only a couple of MATLAB standard statements to obtain a state space model having as an input additive disturbances.

This model can be used in an iterative design of the system with unmatched additive disturbance. Since the formulas are simple, and simulations are not necessary, a repetitive design procedure can be devised so that the desired spectrum or matrices defining optimal behaviour are systematically modified in order to improve the sensitivity to specific type of disturbances.

The two design aims considered in this chapter are that of achieving a given spectrum or an optimal behavior in the LQR sense in the SM. Four brief algorithms cover the design procedure. The first two algorithms determine the SM subspace which has a given dynamic for both reduced order and integral SM respectively. Two more algorithms determine the sliding subspace for an optimal SM motion for the reduced order and integral SM.

The description of the content of the rest of the chapter is as follows. In Section 1.2 the operator notation taken from [5] is used to derive a common model of the reduced order and full order SM. In Section 1.3 the algorithms for achieving a given spectrum in reduced and integral SM are presented as well as the design for optimal behavior in SM. The derivation of the new state space model is the topic of Section 1.4. Section 1.5 contains examples including MATLAB codes and some simulations. The conclusion ends the chapter.

1.2 Common Model of Continuous and Discrete-Time SM Dynamics

Both CT (1.1) and DT (1.3) linear time invariant systems are described by the same equation

$$\delta x = Ax + Bu. \tag{1.5}$$

In CT systems the state $x \in \Re^n$ is a function of continuous time t denoted as $x(t)$, where δ represents the differential operator d/dt. In DT systems, the state x is a function of discrete time k, denoted as x_k while δ denotes the forward shift operator. The control is represented by $u \in \Re^m$, while $A \in \Re^{n\times n}$ and $B \in \Re^{n\times m}$ are system matrices. In the following it will be assumed that rank$(B) = m$, while (A,B) is a controllable pair. The derivations of the SM equations of reduced order and full order follow.

It is assumed that in the reduced order case, a suitable switching type of control places the SM in a subspace defined by the sliding subspace matrix C. SM motion is defined by a pair of equations appended to (1.5): The equation (1.6) says that the state at a moment belongs to a subspace, and (1.7) states that in its further motion it stays in the same subspace:

$$g = Cx = 0, \tag{1.6}$$

$$\delta g = C\delta x = C(Ax + Bu) = 0. \tag{1.7}$$

Here g denotes the so-called vector switching function. The matrix (CB) must have full rank m, and therefore matrix C must also have full rank as well. The solution of (1.7) for u leads to the equivalent control u_{eq} given by

$$u_{eq} = -(CB)^{-1}CAx = -Kx \tag{1.8}$$

where $K \in \Re^{m\times n}$ is referred to as the equivalent feedback matrix. In the equivalent system (1.9), the real switching control is replaced by the equivalent control. Eqs. (1.9) and (1.10) together define a linear system of order $(n-m)$

$$\delta x = (A - BK)x, \tag{1.9}$$

$$Cx = 0. \tag{1.10}$$

To obtain an integral SM, m integrators having outputs represented by the vector σ are connected to the system (1.5). This makes an extended system of order $(n+m)$. The integrator inputs are equal to Ex. As proposed in [12], the SM occurs in the subspace given by

$$g = Dx + \sigma = 0 \tag{1.11}$$

provided DB is a nonsingular matrix.

Since the models of CT and DT integrators differ in the adopted operator notation, the application of the equivalent control concept leads to two different expressions for CT and DT full order SM. CT integrators connected to system (1.5) are represented by

$$\delta\sigma = Ex, \tag{1.12}$$

while DT integrators are modeled as

$$\delta\sigma = Ex + \sigma \tag{1.13}$$

The introduction of a qualifier q, which is equal to 1 for DT systems and to 0 for CT systems, results in an unique triple of equations describing the SM motion of the extended system

$$\delta x = (A - BK)x, \tag{1.14}$$

$$\sigma + Dx = 0, \tag{1.15}$$

$$K = (DB)^{-1}(D(A - qI) + E). \tag{1.16}$$

An important advantage of the full order system is that the state variables in the SM do not depend on the integrator output. Thus, the equation defining the state variables (1.14) represents at the same time the SM motion. This equation mimics a LTI system with a linear feedback. Therefore, the design of the SM may use numerous methods developed for control design of linear systems. The application of the switching control to obtain a system that behaves as an ordinary linear system with linear feedback may sppear strange. However, recall that the sliding mode system completely rejects matched disturbances and may reduce the effect of unmatched disturbances, as illustrated by simulations.This is a significant advantage when compared to linear feedback.

1.3 The Design of SM Subspace

The first part of this section explains the underlying principle of design: the similarity of the SM equations to the equations of first integrals. It specifies the two design goals considered in this chapter: to obtain a desired dynamics, and to have an optimal behaviour in the LQR sense. The second part deals with reduced order SM design, and the third part with full order SM design.

1.3.1 Design Aims and Philosophy

Notice that the pair of equations (1.9) and (1.10) taken together may have the following interpretation: at the subspace (1.10), the motion of the dynamic system (1.9) is described by a first integral. A first integral is a solution of lower order of a higher order system. This feature is the starting point of the proposed design. The design idea in this chapter is to create first a linear autonomous system having some desired properties by choosing a feedback matrix K. This phase of the design may use the rich body of methods developed for linear systems. Then a first integral is to be found such that the desired property of the autonomous system is maintained. Finally the sliding subspace matrix where first integral motion occurs is calculated.

This idea is applied to the two most common design goals: achieving a desired spectrum or determining optimal behavior in the LQR sense. The determination of a feedback matrix K and matrix C for reduced order SM control, and matrices K, D and E for the case of integral SM control, whereby a desired spectrum is achieved will be presented in the next two subsections.

1.3.2 SM with Given Spectrum for Reduced and Full Order Dynamics

Reduced order design will be considered first. In principle, the obtained method may be extended easily to full order systems by treating full order SM as a reduced order SM of the extended system. However, the structure of the extended system enables a more simple design. Accordingly, separate formulas are used for the reduced and full order cases.

If a given set of $(n-m)$ eigenvalues defines a given spectrum in the SM, then the same eigenvalues must be a subset of the equivalent system spectrum. The remaining m eigenvalues should not appear in the SM. If the initial state belongs to the subspace spanned by the $(n-m)$ right eigenvectors corresponding to the desired $(n-m)$ eigenvalues, this aim will be accomplished. These eigenvectors must be mutually linearly independent to ensure the required system order of $(n-m)$ in the SM. The m eigenvalues that will be eliminated may have arbitrary values, except that complex eigenvalues must appear only in conjugate pairs. That idea was implemented in [6]. In the approach adopted here, the required calculation is significantly simplified due to suitable choice of removed eigenvalues.

The first step of the algorithm is the calculation of the feedback matrix K fulfilling two requirements:

a) the matrix $(A-BK)$ is simple, that is, all its eigenvectors are linearly independent;

b) the matrix $(A-BK)$ has m zero eigenvalues and $(n-m)$ given eigenvalues. The obvious consequence of a) and b) is that $\text{rank}(A-BK) = (n-m)$. The calculation of K is a pole placement problem, having no unique solution in MIMO systems. Kautsky's algorithm [5] suits the purpose well since it provides a simple matrix $(A-BK)$. Its only impediment is that multiplicity of eigenvalues can not exceed $\text{rank}(B)$.

The next step is the calculation of the matrix C. It will be shown that this matrix can be obtained as the solution of the following pair of matrix equations:

$$C(A-BK)=0, \tag{1.17}$$

$$CB=I. \tag{1.18}$$

Eq. (1.18) ensures CB is full rank. Since $CB=I$, (1.17) may be replaced by $CA=K$.

The existence of solution will be considered first. The number of independent scalar equations represented by matrix equation (1.17) is at most $m(n-m)$, due to the reduced rank of $(A-BK)$, while the number of independent scalar equations represented by the matrix equation (1.18) is m^2. The total number of independent scalar equations does not exceed $m(n-m)+m^2 = nm$. Since the number of unknown elements of the matrix C is nm, there exists a solution for C. To obtain the solution, (1.17) and (1.18) are rewritten as

$$C\begin{bmatrix} A & B \end{bmatrix} = \begin{bmatrix} K & I_m \end{bmatrix}. \tag{1.19}$$

The solution of this equation comes out by help of application of matrix pseudo-inverse denoted by the superscript +:

$$C = \begin{bmatrix} K & I_m \end{bmatrix} \begin{bmatrix} A & B \end{bmatrix}^+. \tag{1.20}$$

Next, it is shown that this solution provides the desired spectrum. Eq. (1.9) represents an autonomous system with eigenvalues defined by the requirement b). Eq. (1.10) indicates that the rows of C are spanned by m left eigenvectors corresponding to zero eigenvalues. Therefore the components of the solution space of $(A-BK)$ corresponding to zero eigenvalues are not present in the SM motion. It follows that the SM dynamics is defined by the remaining $(n-m)$ eigenvalues that constitute the desired spectrum.

The adoption of the condition $CB=I$ has two benefits. In integral SM this guarantees that the influence of unmatched disturbances will not increase in the SM [13]. The other advantage is related to reaching control design. Although this design is not the topic of this chapter, the significance of the adopted condition should be pointed out. Basically, the prevalent reasoning in the design of the reaching control is to ensure that the amplitudes of the scalar switching functions decrease all the time with a nonzero speed until all scalar switching functions became equal to zero. In CT systems this is achieved by maintaining opposing signs of the scalar switching function and its derivative. In DT systems the sliding subspace may be attained in one step, therefore the future value of the switching functions should be zero. Thus, for both systems the design is based on δg. Combining equations (1.5), (1.17) and (1.18) one obtains the following equation in the reaching stage

$$\delta g = Kx + u = -u_{eq} + u. \tag{1.21}$$

This expression allows one switching function and one component of the control to be paired. This is a useful feature in the overall control design, since the coordination of controls is an important issue in reaching phase design. However, some other design aim may impose a broader restriction as $CB = Z$, as in ⟦2⟧. Then Z may be put into (1.20) instead of I.

The dynamics in integral SM depends on the matrices E and D. A very important advantage of integral systems is that there are more free parameters in design, since the number of available parameters in the matrices D and E is two times larger than in reduced order systems. Since DB must be nonsingular, the condition $DB = I_m$ is set. From this equation, one obtains D using the pseudo-inverse as $D = B^+$. Such a choice of D provides the equation of the reaching mode in the same form as (1.21). Eqs. (1.14) and (1.16) indicate that the SM eigenvalues are determined by the matrix $(A - BK)$. The needed matrix K may be found by using a pole placement procedure. From equation (1.16), one obtains

$$E = (K - B^+(A - qI_n)). \tag{1.22}$$

It should be mentioned that the expression for the matrix E obtained in ⟦11⟧ for discrete systems differs from (1.22) in the value of the matrix D. In ⟦12⟧ D is taken as B^T, and thus there is no advantage of easy reaching control design.

1.3.3 SM with Optimal Behavior for Reduced and Full Order Dynamics

As already explained, two different formulas for integral and reduced order SM will be used, and reduced order design will be considered first. Let the feedback matrix K in (1.9) be obtained by using LQR design. The closed loop system is then an asymptotically stable system. The matrix $Q \geq 0$ defines priorities regarding the states in the system dynamic, while the matrix $R > 0$ affects the amplitude of control Kx. If the value of control amplitude is not crucial, and system (1.5) is minimum phase, the cheap control concept with zero matrix R may be applied. However, MATLAB does not currently have a routine for this case. Therefore, it is practical to assign a nonzero value to the matrix R. A reasonable compromise is to take R in a sense 'smaller' than Q. The resulting matrix $(A-BK)$ is fully represented by its right eigenvectors which form the columns of the matrix V, and the corresponding eigenvalues form the diagonal elements of the diagonal matrix D_g. If all the eigenvalues are different, this matrix is simple. If this condition is not fulfilled for a given choice of R and Q, modifying these matrices may produce a simple matrix.

Trajectories starting in a subspace spanned by any subset of $(n-m)$ right eigenvectors of the closed-loop V_s will be optimal for given A, B, Q and R. If the SM of the system (1.5) is made in such a subspace, all its trajectories will be optimal in the same sense, as well. The obvious choice to construct C is to use the method given in [[8]].

The matrix C defining this subspace must satisfy the equation

$$CV_s = 0 \tag{1.23}$$

where V_s is a matrix that has as columns the selected $(n-m)$ right eigenvectors. The matrix equation (1.23) is equivalent to $m(n-m)$ independent scalar equations. A unique solution may be obtained by adding the equation (1.18). The matrix C is then defined by the matrix equation

$$C\begin{bmatrix} V_s & B \end{bmatrix} = \begin{bmatrix} 0_{m,n-m} & I_m \end{bmatrix}. \tag{1.24}$$

A reasonable way to choose particular right eigenvectors in V_s is based on their corresponding eigenvalues. The question is which eigenvalues/eigenvectors should be dropped. Some obvious options are to delete the dominant eigenvalues often met in LQR design to eliminate overshoot, or to eliminate some real eigenvalues close to zero to improve the stability margin. The calculation of C is eased by creating the selection matrix S, obtained from D_g by replacing desired $(n-m)$ eigenvalues by 1, and other m eigenvalues with 0. The matrix product VS then has as its columns $(n-m)$ selected eigenvectors and m zero columns. Using this matrix, (1.24) may be rewritten as

$$C\begin{bmatrix} VS & B \end{bmatrix} = \begin{bmatrix} 0_{m,n} & I_m \end{bmatrix}. \tag{1.25}$$

The unique solution of this system is

$$C = \begin{bmatrix} 0_{m,n} & I_m \end{bmatrix}\begin{bmatrix} VS & B \end{bmatrix}^+. \tag{1.26}$$

The paper by [[1]] presented a way to obtain an optimal motion in the SM with $R=0$ using the regular form of the system (1.1). The approach applied here is different: it only guarantees that the behavior in the SM is not worse than the behavior of an LQR optimal system. Its advantage is the possibility to remove undesired components from the solution space. Also, the condition $CB=I$, convenient for the design of switching control, is not built into the regular form approach.

Application of integral SM results in a system that mimics a linear LQR system, and also suppresses matched disturbances. In integral SM, the integrator outputs σ need not be optimized. The feedback matrix K in (1.9) or (1.16) may be obtained by the LQR technique. The calculation of D and E is the same as in the design for desired spectrum.

1.4 SM State Space Equations

A dynamical system of n-th order in the reduced order SM with $\mathrm{rank}(C) = m$ behaves as a dynamical system of $n-m$th order. If the design is performed using a state space model of the plant, the motion is described by the equivalent system dynamic equation of n-th order and the sliding subspace algebraic equation of m-th order. This is not a state space model in the strict sense, and numerous tools available in control software cannot be applied for its analysis.

The first way to obtain a state space model for a system of reduced order was proposed in [1] and described in the Introduction. The controlled system state space model is first transformed into regular form. The action of any unmatched disturbances was not considered in this approach. Since their presence may deteriorate system performance, it is very important to include them as an input in the system model.

In further derivations of a new way to obtain such a state space model, a matrix denoted by 0 has all its elements equal to zero, and I is the unity matrix. The matrix dimensions are explicitly stated only if necessary. The reduced SM model will be treated first.

Consider a system given in operator notation having additive disturbances

$$\delta x = Ax + Bu + Gf. \tag{1.27}$$

The system output is

$$y = Hx. \tag{1.28}$$

$G \in \Re^{n \times r}$ is the constant disturbance matrix and $f \in \Re^{r}$ is the disturbance vector. $H \in \Re^{q \times n}$ is the output matrix. Sliding occurs at $g = Cx = 0$, where $g \in \Re^{m}$ is the vector switching function. In the sliding subspace, the equivalent system equation is

$$\delta x = P(Ax + Gf), \tag{1.29}$$

where P denotes the projector matrix:

$$P = I - B(CB)^{-1}C. \tag{1.30}$$

It is easy to verify that the matrix C satisfies the equation $CP = 0$. Since C is of full rank, $\mathrm{rank}(P) = n - m$. Represent P with its SVD:

$$P = USV^{\mathrm{T}}. \tag{1.31}$$

Since P is a quadratic matrix of rank equal to $(n-m)$, $S \in \Re^{n \times n}$ has the following form:

$$S=\begin{bmatrix} S_1 & 0 \\ 0 & 0 \end{bmatrix}, S_1 \in \Re^{(n-m)\times(n-m)} \tag{1.32}$$

The elements of the diagonal matrix S_1 are $(n-m)$ nonzero singular values of P, and thus S_1 is invertible. Since S has the last m rows equal to zero, and the last m columns equal to zero, for any matrix T of appropriate dimensions, the following two properties hold:

(A) product ST has its last m rows equal to zero;

(B) product TS has the last m columns equal to zero.

Matrices U and V are square and unitary, that is their inverse is equal to their transpose. Therefore, U^{T} qualifies as a transform matrix. Introduce new state variables

$$\begin{bmatrix} z \\ w \end{bmatrix} = U^{\mathrm{T}}x, \text{ i.e } x = U \begin{bmatrix} z \\ w \end{bmatrix}, \text{ where } z \in \Re^{n-m}, w \in \Re^{m}. \tag{1.33}$$

The equation of the equivalent system after some matrix manipulations becomes

$$\begin{bmatrix} \delta z \\ \delta w \end{bmatrix} = SV^{\mathrm{T}}AU \begin{bmatrix} z \\ w \end{bmatrix} + SV^{\mathrm{T}}Gf. \tag{1.34}$$

Due to the property (A) matrices $SV^{\mathrm{T}}AU$ and $SV^{\mathrm{T}}G$ may be broken into blocs as follows

$$\begin{bmatrix} \delta z \\ \delta w \end{bmatrix} = \begin{bmatrix} A_z & A_w \\ 0 & 0 \end{bmatrix} \begin{bmatrix} z \\ w \end{bmatrix} + \begin{bmatrix} G_z \\ 0 \end{bmatrix} f. \tag{1.35}$$

Since $\delta w = 0$, in the SM the value of the vector w is constant. Now consider the sliding subspace equation:

$$Cx = CU \begin{bmatrix} z \\ w \end{bmatrix} = 0. \tag{1.36}$$

To find CU, start from $CP = CUSV^{\mathrm{T}}$. Since V is invertible, this reduces to $CUS = 0$. Break also C and U into blocs to obtain

$$CUS = \begin{bmatrix} C_z & C_w \end{bmatrix} \begin{bmatrix} U_{zz} & U_{zw} \\ U_{wz} & U_{ww} \end{bmatrix} \begin{bmatrix} S_1 & 0 \\ 0 & 0 \end{bmatrix} = \begin{bmatrix} (C_zU_{zz} + C_wU_{wz})S_1 & 0 \end{bmatrix} = 0. \tag{1.37}$$

Since S_1 is an invertible matrix, $C_zU_{zz} + C_wU_{wz} = 0$. Now consider the sliding subspace equation (1.36)

$$\begin{bmatrix} C_z & C_w \end{bmatrix} \begin{bmatrix} U_{zz} & U_{zw} \\ U_{wz} & U_{ww} \end{bmatrix} \begin{bmatrix} z \\ w \end{bmatrix} = \begin{bmatrix} C_zU_{zz} + C_wU_{wz} & C_zU_{zw} + C_wU_{ww} \end{bmatrix} \begin{bmatrix} z \\ w \end{bmatrix} = 0. \tag{1.38}$$

Taking into consideration $C_zU_{zz} + C_wU_{wz} = 0$ it follows that

$$\begin{bmatrix} 0 & C_zU_{zw} + C_wU_{ww} \end{bmatrix} \begin{bmatrix} z \\ w \end{bmatrix} = (C_zU_{zw} + C_wU_{ww})w = 0. \tag{1.39}$$

Matrix C has full rank, and U is nonsingular, therefore CU has also full rank. It follows that the quadratic matrix $(C_zU_{zw} + C_wU_{ww})$ is nonsingular, and the above equation reduces to $w = 0$. Therefore in the SM the constant value of w is zero. Hence, the SM is defined by

$$\delta z = A_z z + G_z f, \tag{1.40}$$

$$w = 0. \tag{1.41}$$

In these equations, the coordinates w and z are completely decoupled. Eq. (1.41) defines the sliding subspace, and (1.40) is a state space equation of $(n-m)$ order, describing the SM motion under the influence of additive disturbances f seen as an input. The output of the system (1.41) is

$$y = Hx = HU \begin{bmatrix} z \\ w \end{bmatrix} = \begin{bmatrix} H_z \ H_w \end{bmatrix} \begin{bmatrix} z \\ w \end{bmatrix} = \begin{bmatrix} H_z \ H_w \end{bmatrix} \begin{bmatrix} z \\ 0 \end{bmatrix} = H_z z. \tag{1.42}$$

Now consider the full order SM motion of system (1.27). The extended system equations with additive disturbances are

$$\delta x = Ax + Bu + Gf, \tag{1.43}$$

$$\delta \sigma = Ex + q\sigma. \tag{1.44}$$

The sliding subspace is defined by the equation $\delta\sigma = 0$. If D guarantees that DB is full rank, the equivalent control is

$$u_{eq} = -(DB)^{-1}((D(A-qI)+E)x + DGf). \tag{1.45}$$

The equation of the SM is

$$\delta x = A_{sm}x + G_{sm}f, \tag{1.46}$$

where $A_{sm} = A - B(DB)^{-1}(D(A-qI)+E)$ and $G_{sm} = G - B(DB)^{-1}DG$. Since the output in the SM is $y = Hx$, $H_{sm} = H$.

This completes the construction of the state space model.

1.5 Examples and Simulations

This section contains four design examples and two simulations. The examples illustrate design of the four basic types of SM given in Section 1.3 and generation of SM state space models. The aim of the first simulation is to compare the trajectories of a standard LQR design, and SM design in a system supplied with matched disturbance, with a trajectory having a reaching phase and a sliding phase. The second simulation uses integral SM. Since there is no reaching phase, the trajectory is obtained using a SM state space model. The simulations use designs performed in corresponding examples of the presented theory. MATLAB implementations are supplied to help the reader to apply the results of this chapter in practice. All examples and simulations use the same linearized CT model of an aircraft given in [14] to illustrate the application of the proposed design. The system matrices are:

$$A = \begin{bmatrix} 0 & 0 & 1.132 & 0 & -1 \\ 0 & -0.0638 & -0.1712 & 0 & 0.0705 \\ 0 & 0 & 0 & 1 & 0 \\ 0 & 0.0468 & 0 & -0.8556 & -1.013 \\ 0 & -0.2908 & 0 & 1.0532 & -0.6059 \end{bmatrix}, B = \begin{bmatrix} 0 & 0 & 0 \\ -0.12 & 1 & 0 \\ 0 & 0 & 0 \\ 4.419 & 0 & -1.656 \\ 1.575 & 0 & -0.0732 \end{bmatrix}.$$

The data for the DT version of the same model with $T = 0.1\,\mathrm{s}$, is obtained using the MATLAB command `[Ad,Bd]=c2d(A,B,T)`.

Reduced order CT SM design: A matrix C with desired poles $P = \begin{bmatrix} -1.2+1.6\imath & -1.2-1.6\imath \end{bmatrix}$ is obtained using a single statement according to (1.20)

```
C=[place(A,B,[P zeros(1,rank(B))]) eye(rank(B))]*pinv([A B])
```

$$C = \begin{bmatrix} -0.6332 & 0 & 0.5038 & -0.0318 & 0.7243 \\ -0.076 & 1 & 0.0605 & -0.0038 & 0.0869 \\ -2.5386 & 0 & 0.3831 & -0.6851 & 1.9222 \end{bmatrix}.$$

The state space model for reduced order CT SM: System matrices are A and B. Unmatched disturbance and output matrices are respectively $G = \begin{bmatrix} 1 & 0 & 0 & 0 & -1 \end{bmatrix}^{\mathrm{T}}$ and $H = \begin{bmatrix} 1 & 1 & 0 & 0 & 0 \end{bmatrix}$. The following commands generate the reduced order SM model:

```
[n,m]=size(B)
[n,r]=size(G)
[q,n]=size(H)
P=eye(n)-B*(B*C)^-1*C                     %Projector matrix
[U,S,V]=svd(P)                            %SVD of P
SYSeq=ss(P*A,P*G,H,zeros(q,r))            %Equivalent system
SYStr=ss2ss(SYSeq,U')                     %Transformed system
SYSsm=modred(SYStr,[n-m+1:n],'truncate')  %Trunc. of m variab.
[Asm,Gsm,Hsm,Dsm]=ssdata(SYSsm)           %System matrices:
```

$$A_{sm} = \begin{bmatrix} -1.015 & -1.967 \\ 1.319 & -1.385 \end{bmatrix}, G_{sm} = \begin{bmatrix} -1.56 \\ -1.12 \end{bmatrix}, H_{sm} = \begin{bmatrix} -0.281 & -0.5 \end{bmatrix}, D_{sm} = 0.$$

Full order DT SM design: Let the desired poles be

$$P_d = [e^{\mathrm{T}(-1.2-1.6\imath)}\; e^{\mathrm{T}(-1.2+1.6\imath)}\; e^{-5\mathrm{T}}\; e^{-7\mathrm{T}}\; e^{-10\mathrm{T}}].$$

Based on $DB = I_m$ and (1.20), the following statements give D_d and E_d:

```
Dd=pinv(Bd)
Ed=place(Ad,Bd,Pd)-pinv(Bd)*(Ad-eye(size(Ad)))
```

$$D_d = \begin{bmatrix} -0.3785 & 0.1076 & -0.0303 & -0.7205 & 7.4516 \\ -0.0461 & 10.0397 & -0.0039 & -0.0829 & 0.8585 \\ -1.0064 & 0.2988 & -0.3973 & -8.1575 & 19.4651 \end{bmatrix},$$
$$E_d = \begin{bmatrix} -3.4769 & -0.075 & 0.0898 & -0.5184 & 3.7368 \\ 0.2787 & 6.2641 & 0.5838 & 0.2326 & 0.1806 \\ -13.3457 & -0.2739 & -5.2685 & -5.0862 & 10.7894 \end{bmatrix}.$$

Reduced order CT optimal SM design: The chosen optimization matrices are $R = I_3$, $Q = I_5$. The statement `K=lqr(A,B,Q,R)` gives

$$K = \begin{bmatrix} -0.3004 & -0.0965 & 0.4406 & 0.8111 & 0.6052 \\ 0.2756 & 1.0507 & 0.7903 & 0.1981 & -0.5369 \\ -0.9131 & -0.2837 & -2.679 & -0.8635 & 1.5199 \end{bmatrix}. \tag{1.47}$$

Matrices V and D_g are obtained with `[V,Dg]=eigs(A-B*K)`, where D_g is a diagonal matrix: $\mathrm{diag}(D_g) = (-4.4786, -1.8769, -1.0004, -0.5621 - 0.4812\imath, -0.5621 + 0.4812\imath)$. There are no multiple poles and matrix $(A - BK)$ is simple. The three values nearest to the origin are discarded. The selection of matrices S and VS are then

$$S = \begin{bmatrix} 1&0&0&0&0 \\ 0&1&0&0&0 \\ 0&0&0&0&0 \\ 0&0&0&0&0 \\ 0&0&0&0&0 \end{bmatrix}, VS = \begin{bmatrix} -0.0592 & 0.0416 & 0&0&0 \\ 0.028 & -0.0251 & 0&0&0 \\ 0.2174 & -0.4298 & 0&0&0 \\ -0.9733 & 0.8036 & 0&0&0 \\ -0.0191 & -0.4089 & 0&0&0 \end{bmatrix}.$$

The following statements, based on (1.26), give the matrix C

```
[n,m]=size(B)
C=[zeros(m,n) eye(m)]*pinv([V*S B])
```

$$C = \begin{bmatrix} -3.8119 & 0 & -1.118 & -0.032 & 0.7248 \\ -0.0555 & 1 & -0.1537 & -0.0038 & 0.087 \\ -1.2012 & 0 & -3.2446 & -0.6894 & 1.9341 \end{bmatrix}. \tag{1.48}$$

The matrix of the corresponding equivalent system $(A - B(CB)^{-1}CA)$ has eigenvalues (-4.4789,-1.8695,0,0,0), which confirms the design procedure.

Full order DT optimal SM design: Since MATLAB calculates a DT optimal feedback matrix based on the continuous-time system model, to avoid solving a discrete Riccati equation the following is used `Kd=lqrd(A,B,Q,R,0.1)`

$$K_d = \begin{bmatrix} -0.2936 & -0.0958 & 0.1956 & 0.6245 & 0.5381 \\ 0.2481 & 0.9981 & 0.7221 & 0.1825 & -0.4901 \\ -0.8712 & -0.2846 & -0.5302 & -0.779 & 1.4928 \end{bmatrix}.$$

By further application of (1.20)

```
Dd=pinv(Bd)
Ed=Kd-pinv(Bd)*(Ad-eye(size(Ad)))
```

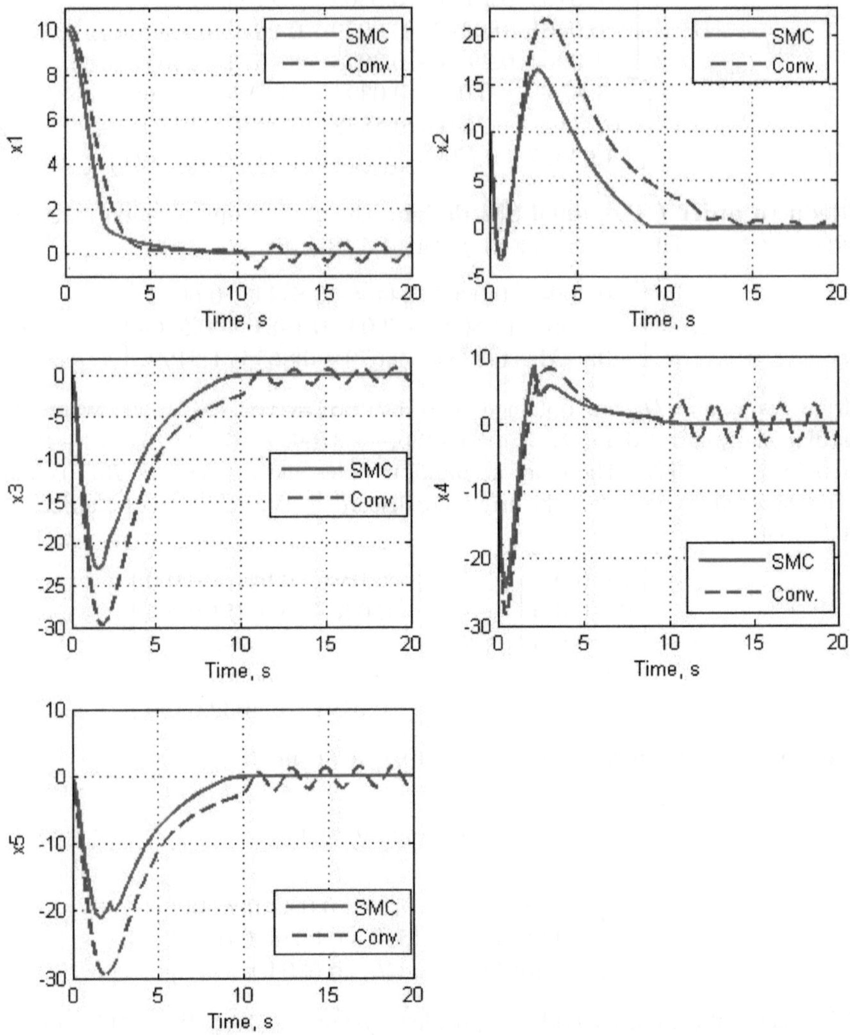

Fig. 1.1 Comparative study of conventional LQR control (Conv.) and CT SM LQR control (SMC). Initial state: $x(0) = [10, 10, 0, 0, 0]^T$. Disturbance $f(t) = 5h(t-10)\sin(\pi t)$ is applied only at the first input. Control is $u = u_{LQR} - [6\text{sgn}(g_1), \text{sgn}(g_2), \text{sgn}(g_3)]^T$.

$$D_d = \begin{bmatrix} -0.3785 & 0.1075 & -0.0304 & -0.7224 & 7.4277 \\ -0.0459 & 10.0447 & -0.0039 & -0.0831 & 0.8557 \\ -1.0074 & 0.2996 & -0.3997 & -8.2068 & 19.5076 \end{bmatrix},$$
$$E_d = \begin{bmatrix} -0.2936 & 0.1165 & 0.2383 & -0.1611 & 0.9072 \\ 0.2481 & 1.0873 & 0.8985 & 0.0969 & -0.5164 \\ -0.8712 & 0.3101 & -2.4162 & -3.3601 & 1.8656 \end{bmatrix}.$$

Reduced order CT optimal SM simulation: Trajectories of a conventional (1.47) (Fig.1.1, dashed lines) and SM version (1.48) (Fig.1.1, solid lines) of an LQR optimal system show that, due to the elimination of slow solution space components, the trajectories supplied with SM control generally reach steady state in a shorter time. Moreover, in these systems outer disturbances are completely suppressed in the sliding mode phase, while the states of systems with conventional control oscillate around the origin.

Full order CT optimal SM simulation: This simulation, Fig.1.2, has only the sliding mode part of the trajectories, since the reaching phase in these systems may be eliminated. Moreover SIMULINK is not used, as in the above simulation, but the SM state space model derived in Section 1.4. The disturbances are matched.

System matrices are A, B, along with $H = [2,0,0,0,0]$, $G = [0\ -0.360\ 13.257\ 4.725]^T$, $Q = I_5$, $R = 0.2I_3$. The disturbance is $f(t) = 5\sin(5t)$, the initial state $x_0 = [2,0,0,0,0]^T$. The program follows:

```
[n,r]=size(G)
[k,n]=size(H)
[n,m]=size(B)
Kopt=lqr(A,B,Q,R)
SYSopt=ss(A-B*Kopt,G,zeros(k,r))       %Conventional optimal system
D=pinv(B)                               %Parameters of integral SM
E=Kopt-D*A
q=0                                     %Continuous system qualifier
Asm=A-B*(D*B)^-1*(D*(A-q*eye(n))+E)    %State space model matrices
Gsm=G-B*(D*B)^-1*D*G
SYSsm=ss(Asm,Gsm,H,zeros(k,r))
```

The response in Fig.1.2 is obtained using:

```
[u,t]=gensig('sin',1,8,0.001)
x0=[2 0 0 0 0]'
lsim(SYSopt,SYSsm,5*u,t,x0)
```

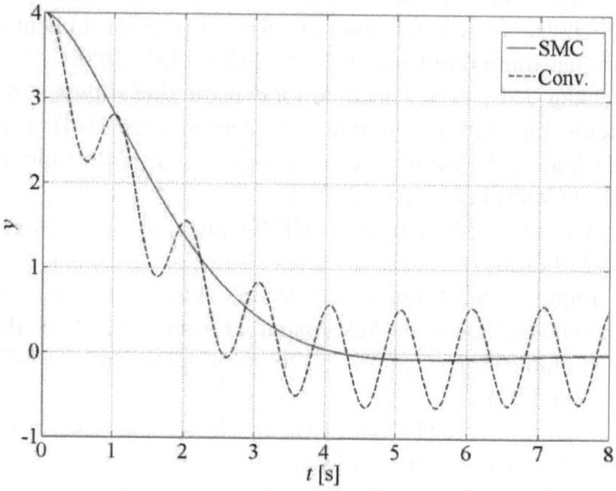

Fig. 1.2 Output trajectories of conventional LQR (dashed line) and integral SM optimal system (solid line). SM system rejects completely a strong disturbance.

1.6 Conclusions

This chapter presents four simple algorithms fulfilling the following goals: the desired spectrum is achieved in both reduced order and integral SM, approximately optimal behavior is determined in reduced order SM and optimal behavior in integral SM. Due to the convenient operator representation of the controlled system, these algorithms work both for CT and DT systems. The switching control design is easy, since scalar controls and scalar switching functions are paired, and thus the issue of coordination of controls in the reaching phase is resolved. A new and simple way to obtain the state space equation of the SM system using SVD is provided. The challenging problem is to extend the proposed approach based on the similarity of SM equations and first integral equations to other issues such as robustness and attenuation of unmatched disturbances, descriptor systems and possibly to some classes of nonlinear systems.

Acknowledgements. The third author acknowledges support from the Ministry of Education and Science of the Republic of Serbia under Project Grant III44004.

References

1. Utkin, V.I., Yang, K.D.: Methods for constructing discontinuity planes in multidimensional variable systems. Automation and Remote Control 39(10), 1466–1470 (1978)
2. El-Ghezawi, O.M.E., Zinober, A.S.I.: Analysis and design of variable structure systems using a geometric approach. International Journal of Control 38(3), 657–671 (1983)
3. Dorling, C.M., Zinober, A.S.I.: Two approaches to hyperplane design in multivariable variable structure control systems. International Journal of Control 44(1), 65–82 (1986)
4. Dorling, C.M., Zinober, A.S.I.: Robust hyperplane design in multivariable variable structure control systems. International Journal of Control 48(5), 2043–2054 (1988)
5. Kautsky, J., Nichols, N.K., Van Dooren, P.: Robust pole assignment in linear state feedback. International Journal of Control 41(2), 1129–1155 (1985)
6. Chen, Y.-P., Chang, J.-L.: A new method for constructing sliding surfaces of linear time-invariant systems. International Journal of System Science 31(4), 417–420 (2000)
7. Chang, J.-L., Chen, Y.-P.: Sliding vector design based on the pole-assignment method. Asian Journal of Control 2(1), 10–15 (2000)
8. Chang, J.-L.: Discrete SM control of MIMO linear systems. Asian Journal of Control 4(2), 217–222 (2002)
9. Huang, J.Y., Yeung, K.S.: Arbitrary eigenvalue assignment via switching hyperplanes design scheme and extension of Ackermann's formula. In: Proc. IEEE Conference on Computer, Communication, Control and Power Engineering, TENCON 1993, Beijing, vol. 4, pp. 17–20 (1993)
10. Ackermann, J., Utkin, V.: SM Control Design Based on Ackermann's Formula. IEEE Transactions of Automatic Control 43(2) (1998)
11. Utkin, V., Shi, J.: Integral SM in Systems Operating under Uncertainty Conditions. In: Proceedings of the 35th Conference on Decision and Control, Kobe, Japan (1996)

12. Abidi, K., Xu, J.-X., Xinghuo, Y.: On the Discrete-Time Integral Sliding-mode Control. IEEE Transactions on Automatic Control 52(4), 709–715 (2007)
13. Castanos, F., Xu, J.-X., Fridman, L.: Integral SMs for Systems with Matched and Unmatched Uncertainties. In: Edwards, C., Fossas Colet, E., Fridman, L. (eds.) Advances Variable Structure and SM Control, pp. 227–246. Springer (2006)
14. Hung, Y.S., MacFarlane, A.G.J.: Multivariable Feedback: A Quasi-Classical Approach. LNCIS, vol. 40. Springer (1982)

Chapter 2
Adaptive Sliding Mode Control

Vadim I. Utkin and Alexander S. Poznyak

Abstract. The main obstacles for application of Sliding Mode Control are two interconnected phenomena: chattering and high activity of control action. It is well known that the amplitude of chattering is proportional to the magnitude of a discontinuous control. These two problems can be handled simultaneously if the magnitude is reduced to a minimal admissible level defined by the conditions for the sliding mode to exist. Here an adaptation methodology is discussed for obtaining the minimum possible value of control based on two approaches developed in recent publications:

- *The σ - adaptation*, providing adequate adjustments of the magnitude of a discontinuous control within the "reaching phase", that is, when the state trajectories are out of a sliding surface;
- *Dynamic adaptation* or the adaptation within a sliding mode (on a sliding surface), based on the, so-called, *equivalent control* obtained by the direct measurements of the output signals of a first-order low-pass filter containing in the input the discontinuous control with the specially adapted magnitude value.

The application of these methodologies is illustrated by the corresponding adaptive versions of the *Super Twist* controller. It is shown that the adaptation based on the equivalent control application enables reducing of the control action magnitude to the minimum possible value keeping the property of finite-time convergence.

Vadim I. Utkin
Department of Electrical Engineering, The Ohio State University, USA
e-mail: utkin@ece.osu.edu

Alexander S. Poznyak
Dept. Control Automatico, CINVESTAV-IPN, Mexico
e-mail: apoznyak@ctrl.cinvestav.mx

B. Bandyopadhyay et al. (Eds.): *Advances in Sliding Mode Control*, LNCIS 440, pp. 21–53.
DOI: 10.1007/978-3-642-36986-5_2

2.1 Introduction

2.1.1 Brief Survey

The *Sliding Mode Control* is a very popular strategy for control of nonlinear uncertain systems, with a very large frame of applications fields ⟦26⟧, ⟦30⟧. Due to the use of the discontinuous function, its main features are

- the robustness of closed-loop system
- and the finite-time convergence.

However, its design requires the knowledge of the bound on the uncertainties, which could be, from a practical point of view, a hard task: it often follows that this bound is overestimated, which yields excessive gain. Then, the main drawback of the sliding mode control, the well-known *chattering phenomenon* (for its analysis, see ⟦4⟧, ⟦5⟧), is important and could damage actuators and systems. A first way to reduce the chattering is the use of a boundary layer: in this case, many approaches have proposed adequate controller gains tuning ⟦26⟧. A second way to decrease the effect of the chattering phenomenon is the use of higher order sliding mode controller ⟦16⟧, ⟦17⟧, ⟦19⟧, ⟦7⟧, ⟦14⟧, ⟦22⟧. However, in both these control approaches, knowledge of the bound on the uncertainties is required. As the objective is the non-requirement of the uncertainties bound, another way consists in using adaptive sliding mode, the goal being to ensure a dynamic adaptation of the control gain to be as small as possible whereas sufficient to counteract the uncertainties and perturbations.

The basic idea of the *Adaptive Control Approach* consists in designing the systems exhibiting the same dynamic properties under uncertainty conditions based on utilization of current information. It involves modifying the control law used by a controller to cope with the fact that the parameters of the system being controlled are slowly time-varying or uncertain. Even more, adaptive control implies improving dynamic characteristics while properties of a controlled plant or environment are varying ⟦1⟧, ⟦25⟧. Without adaptation the original SMC demonstrates *robustness* properties with respect to parameter variations and disturbances ⟦29⟧. The first attempts to apply the ideas of adaptation in Sliding Mode Control (SMC) were made in the 60's ⟦8⟧, ⟦9⟧ and ⟦10⟧: the control efficiency was improved by changing the position or equation of the discontinuity surfaces without any information on a plant parameters. The design idea might be formulated as follows: if sliding mode exists, then the coefficients of switching plane can be varied to improve the system dynamics. However those early publications did not take into account the main obstacle for SMC application - the *chattering* phenomenon which is inherent in sliding motions (see, for example, ⟦7⟧, ⟦4⟧ and ⟦5⟧). This phenomenon is well-known from literature on power converters and referred to as "*ripple*" ⟦21⟧. Then the efforts of the researchers were oriented to the application of *adaptability principles* to reduce the effect of chattering. Since the amplitude of chattering is proportional to discontinuity magnitude in control, one of possible adaptation methods is related to reducing this magnitude to the minimum admissible value dictated by the conditions for SM to exist. So, in [23] the control gain depended on the distance of system state to a discontinuity surface. The tracks of adaptability can be found in the first publications about variable structure systems with SM (see [29], [30]) with the control gain proportional the system state. As recalled previously, this problem is an exciting challenge for applications given that, in many cases, gains are also overestimated, which gives larger control magnitude and larger chattering. In order to adapt the gain, many controllers based on fuzzy tools (see [20], [28]) have been published; however, these papers do not guarantee the tracking performances. In [13] gain dynamics directly depends on the tracking error (sliding variable): the control gain is increasing since sliding mode is not established. Once this is the case, gain dynamics equals zero. The main drawback of this approach is the gain over-estimation with respect to uncertainties bound. Furthermore, this approach is not directly applicable, but requires modifications for its application to real systems: thus, the sign function is replaced by a saturation function where the boundary layer width affects accuracy and robustness. Furthermore, no boundary layer width tuning methodology is provided. A method proposed in [15] in order to limit the switching gain must be mentioned. The idea is based on use of equivalent control: once sliding mode occurs, disturbance magnitude is evaluable and allows an adequate tuning of control gain. However, this approach requires the knowledge of uncertainties/perturbations bounds and the use of low-pass filter, which introduces signal magnitude attenuation, delay and transient behavior when disturbances are acting. In [12] a gain-adaptation algorithm is proposed by using sliding mode disturbance observer. The main drawback is that the knowledge of uncertainties bounds is required to design observer-based controller. There exist also adaptive SMC (ASMC) algorithms that allow adjusting dynamically the control gains without knowledge of uncertainties/perturbations bounds. In particular, several adaptive fuzzy SMC algorithms were proposed. However, they do not guarantee the tracking performance or overestimate the switching control gains as in [13]. Of course, another efficient tool to suppress chattering is the application of state observers [6], but for this method the plant parameters are assumed to be known.

2.1.2 *Objective and the Design Idea*

The objective of the current article is to discuss two new methodologies for control of a class of uncertain nonlinear systems:

1) the first one is referred to as *Sigma Adaptation* of Sliding Mode Controllers and deals with models for which uncertainties are bounded, but this bound (which is finite) is not known; this adaptation process with the varying magnitude of the control gain terminates in the moment when the sliding mode starts (see [23] and [27]).
2) the second methodology is associated with *Dynamic Adaptation* of Sliding Mode Controllers (under known uncertainty bounds) where the adaptation

process is continued during sliding mode, using the current estimates of the corresponding *equivalent control* (see [[31]]), that leads to minimization of chattering effect.

2.1.3 Main Contribution

The first contribution of this chapter is in developing the adaptation methodology for finding the control gain $k(t)$ providing a minimum value of discontinuity resulting in minimization of the chattering effect.

The second contribution consists in developing the modified version of the suggested methodology for the super-twisting algorithm with a SOSM (Second Order Sliding Mode).

2.2 The σ-Adaptation Method

2.2.1 System Description, Main Assumptions and Restrictions

Consider the nonlinear uncertain system

$$\left.\begin{aligned} \dot{x}(t) &= f(x(t)) + g(x(t))\,u \\ x(0) &= x_0,\ t \geq 0 \end{aligned}\right\} \tag{2.1}$$

where $x(t) \in \mathscr{X} \subset \mathbb{R}^n$ the state vector and $u \in \mathbb{R}$ the control input to be designed. Function $f(x)$ and $g(x)$ are supposed to be smooth uncertain functions and are bounded for all $x \in \mathscr{X}$; furthermore, $f(x)$ contains unmeasured perturbations term and $g(x) \neq 0$ for all $x \in \mathscr{X}$.

The control objective consists in forcing the continuous function $\sigma(x,t)$, named sliding variable, to 0. Supposing that σ admits the relative degree equal to 1 with respect to u, one gets

$$\begin{gathered} \dot{\sigma}(x,t) = \left(\frac{\partial \sigma(x,t)}{\partial x}\right)^{\mathsf{T}} \dot{x} + \frac{\partial \sigma(x,t)}{\partial t} = \frac{\partial \sigma(x,t)}{\partial t} + \\ \left(\frac{\partial \sigma(x,t)}{\partial x}\right)^{\mathsf{T}} f(x) + \left(\frac{\partial \sigma(x,t)}{\partial x}\right)^{\mathsf{T}} g(x)\,u = \Psi(x,t) + \Gamma(x,t)\,u \end{gathered} \tag{2.2}$$

where

$$\Psi(x,t) := \frac{\partial \sigma(x,t)}{\partial t} + \left(\frac{\partial \sigma(x,t)}{\partial x}\right)^{\mathsf{T}} f(x)$$

$$\Gamma(x,t) := \left(\frac{\partial \sigma(x,t)}{\partial x}\right)^{\mathsf{T}} g(x)$$

Functions $\Psi(x,t)$ and $\Gamma(x,t)$ are supposed to be bounded such that for all $x \in \mathscr{X}$ and all $t \geq 0$

$$|\Psi(x,t)| \leq \Psi_M, \; 0 < \Gamma_m \leq \Gamma(x,t) \leq \Gamma_M \tag{2.3}$$

It is assumed that Ψ_M, Γ_m and Γ_M exist but are *not known*. The objective for a designer is to propose a sliding mode controller $u(\sigma,t)$ with the same features as classical SMC, namely, robustness and finite-time convergence but without any information on uncertainties and perturbations (appearing in $f(x)$); this latter is only known to be bounded. Furthermore, this objective allows to ensure a global stability of closed-loop system whereas the classical way (with knowledge of uncertainties bounds) only ensures its semi-global stability.

2.2.2 *Real and Ideal Sliding Modes*

In the sequel, the definitions of *ideal* and *real* sliding mode are recalled.

Definition 2.1. [2] We say that an **ideal sliding mode** exists

$$\lim_{\sigma \to +0} \dot{\sigma} < 0 \text{ and } \lim_{\sigma \to -0} \dot{\sigma} > 0$$

In real applications, an 'ideal' sliding mode, as defined in Definition 2.1, cannot be established. Then, it is necessary to introduce the concept of 'real' sliding mode.

Definition 2.2. [2], [16] If, due to some small positive parameter μ, the state trajectories belong to domain

$$|\sigma(t)| \leq \Delta(\mu), \; \lim_{\mu \to 0} \Delta(\mu) = 0$$

then the motion is called a **real sliding mode**.

As it is common for Sliding Mode Theory [29], we will consider the scalar discontinuous control action $u = u(\sigma,t)$ at time t as

$$u(\sigma,t) = -K(t)\,\mathrm{sign}(\sigma)$$

$$\mathrm{sign}(\sigma) := \begin{cases} 1 & if\ \sigma > 0 \\ -1 & if\ \sigma < 0 \\ \in [-1;1] & if\ \sigma = 0 \end{cases} \tag{2.4}$$

Below we consider two different sliding mode control laws:

- the *first* one combines the adaptive schemes formulated in [13] and [15], while mitigating the deficiencies of the combined gain-adaptation schemes;
- the *second* the adaptive scheme is suggested in [23]; it does not require the estimation of the perturbations via equivalent control as in [15] and does not overestimate the control gain as in [13]. Furthermore, the second adaptive-gain

sliding mode control algorithm requires smaller amount of tuning parameters than the first algorithm, and is developed in the real sliding mode context.

2.2.3 First Adaptive Sliding Mode Control Law

Consider the controller (2.4) with the gain $K(t)$ defined as follows:

- If $|\sigma(x(t),t)| > \varepsilon > 0$, then $K(t)$ is the solution of the following differential equation
$$\dot{K}(t) = \bar{K}_1 |\sigma(x(t),t)| \tag{2.5}$$
with $\bar{K}_1 > 0$ and $K(0) > 0$;
- If $|\sigma(x(t),t)| \leq \varepsilon$, then $K(t)$ reads as
$$\begin{aligned} K(t) &= \bar{K}_2 |\eta(t)| + \bar{K}_3 \\ \tau\dot{\eta}(t) + \eta(t) &= \mathrm{sign}(\sigma(x(t),t)) \end{aligned} \tag{2.6}$$
with $\bar{K}_2 = K(t^*)$, $\bar{K}_3 > 0$ and $\tau > 0$, where t^* is the largest value of t such that
$$|\sigma(x(t^*-0),t^*-0)| > \varepsilon, \ |\sigma(x(t^*),t^*)| \leq \varepsilon$$

Obviously, this controller is based on the real sliding mode concept. By supposing that
$$|\sigma(x(0),0)| > \varepsilon$$
the adaptive sliding mode control law (2.5) - (2.6) works as follows:

- The gain $K(t)$ is increasing due to the adaptation law (2.5) up to a value large enough to counteract the bounded uncertainty with unknown bounds in (2.1) until the real sliding mode starts. Denote the time instant when the real sliding mode starts for the first time as t_1.
- As sliding mode has started, i.e.,
$$|\sigma(x(t),t)| \leq \varepsilon$$
from $t = t_1$, $K(t)$ follows the gain-adaptation law (2.6). Then, gain $K(t)$ is adapted through (2.6) with $\bar{K}_2 = K(t_1)$. Note that this strategy will allow to decrease the gain and then to adjust it with respect to the current uncertainties and perturbations.
- However, if the varying uncertainty and perturbation exceeds the value $\bar{K}_2 = K(t_1)$, then the real sliding mode will be destroyed and we get that
$$|\sigma(x(t),t)| > \varepsilon$$
- Next, the gain adaptation will happen in accordance with (2.5). The gain $K(t)$ will be increasing until the sliding mode occurs again at the reaching time instant t_2.

- As sliding mode has occurred and $|\sigma(x(t),t)| \leq \varepsilon$ starting from $t = t_2$, $K(t)$ now follows the gain-adaptation law (2.6) with $\bar{K}_2 = K(t_2)$. And so on.

The following theorem confirms the workability of the first adaptive sliding mode control law (2.5) - (2.6).

Theorem 2.1 ([23]). *Given the nonlinear uncertain system (2.1) with the sliding variable $\sigma(x(t),t)$, satisfying (2.2), controlled by (2.4), (2.5) and (2.6), there exists a finite time t_F so that a real sliding mode is established for all $t \geq t_F$, i.e.,$|\sigma(x(t),t)| \leq \varepsilon$ for all $t \geq t_F$.*

2.2.4 *Second Adaptive Sliding Mode Control Law*

The first adaptive sliding mode control law (2.4), (2.5) and (2.6) uses the concept of "equivalent control" [30]. It can be obtained by a low-pass filter, although it is not easy to tune its parameter. The controller displayed in this subsection does not estimate the boundary of perturbation and uncertainties. But, there is an eminent price to do that: the new strategy guarantees a real sliding mode only.

Consider the following controller (2.4)

$$u(\sigma,t) = -K(t)\,\mathrm{sign}\,(\sigma(x(t),t))$$

where the gain coefficient $K(t)$ satisfies

$$\dot{K} = \begin{cases} \bar{K}|\sigma(x(t),t)|\,\mathrm{sign}\,(|\sigma(x(t),t)| - \varepsilon) \text{ if } K > \mu \\ 0 \text{ if } K \leq \mu \end{cases} \tag{2.7}$$

with $\bar{K} > 0$, $\varepsilon > 0$ and a small enough positive μ. The parameter μ is introduced in order to get only positive values for K.

Once sliding mode with respect to $\sigma(x(t),t)$ is established, the proposed gain-adaptation law (2.7) allows the gain K declining (while $|\sigma(x(t),t)| < \varepsilon$). In other words, the gain K will be kept at the smallest level that allows a given accuracy of σ - stabilization. Of course, as described in the sequel, this adaptation law allows to get an adequate gain with respect to uncertainties/perturbations magnitude.

First, let us prove the following auxiliary result.

Lemma 2.1. *Given the nonlinear uncertain system (2.1) with the sliding variable $\sigma(x(t),t)$ dynamics (2.2) controlled by (2.4) and (2.7), the gain $K(t)$ has an upper-bound, i.e. there exists a positive constant K^* so that $K(t) \leq K^*$ for all $t \geq 0$.*

Proof. Suppose that $|\sigma(x(0),0)| \geq \varepsilon$. From K - dynamics, and given that functions Ψ and Γ are supposed bounded, it follows that K is increasing and there exists a time t_1 (see 2.1) such that

$$K(t_1) = \left|\frac{\Psi(t_1)}{\Gamma(t_1)}\right|$$

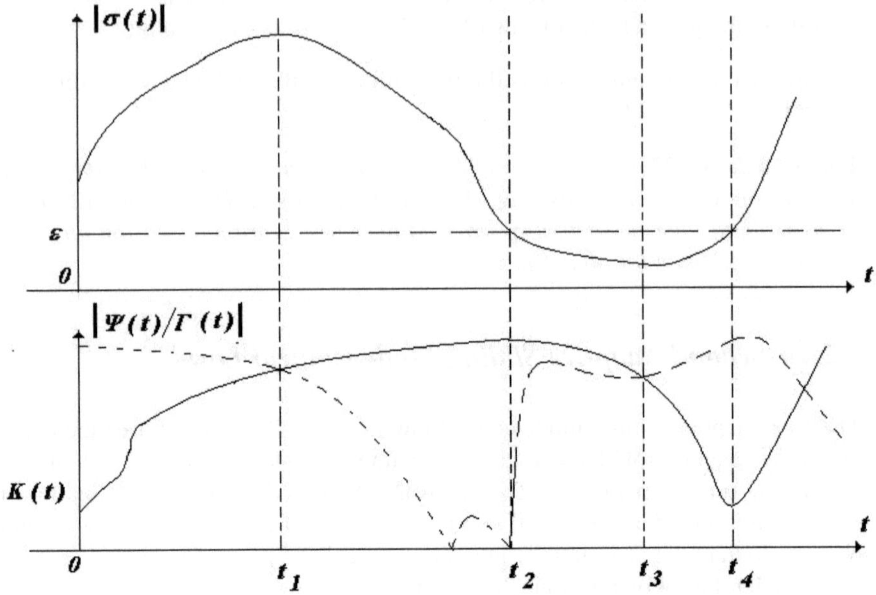

Fig. 2.1 Scheme describing the behaviour of σ (top) and K (bottom) versus time

From $t = t_1$, given K - dynamics, the gain K is large enough to make the sliding variable decreasing. Then, it yields that, in a finite time t_2 (2.1), $|\sigma(x(t),t)| < \varepsilon$. It yields that gain K is decreasing from t_2, the gain K being at a maximum value at $t = t_2$. From K - dynamics, it yields that there exists a time instant $t_3 > t_2$ (2.1) such that

$$K(t_3) = \left|\frac{\Psi(t_3)}{\Gamma(t_3)}\right|$$

From $t = t_3$, the gain K is not large enough to counteract perturbations and uncertainties as it is decreasing. It yields that there exists a time instant $t_4 > t_3$ such that

$$|\sigma(x(t_4),t_4)| \geq \varepsilon$$

The process then restarts from the beginning. By the adaptation law (2.7), the gains $K(t_i)$ remain bounded uniformly on t_i. In fact,

$$K(t_i) = \left|\frac{\Psi(t_i)}{\Gamma(t_i)}\right| \leq \frac{\Psi^{**}}{\Gamma^{**}} := K^{**}$$

and, hence, there always exists a finite constant K^* such that $K^* < K^{**}$, which proves the desired result. ■

Theorem 2.2 ([[23]]). *Given the nonlinear uncertain system (2.1) with the sliding variable $\sigma(x(t),t)$ dynamics (2.2) controlled by (2.4) and (2.7), there exists a finite time t_F so that a real sliding mode is established for all $t \geq t_F$, i.e.,*

$$|\sigma(x(t),t)| < \delta$$

for all $t \geq t_F$ with

$$\delta = \sqrt{\varepsilon^2 + \frac{\Psi_M^2}{\bar{K}\Gamma_m}} \tag{2.8}$$

Proof. The proof is based on the Lyapunov approach and shows that, when $|\sigma(x(t),t)| > \varepsilon$, then control strategy ensures that $|\sigma(x(t),t)| \leq \varepsilon$ in a finite time. Furthermore, it is proven that as soon as σ reaches the domain $|\sigma(x(t),t)| \leq \varepsilon$, it stays in the domain $|\sigma(x(t),t)| \leq \delta$ defined by (2.8) for all consecutive time. Therefore, the proof shows that the real sliding mode is established in finite time in the domain $|\sigma(x(t),t)| \leq \delta$. Consider the following Lyapunov candidate function

$$V = \frac{1}{2}\sigma^2 + \frac{1}{2\gamma}(K-K^*)^2,\ \gamma > 0 \tag{2.9}$$

So one has

$$\dot{V} = \sigma\Psi - \sigma\Gamma K \text{sign}(\sigma) + \frac{1}{\gamma}(K-K^*)\,|\sigma|\,\text{sign}\,(|\sigma|-\varepsilon) \leq$$

$$|\sigma|\,\Psi_M - |\sigma|\,\Gamma_m K + \frac{1}{\gamma}(K-K^*)\,\bar{K}\,|\sigma|\,\text{sign}\,(|\sigma|-\varepsilon) =$$

$$|\sigma|\,(\Psi_M - \Gamma_m K^*) + (K-K^*)\left(-\Gamma_m|\sigma| + \frac{\bar{K}}{\gamma}|\sigma|\,\text{sign}\,(|\sigma|-\varepsilon)\right)$$

By Lemma 2.1 there always exists $K^* > 0$ such that $K(t) - K^* \leq 0$ and hence, for some $\beta_K > 0$ it follows

$$\dot{V} \leq -\beta_\sigma|\sigma| - \beta_K|K-K^*| - \xi$$

$$\xi := |K-K^*|\left(-\Gamma_m|\sigma| + \frac{\bar{K}}{\gamma}|\sigma|\,\text{sign}\,(|\sigma|-\varepsilon) - \beta_K\right)$$

$$\beta_\sigma := (\Gamma_m K^* - \Psi_M) > 0$$

which results to

$$\dot{V} \leq -\beta\sqrt{V} - \xi \tag{2.10}$$

for

$$\beta := \sqrt{2}\min\{\beta_\sigma; \beta_K\sqrt{\gamma}\}$$

Case 1: $|\sigma| > \varepsilon$. In this case it is possible to obtain $\xi > 0$ selecting γ

$$\frac{\bar{K}}{\Gamma_m + \beta_K \varepsilon^{-1}} > \gamma$$

which provides

$$-\Gamma_m |\sigma| + \frac{\bar{K}}{\gamma} |\sigma| - \beta_K > 0$$

for all σ satisfying $|\sigma| > \varepsilon$. From (2.10) we get

$$\dot{V} \leq -\beta\sqrt{V} - \xi \leq -\beta\sqrt{V}$$

Therefore, the finite time convergence to a domain $|\sigma| \leq \varepsilon$ is guaranteed from any initial condition corresponding $|\sigma(x(0), 0)| > \varepsilon$.

Case 2: $|\sigma| < \varepsilon$. Function ξ in (2.10) can be negative. It means that $\dot{V}$ would be sign indefinite, and it is not possible to conclude on the closed-loop system stability. Therefore, $|\sigma|$ can increase over ε. As soon as $|\sigma|$ becomes greater than ε , we return to the previous case and V starts decreasing. Apparently, the decrease of V can be achieved via increase of K allowing $|\sigma|$ to increase before it starts decreasing down to $|\sigma| \leq \varepsilon$. Without loss of generality, let us estimate the overshoot when $\sigma(x(0), 0) = \varepsilon^+$ and $K(0) > 0$: considering the 'worst' case (with respect to uncertainties/perturbations), one has

$$\dot{\sigma} = \Psi_M - K\Gamma_m$$

$$\dot{K} = \bar{K}|\sigma| = \bar{K}\sigma$$

leading tot

$$\sigma(x(t), t) = \varepsilon^+ \cos\left(\sqrt{\bar{K}\Gamma_m t}\right) + \frac{\Psi_M - \bar{K}\Gamma_m}{\sqrt{\bar{K}\Gamma_m}} \sin\left(\sqrt{\bar{K}\Gamma_m t}\right)$$

$$K(t) = \varepsilon^+ \sqrt{\frac{\bar{K}}{\Gamma_m}} \sin\left(\sqrt{\bar{K}\Gamma_m t}\right) + \left(K(0) - \frac{\Psi_M}{\Gamma_m}\right) \cos\left(\sqrt{\bar{K}\Gamma_m t}\right) + \frac{\Psi_M}{\Gamma_m}$$

Taking $\varepsilon^+ \to \varepsilon$, for $\delta = \max_t \sigma(x(t), t)$ (maximization is taken over all t within the considered interval), we obtain (2.8). Theorem is proven. ■

So, the convergence to the domain $|\sigma| \leq \varepsilon$ is in a finite time, but could be sustained in the bigger domain $|\sigma| \leq \delta$. Therefore, the real sliding mode exists in the domain $|\sigma| \leq \delta$.

2.2.5 *On the ε - Parameter Tuning*

The choice of parameter ε has to be made by an adequate way because a 'bad' tuning could provide either instability and control gain increasing to infinity, or bad accuracy for closed-loop system as described in the sequel.

- If the parameter ε is too small, and due to large gain K and sampling period T_e, system trajectories are such that $|\sigma|$ never stays lower than ε. From K-dynamics ((2.7)) it yields that the gain K is increasing, which induces larger oscillation, and so on.
- If the parameter ε is too large, system trajectories are such that, in spite of large gain K and sampling period T_e, $|\sigma|$ is evolving around ε, it follows that controller accuracy is not as good as possible.

In [[23]] there is suggested to select ε adjusted in time as

$$\varepsilon(t) = 4K(t)T_e$$

2.3 The Dynamic Adaptation Based on the Equivalent Control Method

In the previous section, following to [[23]] and [[27]], the adaptation process with the varying magnitude of the control gain terminates in the moment when the sliding mode starts. In [[15]] the authors tried to continue the adaptation process during sliding mode estimating the corresponding *equivalent control*. However, none of the above algorithms resulted in minimum possible value of the discontinuous control. Finding the solution of this problem under uncertainty conditions is the objective of this section.

2.3.1 *Simple Illustrative Example Explaining the Main Idea of the Method*

We start with a simple example. It is assumed that for the first-order system

$$\begin{gathered} \dot{x}(t) = a(t) + u \\ u = -k\text{sign}(x(t)),\ k > 0 \end{gathered} \tag{2.11}$$

The ranges of a time varying parameter

$$0 < |a(t)| \leq a_+$$

and the upper bound A for its time derivative

$$|\dot{a}(t)| \leq A$$

are known only.

The sliding mode with $x(t) \equiv 0$ exists for all values of unknown parameter $a(t)$ if

$$k > a_+$$

However if parameter $a(t)$ is varying, the gain k can be decreased and, as a result, chattering amplitude can be reduced. The objective of adaptation is decreasing k to the minimal value preserving sliding mode, if parameter a is unknown.

If the condition $k > a_+$ holds, then sliding mode with $x(t) \equiv 0$ occurs and control in (2.11) should be replaced by the, so-called, *equivalent control* u_{eq} [[29]] for which the right-hand side in (2.11) is equal to zero, namely,

$$\dot{x}(t) = 0 = a(t) + u_{eq} \tag{2.12}$$

that leads to

$$k(t)\left[\operatorname{sign}(x(t))\right]_{eq} = a(t) \tag{2.13}$$

If $k < a$, the set $x(t) \equiv 0$ is of zero measure in time and can be disregarded. The function $\left[\operatorname{sign}(x(t))\right]_{eq}$ is an average value, or a slow component of discontinuous function $\operatorname{sign}(x(t))$ switching at high frequency and can be easily obtained by a low pass filter filtering out the high frequency component [[29]]. Of course, the average value is in the range $(-1, 1)$.

Then the design idea of adaptation seems to be evident:

after sliding mode occurs the control parameter $\left[\operatorname{sign}(x(t))\right]_{eq}$ *should be decreased until and becomes close to* 1.

On one hand, the condition $k(t) > a(t)$ should hold. But the chattering amplitude is proportional to $k(t)$. The objective of adaptation process looks now transparent:

the gain $k(t)$ should tend to $a(t)/\alpha$ with $\alpha \in (0,1)$ which is very close to 1.

As a result, the minimal possible value of discontinuity magnitude is found for the current value of parameter $a(t)$ to reduce the amplitude of chattering. For that purpose select the *adaptation algorithm* in the form

$$\begin{gathered}
\dot{k}(t) = \rho k(t)\operatorname{sign}(\delta(t)) - M\left[k(t) - k^+\right]_+ + M\left[\mu - k(t)\right]_+ \\
\delta(t) := \left|\left[\operatorname{sign}(x(t))\right]_{eq}\right| - \alpha,\ \alpha \in (0,1) \\
[z]_+ := \begin{cases} 1 \text{ if } z \geq 0 \\ 0 \text{ if } z < 0 \end{cases},\ M > \rho k^+,\ k^+ > a^+, \rho > 0
\end{gathered} \tag{2.14}$$

The gain k can vary in the range $[\mu, k^+]$, $\mu > 0$ is a preselected minimal value of k. For the adaptation algorithm (2.14) sliding mode will occur after a finite time interval. Indeed, if it does not exist, then

$$\left|\left[\operatorname{sign}(x(t))\right]_{eq}\right| = 1$$

that leads to $\delta > 0$, and the increasing gain $k(t)$ will reach the value k^+ which is sufficient for enforcing sliding mode for any value of parameter $a(t)$.

Show that in sliding mode the adaptation process (2.14) with

$$\delta(t) = 0 \text{ or } k = \mu$$

is over after a finite time t_f. To do that calculate the time derivative of the Lyapunov function

$$V(\delta) = \delta^2/2$$

First, assume that during the adaptation process $k(t) \in [\mu, k^+]$ which means that

$$|a(t)|/\alpha > \mu \text{ or } (|a(t)| > \alpha\mu)$$

the time derivatives of $\left|[\mathrm{sign}(x(t))]_{eq}\right|$ (2.13) and $|a(t)|$ exist and the terms depending on M in the adaptation algorithm (2.14) are equal to zero. Calculate the time derivative of the Lyapunov function $V(\delta)$ by virtue of (2.13) and (2.14):

$$\dot{V}(\delta) = \delta\dot{\delta} = \delta\frac{d}{dt}\left|[\mathrm{sign}(x)]_{eq}\right| =$$

$$\begin{aligned} &\delta\frac{d}{dt}(|a|/k) = -|a|\,\delta k^{-2}\dot{k} + \delta k^{-1}\frac{d}{dt}(|a|) = \\ &-|a|\,\delta k^{-1}\rho\,\mathrm{sign}\left(\delta - M[k-k^+]_+ + M[\mu-k]_+\right) \\ &+\delta k^{-1}\dot{a}\,\mathrm{sign}(a) \le -|a|\,\delta k^{-1}\rho\,\mathrm{sign}(\delta) + |\delta|\,k^{-1}A \\ &\le -\alpha\mu\rho k^{-1}|\delta| + |\delta|\,k^{-1}A = -|\delta|\,k^{-1}(\alpha\mu\rho - A) \end{aligned} \tag{2.15}$$

and if $\rho > A/\alpha\mu$ it follows

$$\dot{V}(\delta) \le -\sqrt{2}\frac{(\alpha\mu\rho - A)}{k^+}\sqrt{V(\delta)}$$

It is evident from the solution

$$0 \le \sqrt{V(\delta(t))} \le \sqrt{V(\delta(0))} - \frac{(\alpha\mu\rho - A)}{\sqrt{2}k^+}t$$

of the differential inequality (2.15) that $\sqrt{V(\delta(t))} = 0$ at least after

$$t_f = \frac{k^+}{(\alpha\mu\rho - A)}\sqrt{2V(\delta(0))} = \frac{k^+}{(\alpha\mu\rho - A)}|\delta(0)|$$

and, as a result, $\delta(t)$ becomes equal to zero identically after the finite time t_f.

After the adaptation process is over ($t > t_f$) we have

$$\left|[\mathrm{sign}(x(t))]_{eq}\right| = \frac{|a|}{k} = \alpha$$

So, $k = |a| / \alpha$. If in the course of motion $|a(t)| / \alpha < \mu$, then the gain $k(t)$ decreases until $k(t) = \mu$ and, as it follows from (2.14), it will be maintained at this level. Since the gain $a(t)$ is time varying its increase can can result in $|a(t)| / \alpha = \mu$ and $\delta(t) = 0$ at a time t_f. As it follows from the above analysis, for the further motion in the domain $k(t) \in (\mu, k^+]$ with the initial condition $\delta(t_f) = 0$ the time function $\delta(t)$ will be equal to zero with $\alpha = |a(t)| / k(t)$.

Remark 2.1. The function $[\text{sign}(x(t))]_{eq}$ is needed here for the implementation of the adaptation algorithm (2.14). As it was mentioned above, it can be derived by filtering out a high frequency component of the discontinuous function $\text{sign}(x(t))$ by a low pass filter

$$\tau \dot{z} + z = \text{sign}(x(t)),\ z(0) = 0$$

with a small time constant $\tau > 0$ and the output $z(t)$ which is, in fact, an estimate of $[\text{sign}(x(t))]_{eq}$ satisfying

$$\left| z(t) - [\text{sign}(x(t))]_{eq} \right| \le O(\tau) \underset{\tau \to 0}{\to} 0$$

Then the convergence analysis of (2.14)-(2.15) with $\delta(t) = z(t) - \alpha$ is valid beyond the domain $|\delta(t)| \le O(\tau)$. This inequality defines the accuracy of adaptation. Note that the switching frequencies of the modern power converters are of order dozens of kHz, and very small time constant τ can be selected to get a high accuracy of adaptation.

Notice also that, as follows from [[29]],

$$z(t) = \psi(t) + O(\sup|x(t)| + \tau) + O(\sup|x(t)|/\tau)$$

where $\psi(t)$ is the fast rate exponentially decreasing function. The term $\sup|x(t)|$ is inverse proportional to the sliding mode frequency f. It is of order of dozen kHz in the modern switching devices. Therefore it is not a problem to make the term

$$O(\sup|x| + \tau) + O(\sup|x|/\tau)$$

negligible. Of course, this engineering language can be translated into mathematical one, for example as follows: *for any* $\varepsilon > 0$ *there exists a switching frequency* f_0 *such that*

$$|z - u_{eq}| < \varepsilon \text{ if } f > f_0$$

implying

$$\text{sign}[\left| [\text{sign}(x(t))]_{eq} \right| - \alpha] = \text{sign}[|z(t)| - \alpha]$$

Certainly, we have described the idea only. The generalization of the adaptive procedure (2.14) for the vector-state models with uncertainties $a = a(t,x)$ constitutes the main result given below.

2.3.2 Main Assumptions

Here we consider an arbitrary order system

$$\dot{x}(t) = f(t,x(t)) + b(t,x(t))\,u(t,x(t))$$

$$\begin{gathered} x(t) \in R^n,\ f: R^+ \times R^n \to R^n \\ u: R^+ \times R^n \to R,\ b: R^+ \times R^n \to R^n \end{gathered} \tag{2.16}$$

for which we assume that

A1 the control $u = u(t,x)$ enforces siding mode on some surface

$$\sigma(x) = 0\ (\sigma \in C^1)$$

and is in the following form

$$\begin{gathered} u(t,x) = -k(t)\left(1 + \lambda\sqrt{\|x\|^2 + \varepsilon}\right)\text{sign}(\sigma(x)) \\ \lambda \geq 0,\ \varepsilon > 0,\ k(t) \in [\mu, k^+],\ \mu > 0 \end{gathered} \tag{2.17}$$

Similarly to the example (2.11) the control gain $k(t)$ is a time varying function governed by the adaptation procedure described below.

A2 the uncertain functions $f(t,x)$ and $b(t,x)$ satisfy the commonly accepted conditions (which are much more general then in (2.3)):

$$\begin{gathered} \|f(t,x)\| \leq f_0 + f_1\|x\| \\ 0 < b_0 \leq \nabla^\intercal\sigma(x)\,b(t,x) \\ \|b(t,x)\| \leq b^+,\ \|\nabla\sigma(x)\| \leq \sigma^+ \end{gathered} \tag{2.18}$$

$$\begin{gathered} \Phi(t,x) := \frac{\nabla^\intercal\sigma(x)\,f(t,x)}{\nabla^\intercal\sigma(x)\,b(t,x)} \\ \|\nabla^\intercal\Phi(t,x)\| \leq \Phi_0 + \Phi_1\|x\| \\ \left|\frac{\partial}{\partial t}\Phi(t,x)\right| \leq \varphi_0 + \varphi_1\|x\| \end{gathered} \tag{2.19}$$

All coefficients in the right-hand sides of these inequalities are constant and positive. The function $\sigma(x)$ and its time derivative

$$\begin{gathered} \dot{\sigma}(x) = \nabla^\intercal\sigma(x)\,f(t,x) - \\ \nabla^\intercal\sigma(x)\,b(t,x)\,k(t)\left(1 + \lambda\sqrt{\|x\|^2 + \varepsilon}\right)\text{sign}(\sigma(x)) \end{gathered} \tag{2.20}$$

should have opposite signs ($\sigma(x)\dot{\sigma}(x) < 0$ if $\sigma(x) \neq 0$) for sliding mode to exist on the surface $\sigma(x) = 0$. The sufficient condition for this follows from (2.18),(2.19) and (2.20):

$$\sigma(x)\dot{\sigma}(x) = \sigma(x)\nabla^{\mathsf{T}}\sigma(x) f(t,x) -$$

$$\nabla^{\mathsf{T}}\sigma(x) b(t,x) k(t)\left(1+\lambda\sqrt{\|x\|^2+\varepsilon}\right)|\sigma(x)|$$

$$\leq [\nabla^{\mathsf{T}}\sigma(x) b(t,x)]\,|\sigma(x)| \times$$

$$\left(|\Phi(t,x)| - k(t)\left(1+\lambda\sqrt{\|x\|^2+\varepsilon}\right)\right) < 0$$

if

$$|\Phi(t,x)| - k(t)\left(1+\lambda\sqrt{\|x\|^2+\varepsilon}\right) < 0 \tag{2.21}$$

which is always holds when

$$\lambda \geq f_1/f_0,\ \mu > f_0\sigma^+/b_0,\ k(t) \in \left(\mu, k^+\right] \tag{2.22}$$

in view of the relation

$$|\Phi(t,x)| - k(t)\left(1+\lambda\sqrt{\|x\|^2+\varepsilon}\right) \leq$$

$$f_0\frac{\sigma^+(1+\|x\| f_1/f_0)}{b_0} - \mu(1+\lambda\|x\|)$$

To derive the sliding mode equation the function $\operatorname{sign}(\sigma(x))$ should be replaced by the solution of the equation $\dot{\sigma}(x) = 0$ with respect to the term $\operatorname{sign}(\sigma(x))$, called *the equivalent control*:

$$[\operatorname{sign}(\sigma(x))]_{eq} := \begin{cases} \dfrac{\Phi(t,x)}{k(t)\left(1+\lambda\sqrt{\|x\|^2+\varepsilon}\right)} & \text{if } \sigma(x(t)) = 0 \\ \operatorname{sign}(\sigma(x(t))) & \text{if } \sigma(x(t)) \neq 0 \end{cases} \tag{2.23}$$

satisfying (in view of (2.21)) in the sliding mode ($\sigma(x(t)) = 0$)

$$\left|[\operatorname{sign}(\sigma(x))]_{eq}\right| < 1 \tag{2.24}$$

Note that the state-depended magnitude of discontinuity is the conventional tool to minimize chattering. Indeed, in the course of approaching the origin $x = 0$ it is decreasing automatically [[15]]. Similarly the term

$$\left(1+\lambda\sqrt{\|x\|^2+\varepsilon}\right)$$

may also affect the amplitude of chattering appearing on sliding mode phase. The necessity of this term in (2.17) is related with the considered class of nonlinear functions satisfying

$$\|f(t,x)\| \leq f_0 + f_1 \|x\|$$

If $f_1 = 0$ (nonlinear function is bounded satisfying $\|f(t,x)\| \leq f_0$) similarly to the example we may take $\lambda = 0$, and, in this case, the term $\left(1+\lambda\sqrt{\|x\|^2+\varepsilon}\right)$ does not affect a chattering amplitude. It is important, that this methodology is oriented to the worst case - sliding mode should exist for all values of unknown functions or parameters from some range. The method of the paper guarantees the minimal magnitude for their current values of unknown functions and parameters. In general, we may add the term $\left(1+\lambda\sqrt{\|x\|^2+\varepsilon}\right)$ to the gain multiplying the discontinuous function $\text{sign}(\sigma(x))$ to enforce sliding mode when $f(x)$ is unbounded. This term with $\varepsilon = 0$ in the form $(1+\lambda\|x\|)$ can solve this problem as well. However, the adaptation algorithm implies existence of the gradient of this term, but it does not exist for the last case.

Below we will show that in the general case the adaptation of the gain-parameter $k(t)$ only is sufficient to minimize the chattering effect on sliding mode phase since the suggested "learning law" for $k(t)$ variation automatically takes into account the presence of this term.

2.3.3 *Adaptation Algorithm in Sliding Mode*

2.3.3.1 Description of the Adaptation Procedure

The idea of the *adaptation law* for the control gain $k(t)$ is similar to that for our first-order system in the previous subsection:

$$\dot{k}(t) = \begin{cases} (\gamma_0 + \gamma_1 \|x\|)\, k(t)\text{sign}(\delta(t)) \\ -M[k(t)-k^+]_+ + M[\mu - k(t)]_+ \end{cases} \tag{2.25}$$

where

$$\delta(t) := \left|[\text{sign}(\sigma(x))]_{eq}\right| - \alpha$$
$$\alpha \in (0,1),\ \lambda > 0,\ \gamma_0, \gamma_1 > 0 \tag{2.26}$$

Notice that in (2.23)

$$\frac{|\Phi(t,x)|}{\left(1+\lambda\sqrt{\|x\|^2+\varepsilon}\right)} < k$$

and, moreover,

$$\begin{aligned}&\frac{|\Phi(t,x)|}{\left(1+\lambda\sqrt{\|x\|^2+\varepsilon}\right)} \leq \frac{\sigma^+ (f_0 + f_1 \|x\|)}{b_0 (1+\lambda \|x\|)} = \\ &\sigma^+ \frac{f_0}{b_0}\left(1 + \frac{(f_1 f_0^{-1} - \lambda) \|x\|}{1+\lambda \|x\|}\right) \leq \sigma^+ \frac{f_0}{b_0}\end{aligned} \tag{2.27}$$

Select in (2.25)

$$k^+ > \sigma^+ \frac{f_0}{b_0} \tag{2.28}$$

If sliding mode does not exist, then

$$\left|[\mathrm{sign}\,(\sigma\,(x))]_{eq}\right| = 1$$

and the gain $k(t)$ will be equal to k^+ which results in the occurrence of this motion in the surface $\sigma\,(x\,(t)) = 0$.

2.3.3.2 Analysis of the δ-Stability for the Adaptive Version

The following theorem describes the main stability property of the sliding mode controller with the gain adaptation based on the "equivalent control method".

Theorem 2.3 (on the adaptive sliding mode controller). *For the dynamic system (2.16) closed by the control (2.17) with the gain adaptation law (2.25) - (2.26) with the parameters satisfying*

$$\begin{aligned}k^+ > \sigma^+ \frac{f_0}{b_0},\ \mu > f_0\sigma^+/b_0,\ 0 < \varepsilon << 1\\ \gamma_0 > \alpha^{-1}\left[\left(\frac{f_0}{\mu}+b^+\right)\Phi_0 + \frac{\varphi_0}{\mu} + f_0 + b^+k^+\right]\\ \gamma_1 \geq \alpha^{-1}\left(\frac{f_0}{\mu}+b^+\right)\Phi_1,\ M > \gamma_0 k^+\end{aligned} \tag{2.29}$$

there exist

$$\theta := \alpha\gamma_0 - \left[\left(\frac{f_0}{\mu}+b^+\right)\Phi_0 + \frac{\varphi_0}{\mu} + f_0 + b^+k^+\right] > 0 \tag{2.30}$$

and

$$t_f = \theta^{-1} |\delta(0)|$$

(where $\delta(0)$ is defined by (2.26)) such that for all $t \geq t_f$ the condition

$$\left|[\text{sign}(\sigma(x(t)))]_{eq}\right| = \alpha \tag{2.31}$$

holds. It means that the sliding surface $\sigma(x) = 0$ is attained in a finite time t_f, and for

$$\alpha = 1 - \varepsilon_0$$

($\varepsilon_0 > 0$ is a small enough positive number) the suggested adaptation procedure provides $k(t)$ tending to a vicinity of the minimum possible value $k_{\min}(t)$, that is, as it follows from (2.23), in sliding mode

$$\begin{gathered} k(t) = \begin{cases} \dfrac{1}{1-\varepsilon_0} k_{\min}(t) & if\ k_{\min}(t) \geq \mu \\ \mu & if\ k_{\min}(t) < \mu \end{cases} \\ k_{\min}(t) := \frac{|\Phi(t, x(t))|}{1 + \lambda\sqrt{\|x(t)\|^2 + \varepsilon}} \end{gathered} \tag{2.32}$$

Proof. Consider the Lyapunov function candidate as

$$V(\delta) := \frac{1}{2}\delta^2 \tag{2.33}$$

and it is assumed that during adaptation process $k(t) \in [\mu, k^+]$ which means that

$$\frac{|\Phi(t,x)|}{k(t)\left(1 + \lambda\sqrt{\|x\|^2 + \varepsilon}\right)} > \alpha$$

$$|\Phi(t,x)| > \alpha\mu\left(1 + \lambda\|x\|\right)$$

taking into account that

$$[\text{sign}(\sigma(x))]_{eq} \neq 0$$

and the time derivative of $\left|[\text{sign}(\sigma(x))]_{eq}\right|$ exists. If in the course of adaptation process the above inequality does not hold, the gain $k(t)$ will decrease and after the $k(t) = \mu$ it will remain constant. Note that the similar consideration was given for the simple example in the previous subsection. Calculate the time derivative of (2.9) (it exists with any $\varepsilon > 0$, while it does not exist with $\varepsilon = 0$):

$$\dot{V}(\delta) = \delta\dot{\delta} = \delta \frac{d}{dt}\left[\frac{|\Phi(t,x)|}{k(t)\left(1+\lambda\sqrt{\|x\|^2+\varepsilon}\right)} - \alpha\right]$$

$$= \delta\left[\frac{\nabla^{\mathsf{T}}\Phi(t,x)(f+bu)+\frac{\partial}{\partial t}\Phi(t,x)}{k(t)\left(1+\lambda\sqrt{\|x\|^2+\varepsilon}\right)}\operatorname{sign}(\Phi(t,x)) - \right. \tag{2.34}$$

$$\left. |\Phi(t,x)| \frac{\dot{k}(t)\left(1+\lambda\sqrt{\|x\|^2+\varepsilon}\right)+k(t)\lambda\frac{x^{\mathsf{T}}(f+bu)}{\sqrt{\|x\|^2+\varepsilon}}}{k^2(t)\left(1+\lambda\sqrt{\|x\|^2+\varepsilon}\right)^2}\right]$$

Applying here the estimates (2.18), (2.19) and using (2.17), we derive

$$\dot{V}(\delta) \leq \delta\left[\frac{\|\nabla^{\mathsf{T}}\Phi(t,x)\|\,\|f\| + \left|\frac{\partial}{\partial t}\Phi(t,x)\right|}{k(t)\left(1+\lambda\sqrt{\|x\|^2+\varepsilon}\right)} + \right.$$

$$\|\nabla^{\mathsf{T}}\Phi(t,x)\|\,\|b\| - |\Phi(t,x)|\frac{(\gamma_0+\gamma_1\|x\|)\operatorname{sign}(\delta(t))}{k(t)\left(1+\lambda\sqrt{\|x\|^2+\varepsilon}\right)}$$

$$\left. + \ |\Phi(t,x)|\frac{\lambda x^{\mathsf{T}}(f+bu)}{k(t)\left(1+\lambda\sqrt{\|x\|^2+\varepsilon}\right)\sqrt{\|x\|^2+\varepsilon}}\right]$$

which (in view of (2.24)) implies

$$\dot{V}(\delta) \leq \delta\left[\frac{(\Phi_0+\Phi_1\|x\|)(f_0+f_1\|x\|)+(\varphi_0+\varphi_1\|x\|)}{k(t)\left(1+\lambda\sqrt{\|x\|^2+\varepsilon}\right)}\right.$$

$$+\|\nabla^{\mathsf{T}}\Phi(t,x)\|\,\|b\| - |\Phi(t,x)|\frac{(\gamma_0+\gamma_1\|x\|)\operatorname{sign}(\delta(t))}{k(t)\left(1+\lambda\sqrt{\|x\|^2+\varepsilon}\right)}$$

$$+\left|\Phi(t,x)\right|\frac{\lambda x^{\mathsf{T}}(f+bu)}{k(t)\left(1+\lambda\sqrt{\|x\|^{2}+\varepsilon}\right)^{3}}\Bigg]\leq$$

$$|\delta|\frac{f_0\left(\Phi_0+\Phi_1\|x\|\right)\left(1+\frac{f_1}{f_0}\|x\|\right)+\varphi_0\left(1+\frac{\varphi_1}{\varphi_0}\|x\|\right)}{\mu\left(1+\lambda\|x\|\right)}$$

$$+\ |\delta|\left(\Phi_0+\Phi_1\|x\|\right)b^{+}-|\delta|\,\alpha\left(\gamma_0+\gamma_1\|x\|\right)$$

$$+|\delta|\frac{\lambda\|x\|\left(f_0+f_1\|x\|+b^{+}k^{+}\left(1+\lambda\sqrt{\|x\|^{2}+\varepsilon}\right)\right)}{\left(1+\lambda\sqrt{\|x\|^{2}+\varepsilon}\right)^{2}}$$

Selecting $\lambda\geq\max\left\{\frac{f_1}{f_0},\frac{\varphi_1}{\varphi_0}\right\}$ we get

$$\dot{V}(\delta)\leq|\delta|\left[\left(\frac{f_0}{\mu}+b^{+}\right)\left(\Phi_0+\Phi_1\|x\|\right)+\frac{\varphi_0}{\mu}\right]$$

$$-|\delta|\,\alpha\left(\gamma_0+\gamma_1\|x\|\right)+|\delta|\left[f_0\frac{1+\frac{f_1}{f_0}\|x\|}{1+\lambda\|x\|}+b^{+}k^{+}\right]$$

$$\leq|\delta|\left[\left(\frac{f_0}{\mu}+b^{+}\right)\Phi_0+\frac{\varphi_0}{\mu}+f_0+b^{+}k^{+}-\alpha\gamma_0\right.$$

$$\left.+\|x\|\left(\left(\frac{f_0}{\mu}+b^{+}\right)\Phi_1-\alpha\gamma_1\right)\right]$$

Taking γ_0 and γ_1 satisfying (2.29) we finally get

$$\dot{V}(\delta)\leq-|\delta|\,\theta=-\theta\sqrt{2V(\delta)}$$

where θ is given by (2.30). Then similarly to the primitive example it can be shown that the adaptation process will be over after time instant

$$t_f=\theta^{-1}\left|\delta(0)\right|$$

and if in the course of motion $k(t)$ decreases and becomes equal to μ, then it will be maintained at this level.

2.4 Adaptive Super-Twist Control

In this section we consider the application of the presented adaptation concept for, the adaptive version of the, so-called, Super-Twist Control.

2.4.1 *Main Properties of the Standard Super Twist without Adaptation*

Consider the simple two dimensional nonlinear system containing discontinuous nonlinearity in the right-hand side of the second component:

$$\begin{cases} \dot{x}(t) = y(t) - \bar{\alpha}\sqrt{|x(t)|}\operatorname{sign}(x(t)) \\ \dot{y}(t) = \phi(t) + u(t) \\ u(t) := -\bar{\beta}\operatorname{sign}(x(t)) \end{cases} \tag{2.35}$$

referred below to as the "super-twist" controller [[16]], [[17]] and [[19]].

Remark 2.2. Notice that if $y(0) = 0$, then (2.35) can be represented as

$$\dot{x}(t) = -\bar{\alpha}\sqrt{|x(t)|}\operatorname{sign}(x(t)) - \bar{\beta}\int_{s=0}^{t} \operatorname{sign}(x(s))ds$$

which is exactly a *PI*-controller (with the *P*-part modulation) with respect to the $\operatorname{sign}(x)$ - term. Recall that standard *PI*-controllers contain the same Proportional and Integration terms (PI terms), but with respect to the state variable x.

In (2.35) it is supposed that

$$\bar{\alpha} > 0 \text{ and } |\phi(t)| \leq \phi_0 < \bar{\beta} \tag{2.36}$$

Of course, similarly [[16]], [[17]], [[19]] it is assumed that the super twisting algorithm is applied for a system of an arbitrary order with scalar control; x and y are state variable of the controller only, while the function $\phi(t)$ depends on both x, y and the system state.

2.4.1.1 Convergence Analysis of a Standard Super - Twist Controller without Adaptation

a) The both state variable are sign-varying, therefore the initial conditions can be selected as follows

$$x(0) = -x_0, x_0 > 0, y(0) = 0$$

In our case $\dot{x}(0) > 0$ and, hence, $x(t)$ is increasing. Denote

$$t_1^* := \inf\{t > 0 : x(t_1^*) = 0, x(t) < 0 \text{ for } t \in [0, t_1^*)\}$$

Next, compare two ODE's:

$$\dot{x}(t) = y(t) - \bar{\alpha}\frac{x(t)}{\sqrt{|x(t)|}},\ x(0) = -x_0 < 0 \tag{2.37}$$

where

$$y(t) = \int_{\tau=0}^{t} \left[\bar{\beta} + \phi(\tau)\right] d\tau$$

satisfying

$$\begin{aligned} y(t) &\geq \int_{\tau=0}^{t} \left[\bar{\beta} - \phi_0\right] d\tau = m \cdot t,\ m := \bar{\beta} - \phi_0 > 0 \\ y(t) &\leq \int_{\tau=0}^{t} \left[\bar{\beta} + \phi_0\right] d\tau = Mt,\ M := \bar{\beta} + \phi_0 \end{aligned} \tag{2.38}$$

and

$$\dot{z}(t) = m \cdot t - \bar{\alpha}\frac{z(t)}{\sqrt{x_0}},\ z(0) = x(0) = -x_0 < 0 \tag{2.39}$$

Obviously that the ODE (2.37) is equivalent to (2.35). Since $|x(t)|$ is decreasing it follows that

$$\frac{1}{\sqrt{|x(t)|}} > \frac{1}{\sqrt{x_0}} := k_0 \text{ for } t > 0$$

For any $t \in [0, t_1^*)$ we have $x(t) < 0$ and $z(t) < 0$ which implies $\dot{x}(t) > \dot{z}(t)$ and, as the result, $x(t) > z(t)$. So that

$$t_1^* > t' := \inf\{t > 0 : z(t) = 0\}$$

The solution to (2.39) is

$$z(t) = \frac{m}{\bar{\alpha}k_0}\left(t - \frac{1}{\bar{\alpha}k_0}\right) + \left[\frac{m}{(\bar{\alpha}k_0)^2} - x_0\right] e^{-\bar{\alpha}k_0 t}$$

and one can conclude that

$$t' = (k_0)^{-1} q\left(\frac{1}{\bar{\alpha}}\right) = \sqrt{x_0} q\left(\frac{1}{\bar{\alpha}}\right) < t_1^*$$

for a large enough $\bar{\alpha}$. Here t' is the solution of the transcendent algebraic equation $z(t') = 0$ and $q(s) \to 0$ when $s \to 0$. By (2.38) it follows that $y(t) \leq Mt$ and hence

$$y(t_1^*) \leq Mt_1^* \leq Mt' = M\sqrt{x_0} q\left(\frac{1}{\bar{\alpha}}\right)$$

b) For $t > t_1^*$ we already have that $x(t) > 0$ and $y(t)$ is a decaying function since

$$\dot{y}(t) = \phi(t) - \bar{\beta}\,\text{sign}(x(t)) = \phi(t) - \bar{\beta} \leq 0$$

that implies

$$y(t) \leq y(t_1^*) - m \cdot t \tag{2.40}$$

For the instant t_1^{**}

$$t_1^{**} := \inf\{t > t_1^* : y(t) = 0\}$$

we have

$$\begin{aligned} t_1^{**} &\leq t_1^* + y(t_1^*)/m = t_1^* + \frac{M}{m}\sqrt{x_0}q\left(\frac{1}{\bar{\alpha}}\right) \\ &= \left(1 + \frac{M}{m}\right)\sqrt{x_0}q\left(\frac{1}{\bar{\alpha}}\right) \end{aligned} \tag{2.41}$$

So, by (2.35) and (2.40)

$$\begin{aligned} x(t_1^{**}) &= \int_{\tau=t_1^*}^{t_1^{**}} \left[y(\tau) - \bar{\alpha}\sqrt{|x(\tau)|}\operatorname{sign}(x(\tau))\right] d\tau \\ &= \int_{\tau=t_1^*}^{t_1^{**}} \left[y(\tau) - \bar{\alpha}\sqrt{|x(\tau)|}\right] d\tau \leq \int_{\tau=t_1^*}^{t_1^{**}} y(t_1^*) d\tau = \\ &\quad y(t_1^*)\left(t_1^{**} - t_1^*\right) \leq y^2(t_1^*)/m \leq \gamma x_0 \end{aligned} \tag{2.42}$$

where

$$\gamma := \frac{M^2}{m}\left[q\left(\frac{1}{\bar{\alpha}}\right)\right]^2 \tag{2.43}$$

Selecting $\bar{\alpha}$ large enough we may conclude that $\gamma \in (0,1)$ and

$$x(t_1^{**}) \leq \gamma x_0$$

Here $x(t_1^{**})$ is an initial value of (2.35) for the second interval $\Delta t_2 := t_2^{**} - t_1^{**}$ where

$$t_2^{**} := \inf\{t > t_1^{**} : y(t) = 0\}$$

Similarly to (2.42)

$$|x(t_2^{**})| \leq \gamma x(t_1^{**}) \leq \gamma^2 x_0 \tag{2.44}$$

c) Iterating this process we may conclude that

$$|x(t_i^{**})| \leq \gamma \left|x(t_{i-1}^{**})\right| \leq \cdots \leq \gamma^i x_0 \tag{2.45}$$

and

$$\begin{aligned} \Delta t_i := t_i^{**} - t_{i-1}^{**} &\leq \sqrt{\left|x(t_{i-1}^{**})\right|}\left(1 + \frac{M}{m}\right) q\left(\frac{1}{\bar{\alpha}}\right) \\ &\leq \gamma^{i/2}\sqrt{x_0}\left(1 + \frac{M}{m}\right) q\left(\frac{1}{\bar{\alpha}}\right) \end{aligned} \tag{2.46}$$

Last two inequalities permit to formulate the following result.

Proposition 2.1. *If*

$$|\phi(t)| \leq \phi_0 < \beta$$

then for any initial value $x(0)$ from bounded domain there exists large enough $\bar{\alpha} > 0$, such that the super-twist procedure (2.35) has a finite time convergence or reaching time proceeding ***the second order sliding mode****, and the following properties holds:*

1)

$$|x(t)| \simeq q\left(\gamma^{i/2}\right) \underset{t\to\infty}{\to} 0$$

$$\gamma := \frac{M^2}{m}\left[q\left(\frac{1}{\bar{\alpha}}\right)\right]^2 \in (0,1)$$

2) The reaching time

$$t_{reach} := \inf_{\bar{t}\geq 0}\{\bar{t} : x(t) = 0 \; for\ all\ t \geq \bar{t}\}$$

is estimated by

$$\begin{aligned} t_{reach} &\leq \sum_{i=1}^{\infty} \Delta t_i \leq \sqrt{x_0}\left(1+\frac{M}{m}\right) q\left(\frac{1}{\bar{\alpha}}\right) \sum_{i=1}^{\infty} \gamma^{i/2} \\ &\leq \frac{\sqrt{\gamma}}{1-\sqrt{\gamma}}\sqrt{x_0}\left(1+\frac{M}{m}\right) q\left(\frac{1}{\bar{\alpha}}\right) \end{aligned} \tag{2.47}$$

The important comments can be done:

- the reaching time tends to zero with gain $\bar{\alpha} \to \infty$;
- a finite-time convergence takes place for any small $m := \bar{\beta} - \phi_0 > 0$;
- the sufficient convergence conditions, derived in previous publications(see, for example, [[18]], [[24]]) led to upper estimate of admissible disturbance less than $0.5\bar{\beta}$. Note that the system is not even asymptotically stable for $\phi_0 \geq \bar{\beta}$. As it follows from (2.35) in this case $y(t)$ is constant or diverging, if the disturbance ϕ is such that $|\phi(t)| \geq \bar{\beta}$, and has sign opposite to control $u(t)$.
- the upper bound (2.47) for the reaching time t_{reach} is proportional to the root of the initial state, namely, $\sqrt{|x_0|}$ (since, in our case $y_0 = 0$) and inverse-proportinal to the parameter $\bar{\alpha}$, i.e., $q\left(\frac{1}{\bar{\alpha}}\right)$, which coincides with the estimates in [[18]] proportional also to ($y_0 = 0$)

$$\sqrt{|x_0|} + |y_0| = \sqrt{|x_0|}$$

2.4.1.2 Simulations of a Super-Twist Control without Adaptation

The figure 2.2 illustrates the dynamics of the super-twist controller with

$$\phi(t) = \phi_0 \sin(\omega t)$$

the following parameters:

$$\bar{\alpha} = 1.2,\ \bar{\beta} = 0.6,\ \phi_0 = 1,\ \omega = 0.09 \text{ and } x(0) = \begin{bmatrix} -0.3\ 0.25 \end{bmatrix}^{\mathsf{T}}$$

One can see a finite-time convergence to zero (approximately in 3.5 sec.) of the first state variable $x(t)$ and the corresponding discontinuous control of the amplitude $\bar{\beta}$.

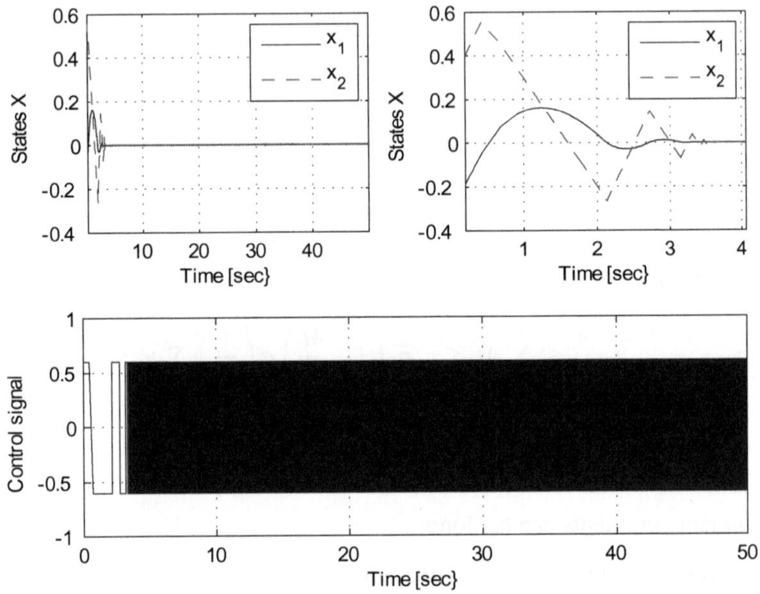

Fig. 2.2 The states and the control signal for the super-twist controller without adaptation of the gain parameter $\bar{\beta}$

2.4.2 *Super-Twist Control with Adaptation*

Denoting

$$x_1 = x,\ \ x_2 = y,\ k := \bar{\beta}$$

the system (2.35) can be represented as

$$\begin{cases} \dot{x}_1 = x_2 - \bar{\alpha}\sqrt{|x_1|}\text{sign}(x_1) \\ \dot{x}_2 = \phi(t) + u \\ u := -k\text{sign}(x_1) \end{cases}$$

or, in the vector format (2.16)

$$\dot{x} = f(t,x) + b(t,x)u$$

with

$$f(t,x) := \begin{pmatrix} x_2 - \bar{\alpha}\sqrt{|x_1|}\mathrm{sign}(x_1) \\ \phi(t) \end{pmatrix}, \; b(t,x) := \begin{pmatrix} 0 \\ 1 \end{pmatrix}$$

Taking

$$\sigma(x) = x_1$$

and permitting for the gain parameter to be time-varying, i.e.,

$$k(t) = \beta(t)$$

we may apply the adaptation procedure (2.25)-(2.26) in spite of the fact that $\left\| \frac{\partial}{\partial x} f(t,x) \right\|^2$ is unbounded since in this case it does not participate directly in the construction of $[\mathrm{sign}(\sigma(x(t)))]_{eq}$. Below we will consider this with more details.

2.4.2.1 The σ-Adaptation Method

Here, following to [[23]], we apply the adaptation law given by

$$\begin{aligned} u(t) &= -k(t)\mathrm{sign}(x_1(t)) \\ \dot{k}(t) &= \begin{cases} k(t)\,|\sigma(x(t))|\,\mathrm{sign}(|\sigma(x(t)| - \varepsilon) & \text{if } k(t) > \bar{\mu} \\ \bar{\mu} & \text{if } k(t) \le \bar{\mu} \end{cases} \end{aligned} \tag{2.48}$$

referred to as "σ–*adaptation*". In (2.48)

$$k(0) = 2.8, \varepsilon = 0.0003 \text{ and } \bar{\mu} = 0.04$$

The specific feature of this procedure is that the adaptation process practically stops after the reaching time t_{reach} when

$$\sigma(x(t)) = x_1(t) = 0$$

for any $t \ge t_{reach}$, and, as the result, the gain parameter $k(t) = \beta(t)$, defining the size of the discontinuous control (or a chattering amplitude) may be still too far from the disturbance level $|\phi(t)| \le \phi_0$ which is minimal possible one guarantying the finite-time convergence. This effect is clearly seen in the Figure 2.3: the reaching time $t_{reach} \simeq 1$ sec., but the gain parameter (the chattering amplitude) remains around the initial level 2.8 (in fact, 3.5) which is too high comparing with $\phi_0 = 1$. So, the adaptation period is to short to decrease significantly the gain parameter $k(t) = \bar{\beta}(t)$, and in sliding mode regime there is no adaptation.

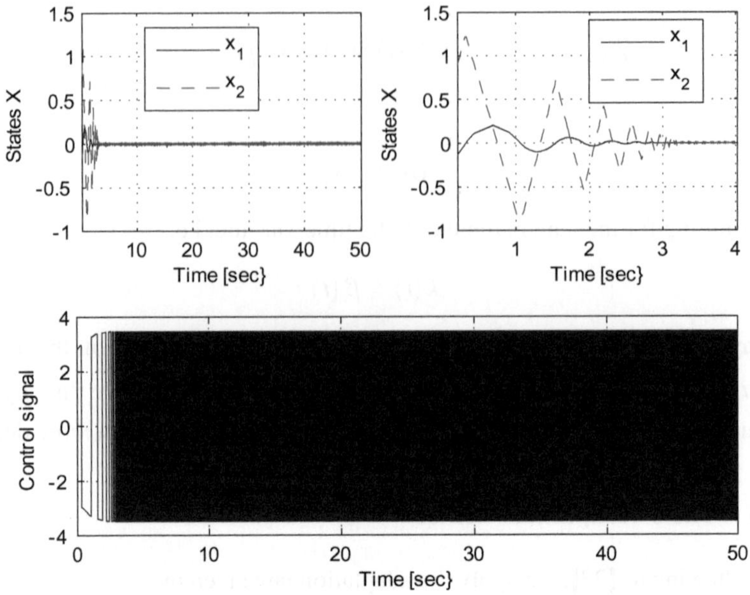

Fig. 2.3 The states and the control signal (with the zooms in the right column) for the super-twist controller with σ-adaptation of the gain parameter $\bar{\beta}$

2.4.2.2 Adaptation Based on the "Equivalent Control"

Theoretical Analysis

The adaptation procedure (2.25)-(2.26) suggested here is applied to minimize the magnitude of discontinuous input $\bar{\beta}\text{sign}(x(t))$ in (2.35). In sliding mode $y(t) \equiv 0$ therefore

$$[\text{sign}(\sigma(x(t)))]_{eq} = \phi(t)/\bar{\beta}(t) = \phi(t)/k(t) \tag{2.49}$$

and the algorithm (2.25)-(2.26) with

$$\lambda = \gamma_1 = 0$$

can be used directly for this case if time derivative of $|\phi(t)|$ is bounded, namely, if

$$\frac{d}{dt}|\phi(t)| \leq L \tag{2.50}$$

Indeed, following to (2.26) and (2.49) for $V(\delta) = \delta^2/2$ we have

$$\dot{V}(\delta(t)) = \delta(t)\dot{\delta}(t) = \delta(t)\frac{d}{dt}\left(\left|[\text{sign}(\sigma(x))]_{eq}\right|\right)$$

If $|\phi(t)|$ is differentiable then

$$\frac{d}{dt}\left(\left|[\text{sign}(\sigma(x))]_{eq}\right|\right) = \frac{1}{k}\frac{d}{dt}|\phi| - \frac{|\phi|}{k^2}\dot{k}$$

and

$$\dot{V}(\delta(t)) = \delta(t)\left(\frac{1}{k}\frac{d}{dt}|\phi| - \frac{|\phi|}{k^2}\dot{k}\right) \tag{2.51}$$

Substitution

$$\dot{k}(t) = \gamma_0 k(t)\,\text{sign}(\delta(t)) \tag{2.52}$$

in (2.51) and using (2.50) imply

$$\dot{V}(\delta(t)) = \delta(t)\left[\frac{1}{k}\frac{d}{dt}|\phi| - \right.$$

$$\left.\frac{|\phi|}{k^2}k\gamma_0\text{sign}(\delta(t))\right] \leq \frac{|\delta(t)|}{k}(L - \phi_0\gamma_0)$$

Taking $\gamma_0 > L/\phi_0$ and denoting

$$\varkappa := \phi_0\gamma_0 - L > 0$$

from the last inequality we get

$$\dot{V}(\delta(t)) \leq -|\delta(t)|\varkappa/k^+ = -\varkappa/k^+\sqrt{2V(\delta(t))}$$

which proofs the finite convergence before $t_f = |\delta(0)|k^+/\varkappa$. So, the following statement can be formulated.

Theorem 2.4 (on adaptive super-twist [[31]]). *The system (2.35) with disturbances $\phi(t)$ having a bounded derivative (fulfilling (2.50)), and with the parameter $\bar{\beta}(t) = k(t)$ adapted on-line according to the adaptation law*

$$\dot{k}(t) = \begin{cases} \gamma_0 k(t)\text{sign}(\delta(t)) - M[k(t) - k^+]_+ + M[\mu - k(t)]_+ \\ \qquad \textit{if}\; 0 < \mu \leq k(t) \leq k^+ \\ 0 \;\textit{otherwise} \end{cases}$$

where $\gamma_0 > L/\mu$ converges in the finite time

$$t_f = |\delta(0)|k^+/(\mu\gamma_0 - L)$$

to the sliding mode regime

$$\sigma(x) = x_1 = 0$$

maintaining within the relation

$$\phi(t)/k(t) = \alpha = 1 - \varepsilon_0$$

for small enough $\varepsilon_0 > 0$.

Numerical Illustration

To demonstrate the properties of the adaptation procedure (2.25)-(2.26), simulation was performed for the case $\sigma(x) = x_1$ with the following parameters:

$$\gamma_0 = 2,\ \phi_0 = 1,\ \mu = 0.04$$

$$\alpha = 0.95,\ k^+ = 10\ ,\ k(0) = 2.8$$

In the simulations, the filter (given in Remark 1) with $\tau = 0.5$ was used to calculate the function $\delta(t)$. We obtained the following dynamics (see the figure):

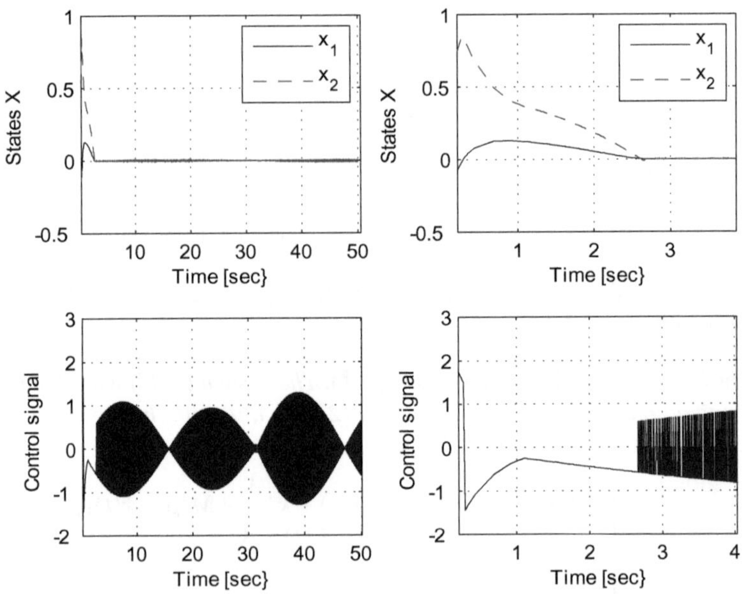

Fig. 2.4 The states and the control signal (with the zooms in the right column) for the super-twist controller with adaptation of the gain parameter $\bar{\beta}$ based on the "equivalent control" signal

Here is clearly seen from Fig.2.5 and Fig.2.4 that gain parameter $k(t) = \beta(t)$, defining the chattering amplitude in the sliding mode (after the reaching time $t_{reach} \simeq 1$ sec.), continues to decrease attaining after 1.3 sec.the level 0.1 and after follows the amplitude of the external perturbation signal $\phi(t)$. Notice that the simulation with lower frequency demonstrates perfect adaptation process.

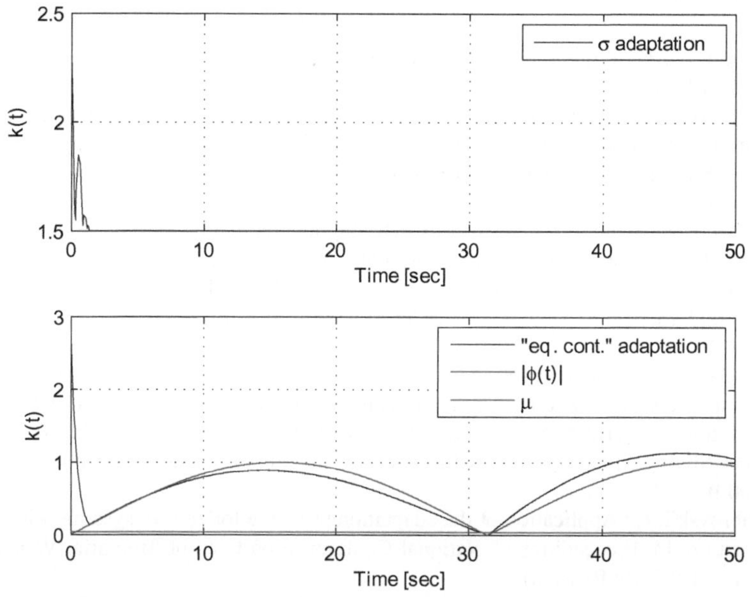

Fig. 2.5 The gain parameter $k(t)$ for σ-adaptation and the adaptation process based on "equivalent control method"

Remark 2.3. The main source of chattering of the super-twisting controller (2.35) is the square root, and not the relay term. A large $\bar{\alpha}$ is really bad selection. If the disturbance is close to the magnitude of the relay function, convergence takes place only for high enough value of $\bar{\alpha}$. So, the magnitude of the relay should be increased to decrease $\bar{\alpha}$. The open problem is to find the trade off to minimize chattering.

2.4.3 Conclusions

In this work an adaptation methodology is developed to find the control gain of a sliding-mode control providing a minimum value of discontinuity resulting in minimization of the chattering effect. The application of this methodology to the super-twist control enables reducing of the control action magnitude to minimum possible value along with a finite-time convergence. The numerical examples clearly illustrate the positive effect of the gain coefficient adaptation being applied to the SOSM controllers (in particularly, to the super-twist controller).

Acknowledgements. The authors would like to express their appreciation to the Ph.D. student of CINVESTAV-IPN M.S. Patricio Ordaz for his help in the numerical realization of the presented methodologies.

References

1. Astrom, K.J., Wittenmark, B.: Adaptive Control, 2nd edn. Addison-Wesley, NY (1994,1989)
2. Barbashin, E.: Introduction in to Stability Theory, Nauka, Moscow (1967) (in Russian)
3. Bartolini, G., Ferrara, A., Usai, E., Utkin, V.I.: On Multi-input Chattering-free Second-order Sliding Mode Control. IEEE Transactions on Automatic Control 45, 1711–1717 (2000)
4. Boiko, I., Fridman, L.: Analysis of Chattering in Continuous Sliding-mode Controllers. IEEE Transaction on Automatic Control 50, 1442–1446 (2005)
5. Boiko, I., Fridman, L., Pisano, A., Usai, E.: Analysis of Chattering in Systems with Second Order Sliding Modes. IEEE Transactions on Automatic Control 52, 2085–2102 (2007)
6. Bondarev, A., Bondarev, S., Kostyleva, N., Utkin, V.: Sliding modes in the systems asymptotic state observers. Automation and Remote Control 46, 49–64 (1985)
7. Bartolini, G., Ferrara, A., Usai, E., Utkin, V.I.: On Multi-input Chattering-free Second-order Sliding Mode Control. IEEE Transactions on Automatic Control 45, 1711–1717 (2000)
8. Dubrovskii, E.: Application of the adaptation principle for control systems with variable structure. In: Proceedings of National Conference on Control, Bulgaria, Varna, vol. 1, part 1 (1967) (in Russian)
9. Dubrovskii, E., Kortnev, A.: Adaptative control systems with variable structure with bounded control actions, in the set of papers. Systems with variable structure for flight automation, Nauka, 34–44 (1968) (in Russian)
10. Emelyanov, S.: Theory of systems with variable structure, Nauka, Moscow (1970) (in Russian)
11. Filippov, A.F.: Differential Equations with Discontinuous Right-hand Side. Kluwer Academic Publishers, Dordrecht (1988)
12. Hall, C., Shtessel, Y.: Sliding mode disturbance observer-based control for a reusable launch vehicle. AIAA Journal of Guidance, Control, and Dynamics 29, 1315–1328 (2006)
13. Huang, Y.-J., Kuo, T.-C., Chang, S.-H.: Adaptive Sliding-mode Control for Nonlinear Systems with Uncertain Parameters. IEEE Transactions on System, Man, and Cybernetics – Part B: Cybernetics 38, 534–539 (2008)
14. Laghrouche, S., Plestan, P., Glumineau, A.: Higher order sliding mode control based on integral sliding surface. Automatica 43, 531–537 (2007)
15. Lee, H., Utkin, V.I.: Chattering Suppression Methods in Sliding Mode Control Systems. Annual Reviews in Control 31, 179–188 (2007)
16. Levant, A.: Sliding Order and Sliding Accuracy in Sliding Mode Control. International Journal of Control 58, 1247–1263 (1993)
17. Levant, A.: Universal SISO Sliding-mode Controllers with Finite-time Convergence. IEEE Transactions on Automatic Control 49, 1447–1451 (2001)
18. Levant, A.: Exact Differentiation of Signals with Unbounded Higher Derivatives. In: Proceedings of 45-th IEEE Conference on Decision and Control (CDC 2006), CA, USA, pp. 5585–5590 (2006)
19. Levant, A.: Principles of 2-sliding Mode Design. Automatica 43, 576–586 (2007)
20. Munoz, D., Sbarbaro, D.: An Adaptive Sliding Mode Controller for Discrete Nonlinear Systems. IEEE Transactions on Industrial Electronics 47, 574–581 (2000)
21. Perreault, D.J., Selders Jr., R.L.: Switching-Ripple-Based Current Sharing for Paralleled Power Converters. IEEE Transaction on Circuit and Systems 46(10), 1264–1274 (1999)

22. Plestan, F., Glumineau, A., Laghrouche, S.: A new algorithm for high- order sliding mode control. International Journal of Robust and Nonlinear Control 18, 441–453 (2008)
23. Plestan, F., Shtessel, Y., Brégeault, V., Poznyak, A.: New methodologies for adaptive sliding mode control. International Journal of Control 83(9), 1907–1919 (2010)
24. Polyakov, A., Poznyak, A.: Reaching time estimation for super-twisting second order sliding mode controller via Lyapunov function designing. IEEE Trans. Automatic Control 54(8), 1951–1955 (2009)
25. Sastry, S., Bodson, M.: Adaptive Control: Stability, Convergence, and Robustness. Prentice-Hall, NY (1994)
26. Slotine, J., Sastry, S.: Tracking control of nonlinear system using sliding surfaces with application to robot manipulators. International Journal of Control 38, 465–492 (1983)
27. Shtessel, Y.B., Moreno, J.A., Plestan, F., Fridman, L.M., Poznyak, A.S.: Super-twisting Adaptive Sliding Mode Control: a Lyapunov Design. In: Proceedings of the 49th IEEE Conference on Decision and Control, Atlanta, GA, USA, pp. 5109–5113 (2010)
28. Tao, C.W., Chan, M.L., Lee, T.T.: Adaptive Fuzzy Sliding Mode Controller for Linear Systems with Mismatched Time-varying Uncertainties. IEEE Transactions on System, Man, and Cybernetics – Part B: Cybernetics 33, 283–294 (2003)
29. Utkin, V.I.: Slides Modes in Control and Optimization. Springer, Berlin (1992)
30. Utkin, V.I., Guldner, J., Shi, J.: Sliding Mode in Control in Electromechanical Systems. Taylor & Francis, London (1999)
31. Utkin, V.I., Poznyak, A.S.: Adaptive sliding mode control with application to super-twist algorithm: Equivalent control method. Automatica 49(1), 39–47 (2013)

Chapter 3
Decentralised Variable Structure Control for Time Delay Interconnected Systems

Xing-Gang Yan and Sarah K. Spurgeon

Abstract. A class of multiple time varying delay interconnected systems with nonlinear disturbances is considered in this Chapter, where both the known and uncertain interconnections involve time delay. A decentralised static output feedback variable structure control is synthesised, which is independent of the time delays, to stabilise the system globally uniformly asymptotically. The stability of the closed loop system is analysed based on the Lyapunov Razumikhin approach. Then, for interconnected systems where each subsystem is square, it is shown that the effects of the uncertain interconnections can be largely rejected by appropriate controllers if the delays are known and the uncertain interconnections are bounded by a class of functions of the outputs and delayed outputs. A case study relating to a river pollution control problem is presented to illustrate the proposed approach.

3.1 Introduction

Interconnected systems exist widely in the real world. Examples include power networks, cellular systems, ecological systems and financial systems. Such systems are often widely distributed in space. A fundamental characteristic of interconnected systems, which holds for both natural and engineered systems, is that they tend to operate in a decentralised manner. For interconnected systems, the presupposition of centrality generally fails to hold due to the lack of centralised information or the lack of a centralised decision making focus. Even with engineered systems, issues such as the economic cost and reliability of communication links, particularly when systems are characterised by geographical separation, limit the appetite

Xing-Gang Yan · Sarah K. Spurgeon
Instrumentation, Control and Embedded Systems Research Group,
School of Engineering and Digital Arts, University of Kent, Canterbury,
Kent CT2 7NT, United Kingdom
e-mail: {X.Yan, S.K.Spurgeon}@kent.ac.uk

B. Bandyopadhyay et al. (Eds.): *Advances in Sliding Mode Control*, LNCIS 440, pp. 55–74.
DOI: 10.1007/978-3-642-36986-5_3

to develop centralised systems. This has motivated the development of a wide literature in the area of decentralised control for interconnected systems, see, for example, [12, 17, 20, 26, 27].

3.1.1 Interconnected Systems

Interconnected systems are often modelled as dynamical equations composed of interconnections between a collection of lower-dimensional subsystems. A fundamental property of any interconnected system is that a perturbation of one subsystem can affect the other subsystems as well as the overall performance of the network. The purpose of control and monitoring paradigms from the domain of engineering within an interconnected system's architecture is thus to minimise the effect of any perturbation or uncertainty on the overall system behaviour.

Large scale interconnected systems were studied from the engineering perspective as early as the 1970's [24]. This early work focussed primarily on linear interconnected systems. The dynamics of large scale natural and engineered interconnected systems are usually highly nonlinear, and thus it is not only the structure of the system which produces complexity but also the nonlinearity of the dynamics. It is clear that although a linear dynamics may approximate the orbit of a nonlinear system locally, it does not permit the existence of the multiple states observed in real networks and does not accommodate global properties of the system. Such global properties can be crucial because they may become significant when the system is perturbed or a subsystem enters a failure state. Increasing requirements on system performance coupled with the ability to model and simulate reality by means of complex, possibly nonlinear, interconnected systems models have motivated increasing contributions to the study of such systems. This interest has been further stimulated by the simultaneous development of nonlinear systems theory and the emergence of powerful mathematical and computational tools which render the formal and constructive study of nonlinear large scale systems increasingly possible.

3.1.2 Decentralised Output Feedback Control

Decentralised output feedback control, where only limited local system state information is available to design any corrective action, has received much attention in the literature and many interesting results have been obtained. Many of these methods are based on Lyapunov approaches or involve adaptive control. In the contributions of Saberi and Khalil [23] and Yan et al. [31], Lyapunov methods are used to form the control scheme and strict structural conditions are imposed on each of the nominal subsystem models. The work also includes some strong limitations on the admissible interconnections. Adaptive control techniques have been employed by Zhou and Wen [34], and Jain and Khorrami [11], but only parametric uncertainty is dealt with due to the limitation of the approach; this is clearly a strong limitation as uncertainty in the possibly nonlinear system structure as well as external perturbations are key factors in any interconnected system of practical significance. The corresponding results can thus only be applied to certain systems with special structure. An appropriate methodology must be able to deal with a broad class of nonlinear subsystems where the subsystems themselves as well as the possibly nonlinear interconnections between them will be uncertain and only limited system state variables will be measurable. So called sliding mode control has been used successfully by many authors in such uncertain, nonlinear scenarios ⟦2, 3, 25, 32⟧. However, the primary focus in the literature has been on centralised control which is problematic to implement in large-scale interconnected systems using decentralised control.

Sliding mode control schemes for large-scale systems have also been proposed in the literature, see for example ⟦7, 13⟧. However, in such contributions it is required that the uncertainties and the interconnections have special structure, or else have linear or polynomial bounds. In addition, most methods focus on the so-called state feedback control case where all state information is assumed available to the control design. Much less attention has been paid to the output feedback, or limited information, case. Lee proposed a decentralised output feedback control scheme using sliding mode techniques ⟦13⟧, where not only were the isolated subsystems required to be linear, but the interconnections were restricted to the linear case as well. Also all of the uncertainties and interconnections are required to adopt a specific structure, i.e. satisfy the so-called matching conditions whereby all the perturbations and interconnection effects are assumed to be implicit in the control injection channels. Recent work has made significant contributions to alleviating these constraints and has developed constructive frameworks for the development of output feedback control strategies based on sliding mode techniques ⟦27–29⟧. This work encompasses nonlinear system representations, uncertainty and unknown perturbations as well as limited available measurements of the system state. A class of nonlinear, large-scale interconnected systems incorporating a broad range of uncertainties has been considered where no statistical information about the uncertainties is imposed.

3.1.3 Time Delay in Interconnected Systems

Interconnections between two or more subsystems in a network are often accompanied by phenomena such as material transfer, energy transfer and information transfer. From a mathematical point of view, transfer phenomena can be represented by delay elements ⟦19⟧. However, for such a time delay interconnected system, the future evolution frequently depends not only on the present state but also on the past history of the system. The presence of even a small delay may greatly affect the performance of a system; a stable system may become unstable, or chaotic behaviour may result ⟦19⟧. This has motivated the importance of the study of interconnected systems in the presence of delay ⟦1⟧.

It should be noted that time delay is another important factor which makes the study of interconnected systems complex ⟦21⟧. Mahmoud and Bingulac ⟦18⟧ considered a class of interconnected systems where delay does not appear in the interconnection terms. Although time delay interconnected systems have been considered, and many results have been achieved [1, 8], most of the existing results are based on the fact that the system states are available. The associated decentralised output feedback results for time-delayed interconnected systems are few [10,15,33]. An output feedback decentralised control scheme is given in [16] where discrete interconnected systems are considered. A class of nonlinear interconnected systems with triangular structure is considered in [10], and an interconnected system composed of a set of single input single output subsystems with dead zone input is considered in [33]. In both [10] and [33], the control schemes are based on dynamical output feedback which increases the computation greatly due to the associated closed-loop system possessing possibly double the order of the actual plant. A decentralised model reference adaptive control scheme is proposed in [15] where the considered interconnections are linear and matched.

Some work considers systems of particular structure, such as the work of Hua and Guan [9] where a triangular structure is assumed. In all of the existing output feedback control strategies for interconnected time delay systems, the nominal isolated subsystems are required to be linear, and the bounds on the disturbances are functions of the outputs and/or largely linear [9, 16, 33]. A class of interconnected systems with time delay is considered in [8] where a model following problem is considered and state feedback is employed. Building on a strong track record of work in the area of control of delay systems [29] and interconnected systems [27,28], recent work has sought to develop a global decentralised static output feedback robust control scheme for interconnected systems [30] where it is assumed that all the time delays are known.

3.1.4 Contribution

In this Chapter, a variable structure control is synthesised to stabilise a class of time delay interconnected systems with nonlinear disturbances. The bounds on the uncertainties are nonlinear and time delayed. Both the isolated subsystems and the interconnections involve multiple time varying delays. A decentralised variable structure control scheme using only output information is proposed firstly which is independent of time delay. Based on the Lyapunov Razumikhin approach, sufficient conditions are derived such that the closed-loop systems formed by the designed control and the considered interconnected systems are globally uniformly asymptotically stable. Then, for interconnected systems composed of a set of square subsystems, it is shown that the effects of the nonlinear interconnections can be largely rejected if their bounds are nonlinear functions of only the outputs and delayed outputs, and the delays are available for design. The limitation that the rate of change of the time delay is less than one, is not required. A compensator, which increases the required computation levels for large-scale interconnected systems, is not required either. A case study on the river pollution problem is presented to demonstrate the work.

3.2 Preliminaries

This section will provide the required notation and some basic results which will be used later in this Chapter.

3.2.1 Notation

In this Chapter, $\mathscr{R}^+$ denotes the nonnegative set of real numbers $\{t \mid t \geq 0\}$. The symbol $\mathscr{C}_{[a,b]}$ represents the set of $\mathscr{R}^n$-valued continuous function on $[a,b]$ and I_n denotes the unit matrix with dimension n. For a matrix A, the expression $A > 0$ ($A < 0$) means that A is symmetric positive (negative) definite and $\lambda_{\max}(A)$ ($\lambda_{\min}(A)$) represents its maximum (minimum) eigenvalue. The symbol $\text{diag}\{A_1, A_2, \cdots, A_n\}$ represents diagonal/block-diagonal matrix with diagonal entries $A_1, A_2, \cdots, A_n$. For vectors $x = (x_1, x_2, \ldots, x_{n_1})^T \in \mathscr{R}^{n_1}$ and $y = (y_1, y_2, \ldots, y_{n_2})^T \in \mathscr{R}^{n_2}$, the expression $f(x,y)$ denotes a function $f(x_1, x_2, \ldots, x_{n_1}, y_1, y_2, \ldots, y_{n_2})$ defined on $\mathscr{R}^{n_1+n_2}$. Finally, $\|\cdot\|$ denotes the Euclidean norm or its induced norm.

3.2.2 Basic Results

Definition 3.1. (see, [6]) A continuous function $\alpha : [0,a) \mapsto [0,\infty)$ is said to belong to class $\mathscr{K}$ if it is strictly increasing and $\alpha(0) = 0$. Further, it is said to belong to class $\mathscr{K}_\infty$ if $a = \infty$ and $\lim_{r\to\infty} \alpha(r) = \infty$.

Consider a time-delay system

$$\dot{x}(t) = f(t, x(t - d(t))) \tag{3.1}$$

with initial condition

$$x(t) = \phi(t), \qquad t \in [-\overline{d}, 0]$$

where $f : \mathscr{R}^+ \times \mathscr{C}_{[-\overline{d},0]} \mapsto \mathscr{R}^n$ takes $\mathscr{R} \times$ (bounded sets of $\mathscr{C}_{[-\overline{d},0]}$) into bounded sets in $\mathscr{R}^n$; $d(t)$ is the time-varying delay and $\overline{d} := \sup_{t\in\mathscr{R}^+}\{d(t)\} < \infty$.

Lemma 3.1. *If there exist class $\mathscr{K}_\infty$ functions $\zeta_1(\cdot)$ and $\zeta_2(\cdot)$, a class $\mathscr{K}$ function $\zeta_3(\cdot)$ and a continuous function $V_1(\cdot) : [-\overline{d}, \infty] \times \mathscr{R}^n \mapsto \mathscr{R}^+$ satisfying*

$$\zeta_1(\|x\|) \leq V_1(t,x) \leq \zeta_2(\|x\|), \qquad t \in \mathscr{R}^+, \quad x \in \mathscr{R}^n$$

such that the time derivative of V_1 along the solution of system (3.1) satisfies

$$\dot{V}_1(t,x) \leq -\zeta_3(\|x\|) \tag{3.2}$$

whenever

$$V_1(t+d, x(t+d)) \leq V_1(t, x(t))$$

for any $d \in [-\overline{d}, 0]$, *then the system (3.1) is uniformly stable. If in addition,* $\zeta_3(\tau) > 0$ *for* $\tau > 0$ *and there exists a continuous nondecreasing function* $\xi(\tau) > \tau$ *for* $\tau > 0$ *such that (3.2) is strengthened to*

$$\dot{V}_1(t,x) \leq -\zeta_3(\|x\|) \quad \text{if} \quad V_1(t+d, x(t+d)) \leq \xi(V_1(t,x(t))) \tag{3.3}$$

for $d \in [-\overline{d}, 0]$, *then the system (3.1) is uniformly asymptotically stable. Further, if in addition* $\lim_{\tau \to \infty} \zeta_1(\tau) = \infty$, *then, the system (3.1) is globally uniformly asymptotically stable.*

Proof. See pages 14-15 in [6].

Lemma 3.1 is the well known Razumikhin Theorem [6]. From Lemma 3.1, the following result can be obtained.

Lemma 3.2. *Consider system (3.1). If there exists a function* $V_0(x) = x^T P x$ *with* $P > 0$ *such that for* $d \in [-\overline{d}, 0]$, *the time derivative of* V_0 *along the solution of system (3.1) satisfies*

$$\dot{V}_0(x) \leq -q_1 \|x\|^2 \quad \text{if} \quad V_0(x(t+d)) \leq q_2 V_0(x(t)) \tag{3.4}$$

for some $q_1 > 0$ *and* $q_2 > 1$, *then system (3.1) is globally uniformly asymptotically stable.*

Proof. See Lemma 1 of Appendix 1 in [30].

Lemma 3.3. *Assume the matrix/vector functions* $H_{ij}(t,x_j) \in \mathscr{R}^{n_i \times m_j}$ *with* n_i *and* m_j *positive integral numbers, and* $x = col(x_1, x_2, \cdots, x_n)$ *where* $x_i \in \mathscr{R}^{n_i}$ *for* $i = 1, 2, \ldots, n$. *Then*

$$\sum_{i=1}^{n} \sum_{\substack{j=1 \\ j \neq i}}^{n} H_{ij}(t,x_j) = \sum_{i=1}^{n} \sum_{\substack{j=1 \\ j \neq i}}^{n} H_{ji}(t,x_i)$$

Proof. From the fact that

$$\sum_{i=1}^{n} \sum_{j=1}^{n} H_{ij}(t,x_j) = \sum_{j=1}^{n} \sum_{i=1}^{n} H_{ij}(t,x_j)$$

it follows that

$$\begin{aligned}
&\textstyle\sum_{i=1}^{n} \sum_{\substack{j=1 \\ j \neq i}}^{n} H_{ij}(t,x_j) \\
&= \textstyle\sum_{i=1}^{n} \sum_{j=1}^{n} H_{ij}(t,x_j) - H_{11}(t,x_1) - H_{22}(t,x_2) - \cdots - H_{nn}(t,x_n) \\
&= \textstyle\sum_{j=1}^{n} \sum_{i=1}^{n} H_{ij}(t,x_j) - \sum_{j=1}^{n} H_{jj}(t,x_j) \\
&= \textstyle\sum_{j=1}^{n} (\sum_{i=1}^{n} H_{ij}(t,x_j) - H_{jj}(t,x_j)) \\
&= \textstyle\sum_{j=1}^{n} \sum_{\substack{i=1 \\ i \neq j}}^{n} H_{ij}(t,x_j) \\
&= \textstyle\sum_{i=1}^{n} \sum_{\substack{j=1 \\ j \neq i}}^{n} H_{ji}(t,x_i)
\end{aligned}$$

Hence the conclusion follows.

The results presented in this section will be used in the later analysis.

3.3 System Description and Basic Assumptions

In this section, the systems considered in this chapter will be presented and basic assumptions will be imposed.

3.3.1 *Interconnected System Description*

Consider a time-varying delayed interconnected system composed of n n_i-th order subsystems

$$\dot{x}_i = A_i x_i + B_i \left(u_i + G_i(t, x_i, x_{id_i})\right) + \sum_{\substack{j=1 \\ j\neq i}}^{n} \left(E_{ij} x_{jd_j} + F_{ij} x_j + \Phi_{ij}(t, x_j, x_{jd_j})\right) \quad (3.5)$$

$$y_i = C_i x_i, \qquad i = 1, 2, \ldots, n, \quad (3.6)$$

where $x := col(x_1, \ldots, x_n)$, $x_i \in \mathscr{R}^{n_i}$, $u_i \in \mathscr{R}^{m_i}$ and $y_i \in \mathscr{R}^{p_i}$ are the state variables, inputs and outputs of the i-th subsystem respectively. The triple (A_i, B_i, C_i) and $E_{ij}, F_{ij} \in \mathscr{R}^{n_i \times n_j}$ with $i \neq j$ represent constant matrices of appropriate dimensions with B_i and C_i of full rank. The functions $G_i(\cdot)$ are matched nonlinear uncertainties in the i-th subsystem. The terms

$$\sum_{\substack{j=1 \\ j\neq i}}^{n} (E_{ij} x_{jd_j} + F_{ij} x_j) \qquad \text{and} \qquad \sum_{\substack{j=1 \\ j\neq i}}^{n} \Phi_{ij}(t, x_j, x_{jd_j})$$

are, respectively, the known and uncertain interconnections of the i-th subsystem; $x_{id_i} := x_i(t - d_i)$ are the delayed states, and the symbols $d_i := d_i(t)$ denote the time-varying delays which are assumed to be known, nonnegative and bounded in $\mathscr{R}^+$, that is

$$\overline{d}_i := \sup_{t \in \mathscr{R}^+} \{d_i(t)\} < \infty, \qquad i = 1, 2, \ldots, n$$

The initial conditions associated with the time delays are given by

$$x_i(t) = \phi_i(t), \qquad t \in [-\overline{d}_i, 0]$$

where $\phi_i(\cdot)$ are continuous in $[-\overline{d}_i, 0]$ for $i = 1, 2, \ldots, n$. It is assumed that all the nonlinear functions are smooth enough such that the unforced interconnected system has a unique continuous solution.

Definition 3.2. Consider system (3.5)–(3.6). The systems

$$\dot{x}_i = A_i x_i + B_i(u_i + G_i(t, x_i, x_{id_i}))$$
$$y_i = C_i x_i, \qquad i = 1, 2, \ldots, n,$$

are called the i-th isolated subsystems of the system (3.5)–(3.6), and the systems

$$\dot{x}_i = A_i x_i + B_i u_i \tag{3.7}$$
$$y_i = C_i x_i, \qquad i = 1, 2, \ldots, n, \tag{3.8}$$

are said to be the i-th nominal isolated subsystems of the system (3.5)–(3.6).

3.3.2 Assumptions

For the interconnected system (3.5)–(3.6), it is required to impose the following conditions.

Assumption 3.1. There exist known continuous functions $\rho_i(\cdot)$ and $\varpi_i(\cdot)$ and constants α_{ij} and β_{ij} such that for $i \neq j,\ i, j = 1, 2, \ldots, n$

$$\|G_i(t, x, x_{id_i})\| \leq \rho_i(t, y_i) + \varpi_i(t, y_i)\|x_{id_i}\| \tag{3.9}$$
$$\|\Phi_{ij}(t, x_j, x_{jd_j})\| \leq \alpha_{ij}\|x_j\| + \beta_{ij}\|x_{jd_j}\| \tag{3.10}$$

Remark 1. Assumption 3.1 is a limitation on the uncertainties that can be tolerated by the system. It is not required that the interconnections are described or bounded by functions of the system outputs. Unlike [22, 33], time delays are involved in the interconnections; and the result obtained in this chapter will be global.

Assumption 3.2. There exist matrices K_i, D_i and $P_i > 0$ such that for $i = 1, 2, \ldots, n$

$$-Q_i := (A_i - B_i K_i C_i)^T P_i + P_i(A_i - B_i K_i C_i) < 0 \tag{3.11}$$
$$B_i^T P_i = D_i C_i \tag{3.12}$$

Remark 2. Assumption 3.2 describes a structural property associated with the triple (A_i, B_i, C_i) which is the standard Constrained Lyapunov Problem (CLP) [5]. A similar limitation has been imposed by many authors (see e.g, [5, 31]). Necessary and sufficient conditions for solving the CLP can be found in [4, 5].

3.3.3 Problem Statement

In this chapter, it is assumed that all the isolated subsystems (3.7) and (3.8) are output feedback stabilisable. The objective is to design a variable structure control law of the form

$$u_i = u_i(t, y_i), \qquad i = 1, 2, \ldots, n \tag{3.13}$$

such that the associated closed-loop system formed by applying the control law in (3.13) to the interconnected system (3.5)–(3.6), is globally uniformly asymptotically stable even in the presence of the uncertainties and time delays. Since the control elements u_i in (3.13) are only dependent on the time t and output y_i, and are independent of time delay, they are called a memoryless decentralised static output feedback control. Then, for interconnected systems with square subsystems, delay dependent decentralised output feedback control elements

$$u_i = u_i(t, y_i, y_{id_i}), \qquad i = 1, 2, \ldots, n$$

are proposed such that the effects of the uncertain interconnections are largely rejected.

3.4 Decentralised Delay Independent Control

In this section, a decentralised output feedback controller which is independent of the time delay will be designed for the interconnected systems (3.5)–(3.6).

3.4.1 *Designed Control*

Consider the control

$$u_i = -K_i y_i - \frac{1}{2\varepsilon_i} D_i y_i \varpi_i^2(t, y_i) + u_i^a(t, y_i), \qquad i = 1, 2, \ldots, n \tag{3.14}$$

where $K_i \in \mathscr{R}^{m_i \times p_i}$ are design parameters satisfying Assumption 3.2, $\varepsilon_i > 0$ are constant and $u_i^a(\cdot)$ are defined by

$$u_i^a(\cdot) := \begin{cases} -\frac{D_i y_i}{\|D_i y_i\|} \rho_i(t, y_i), & D_i y_i \neq 0 \\ 0, & D_i y_i = 0 \end{cases} \tag{3.15}$$

where D_i satisfy (3.12). Since the structure of the control u_i in (3.14) are variable due to $u_i^a(\cdot)$ in (3.15), they are called a variable structure control. Clearly it is decentralised because u_i is only dependent on time t and local output information y_i. Thus u_i in (3.14) are called decentralised output feedback variable structure controllers.

3.4.2 *Main Result*

The following result can now be presented:

Theorem 3.1. *Assume that Assumptions 3.1–3.2 hold. Then, the closed-loop system formed by applying the control (3.14)–(3.15) into system (3.5)–(3.6) is globally* *uniformly asymptotically stable if* $W^T + W > 0$ *where the matrix* $W = [w_{ij}]_{2n\times 2n}$ *is defined by*

$$w_{ij} = \begin{cases} \lambda_{\min}(Q_i) - q\lambda_{\max}(P_i), & 1 \le i = j \le n \\ \lambda_{\min}(P_i) - \varepsilon_i, & n+1 \le i = j \le 2n \\ -2\|P_iF_{ij}\| - 2\alpha_{ij}\|P_i\|, & i \ne j \ \ and \ \ 1 \le i, j \le n \\ -2\|P_iE_{i(j-n)}\| - 2\beta_{i(j-n)}\|P_i\|, & 1 \le i \le n, \ j > n \ and \ j-n \ne i \\ -2\|P_{i-n}E_{(i-n)j}\| - 2\beta_{(i-n)j}\|P_{i-n}\|, & i > n, \ 1 \le j \le n \ and \ i-n \ne j \\ 0, & otherwise \end{cases}$$

for constants $q > 1$ *and* $\varepsilon_i > 0$, *where* α_{ij} *and* β_{ij} *are defined in (3.10) for* $i, j = 1, 2, \ldots, n$, $i \ne j$.

Proof. Applying the control (3.14)–(3.15) into system (3.5)–(3.6), it follows that the closed-loop system is described by

$$\dot{x}_i = A_ix_i + B_i\left(-K_iC_ix_i - \frac{1}{2\varepsilon_i}D_iy_i\varpi_i^2(t,y_i) + u_i^a(t,y_i) + G_i\big(t,x_i,x_{id_i}\big)\right) \\ + \sum_{\substack{j=1\\ j\ne i}}^{n}\Big(E_{ij}x_{jd_j} + F_{ij}x_j + \Phi_{ij}(t,x_j,x_{jd_j})\Big) \tag{3.16}$$

where $u_i^a(\cdot)$ are given by (3.15) for $i = 1, 2, \ldots, n$. For system (3.16), consider the Lyapunov function candidate

$$V(x(t)) = \sum_{i=1}^{n} x_i^T(t)P_ix_i(t) \tag{3.17}$$

where $P_i > 0$ satisfy Assumption 3.2 for $i = 1, 2, \ldots, n$. Then, the time derivative of $V(\cdot)$ along the trajectories of system (3.16) is given by

$$\dot{V} = -\sum_{i=1}^{n} x_i^TQ_ix_i + 2\sum_{i=1}^{n} x_i^TP_iB_i\left(-\frac{1}{2\varepsilon_i}D_iy_i\varpi_i^2(t,y_i) + u_i^a(t,y_i)\right) \\ +2\sum_{i=1}^{n} x_i^TP_iB_iG_i(t,x_i,x_{id_i}) + 2\sum_{i=1}^{n}\sum_{\substack{j=1\\ j\ne i}}^{n} x_i^TP_iE_{ij}x_{jd_j} \\ +2\sum_{i=1}^{n}\sum_{\substack{j=1\\ j\ne i}}^{n} x_i^TP_iF_{ij}x_j + 2\sum_{i=1}^{n}\sum_{\substack{j=1\\ j\ne i}}^{n} x_i^TP_i\Phi_{ij}(t,x_j,x_{jd_j}) \tag{3.18}$$

From (3.9), (3.12) and Young's inequality, it follows that for any $\varepsilon_i > 0$

$$
\begin{aligned}
x_i^T P_i B_i G_i(t,x_i,x_{id_i}) &= (D_i y_i)^T G_i(t,x_i,x_{id_i}) \\
&\le \|D_i y_i\|\rho_i(t,y_i) + \|D_i y_i\|\varpi_i(t,y_i)\|x_{id_i}\| \\
&\le \|D_i y_i\|\rho_i(t,y_i) + \frac{1}{2\varepsilon_i}\|D_i y_i\|^2\varpi_i^2(t,y_i) + \frac{\varepsilon_i}{2}\|x_{id_i}\|^2 \qquad (3.19)
\end{aligned}
$$

From (3.12) and the definition of $u_i^a(\cdot)$ in (3.15), it follows that
i) if $D_i y_i = 0$, then $u_i^a(\cdot) = 0$, and thus

$$
x_i^T P_i B_i u_i^a(t,y_i) + \|D_i y_i\|\rho_i(t,y_i) = \|D_i y_i\|\rho_i(t,y_i) = 0
$$

ii) if $D_i y_i \neq 0$, from the definition of $u_i^a(\cdot)$ in (3.15),

$$
\begin{aligned}
& x_i^T P_i B_i u_i^a(t,y_i) + \|D_i y_i\|\rho_i(t,y_i) \\
& \le -(D_i y_i)^T \frac{D_i y_i}{\|D_i y_i\|}\rho(t,y_i) + \|D_i y_i\|\rho_i(t,y_i) \\
& = 0
\end{aligned}
$$

Thus, from i) and ii) above,

$$
x_i^T P_i B_i u_i^a(t,y_i) + \|D_i y_i\|\rho_i(t,y_i) \le 0, \qquad i = 1,2,\cdots,n \tag{3.20}
$$

Further, from (3.12),

$$
\begin{aligned}
& -\frac{1}{2\varepsilon_i}x_i^T P_i B_i D_i y_i \varpi_i^2(t,y_i) + \frac{1}{2\varepsilon_i}\|D_i y_i\|^2\varpi_i^2(t,y_i) \\
&= -\frac{1}{2\varepsilon_i}x_i^T C_i^T D_i^T D_i y_i \varpi_i^2(t,y_i) + \frac{1}{2\varepsilon_i}\|D_i y_i\|^2\varpi_i^2(t,y_i) \\
&= -\frac{1}{2\varepsilon_i}(D_i y_i)^T D_i y_i \varpi_i^2(t,y_i) + \frac{1}{2\varepsilon_i}\|D_i y_i\|^2\varpi_i^2(t,y_i) = 0 \qquad (3.21)
\end{aligned}
$$

Therefore, from (3.19), (3.20) and (3.21)

$$
\begin{aligned}
2\sum_{i=1}^{n} x_i^T P_i B_i \left(-\frac{1}{2\varepsilon_i} D_i y_i \varpi_i^2(t,y_i) + u_i^a(t,y_i)\right) &+ 2\sum_{i=1}^{n} x_i^T P_i B_i G_i(t,x_i,x_{id_i}) \\
&\le \sum_{i=1}^{n} \varepsilon_i \|x_{id_i}\|^2 \qquad (3.22)
\end{aligned}
$$

From (3.10),

$$
\begin{aligned}
x_i^T P_i \Phi_{ij}(t,x_j,x_{jd_j}) &\le \|x_i\|\,\|P_i\|\left(\alpha_{ij}\|x_j\| + \beta_{ij}\|x_{jd_j}\|\right) \\
&= \alpha_{ij}\|P_i\|\,\|x_i\|\,\|x_j\| + \beta_{ij}\|P_i\|\,\|x_i\|\,\|x_{jd_j}\| \qquad (3.23)
\end{aligned}
$$

Applying (3.22) and (3.23) to equation (3.18) yields

$$\dot{V} \leq -\sum_{i=1}^{n} x_i^T Q_i x_i + \sum_{i=1}^{n} \varepsilon_i \|x_{id_i}\|^2 + 2\sum_{i=1}^{n} \sum_{\substack{j=1\\j\neq i}}^{n} x_i^T P_i E_{ij} x_{jd_j} + 2\sum_{i=1}^{n} \sum_{\substack{j=1\\j\neq i}}^{n} x_i^T P_i F_{ij} x_j$$
$$+2\sum_{i=1}^{n} \sum_{\substack{j=1\\j\neq i}}^{n} \Big(\alpha_{ij}\|P_i\|\,\|x_i\|\,\|x_j\| + \beta_{ij}\|P_i\|\,\|x_i\|\|x_{jd_j}\|\Big) \tag{3.24}$$

From the definition of $V(\cdot)$ in (3.17), it is clear that

$$V(x_{1d_1}, x_{2d_2}, \ldots, x_{nd_n}) \leq qV(x_1, x_2, \ldots, x_n), \qquad (q > 1)$$

implies that

$$q\sum_{i=1}^{n} \lambda_{\max}(P_i)\|x_i\|^2 - \sum_{i=1}^{n} \lambda_{\min}(P_i)\|x_{id_i}\|^2 \geq q\sum_{i=1}^{n} x_i^T P_i x_i - \sum_{i=1}^{n} x_{id_i}^T P_i x_{id_i} \geq 0 \tag{3.25}$$

Therefore, from (3.25) and (3.24), it follows that when $V(x_{1d_1}, \ldots, x_{nd_n}) \leq qV(x_1, \ldots, x_n)$,

$$\dot{V} \leq -\sum_{i=1}^{n} x_i^T Q_i x_i + \sum_{i=1}^{n} \varepsilon_i \|x_{id_i}\|^2 + 2\sum_{i=1}^{n} \sum_{\substack{j=1\\j\neq i}}^{n} x_i^T P_i E_{ij} x_{jd_j} + 2\sum_{i=1}^{n} \sum_{\substack{j=1\\j\neq i}}^{n} x_i^T P_i F_{ij} x_j$$
$$+2\sum_{i=1}^{n} \sum_{\substack{j=1\\j\neq i}}^{n} \Big(\alpha_{ij}\|P_i\|\,\|x_i\|\,\|x_j\| + \beta_{ij}\|P_i\|\,\|x_i\|\|x_{jd_j}\|\Big)$$
$$+q\sum_{i=1}^{n} \lambda_{\max}(P_i)\|x_i\|^2 - \sum_{i=1}^{n} \lambda_{\min}(P_i)\|x_{id_i}\|^2$$

$$\leq -\sum_{i=1}^{n} \Big(\lambda_{\min}(Q_i) - q\lambda_{\max}(P_i)\Big)\|x_i\|^2 - \sum_{i=1}^{n} \Big(\lambda_{\min}(P_i) - \varepsilon_i\Big)\|x_{id_i}\|^2$$
$$+2\sum_{i=1}^{n} \sum_{\substack{j=1\\j\neq i}}^{n} \Big(\|P_i F_{ij}\| + \alpha_{ij}\|P_i\|\Big)\|x_i\|\,\|x_j\|$$
$$+2\sum_{i=1}^{n} \sum_{\substack{j=1\\j\neq i}}^{n} \Big(\|P_i E_{ij}\| + \beta_{ij}\|P_i\|\Big)\|x_i\|\,\|x_{jd_j}\|$$
$$= -\frac{1}{2} Y(W^T + W)Y^T$$
$$\leq -\frac{1}{2}\lambda_{\min}(W^T + W)(\|x\|^2 + \|x_d\|^2)$$
$$\leq -\frac{1}{2}\lambda_{\min}(W^T + W)\|x\|^2$$

where $Y := \left[\|x_1\| \cdots \|x_n\| \; \|x_{1d_1}\| \cdots \|x_{nd_n}\| \right]$. From $W^T + W > 0$, it follows that $\lambda_{\min}(W^T + W) > 0$. Hence the conclusion follows from Lemma 3.2.

Remark 3. Consider (3.10) in Assumption 3.1. The bounds on the uncertain interconnections in system (3.5) are dependent on the systems states, and thus they cannot be employed in the control design since static output feedback is used in this chapter. The effects of such interconnections have been reflected through α_{ij} and β_{ij} in the matrix W.

3.5 Decentralised Control Synthesised for Square Case

Consider interconnected systems where all of the subsystems are square (each subsystem has the same number of outputs as the number of inputs). In this case, it is possible to design decentralised controllers such that the effect of the uncertain interconnections can be largely rejected if delay is available for design. This problem is far from trivial because the uncertain interconnections in each subsystem involve all subsystems' output information while the decentralised control is only allowed to use local output information.

3.5.1 *Controller Design*

The following assumption is imposed on the system (3.5)–(3.6).

Assumption 3.3. It is assumed that $m_i = p_i$ and the time delays d_i are known. The uncertainties $G_i(\cdot)$ satisfy (3.9) and the uncertainties $\Phi_{ij}(\cdot)$ satisfy

$$\|\Phi_{ij}(t, x_j, x_{jd_j})\| \leq \xi_{ij}(t, y_j, y_{jd_j})\|y_j\| \tag{3.26}$$

where functions $\xi_{ij}(\cdot)$ are known nonnegative and continuous for $i \neq j$ and $i, j = 1, 2, \ldots, n$.

Since both B_i and C_i are of full rank, it follows that under Assumption 3.2 the matrix D_i is nonsingular in square case. Then, consider the following control law

$$u_i = -K_i y_i - \frac{1}{2\varepsilon_i} D_i y_i \varpi_i^2(t, y_i) + u_i^a(\cdot) + u_i^b(\cdot) \tag{3.27}$$

where K_i and $u_i^a(\cdot)$ are the same as in (3.14), and $u_i^b(\cdot)$ is defined by

$$u_i^b(\cdot) = -D_i^{-T} y_i \sum_{\substack{j=1 \\ j \neq i}}^{n} \frac{1}{\varepsilon_{ji}} (\lambda_{\max}(P_j))^2 \xi_{ji}^2(t, y_i, y_{id_i}) \tag{3.28}$$

where $\varepsilon_{ji} > 0$ ($j \neq i$) are constants for $i, j = 1, 2, \ldots, n$. It is obvious that the delays are employed in the control design for u_i^b and thus it is required to be known.The result in section 3.5.2 will show that the controllers in (3.27) can largely reject the effects of the uncertain interconnections.

3.5.2 Main Result

Theorem 3.2. *Under Assumptions 3.2 and 3.3, the closed-loop system formed by applying control (3.27) with $u_i^b(\cdot)$ defined in (3.28) into system (3.5)–(3.6) is globally uniformly asymptotically stable if $\Gamma^T + \Gamma > 0$ where the matrix*

$$\Gamma := \begin{bmatrix} \Gamma_{11} & \Gamma_{12} \\ \Gamma_{21} & \Gamma_{22} \end{bmatrix}$$

is defined by

$$\Gamma_{11} := \begin{bmatrix} \Pi_1^a & -2P_1F_{12} & \cdots & -2P_1F_{1n} \\ -2P_2F_{21} & \Pi_2^a & \ddots & \vdots \\ \vdots & \ddots & \ddots & -2P_{n-1}F_{(n-1)n} \\ -2P_nF_{n1} & \cdots & -2P_nF_{n(n-1)} & \Pi_n^a \end{bmatrix}$$

$$\Gamma_{12} := \begin{bmatrix} 0 & -2P_1E_{12} & \cdots & -2P_1E_{1n} \\ -2P_2E_{21} & 0 & \ddots & \vdots \\ \vdots & \ddots & \ddots & -2P_{n-1}E_{(n-1)n} \\ -2P_nE_{n1} & \cdots & -2P_nE_{n(n-1)} & 0 \end{bmatrix}$$

$$\Gamma_{21} := \begin{bmatrix} 0 & -2P_2E_{21} & \cdots & -2P_nE_{n1} \\ -2P_1E_{12} & 0 & \ddots & \vdots \\ \vdots & \ddots & \ddots & -2P_{n-1}E_{(n-1)n} \\ -2P_nE_{n1} & \cdots & -2P_nE_{n(n-1)} & 0 \end{bmatrix}$$

and

$$\Gamma_{22} := diag\left\{\Pi_1^b, \Pi_2^b, \cdots, \Pi_n^b\right\}$$

where $\Pi_i^a := Q_i - (q\lambda_{\max}(P_i) + \Sigma_{\substack{j=1 \\ j\neq i}}^n \varepsilon_{ij})I_{n_i}$ and $\Pi_i^b := (\lambda_{\min}(P_i) - \varepsilon_i)I_{n_i}$ for $i = 1, \ldots, n$ and $q > 1$.

Proof. Consider the uncertain interconnection terms $\sum_{\substack{j=1\\j\neq i}}^{n} \Phi_{ij}(t,x_j,x_{jd_j})$. From the condition (3.26) and Young's inequality ($ab \le \frac{1}{2\varepsilon}a^2 + \frac{\varepsilon}{2}b^2$ for $\varepsilon > 0$),

$$\begin{aligned}
&2x_i^T P_i \Phi_{ij}(t,x_j,x_{jd_j}) \\
&\le 2\lambda_{\max}(P_i)\|x_i\|\xi_{ij}(t,y_j,y_{jd_j})\|y_j\| \\
&\le \varepsilon_{ij}\|x_i\|^2 + \frac{1}{\varepsilon_{ij}}(\lambda_{\max}(P_i))^2\,\xi_{ij}^2(t,y_j,y_{jd_j})\|y_j\|^2
\end{aligned} \tag{3.29}$$

for constant scalars $\varepsilon_{ij} > 0$. Then, from inequalities (3.29)

$$\begin{aligned}
&2\sum_{i=1}^{n}\sum_{\substack{j=1\\j\neq i}}^{n} x_i^T P_i \Phi_{ij}(t,x_j,x_{jd_j}) \\
&\le \sum_{i=1}^{n}\sum_{\substack{j=1\\j\neq i}}^{n} \varepsilon_{ij}\|x_i\|^2 + \sum_{i=1}^{n}\sum_{\substack{j=1\\j\neq i}}^{n}\frac{1}{\varepsilon_{ij}}(\lambda_{\max}(P_i))^2\,\xi_{ij}^2(t,y_j,y_{jd_j})\|y_j\|^2 \\
&= \sum_{i=1}^{n}\Big(\sum_{\substack{j=1\\j\neq i}}^{n}\varepsilon_{ij}\Big)\|x_i\|^2 + \sum_{i=1}^{n}\sum_{\substack{j=1\\j\neq i}}^{n}\frac{1}{\varepsilon_{ji}}(\lambda_{\max}(P_j))^2\xi_{ji}^2(t,y_i,y_{id_i})\|y_i\|^2
\end{aligned} \tag{3.30}$$

where Lemma 3.3 is used to obtain the last equality. From the definition of $u_i^b(\cdot)$ in (3.28),

$$\begin{aligned}
&x_i^T P_i B_i u_i^b(\cdot) + \sum_{\substack{j=1\\j\neq i}}^{n}\frac{1}{\varepsilon_{ji}}(\lambda_{\max}(P_j))^2\xi_{ji}^2(t,y_i,y_{id_i})\|y_i\|^2 \\
&\le -x_i^T C_i^T D_i^T D_i^{-T} y_i \sum_{\substack{j=1\\j\neq i}}^{n}\frac{1}{\varepsilon_{ji}}(\lambda_{\max}(P_j))^2\xi_{ji}^2(t,y_i,y_{id_i}) \\
&\quad + \sum_{\substack{j=1\\j\neq i}}^{n}\frac{1}{\varepsilon_{ji}}(\lambda_{\max}(P_j))^2\xi_{ji}^2(t,y_i,y_{id_i})\|y_i\|^2 \\
&= 0
\end{aligned} \tag{3.31}$$

Therefore, from (3.30) and (3.31)

$$2\sum_{i=1}^{n} x_i^T P_i B_i u_i^b(t,y_i) + 2\sum_{i=1}^{n}\sum_{\substack{j=1\\j\neq i}}^{n} x_i^T P_i \Phi_{ij}(t,x_j,x_{jd_j}) \le \sum_{i=1}^{n}\sum_{\substack{j=1\\j\neq i}}^{n}\varepsilon_{ij}\|x_i\|^2 \tag{3.32}$$

Consider the same Lyapunov function as given in (3.17). Following the analysis and proof in Theorem 1, it is straightfoward to see that when $V(x_{1d_1},\dots,x_{nd_n}) \le qV(x_1,\dots,x_n)$,

$$
\begin{aligned}
\dot{V} \leq & -\sum_{i=1}^{n} x_i^T Q_i x_i + \sum_{i=1}^{n} \varepsilon_i \|x_{id_i}\|^2 + 2\sum_{i=1}^{n}\sum_{\substack{j=1\\ j\neq i}}^{n} x_i^T P_i E_{ij} x_{jd_j} + 2\sum_{i=1}^{n}\sum_{\substack{j=1\\ j\neq i}}^{n} x_i^T P_i F_{ij} x_j \\
& + \sum_{i=1}^{n}\sum_{\substack{j=1\\ j\neq i}}^{n} \varepsilon_{ij}\|x_i\|^2 + q\sum_{i=1}^{n} \lambda_{\max}(P_i)\|x_i\|^2 - \sum_{i=1}^{n} \lambda_{\min}(P_i)\|x_{id_i}\|^2 \\
\leq & -\sum_{i=1}^{n} x_i^T \Big(Q_i - \Big(q\lambda_{\max}(P_i) + \sum_{\substack{j=1\\ j\neq i}}^{n} \varepsilon_{ij}\Big) I_{n_i}\Big) x_i - \sum_{i=1}^{n} \Big(\lambda_{\min}(P_i) - \varepsilon_i\Big) x_{id_i}^T x_{id_i} \\
& + 2\sum_{i=1}^{n}\sum_{\substack{j=1\\ j\neq i}}^{n} x_i^T P_i E_{ij} x_{jd_j} + 2\sum_{i=1}^{n}\sum_{\substack{j=1\\ j\neq i}}^{n} x_i^T P_i F_{ij} x_j \\
= & -\frac{1}{2} Z^T(\Gamma^T + \Gamma) Z \leq -\frac{1}{2}\lambda_{\min}(\Gamma^T + \Gamma)\|Z\|^2 \leq -\frac{1}{2}\lambda_{\min}(\Gamma^T + \Gamma)\|x\|^2
\end{aligned}
$$

where $Z := col(x_1, \cdots, x_n, x_{1d_1}, \ldots, x_{nd_n})$.

Hence, the conclusion follows from $\Gamma + \Gamma^T > 0$.

Remark 4. From the proof of Theorem 3.2, it is clear to see that the only terms resulting from the uncertain interconnections, are $\sum_{\substack{j=1\\ j\neq i}}^{n} \varepsilon_{ij}$ in the matrix Γ. Compared the matrix W in Theorem 3.1 and the matrix Γ in Theorem 3.2, it is straightforward to see that the effects of the uncertain interconnections have been largely rejected by the control (3.27) because the terms $\sum_{\substack{j=1\\ j\neq i}}^{n} \varepsilon_{ij}$ appeared in the matrix Γ, can be very small if the parameters ε_{ij} are chosen to be small enough although small ε_{ij} usually result in high gain control.

3.6 Case Study—River Pollution Control Problem

Consider a two-reach model of a river pollution control problem [[14]]. It is assumed that the concentration of biochemical oxygen demand (BOD) for the first subsystem is perturbed by a time delay. Then, the system can be described by (See, [[30]])

$$
\dot{x}_1 = \underbrace{\begin{bmatrix} -1.32\delta & 0 \\ -0.32 & -1.2 \end{bmatrix}}_{A_1} x_1 + \underbrace{\begin{bmatrix} 0.1 \\ 0 \end{bmatrix}}_{B_1} \Big(u_1 + \underbrace{(-13.2(1-\delta)) y_{1d_1}}_{G_1(\cdot)}\Big) + \Phi_{12}(\cdot) \tag{3.33}
$$

$$
\begin{aligned}
\dot{x}_2 = & \underbrace{\begin{bmatrix} -1.32 & 0 \\ -0.32 & -1.2 \end{bmatrix}}_{A_2} x_2 + \underbrace{\begin{bmatrix} 0.1 \\ 0 \end{bmatrix}}_{B_2} (u_2 + G_2(\cdot)) + \underbrace{\begin{bmatrix} 0.9\delta & 0 \\ 0 & 0 \end{bmatrix}}_{E_{21}} x_{1d_1} \\
& + \underbrace{\begin{bmatrix} 0.9 & 0 \\ 0 & 0.9 \end{bmatrix}}_{F_{21}} x_1 + \underbrace{\begin{bmatrix} -0.9\delta y_1 \\ 0 \end{bmatrix}}_{\Phi_{21}(\cdot)}
\end{aligned} \tag{3.34}
$$

$$y_1 = \underbrace{[1 \quad 0]}_{C_1} x_1, \qquad y_2 = \underbrace{[1 \quad 0]}_{C_2} x_2 \tag{3.35}$$

where $x_1 := col(x_{11}, x_{12})$ and $x_2 := col(x_{21}, x_{22})$. The variables x_{i1} and x_{i2} represent the concentration of the BOD and the concentration of dissolved oxygen respectively, and the control u_i are the BOD of the effluent discharge into the river for $i = 1, 2$. The constant $\delta \in [0, 1]$ is the retarded coefficient. The uncertainties $G_2(\cdot)$ and $\Phi_{12}(\cdot)$ are added to illustrate the results obtained.

By direct computation, it is obtained that the matrix $W + W^T$ defined in Theorem 3.1 is not positive definite even if it is assumed that both $G_2(\cdot)$ and $\Phi_{12}(\cdot)$ are zero. Therefore, for the system (3.33)–(3.35), the result in Theorem 3.1 does not hold.

Next a decentralised controller based on the result given in section 3.5 will be proposed. Rewrite the system (3.33)–(3.35) as follows

$$\dot{x}_1 = \underbrace{\begin{bmatrix} -1.32\delta & 0 \\ -0.32 & -1.2 \end{bmatrix}}_{A_1} x_1 + \underbrace{\begin{bmatrix} 0.1 \\ 0 \end{bmatrix}}_{B_1} \Big(u_1 + \underbrace{(-13.2(1-\delta))y_{1d_1}}_{G_1(\cdot)}\Big) + \Phi_{12}(\cdot) \tag{3.36}$$

$$\dot{x}_2 = \underbrace{\begin{bmatrix} -1.32 & 0 \\ -0.32 & -1.2 \end{bmatrix}}_{A_2} x_2 + \underbrace{\begin{bmatrix} 0.1 \\ 0 \end{bmatrix}}_{B_2} (u_2 + G_2(\cdot)) + \underbrace{\begin{bmatrix} 0.9\delta & 0 \\ 0 & 0 \end{bmatrix}}_{E_{21}} x_{1d_1}$$
$$+ \underbrace{\begin{bmatrix} 0 & 0 \\ 0 & 0.9 \end{bmatrix}}_{F_{21}} x_1 + \underbrace{\begin{bmatrix} (1-\delta)0.9y_1 \\ 0 \end{bmatrix}}_{\Phi_{21}(\cdot)} \tag{3.37}$$

$$y_1 = \underbrace{[1 \quad 0]}_{C_1} x_1, \qquad y_2 = \underbrace{[1 \quad 0]}_{C_2} x_2 \tag{3.38}$$

It is assumed that

$$|G_2(\cdot)| \le \underbrace{1 + \sin y_2}_{\rho_2} + \underbrace{|y_2|}_{\varpi_2} \|x_{2d_2}\|, \qquad \|\Phi_{12}\| \le \underbrace{|y_2 y_{2d_2}| \sin^2 t}_{\xi_{12}} |y_2| \tag{3.39}$$

Let $\rho_1 = 0$, $\varpi_1(\cdot) = 13.2(1-\delta)$ and $\xi_{21} = 0.9(1-\delta)$. It is clear to see that the Assumption 3.3 hold. Then let $K_1 = 20$, $K_2 = 30$ and

$$Q_1 = \begin{bmatrix} 4.5280 & 0.3200 \\ 0.3200 & 2.4000 \end{bmatrix}, \qquad Q_2 = \begin{bmatrix} 8.6400 & 0.3200 \\ 0.3200 & 2.4000 \end{bmatrix}$$

The solutions to the Lyapunov equations in (3.11) are $P_1 = P_2 = I_2$ and the equations (3.12) are satisfied with $D_1 = D_2 = 0.1$. Comparing system (3.36)–(3.37) with the system (3.5)–(3.6), it is straightforward to see that $E_{12} = F_{12} = 0$. Let $\varepsilon_1 = \varepsilon_2 = 0.5$ and $\varepsilon_{12} = \varepsilon_{21} = 0.1$. By direct computation,

$$\Gamma_{11} = \begin{bmatrix} 3.4180 & 0.3200 & 0 & 0 \\ 0.3200 & 1.2900 & 0 & 0 \\ 0 & 0 & 7.5300 & 0.3200 \\ 0 & -1.8000 & 0.3200 & 1.2900 \end{bmatrix}$$

$$\Gamma_{22} = \begin{bmatrix} 0.5000 & 0 & 0 & 0 \\ 0 & 0.5000 & 0 & 0 \\ 0 & 0 & 0.5000 & 0 \\ 0 & 0 & 0 & 0.5000 \end{bmatrix}$$

$$\Gamma_{12} = \begin{bmatrix} 0 & 0 & 0 & 0 \\ 0 & 0 & 0 & 0 \\ -0.3600 & 0 & 0 & 0 \\ 0 & 0 & 0 & 0 \end{bmatrix}, \quad \Gamma_{21} = \begin{bmatrix} 0 & 0 & -0.3600 & 0 \\ 0 & 0 & 0 & 0 \\ 0 & 0 & 0 & 0 \\ 0 & 0 & 0 & 0 \end{bmatrix}$$

and the matrix $\Gamma^T + \Gamma$ where Γ is defined in Theorem 3.2, is positive definite. Clearly the controllers (3.27)–(3.28) are well defined, and, from Theorem 3.2, they stabilise the system (3.33)-(3.35) globally asymptotically.

For simulation purposes, choose $\sigma = 0.20$ and assume the delays are chosen as $d_1(t) = 3 - 2\sin(t)$ and $d_2(t) = 2 - \cos t$, and the delay related initial conditions are chosen as $\phi_1(t) = col(2\cos t, 1)$ and $\phi_2(t) = col(0, 1 - \sin(t))$. The simulation results shown in Figure 3.1 are as expected.

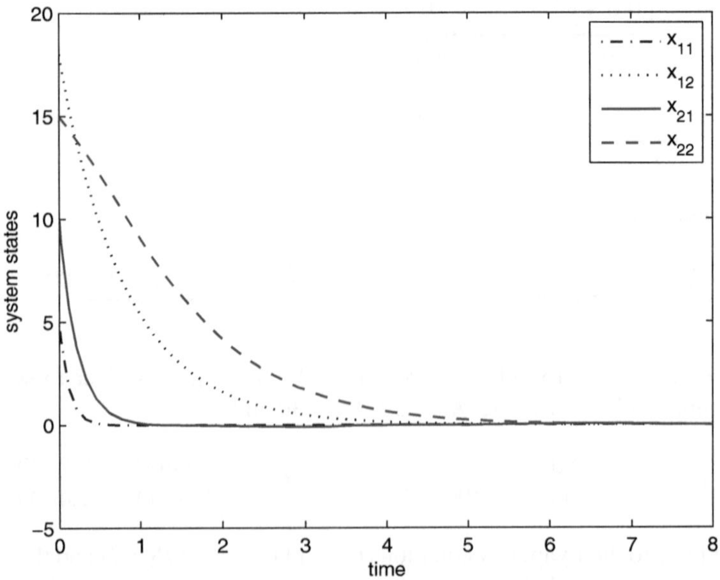

Fig. 3.1 The time responses of the state variables of system (3.33)–(3.34)

3.7 Conclusions

This Chapter has presented two control strategies based on different classes of uncertain interconnections. The proposed control schemes are decentralised and based only on output information, which is convenient for real implementation. The differences between the inaccessible bounds which cannot be used in the control design, and the accessible bounds on the uncertain interconnections which can be employed in the control design, are shown. The proposed approach can be used to accommodate mismatched uncertain interconnections if the delay is known and the bounds on the uncertain interconnections are a class of functions of the outputs and delayed outputs. The limitation on the rate of change of the time varying delay is not required, as is required using the Lyapunov-Krasovskii approach. The case study shows the practicability of the proposed approach.

References

1. Bakule, L.: Decentralized control: an overview. Annual Reviews in Control 32(1), 87–98 (2008)
2. Edwards, C., Akoachere, A., Spurgeon, S.K.: Sliding-mode output feedback controller design using linear matrix inequalities. IEEE Trans. on Automat. Control 46(2), 115–119 (2001)
3. Edwards, C., Spurgeon, S.K.: Sliding mode control: theory and applications. Taylor and Francis Ltd., London (1998)
4. Edwards, C., Yan, X.G., Spurgeon, S.K.: On the solvability of the constrained Lyapunov problem. IEEE Trans. on Automat. Control 52(10), 1982–1987 (2007)
5. Galimidi, A.R., Barmish, B.R.: The constrained Lyapunov problem and its application to robust output feedback stabilization. IEEE Trans. on Automat. Control 31(5), 410–419 (1986)
6. Gu, K., Kharitonov, V.L., Chen, J.: Stability of time-delay systems. Birkhäuser, Boston (2003)
7. Hsu, K.C.: Decentralized variable-structure control design for uncertain large-scale systems with series nonlinearities. Int. J. Control 68(6), 1231–1240 (1997)
8. Hua, C., Ding, S.X.: Model following controller design for large-scale systems with time-delay interconnections and multiple dead-zone inputs. IEEE Trans. on Automat. Control 56(4), 962–968 (2011)
9. Hua, C., Guan, X.: Output feedback stabilization for time-delay nonlinear interconnected systems using neural networks. IEEE Trans. on Neural Networks 19(4), 673–688 (2008)
10. Hua, C.C., Wang, Q.G., Guan, X.P.: Memoryless state feedback controller design for time delay systems with matched uncertain nonlinearities. IEEE Trans. on Automat. Control 53(3), 801–807 (2008)
11. Jain, S., Khorrami, F.: Decentralized adaptive output feedback design for large-scale nonlinear systems. IEEE Trans. on Automat. Control 42(5), 729–735 (1997)
12. Jiang, Z.P.: Decentralized disturbance attenuating output-feedback trackers for large-scale nonlinear systems. Automatica 38(8), 1407–1415 (2002)
13. Lee, J.L.: On the decentralized stabilization of interconnected variable structure systems using output feedback. Journal of the Franklin Institute 332(5), 595–605 (1995)

14. Lunze, J.: Feedback control of Large scale systems. Prentice Hall International (UK) Ltd., Hemel Hempstead (1992)
15. Mirkin, B.M., Gutman, P.: Decentralized output-feedback MRAC of linear state delay systems. IEEE Trans. on Automat. Control 48(9), 1613–1619 (2003)
16. Mahmoud, M.S.: Decentralized output-feedback stabilization for interconnected discrete systems with unknown delays. Optimal Control Applications & Methods 31(6), 529–545 (2010)
17. Mahmoud, M.S.: Decentralized Systems with Design Constraints. Springer-Verlag London Limited (2011)
18. Mahmoud, M.S., Bingulac, S.: Robust design of stabilizing controllers for interconnected time-delay systems. Automatica 34(5), 795–800 (1998)
19. Michiels, W., Niculescu, S.I.: Stability and stabilization of time-delay systems: an eigenvalue-based approach. The Society for Industrial and Applied Mathematics, Philadelphia (2007)
20. Panagi, P., Polycarpou, M.M.: Decentralized fault tolerant control of a class of interconnected nonlinear systems. IEEE Trans. on Automat. Control 56(1), 178–184 (2011)
21. Richard, J.P.: Time-delay systems: An overview of some recent advances and open problems. Automatica 39(10), 1667–1694 (2003)
22. Rodellar, J., Leitmann, G., Ryan, E.P.: Output feedback control of uncertain coupled systems. Int. J. Control 58(2), 445–457 (1993)
23. Saberi, A., Khalil, H.: Decentralized stabilization of interconnected systems using output feedback. Int. J. Control 41(6), 1461–1475 (1985)
24. Sandell, N.R., Varaiya, P., Athans, M., Safonov, M.G.: Survey of decentralized control methods for large-scale systems. IEEE Trans. on Automat. Control 23(2), 108–128 (1978)
25. Utkin, V.I.: Sliding modes in control optimization. Springer, Berlin (1992)
26. Xie, S., Xie, L.: Decentralized stabilization of a class of interconnected stochastic nonlinear systems. IEEE Trans. on Automat. Control 45(1), 132–137 (2000)
27. Yan, X.G., Edwards, C., Spurgeon, S.K.: Decentralised robust sliding mode control for a class of nonlinear interconnected systems by static output feedback. Automatica 40(4), 613–620 (2004)
28. Yan, X.G., Spurgeon, S.K., Edwards, C.: Decentralised sliding mode control for nonminimum phase interconnected systems based on a reduced-order compensator. Automatica 42(10), 1821–1828 (2006)
29. Yan, X.G., Spurgeon, S.K., Edwards, C.: Sliding mode control for time-varying delayed systems based on a reduced-order observer. Automatica 46(8), 1354–1362 (2010)
30. Yan, X.G., Spurgeon, S.K., Edwards, C.: Global decentralised static output feedback sliding mode control for interconnected time-delay systems. IET Control Theory and Applications 6(2), 192–202 (2012)
31. Yan, X.G., Wang, J., Lü, X., Zhang, S.: Decentralized output feedback robust stabilization for a class of nonlinear interconnected systems with similarity. IEEE Trans. on Automat. Control 43(2), 294–299 (1998)
32. Zak, S.H., Hui, S.: Output feedback variable structure controllers and state estimators for uncertain/nonlinear dynamic systems. IEE Proc. Part D: Control Theory Appl. 140(1), 41–50 (1993)
33. Zhou, J.: Decentralized adaptive control for large-scale time-delay systems with deadzone input. Automatica 44(7), 1790–1799 (2008)
34. Zhou, J., Wen, C.: Decentralized backstepping adaptive output tracking of interconnected nonlinear systems. IEEE Trans. on Automat. Control 53(10), 2378–2384 (2008)

Chapter 4
On the Second Order Sliding Mode Approach to Distributed and Boundary Control of Uncertain Parabolic PDEs

Y. Orlov, A. Pisano, and E. Usai

Abstract. This chapter addresses the Lyapunov-based design of distributed and boundary second order sliding mode (2-SM) controllers in the domain of distributed-parameters systems (DPSs). New distributed control results are given and, additionally, an already presented boundary control approach is briefly recalled to enhance the tutorial value of the chapter. Non-standard Lyapunov functionals are invoked to establish the relevant stability and convergence results in appropriate functional spaces. As the novelty, the state tracking problem for an uncertain reaction-diffusion process with spatially varying parameters and non-homogeneous mixed boundary conditions is first addressed. The reference profile is both time and space dependent, and the process is affected by a smooth distributed disturbance. The proposed robust synthesis of the distributed control input is formed by linear PI-type feedback design and the "Super-Twisting" second-order sliding-mode control algorithm, suitably combined and re-worked in the infinite-dimensional setting.

In the second part of the chapter, recently achieved results [20] concerning the regulation of an uncertain heat process with collocated boundary sensing and actuation are recalled. The underlying heat process is governed by an uncertain parabolic partial differential equation (PDE) with controlled mixed boundary conditions, it exhibits an unknown spatially varying diffusivity parameter, and it is affected by a smooth boundary disturbance. The proposed robust synthesis is formed by combining linear PD-type feedback design and the "Twisting" second-order sliding-mode control algorithm. The stability of the twisting-based boundary controller can be investigated by means of a totally different family of Lyapunov functionals as compared to that used for the super-twisting-based distributed control scheme, and the

Y. Orlov
CICESE Research Center, Ensenada (Mexico)
e-mail: yorlov@cicese.mx

A. Pisano · E. Usai
Universitá degli studi di Cagliari, Cagliari, Italy,
Dipartimento di Ingegneria Elettrica ed Elettronica (DIEE)
e-mail: {pisano,eusai}@diee.unica.it

B. Bandyopadhyay et al. (Eds.): *Advances in Sliding Mode Control*, LNCIS 440, pp. 75–95.
DOI: 10.1007/978-3-642-36986-5_4

approach is recalled in order to give a more complete overview of the available second-order sliding mode designs for PDEs. The presented control schemes are supported by simulation results.

Keywords: Distributed parameter systems, Parabolic PDEs, Heat equation, Reaction-diffusion equation, Boundary control, Second-order sliding mode control, Lyapunov analysis.

4.1 Introduction

Sliding-mode control has long been recognized as a powerful control method to counteract non-vanishing external disturbances and unmodelled dynamics when controlling dynamical systems of finite and infinite dimension [[27]].

Presently, the discontinuous control synthesis in the infinite-dimensional setting is well documented [[12, 15, 16, 18, 19]], and it is generally shown to retain the main robustness features as those possessed by its finite-dimensional counterpart. Other robust control paradigms have been fruitfully applied in the infinite dimensional setting such as adaptive and model-reference control (see [[4, 10]]), geometric and Lyapunov-based design (see [[2]]), H_∞ and LMI-based design (see [[6]]). It should be noted that the latter paradigms are capable of attenuating vanishing disturbances only, whereas the former discontinuous control is additionally capable of rejecting persistent disturbances with an *a priori* known bound on their L_2 norm.

In the present chapter we basically present some results concerning the tracking and regulation of infinite dimensional processes by means, respectively, of distributed and boundary control techniques based on the second-order sliding mode concept. In some recent authors' publications (see [[22–24]])) two finite dimensional robust control algorithms, namely, the "Super-Twisting" and "Twisting" second-order sliding-mode (2-SM) controllers (see [[5, 13]]) for details on these controllers) have been generalized to the infinite-dimensional setting and applied, in distributed form, for controlling certain heat and wave processes with uncertain constant parameters. In [[25]] the regulation of a heat process with collocated boundary sensing and actuation and uncertain constant diffusivity was tackled by relying on a suitable combination between linear PD design and the Twisting algorithm. In a more recent authors' work [[21]], the regulation problem for an uncertain reaction-diffusion process with spatially varying parameters and non-homogeneous mixed boundary conditions, and affected by a smooth distributed disturbance, was addressed. The proposed robust synthesis of the distributed control input is formed by linear PI-type feedback design and the "Super-Twisting" second-order sliding-mode control algorithm, suitably combined and re-worked in the infinite-dimensional setting. The reference profile is admitted to be space dependent but must be time-invariant.

As the novelty, we shall extend the results of [[21]] by addressing the *tracking* problem with a reference profile which is both time and space dependent. This generalization is far from being trivial as it requires the imposition of an additional assumption concerning the admitted disturbance, that was not required in the regulation case of [21], as well as corresponding significant changes in the Lyapunov analysis procedure. We put the constraint that the distributed control input must be a continuous (although possibly non-smooth) function of time.

The additional concern of the present chapter is to investigate the use of the second order sliding mode approach within the framework of Parabolic PDEs with collocated boundary sensing and actuation. The boundary control problem for heat processes was studied, e.g., in [6, 9, 26] under more strict assumptions on the admitted uncertainties and perturbations compared to those made in the present work. Here we address the boundary control problem for an uncertain heat process with uncertain and spatially varying diffusivity parameter and controlled Robin's boundary conditions.

Summarizing the results previously presented in [20], an appropriate combination between linear PD-type feedback design and the "Twisting" second-order sliding-mode control algorithm (see [13, 16] for details on the application of this algorithm in the finite-dimensional setting) is considered in order to provide system's stability in a suitable Sobolev space, involving spatial state derivatives up to the second order, while rejecting a class of non-vanishing matched perturbations of arbitrary shape, possibly unbounded in magnitude, requiring just the knowledge of a constant upper bound to the magnitude of the disturbance time derivative. In the resulting closed-loop system, the discontinuous 2-SM controller is connected to the plant input through a dynamical filter (an integrator) thereby augmenting the system state with its time derivative. While passing through the filter, the discontinuous signal is smoothed out, and the so-called chattering phenomenon, extremely undesired in practice, is thus attenuated.

The rest of the paper is structured as follows. Some notations are introduced in the remainder of the Introduction. Section 2 deals with the new results concerning the distributed control of an uncertain reaction-diffusion equation. Particularly, Subsection 2.1 deals with the problem formulation, Subsection 2.2 presents the proposed controller and provides the relevant Lyapunov based stability analysis. Subsection 2.3 illustrates some numerical simulation results. Section 3 summarizes a previously presented result concerning the boundary regulation of an uncertain heat process. The three Subsections 3.1, 3.2 and 3.3 deal with, respectively, the problem formulation, the description and stability analysis of the suggested scheme, and the illustration of some simulation results. Finally, Section 4 gives some concluding remarks pointing out possible directions of improvement of the proposed results.

Notation

The notation used throughout is fairly standard. $L_2(0,1)$ stands for the Hilbert space of square integrable functions $z(\zeta)$, $\zeta \in (0,1)$, whose L_2-norm is given by

$$\|z(\cdot)\|_2 = \sqrt{\int_0^1 z^2(\zeta)d\zeta}. \tag{4.1}$$

$W^{0,2}(0,1)$denotes the Hilbert space $L_2(0,1)$. $W^{1,2}(0,1)$ denotes the Sobolev space of absolutely continuous scalar functions $z(\zeta)$ on $(0,1)$ with square integrable derivative $z_\zeta(\zeta)$ and the norm

$$\|z(\cdot)\|_{1,2} = \sqrt{\|z(\cdot)\|_2 + \|z_\zeta(\cdot)\|_2} \tag{4.2}$$

$W^{2,2}(0,1)$ denotes the Sobolev space of absolutely continuous scalar functions $z(\zeta)$ on $(0,1)$ with square integrable derivatives $z^{(i)}(\zeta)$ up to the order $i=2$ and the weighted norm

$$\|z(\cdot)\|_{2,2,\varphi} = \sqrt{\|z(\cdot)\|_2 + \|z_\zeta(\cdot)\|_2 + \|[\varphi(\cdot)z_\zeta(\cdot)]_\zeta\|_2} \tag{4.3}$$

which depends on the weighting function $\varphi(\cdot) \in W^{1,2}(0,1)$. More generally, $W^{\ell,2}(0,1)$ denotes the Sobolev space of absolutely continuous scalar functions $z(\zeta)$, $\zeta \in [0,1]$ with square integrable derivatives $z^{(i)}(\zeta)$ up to the order $\ell \geq 1$. A non-standard notation stands for

$$\Omega^{4,2}(0,1) = \{\omega(\xi) \in W^{4,2}(0,1) : \sqrt{|\omega(\xi)|}\mathtt{sign}(\omega(\xi)) \in W^{2,2}(0,1)\}. \tag{4.4}$$

4.2 Distributed Control of Reaction-Diffusion Processes

4.2.1 Problem Formulation

Consider the space- and time-varying scalar field $Q(\xi,t)$ evolving in a Hilbert space $L_2(0,1)$, where $\xi \in [0,1]$ is the monodimensional (1D) space variable and $t \geq 0$ is time. Let it be governed by the following perturbed Reaction-Diffusion Equation with spatially-varying parameters

$$Q_t(\xi,t) = [\theta_1(\xi)Q_\xi(\xi,t)]_\xi + \theta_2(\xi)Q(\xi,t) + u(\xi,t) + \psi(\xi,t), \tag{4.5}$$

where $\theta_1(\cdot) \in C^1(0,1)$ is a positive-definite spatially-varying parameter called *thermal conductivity* (or, more generally, *diffusivity*), $\theta_2(\cdot) \in C(0,1)$ is another spatially-varying parameter called *dispersion* (or *reaction constant*), $u(\xi,t)$ is the modifiable source term (the distributed control input), and $\psi(\xi,t)$ represents a distributed **uncertain** disturbance source term. This uncertain term is supposed to satisfy the following conditions

$$\psi(\xi,t) \in L_2(0,1), \qquad \psi_t(\xi,t) \in W^{1,2}(0,1) \tag{4.6}$$

The spatially-varying diffusivity and dispersion coefficients $\theta_1(\xi)$ and $\theta_2(\xi)$ are supposed to be uncertain, too. We consider non-homogeneous mixed boundary conditions (BCs)

$$Q(0,t) - \alpha_0 Q_\xi(0,t) = Q_0(t) \in W^{1,2}(0,\infty), \tag{4.7}$$

$$Q(1,t) + \alpha_1 Q_\xi(1,t) = Q_1(t) \in W^{1,2}(0,\infty), \tag{4.8}$$

with some positive **uncertain** constants α_0, α_1. The initial conditions (ICs)

$$Q(\xi,0) = \omega_0(\xi) \in W^{2,2}(0,1) \tag{4.9}$$

are assumed to meet the same BCs. Since nonhomogeneous BCs are in force, a solution of the above boundary-value problem is defined in the mild sense (see [[3]]) as that of the corresponding integral equation, written in terms of the strongly continuous semigroup, generated by the infinitesimal plant operator.

The control task is to make the scalar field $Q(\xi,t)$ to track a given reference $Q^r(\xi,t) \in W^{2,2}(0,1)$ which should be selected in accordance with the BCs (4.7)-(4.8) and which also satisfies the following condition

$$Q_t^r \in W^{3,2}(0,1). \tag{4.10}$$

4.2.2 *Super-Twisting Based Synthesis*

Consider the deviation variable

$$x(\xi,t) = Q(\xi,t) - Q^r(\xi,t) \tag{4.11}$$

whose L_2 norm will be driven to zero by the designed feedback control. The dynamics of the error variable (4.11) are easily derived as

$$x_t(\xi,t) = [\theta_1(\xi)x_\xi(\xi,t)]_\xi + \theta_2(\xi)x(\xi,t) + u(\xi,t) - Q_t^r(\xi,t) + \eta(\xi,t), \tag{4.12}$$

with the "augmented" disturbance

$$\eta(\xi,t) = [\theta_1(\xi)Q_\xi^r(\xi,t)]_\xi + \theta_2(\xi)Q^r(\xi,t) + \psi(\xi,t), \tag{4.13}$$

and the next ICs and homogeneous mixed BCs

$$x(\xi,0) = \omega_0(\xi) - Q^r(\xi,0) \in W^{2,2}(0,1) \tag{4.14}$$

$$x(0,t) - \alpha_0 x_\xi(0,t) = x(1,t) + \alpha_1 x_\xi(1,t) = 0. \tag{4.15}$$

Assume what follows:

Assumption 4.1. *There exist a priori known constants Θ_{1m}, Θ_{1M} and Θ_{2M} such that*

$$0 < \Theta_{1m} \le \theta_1(\xi) \le \Theta_{1M}, \quad |\theta_2(\xi)| \le \Theta_{2M} \quad \textit{for all} \quad \xi \in [0,1]. \tag{4.16}$$

Assumption 4.2. *There exist a priori known constants H_0, ..., H_3, Ψ_0 and Ψ_1 such that the following inequalities hold for all $t \ge 0$*

$$\|\theta_2(\cdot)Q_t^r(\cdot,t)\|_2 \le H_0, \qquad \|[\theta_2(\xi)Q_t^r(\cdot,t)]_\xi\|_2 \le H_1, \tag{4.17}$$

$$\|[\theta_1(\xi)Q_\xi^r(\cdot,t)]_{\xi\, t}\|_2 \le H_2, \qquad \|[\theta_1(\xi)Q_\xi^r(\cdot,t)]_{\xi\xi\, t}\|_2 \le H_3, \tag{4.18}$$

$$\|\psi_t(\cdot,t)\|_2 \le \Psi_0, \qquad \|\psi_{t\xi}(\cdot,t)\|_2 \le \Psi_1 \tag{4.19}$$

By the Assumption 4.2, it follows that the L_2 norm of the augmented disturbance time derivative $\eta_t(\xi,t)$, and that of its spatial derivative, fulfill the next conditions

$$\|\eta_t(\cdot,t)\|_2 \le M, \quad \|\eta_{t\xi}(\cdot,t)\|_2 \le M_\xi, \qquad \forall t \ge 0 \tag{4.20}$$

with

$$M = H_2 + H_0 + \Psi_0, \qquad M_\xi = H_3 + H_1 + \Psi_1 \tag{4.21}$$

The class of admissible "augmented" disturbances is further specified by the following additional restriction, being introduced in [[24]]:

Assumption 4.3. *There exist a priori known constant M_x such that the following restriction holds uniformly beyond the origin $\|x(\cdot,t)\|_2 = 0$ in the state space $L_2(0,1)$:*

$$|\eta_t(\xi,t)| \le M_x \frac{|x(\xi,t)|}{\|x(\cdot,t)\|_2}, \qquad \forall t \ge 0, \forall \xi \in [0,1] \tag{4.22}$$

It is worth noticing that according to the Assumption 4.3 an admissible disturbance has a time derivative which is not necessarily vanishing as $\|x(\cdot,t)\|_2 \to 0$ because the norm of the right-hand side of the disturbance restriction (4.22) remains unit according to relation $\left\| \frac{|x(\cdot,t)|}{\|x(\cdot,t)\|_2} \right\|_2 = 1$. Particularly, with $M_x \le M$ a finite-dimensional counterpart of (4.22) would not impose any further restrictions on admissible disturbances in addition to the first relation of (4.20).

It should also be noted that the assumptions on the ICs and BCs, made above, allow us to deal with strong, sufficiently smooth solutions of the uncertain error dynamics (4.12)-(4.15) in the open-loop when no control input is applied.

In order to stabilize the error dynamics it is proposed a dynamical distributed controller defined as follows

$$u(\xi,t) = Q_t^r(\xi,t) - \lambda_1\sqrt{|x(\xi,t)|}\,\mathtt{sign}(x(\xi,t)) - \lambda_2 x(\xi,t) + v(\xi,t) \tag{4.23}$$

$$v_t(\xi,t) = -W_1 \frac{x(\xi,t)}{\|x(\cdot,t)\|_2} - W_2 x(\xi,t), \quad v(\xi,0) = 0 \tag{4.24}$$

which can be seen as a distributed version of the finite-dimensional "Super-Twisting" second-order sliding-mode controller (see [[5, 13]]) complemented by a feed-forward term $Q_t^r(\xi,t)$ and by the two additional proportional and integral linear terms $-\lambda_2 x(\xi,t)$ and $-W_2 x(\xi,t)$. For ease of reference, the combined Distributed Super-Twisting/PI controller (4.23)-(4.24) will be abbreviated as DSTPI.

The non-smooth nature of the DSTPI controller (4.23)-(4.24), that undergoes discontinuities on the manifold $x = 0$ due to the discontinuous term $\frac{x(\xi,t)}{\|x(\cdot,t)\|_2}$, requires appropriate analysis about the meaning of the corresponding solutions for the resulting discontinuous feedback system. The precise meaning of the solutions of (4.12), (4.14), (4.15) with the piece-wise continuously differentiable control input (4.23)-(4.24) can be defined in a generalized sense (see [[16]]) as a limiting result obtained through a certain regularization procedure, similar to that proposed for finite-dimensional systems (see [[7,27]]). According to this procedure, the strong solutions of the boundary-value problem are only considered whenever they are beyond the discontinuity manifold $x = 0$ whereas in a vicinity of these manifolds the original system is replaced by a related system, which takes into account all possible imperfections (e.g., delay, hysteresis, saturation, etc.) in the new input function $u^{\delta}(x,\xi,t)$, for which there exists a strong solution $x^{\delta}(\xi,t)$ of the corresponding boundary-value problem with the smoothed input $u^{\delta}(x,\xi,t)$. In particular, a relevant approximation occurs when the discontinuous term $U(x) = \frac{x(\xi,t)}{\|x(\cdot,t)\|_2}$ is substituted by the smooth approximation $U^{\delta}(x) = \frac{x(\xi,t)}{\delta+\|x(\cdot,t)\|_2}$. A generalized solution of the system in question is then obtained through the limiting procedure by diminishing δ to zero, thereby making the characteristics of the new system approach those of the original one. As in the finite-dimensional case, a motion along the discontinuity manifold is referred to as a "**sliding mode**".

Remark 4.1. The existence of generalized solutions, thus defined, has been established within the abstract framework of Hilbert space-valued dynamic systems (cf., e.g., [[16 . Theorem 2.4]]) whereas the uniqueness and well-posedness appear to follow from the fact that in the system in question, no sliding mode occurs but in the origin $x = 0$. While being well-recognized for second order sliding mode control algorithms if confined to the finite-dimensional setting, this fact, however, remains beyond the scope of the present investigation.

The performance of the closed-loop system is analyzed in the next theorem.

Theorem 4.1. *Consider the perturbed diffusion/dispersion equation (4.5) along with the boundary conditions (4.7) and with the system parameters, reference trajectory and uncertain disturbance satisfying the Assumptions 4.1-4.3. Then, the distributed control strategy (4.23)-(4.24) with the parameters λ_1, λ_2, W_1 and W_2 selected according to*

$$\lambda_2 \geq \Theta_{2M}, \quad W_1 \geq max\left\{M + \frac{\Theta_{1M}M_{\xi}}{2(\lambda_2 - \Theta_{2M})}, \frac{1}{2}\frac{\Theta_{1M}}{\Theta_{1m}}M_{\xi}, 2M_x\right\},$$

$$\lambda_1 \geq max\left\{2M, \frac{2M_x}{W_1}\right\}, \quad W_2 \geq 0, \quad , \qquad (4.25)$$

guarantees that the L_2-norm $\|x(\cdot,t)\|_2$ of the tracking error tends to zero as t tends to infinity.

Proof of Theorem 4.1. Let us define the auxiliary variable

$$\delta(\xi,t) = v(\xi,t) + \eta(\xi,t) \qquad (4.26)$$

System (4.12) with the control law (4.23)-(4.24) yields the following closed-loop dynamics in the new $x-\delta$ coordinates

$$x_t(\xi,t) = [\theta_1(\xi)x_\xi(\xi,t)]_\xi - \lambda_1\sqrt{|x(\xi,t)|}\,\mathtt{sign}(x(\xi,t)) - (\lambda_2-\theta_2(\xi))x(\xi,t) + \delta(\xi,t) \quad (4.27)$$

$$\delta_t(\xi,t) = -W_1\frac{x(\xi,t)}{\|x(\cdot,t)\|_2} - W_2x(\xi,t) + \eta_t(\xi,t) \quad (4.28)$$

In order to simplify the notation, the dependence of the system coordinates from the space and time variables (ξ,t) is omitted from this point on. Consider the following Lyapunov functional

$$V_1(t) = 2W_1\|x\|_2 + W_2\|x\|_2^2 + \frac{1}{2}\|\delta\|_2^2 + \frac{1}{2}\|s\|_2^2 \quad (4.29)$$

inspired from the finite-dimensional treatment (see [14]), where

$$s = x_t = [\theta_1(\xi)x_\xi]_\xi - \lambda_1\sqrt{|x|}\,\mathtt{sign}(x) - [\lambda_2-\theta_2(\xi)]x + \delta. \quad (4.30)$$

The time derivative of $V_1(t)$ is given by

$$\dot{V}_1(t) = \frac{2W_1}{\|x\|_2}\int_0^1 xs\,d\xi + 2W_2\int_0^1 xs\,d\xi + \int_0^1 \delta\delta_t\,d\xi + \int_0^1 ss_t\,d\xi \quad (4.31)$$

Let us evaluate the time derivative of the auxiliary signal s along the strong solutions of (4.27)-(4.28)

$$s_t = [\theta_1(\xi)s_\xi]_\xi - \frac{1}{2}\lambda_1\frac{s}{\sqrt{|x|}} - [\lambda_2-\theta_2(\xi)]s - W_1\frac{x}{\|x\|_2} - W_2x + \eta_t \quad (4.32)$$

Substituting (4.28) and (4.32) into (4.31) and rearranging it yields

$$\begin{aligned}\dot{V}_1(t) = {} & \frac{2W_1}{\|x\|_2}\int_0^1 xs\,d\xi + 2W_2\int_0^1 xs\,d\xi - \frac{W_1}{\|x\|_2}\int_0^1 \delta x\,d\xi - W_2\int_0^1 \delta x\,d\xi \\ & + \int_0^1 \delta\eta_t\,d\xi + \int_0^1 s[\theta_1(\xi)s_\xi]_\xi\,d\xi - \frac{1}{2}\lambda_1\int_0^1 \frac{s^2d\xi}{\sqrt{|x|}} - \int_0^1 [\lambda_2-\theta_2(\xi)]s^2d\xi \\ & - \frac{W_1}{\|x\|_2}\int_0^1 xs\,d\xi - W_2\int_0^1 xs\,d\xi + \int_0^1 s\eta_t\,d\xi \end{aligned} \quad (4.33)$$

which can be manipulated as follows by virtue of Assumption 4.1

$$\begin{aligned}\dot{V}_1(t) \le {} & -\frac{W_1}{\|x\|_2}\int_0^1 x(\delta-s)\,d\xi - W_2\int_0^1 x(\delta-s)\,d\xi + \int_0^1 s[\theta_1(\xi)s_\xi]_\xi\,d\xi - \\ & -\frac{1}{2}\lambda_1\int_0^1 \frac{s^2d\xi}{\sqrt{|x|}} - [\lambda_2-\Theta_{2M}]\int_0^1 s^2d\xi + \int_0^1 (\delta+s)\eta_t\,d\xi \end{aligned} \quad (4.34)$$

By (4.30), one has

$$\delta - s = \lambda_1 \sqrt{|x|}\,\mathtt{sign}(x) + [\lambda_2 - \theta_2(\xi)]x - [\theta_1(\xi)x_\xi]_\xi \tag{4.35}$$

$$\delta + s = 2s + \lambda_1 \sqrt{|x|}\,\mathtt{sign}(x) + [\lambda_2 - \theta_2(\xi)]x - [\theta_1(\xi)x_\xi]_\xi. \tag{4.36}$$

Due to this, and considering once more the Assumption 4.1, (4.34) can further be manipulated as

$$\begin{aligned}
\dot V_1(t) \le & -\frac{W_1\lambda_1}{\|x\|_2}\int_0^1 x\sqrt{|x|}\,\mathtt{sign}(x)d\xi - \frac{W_1[\lambda_2-\Theta_{2M}]}{\|x\|_2}\int_0^1 x^2 d\xi \\
& + \frac{W_1}{\|x\|_2}\int_0^1 x[\theta_1(\xi)x_\xi]_\xi d\xi - W_2\lambda_1\int_0^1 x\sqrt{|x|}\,\mathtt{sign}(x)d\xi \\
& - W_2[\lambda_2-\Theta_{2M}]\int_0^1 x^2 d\xi + W_2\int_0^1 x[\theta_1(\xi)x_\xi]_\xi d\xi + \int_0^1 s[\theta_1(\xi)s_\xi]_\xi d\xi \\
& - \frac{1}{2}\lambda_1\int_0^1 \frac{s^2 d\xi}{\sqrt{|x|}} - [\lambda_2-\Theta_{2M}]\int_0^1 s^2 d\xi + 2\int_0^1 s\eta_t d\xi \\
& + \lambda_1\int_0^1 \sqrt{|x|}\,\mathtt{sign}(x)\eta_t d\xi + [\lambda_2-\Theta_{2M}]\int_0^1 x\eta_t d\xi - \int_0^1 [\theta_1(\xi)x_\xi]_\xi \eta_t d\xi.
\end{aligned} \tag{4.37}$$

By taking into account the BCs (4.15) and their time derivatives, standard integration by parts yields

$$\begin{aligned}
\int_0^1 x[\theta_1(\xi)x_\xi]_\xi d\xi &= -\int_0^1 \theta_1(\xi)x_\xi^2 d\xi + \theta_1(1)x(1,t)x_\xi(1,t) - \theta_1(0)x(0,t)x_\xi(0,t) \\
&\le -\Theta_{1m}\|x_\xi\|_2^2 - \theta_1(1)\frac{x^2(1,t)}{\alpha_1} - \theta_1(0)\frac{x^2(0,t)}{\alpha_0}
\end{aligned} \tag{4.38}$$

$$\begin{aligned}
\int_0^1 s[\theta_1(\xi)s_\xi]_\xi d\xi &= -\int_0^1 \theta_1(\xi)s_\xi^2 d\xi + \theta_1(1)s(1,t)s_\xi(1,t) - \theta_1(0)s(0,t)s_\xi(0,t) \\
&\le -\Theta_{1m}\|s_\xi\|_2^2 - \theta_1(1)\frac{s^2(1,t)}{\alpha_1} - \theta_1(0)\frac{s^2(0,t)}{\alpha_0}
\end{aligned} \tag{4.39}$$

$$\begin{aligned}
\int_0^1 [\theta_1(\xi)x_\xi]_\xi \eta_t d\xi &= -\int_0^1 \theta_1(\xi)x_\xi \eta_{t\xi} d\xi + \theta_1(1)\eta_t(1,t)x_\xi(1,t) - \theta_1(0)\eta_t(0,t)x_\xi(0,t) \\
&= -\int_0^1 \theta_1(\xi)x_\xi \eta_{t\xi} d\xi - \theta_1(1)\eta_t(1,t)\frac{x(1,t)}{\alpha_1} - \theta_1(0)\eta_t(0,t)\frac{x(0,t)}{\alpha_0}
\end{aligned} \tag{4.40}$$

Additional straightforward manipulations of (4.37) taking into account (4.38) and (4.39) yield

$$
\begin{aligned}
\dot{V}_1(t) \leq & -W_1[\lambda_2-\Theta_{2M}]\|x\|_2 - W_2[\lambda_2-\Theta_{2M}]\|x\|_2^2 - W_1\Theta_{1m}\frac{\|x_\xi\|_2^2}{\|x\|_2} \\
& - \frac{W_1}{\|x\|_2}\theta_1(1)\frac{x^2(1,t)}{\alpha_1} - \frac{W_1}{\|x\|_2}\theta_1(0)\frac{x^2(0,t)}{\alpha_0} - W_2\Theta_{1m}\|x_\xi\|_2^2 \\
& - W_2\theta_1(1)\frac{x^2(1,t)}{\alpha_1} - W_2\theta_1(0)\frac{x^2(0,t)}{\alpha_0} - [\lambda_2-\Theta_{2M}]\|s\|_2^2 - \Theta_{1m}\|s_\xi\|_2^2 \\
& - \theta_1(1)\frac{s^2(1,t)}{\alpha_1} - \theta_1(0)\frac{s^2(0,t)}{\alpha_0} - W_2\lambda_1\int_0^1 |x|^{3/2}d\xi \\
& - \frac{1}{2}\lambda_1\int_0^1 \frac{s^2 d\xi}{\sqrt{|x|}} - \frac{W_1\lambda_1}{\|x\|_2}\int_0^1 \sqrt{|x|}|x|d\xi + 2\int_0^1 s\eta_t d\xi \\
& + \lambda_1\int_0^1 \sqrt{|x|}\,\mathtt{sign}(x)\eta_t d\xi + [\lambda_2-\Theta_{2M}]\int_0^1 x\eta_t d\xi \\
& + \int_0^1 \theta_1(\xi)x_\xi\eta_{t\xi}d\xi + \theta_1(1)\eta_t(1,t)\frac{x(1,t)}{\alpha_1} + \theta_1(0)\eta_t(0,t)\frac{x(0,t)}{\alpha_0}
\end{aligned} \tag{4.41}
$$

It is worth noting that by virtue of the tuning inequality $\lambda_2 > \Theta_{2M}$ in (4.25) all terms appearing in the right hand side of (4.41) are negative definite except those depending on the augmented disturbance term η_t and its spatial derivative. Some estimations involving those sign-indefinite terms are now derived by simple application of the Cauchy-Schwartz and Young's inequalities and by considering the Assumptions 4.1 and 4.2, the BCs (4.15) and the derived conditions (4.20)-(4.21):

$$
\begin{aligned}
2\left|\int_0^1 s\eta_t d\xi\right| & \leq 2\int_0^1 |s||\eta_t|d\xi = 2\int_0^1 \frac{|s|\sqrt{|\eta_t|}\sqrt{|\eta_t|}\sqrt{|x|}}{\sqrt{|x|}}d\xi \\
& \leq \int_0^1 \frac{|\eta_t|s^2 + |\eta_t||x|}{\sqrt{|x|}}d\xi \leq M\int_0^1 \frac{s^2}{\sqrt{|x|}}d\xi + \int_0^1 \eta_t\sqrt{|x|}d\xi
\end{aligned} \tag{4.42}
$$

$$
\left|\int_0^1 x\eta_t d\xi\right| \leq \left[\int_0^1 x^2 d\xi\right]^{1/2}\left[\int_0^1 \eta_t^2 d\xi\right]^{1/2} \leq M\|x\|_2 \tag{4.43}
$$

$$
\begin{aligned}
\left|\int_0^1 \theta_1(\xi)x_\xi\eta_{t\xi}d\xi\right| & \leq \Theta_{1M}\int_0^1 |x_\xi||\eta_{t\xi}|d\xi = \Theta_{1M}\int_0^1 \frac{|x_\xi|\sqrt{|\eta_{t\xi}|}\sqrt{|\eta_{t\xi}|}\|x\|_2}{\|x\|_2}d\xi \\
& \leq \frac{1}{2}\Theta_{1M}\int_0^1 \frac{x_\xi^2|\eta_{t\xi}| + |\eta_{t\xi}|\|x\|_2^2}{\|x\|_2}d\xi \\
& \leq \frac{1}{2}\Theta_{1M}M_\xi\frac{\|x_\xi\|_2^2}{\|x\|_2} + \frac{1}{2}\Theta_{1M}M_\xi\|x\|_2
\end{aligned} \tag{4.44}
$$

Taking into account (4.42)-(4.44), the right-hand side of (4.41) can be estimated as

$$\begin{aligned}
\dot{V}_1(t) \leq & -(\lambda_2-\Theta_{2M})\left[W_1 - M - \frac{\Theta_{1M}M_\xi}{2(\lambda_2-\Theta_{2M})}\right]\|x\|_2 - W_2(\lambda_2-\Theta_{2M})\|x\|_2^2 \\
& - \left[W_1\Theta_{1m} - \frac{1}{2}\Theta_{1M}M_\xi\right]\frac{\|x_\xi\|_2^2}{\|x\|_2} - W_2\Theta_{1m}\|x_\xi\|_2^2 - (\lambda_2-\Theta_{2M})\|s\|_2^2 \\
& - \Theta_{1m}\|s_\xi\|_2^2 - W_2\lambda_1\int_0^1 |x|^{3/2}d\xi - \frac{1}{2}(\lambda_1-2M)\int_0^1 \frac{s^2 d\xi}{\sqrt{|x|}} \\
& - \int_0^1 \sqrt{|x|}\left[\frac{W_1\lambda_1}{2\|x\|_2}|x| - \eta_t\right]d\xi - \lambda_1\int_0^1 \sqrt{|x|}\left[\frac{W_1}{2\|x\|_2}|x| - \eta_t\right]d\xi \\
& - \theta_1(1)\frac{|x(1,t)|}{\alpha_1}\left[\frac{W_1}{\|x\|_2}|x(1,t)| - \eta_t(1,t)\right] \\
& - \theta_1(0)\frac{|x(0,t)|}{\alpha_0}\left[\frac{W_1}{\|x\|_2}|x(0,t)| - \eta_t(0,t)\right] \\
& - W_2\theta_1(1)\frac{x^2(1,t)}{\alpha_1} - W_2\theta_1(0)\frac{x^2(0,t)}{\alpha_0} - \theta_1(1)\frac{s^2(1,t)}{\alpha_1} - \theta_1(0)\frac{s^2(0,t)}{\alpha_0}
\end{aligned} \tag{4.45}$$

By virtue of Assumption 4.3, the next inequalities guarantee that all terms in the right hand side of (4.45) are negative definite

$$\lambda_2 > \Theta_{2M}, \quad W_1 > M + \frac{\Theta_{1M}M_\xi}{2(\lambda_2-\Theta_{2M})}, \quad W_2 > 0, \quad W_1 > \frac{1}{2}\frac{\Theta_{1M}}{\Theta_{1m}}M_\xi, \tag{4.46}$$

$$\lambda_1 > 2M, \quad W_1\lambda_1 > 2M_x, \quad W_1 > 2M_x, \quad W_1 > M_x \tag{4.47}$$

The above inequalities collected together form the tuning conditions (4.25). To complete the proof it remains to demonstrate that

$$\|x(\cdot,t)\|_2 \to 0 \text{ as } t \to \infty. \tag{4.48}$$

For this purpose, let us integrate the relation

$$\dot{\bar{V}}(t) \leq -(\lambda_2-\Theta_{2M})\left[W_1 - M - \frac{\Theta_{1M}M_\xi}{2(\lambda_2-\Theta_{2M})}\right]\|x\|_2, \tag{4.49}$$

straightforwardly resulting from the negative definiteness of all terms in the right hand side of (4.45), to conclude that

$$\int_0^\infty \|x(\cdot,t)\|_2 dt < \infty \tag{4.50}$$

The inequality $\dot{V}_1(t) \leq 0$, which is readily concluded from (4.45) and (4.46)-(4.47) in light of the Assumption 4.3, guarantees that $V_1(t) \leq V_1(0)$ for any $t \geq 0$. From this, and considering (4.29), one can conclude that the L_2 norm of $s = x_t$ fulfills the estimation

$$\|x_t\|_2^2 \leq 2V_1(0), \quad \forall t \geq 0 \tag{4.51}$$

Thus, the integrand $\omega(t) = \|x(\cdot,t)\|_2$ of (4.50) possesses a uniformly bounded time derivative

$$\dot{\omega}(t) = \frac{\int_0^1 x x_t d\xi}{\|x\|_2} \leq \|x_t\|_2 \leq \sqrt{2R} \tag{4.52}$$

on the semi-infinite time interval $t \in [0,\infty)$, where R is any positive constant such that $R \geq V_1(0)$. Convergence (4.48) is then verified by applying the Barbalat lemma (see [[8]]). Since the Lyapunov functional (4.29) is radially unbounded the global asymptotic stability of the closed-loop system (4.12)-(4.15) is thus established in the L_2 space. Theorem 4.1 is proved. □

4.2.3 Numerical Simulations

Consider the perturbed reaction diffusion equation (4.5) with the space-varying parameters:

$$\theta_1(\xi) = 0.1 + 0.02\sin(1.3\pi\xi) \tag{4.53}$$

$$\theta_2(\xi) = 1 + 0.1\sin(3.5\pi\xi) \tag{4.54}$$

Mixed-type BCs

$$Q(0,t) - Q_\xi(0,t) = Q(1,t) + Q_\xi(1,t) = 20 \tag{4.55}$$

and ICs

$$Q(\xi,0) = 20 + 10sin(6\pi\xi) \tag{4.56}$$

The chosen function for $\theta_2(\xi)$ makes the system open-loop unstable (see [[9]]). We choose the following reference profile:

$$Q^r(\xi,t) = 20 + 5sin(\pi\xi)sin(\pi t) \tag{4.57}$$

which meets the actual BCs. A space- and time-varying disturbance term is considered in the form

$$\psi(\xi,t) = 5\mathtt{sin}(2.5\pi\xi)\mathtt{cos}(3\pi t) \tag{4.58}$$

The L_2 norm bounds M and M_ξ of the disturbance derivatives η_t and $\eta_{t\xi}$, which are involved in the controller tuning inequalities (4.25) can be easily estimated as $M = 80$ and $M_\xi = 550$, and the constant M_x is selected as $M_x = 100$. Then, the controller gains are set in accordance with (4.25) to the values

$$W_1 = 400, \quad \lambda_1 = 160, \quad W_2 = 2, \quad \lambda_2 = 2 \tag{4.59}$$

For solving the PDE governing the closed-loop system, standard finite-difference approximation method is used by discretizing the spatial solution domain $\xi \in [0,1]$ into a finite number of N uniformly spaced solution nodes $\xi_i = ih$, $h = 1/(N+1)$, $i = 1,2,...,N$. The value $N = 100$ has been used in the present simulations. The resulting 100-th order discretized system is implemented in Matlab-Simulink and solved by fixed-step Euler integration method with constant step $T_s = 10^{-4}$.

The left plot in Figure 4.1 depicts the solution $Q(\xi,t)$, which converges to the given reference profile, and the right plot depicts the distributed control input $u(\xi,t)$ which, as expected, appears to be a smooth function of both time and space. Figure 4.2 depicts the contractive time evolution of the tracking error L_2-norm $\|x(\cdot,t)\|_2$, which fastly tends to zero. The attained results confirm the validity of the presented analysis.

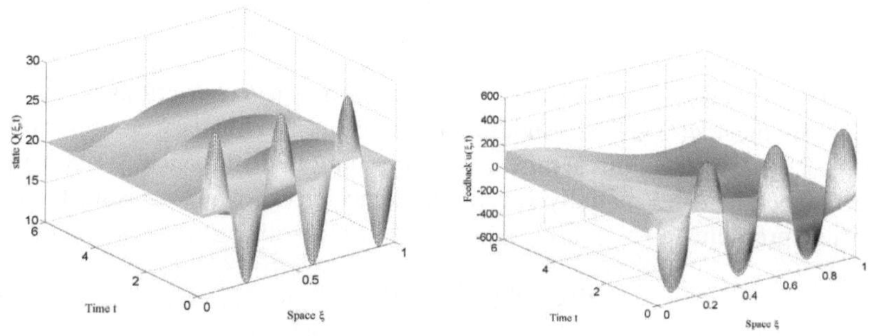

Fig. 4.1 The solution $Q(\xi,t)$ and the distributed control $u(\xi,t)$

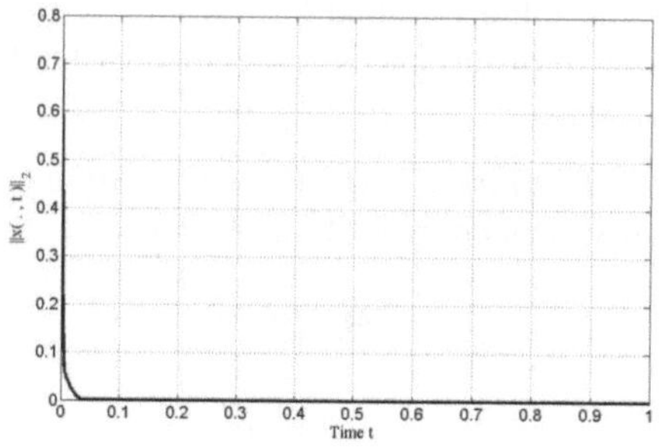

Fig. 4.2 Tracking error L_2 norm $\|x(\cdot,t)\|_2$

4.3 Boundary Control of Uncertain Diffusion Processes

4.3.1 Problem Formulation

Consider the space- and time-varying scalar field $Q(\xi,t)$ with the monodimensional spatial variable $\xi \in [0,1]$ and time variable $t \geq 0$. Let it be governed by a perturbed version of the parabolic PDE which is commonly referred to as the **"Heat Equation"**:

$$Q_t(\xi,t) = [\theta(\xi)Q_\xi(\xi,t)]_\xi, \tag{4.60}$$

where the subscripts t and ξ denote temporal and spatial derivatives, respectively, and $\theta(\cdot) \in C^1(0,1)$ is a positive-definite spatially-varying parameter called *thermal conductivity* (or, more generally, *diffusivity*). The initial condition (IC) is given by

$$Q(\xi,0) = Q^0(\xi) \in W^{2,2}(0,1). \tag{4.61}$$

Throughout, we assume controlled and perturbed Robin's (i.e., mixed) boundary conditions (BCs) of the form

$$Q_\xi(0,t) = \alpha_0 Q(0,t) + \beta_0 \tag{4.62}$$
$$Q_\xi(1,t) = -\alpha_1 Q(1,t) + \beta_1 + u(t) + \psi(t), \tag{4.63}$$

where $\alpha_i \geq 0$ and β_i $(i = 0,1)$, are arbitrary constants (no restrictions on the sign of the β_i's are met), $u(t) \in \mathbb{R}$ is a modifiable source term (boundary control input) and $\psi(t) \in \mathbb{R}$ represents an **uncertain** sufficiently smooth disturbance. We consider the time-independent and spatially varying reference $Q^r(\xi)$ which satisfies the boundary value problem

$$[\theta(\xi)Q^r_\xi(\xi)]_\xi = 0 \tag{4.64}$$
$$Q^r_\xi(0) = \alpha_0 Q^r(0) + \beta_0 \tag{4.65}$$
$$Q^r(1) = Q^r_1 \tag{4.66}$$

for an arbitrary, user-selectable, constant Q^r_1.

The class of admissible disturbances is specified by the following restriction on their time derivative.

Assumption 4.4. *The disturbance $\psi(t)$ is twice continuously differentiable and there exists an a priori known constant M such that*

$$|\psi_t(t)| \leq M \tag{4.67}$$

for almost all $t \geq 0$.

The spatially varying diffusivity is supposed to satisfy the next restriction

Assumption 4.5. *There exist a priori known constants Θ_m, Θ_M such that*

$$0 < \Theta_m \leq \theta(\xi) \leq \Theta_M, \quad \forall \xi \in [0,1]. \tag{4.68}$$

With the assumptions above the evolution of the considered heat process is studied in the Sobolev space $W^{2,2}(0,1)$ and the control objective is to steer the $W^{2,2}$-norm of the deviation

$$x(\xi,t) = Q(\xi,t) - Q^r(\xi) \tag{4.69}$$

of the scalar field $Q(\xi,t)$ from the *a priori* given reference to zero, despite the presence of an uncertain, arbitrarily shaped, smooth boundary disturbance $\psi(t)$ fulfilling the Assumption 4.4. Boundary sensing at $\xi = 1$ of the deviation $x(\xi,t)$ and of its time derivative $x_t(\xi,t)$ is assumed to be the only available information on the state of the system. The deviation variable $x(\xi,t)$ is governed by the heat equation

$$x_t(\xi,t) = [\theta(\xi)x_\xi(\xi,t)]_\xi \tag{4.70}$$

subject to the next Robin-type BCs

$$x_\xi(0,t) - \alpha_0 x(0,t) = 0 \tag{4.71}$$
$$x_\xi(1,t) + \alpha_1 x(1,t) = u(t) + \psi(t) + \gamma_1, \tag{4.72}$$

with the constant

$$\gamma_1 = \beta_1 - Q^r_\xi(1) - \alpha_1 Q^r_1, \tag{4.73}$$

which can be derived by considering (4.69), and its spatial derivative $x_\xi(\xi,t) = Q_\xi(\xi,t) - Q^r_\xi(\xi)$, along with the conditions (4.63) and (4.66). The corresponding ICs are

$$x(\xi,0) = x^0(\xi), \quad x^0(\xi) = Q^0(\xi) - Q^r(\xi) \tag{4.74}$$

It is worth noticing that the disturbance-free system (4.70)-(4.74) in open-loop is only stable, rather than asymptotically stable. Thus, the modifiable control variable $u(t)$ should be designed in order to make the zero solution $x(\xi,t) = 0$ of the closed-loop system (4.70)-(4.74) globally asymptotically stable in the $W^{2,2}$-space despite the presence of an unknown disturbance $\psi(t)$ affecting the state of the system through its boundary. Since non-homogeneous boundary conditions are in force, the meaning of the boundary-value problem (4.70)-(4.74) is subsequently viewed in the mild sense. The mild solutions coincide with those of the following PDE in distributions

$$x_t(\xi,t) = [\theta(\xi)x_\xi(\xi,t)]_\xi + \theta(1)[u(t) + \psi(t) + \gamma_1]\delta(\xi - 1) \tag{4.75}$$

subject to the homogeneous Robin BCs

$$x_\xi(0,t) - \alpha_0 x(0,t) = x_\xi(1,t) + \alpha_1 x(1,t) = 0 \tag{4.76}$$

and to the ICs (4.74). Indeed, (weak) solutions of the boundary-value problem (4.75)-(4.76) are defined by means of the corresponding Green function, yielding the same integral equation. To this end, we note that according to [[3, Theorem 3.3.3]], the unforced system (4.70)-(4.74) with $u(t) \equiv 0$ possesses a unique classical solution in the state space $W^{2,2}(0,1)$ (cf. [[3, Definition 3.2.9]].

4.3.2 Twisting Based Synthesis

To achieve the control goal, the system state is augmented through a *dynamic input extension* by inserting an integrator at the plant input. The control derivative $u_t(t)$ is then regarded as a fictitious control variable to be generated by a suitable feedback mechanism. The next dynamical controller

$$u_t(t) = -\lambda_1 \operatorname{sign} x(1,t) - \lambda_2 \operatorname{sign} x_t(1,t) - W_1 x(1,t) - W_2 x_t(1,t),$$
$$u(0) = 0, \tag{4.77}$$

is currently under study, where the initial condition $u(0)$ is set to zero for certainty. In the above controller description, λ_1, λ_2, W_1 and W_2 are constant tuning parameters subject to the inequalities

$$\lambda_2 > M, \ \lambda_1 > \lambda_2 + M, \ W_1 > \frac{1}{2}\frac{\Theta_M}{\Theta_m}, \ W_2 > 0. \tag{4.78}$$

Since the present work focuses on the stabilizing synthesis we do not analyze the well-posedness of the closed-loop system because of space limitations and due to the parabolic character of the system that ensures it. Thus, in the remainder, we simply assume the following.

Assumption 4.6. *The closed-loop system (4.75)-(4.77) possesses a unique mild solution $x(\cdot,t) \in W^{2,2}(0,1)$ whose time derivative $x_t(\cdot,t) \in W^{2,2}(0,1)$ constitutes a (weak) solution of the distribution boundary-value problem*

$$x_{tt}(\xi,t) = [\theta(\xi)x_{t\xi}(\xi,t)]_\xi + \theta(1)\{u_t[y](t) + \psi_t(t)\}\delta(\xi-1) \tag{4.79}$$

$$x_{t\xi}(0,t) - \alpha_0 x_t(0,t) = x_{t\xi}(1,t) + \alpha_1 x_t(1,t) = 0, \tag{4.80}$$

with respect to $x_t(\xi,t)$, which is formally obtained by differentiating (4.75)-(4.76) in the time variable.

The following relation

$$\|x_t\|_2 = \|[\theta(\xi)x_\xi(\xi,t)]_\xi\|_2 \tag{4.81}$$

is particularly concluded from (4.70). Along with the instrumental lemmas given below, relation (4.81) will be instrumental in our further derivation.

We introduce several technical lemmas that will be instrumental in our next derivations.

Lemma 4.1. *[20] Let $z(\xi) \in W^{1,2}(0,1)$. Then, the following inequality holds:*

$$\|z(\cdot)\|_2^2 \leq 2(z^2(i) + \|z_\xi(\cdot)\|_2^2), \quad i = 0,1. \tag{4.82}$$

Lemma 4.2. *[20] The functional*

$$\tilde{V}(x,x_t) = \lambda_1\theta(1)|x(1,t)| + \frac{1}{2}\theta(1)W_1 x^2(1,t) + \frac{1}{2}\|x_t(\cdot,t)\|_2^2, \tag{4.83}$$

being computed on the mild solutions (x,x_t) of the boundary-value problem (4.79)-(4.80), upper estimates the weighted $W^{2,2}(0,1) \times L_2(0,1)$-norm of these solutions in the sense that

$$\alpha(\|x(\cdot,t)\|^2_{2,2,\theta} + \|x_t(\cdot,t)\|^2_2) \le \tilde{V}(x,x_t) \tag{4.84}$$

for any time instant $t \ge 0$ and for some positive constant α

Lemma 4.3. *[20] Let a set*

$$\mathscr{D}_R^{\tilde{V}} = \{(z(\xi),h(\xi)) \in W^{2,2}(0,1) \times L_2(0,1): \qquad \tilde{V}(z,h) \le R\} \tag{4.85}$$

be determined by means of functional (4.83) and be specified with some positive R. Then the following conditions

$$\int_0^1 z(1)h(\xi)\,d\xi \ge -\frac{1}{2}\left[\frac{R}{\lambda_1\Theta_m}|z(1)| + \|h\|_2^2\right], \tag{4.86}$$

$$\|h\|_2^2 \le 2R, \quad \|h\|_2 \le \sqrt{2R}, \quad \|h\|_2^2 \le \sqrt{2R}\|h\|_2, \tag{4.87}$$

hold for an arbitrary $(z(\xi),h(\xi)) \in \mathscr{D}_R^{\tilde{V}}$.

We are now in a position to state the convergence result while providing a sketch of the proof. A more comprehensive treatment can be found in [20]

Theorem 4.2. *[20] Consider the perturbed heat process (4.60)-(4.63) subject to the dynamic control strategy (4.77), (4.78). Let Assumptions 4-6 be satisfied. Then the solutions (x,x_t) of the resulting error boundary-value problem (4.79)-(4.80) are globally asymptotically stable in the space $W^{2,2}(0,1) \times L_2(0,1)$.*

Proof of Theorem 4.2 (sketch)
By Lemma 4.2, functional (4.83) is positive definite along the mild solutions (x,x_t) of the boundary-value problem (4.79)-(4.80). The time derivative of (4.83) along such solutions is shown, after few manipulations, to meet the next estimation

$$\begin{aligned}\dot{\tilde{V}}(t) \le &-\Theta_m(\lambda_2 - M)|x_t(1,t)| - \Theta_m(W_2+\alpha_1)x_t^2(1,t)\\ &-\Theta_m\alpha_0 x_t^2(0,t) - \Theta_m\|x_{t\xi}\|_2^2.\end{aligned} \tag{4.88}$$

Due to (4.78) and (4.88), the Lyapunov functional $\tilde{V}(t)$, being computed along the mild solutions of the closed-loop system, is a non-increasing function of time, and as a result the domain $\mathscr{D}_R^{\tilde{V}}$, given by (4.85) with an arbitrary $R \ge \tilde{V}(0)$, is **invariant** for the system trajectories. Thus, the subsequent analysis will take into account that the mild solutions (x,x_t) stay in the domain $\mathscr{D}_R^{\tilde{V}}$ forever.

Now consider the "augmented" functional

$$\tilde{V}_R(t) = \tilde{V}(t) + \frac{1}{2}\kappa_R\theta(1)(W_2+\alpha_1)x^2(1,t) + \kappa_R\int_0^1 x(1,t)x_t(\xi,t)\,d\xi \tag{4.89}$$

where κ_R is a sufficiently small positive constant to subsequently be specified.

By Lemma 4.3 specified with $z = x$ and $h = x_t$, and considering the inequality (4.68) and the additional restriction

$$\kappa_R < \min \left\{ \frac{2\lambda_1^2 \Theta_m^2}{R}, 1 \right\} \tag{4.90}$$

the augmented functional (4.89) is lower estimated by functional (4.83) as

$$\tilde{V}_R\,(x, x_t) \geq \mu \tilde{V}(x, x_t) \tag{4.91}$$

$$\mu = \min \left\{ 1 - \frac{\kappa_R R}{2\lambda_1^2 \Theta_m^2}, \frac{W_1 + \kappa_R (W_2 + \alpha_1)}{W_1}, (1 - \kappa_R) \right\} \tag{4.92}$$

It means that along with (4.83), the functional $\tilde{V}_R$ is positive definite on the mild solutions (x, x_t) of the boundary-value problem (4.79)-(4.80) within the invariant set $\mathscr{D}_R^{\tilde{V}}$. Let us now evaluate the time derivative of $\tilde{V}_R(t)$:

$$\dot{\tilde{V}}_R = \dot{\tilde{V}} + \kappa_R \theta(1)(W_2 + \alpha_1) x(1,t) x_t(1,t)$$
$$+ \kappa_R \int_0^1 x_t(1,t) x_t(\xi,t) d\xi + \kappa_R \int_0^1 x(1,t) x_{tt}(\xi,t) \xi. \tag{4.93}$$

After appropriate manipulations (see [20]), (4.93) is shown to fulfill the next estimation

$$\dot{\tilde{V}}_R(t) \leq -\gamma \tilde{V}_R(t) \tag{4.94}$$

for some positive constant parameter γ that establishes the exponential convergence of $\tilde{V}_R(t)$, initialized within (4.85), to zero as $t \to \infty$.

To complete the proof it remains to note that due to the upper estimate (4.91) of the functional $\tilde{V}(t)$ by the functional $\tilde{V}_R(t)$, it follows that $\tilde{V}(t)$, being computed on the mild solutions (x, x_t) of the boundary-value problem (4.79)-(4.80), converges asymptotically to zero, too, and by virtue of Lemma 4.2, the local asymptotic stability of (4.79)-(4.80) with the augmented state (x, x_t) in the $W^{2,2}(0,1) \times L_2(0,1)$-space is established with the initial set (4.85). Since the initial set (4.85) can be specified with an arbitrarily large $R > 0$ the **global** asymptotic stability in the $W^{2,2}(0,1) \times L_2(0,1)$-space is then concluded. Theorem 1 is thus proved. □

4.3.3 Simulations

Consider the perturbed heat equation (4.60) with constant diffusivity $\theta = 1$. The parameters of the uncontrolled Robin's BC (4.62) are set as $\alpha_0 = 1$ and $\beta_0 = -5$.

The boundary value problem (4.64)-(4.66) specialized for a constant diffusivity has a solution $Q^r(\xi) = Q_0^r + \xi(Q_1^r - Q_0^r)$ which linearly depends on the spatial variable, where the reference boundary value $Q^r(1) = Q_1^r$ is arbitrarily selected as $Q_1^r = 15$ and the resulting value for Q_0^r is derived from the other parameters according to $Q_0^r = \frac{Q_1^r - \beta_0}{1+\alpha_0} = 10$, which is obtained by imposing the BC (4.62) on the solution $Q^r(\xi)$. Parameter β_1 is arbitrarily set to the value $\beta_1 = 1$. The disturbance $\psi(t)$ is selected as $\psi(t) = 4\cos(0.5\pi t)$. The magnitude of the disturbance time derivative ψ_t can be easily upper-estimated as $M = 6.5$, as required by (4.67). The initial conditions have been set to $Q^0(\xi) = 3 + 2\sin(4\pi\xi)$.

Controller (4.77) has been implemented with the parameters $\lambda_1 = 15$, $\lambda_2 = 7$, $W_1 = W_2 = 1$ which are selected in accordance with (4.78). The same finite-difference discretization technique and numerical solver used in the previously presented simulations were used.

The two plots in Figure 4.3 show the solution $Q(\xi,t)$ and the applied boundary control $u(t)$. It can be seen that the solution converges to the linear reference $Q^r(\xi)$ along the entire solution domain, and that, as expected, the applied boundary control is a continuous function.

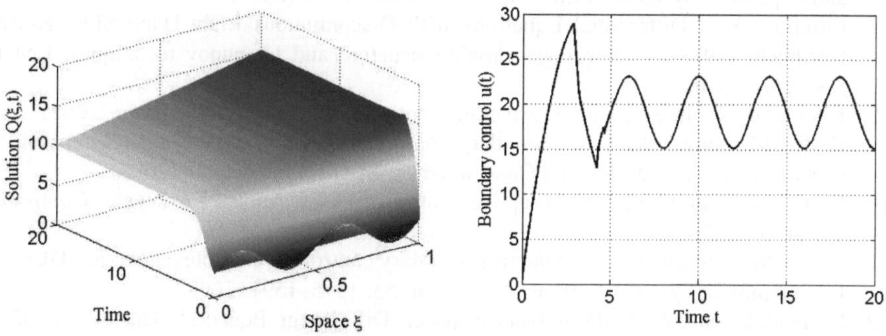

Fig. 4.3 The solution $Q(\xi,t)$ (left plot) and the boundary control $u(,t)$ (right plot)

4.4 Conclusions

The "Super-Twisting" and "Twisting" 2-SMC algorithms have been used in conjunction with linear PI and PD controllers, respectively, to address certain tracking and regulation problems for uncertain parabolic PDEs by means of distributed and boundary control schemes.

Appropriate non-smooth Lyapunov functionals are introduced to demonstrate the convergence properties in the relevant functional spaces. The proposed infinite-dimensional treatments retain robustness features against non-vanishing matching disturbances similar to those possessed by the finite-dimensional counterparts of the considered controllers.

Finite-time convergence of the proposed algorithms, which would be the case if confined to a finite dimensional treatment, cannot be proved using the proposed Lyapunov functionals, and it remains among other problems to be tackled in the future within the present framework.

References

1. Abramowitz, M., Stegun, I.A.: Handbook of Mathematical Functions with Formulas, Graphs, and Mathematical Tables. Dover, New York (1964) ISBN 0-486-61272-4
2. Christofides, P.D.: Nonlinear and Robust Control of PDE Systems. Methods and Applications to Transport-Reaction Processes. Systems & Control: Foundations & Applications series. Birkhäuser, Boston (2001)
3. Curtain, R., Zwart, H.: An Introduction To Infinite-dimensional Linear Systems Theory. Texts in Applied Mathematics. Springer, Berlin (1995)
4. Demetriou, M.A., Ito, K., Smith, R.C.: Adaptive techniques for the MRAC, adaptive parameter identification, and on-line fault monitoring and accommodation for a class of positive real infinite dimensional systems. Int. J. Adaptive Contr. Signal Proc. 23(2), 193–215 (2009)
5. Fridman, L., Levant, A.: Higher Order Sliding Modes as a Natural Phenomenon in Control Theory. In: Garofalo, F., Glielmo, L. (eds.) Robust Control via Variable Structure and Lyapunov Techniques. LNCIS, vol. 217, pp. 107–133. Springer, Heidelberg (1996)
6. Fridman, E., Orlov, Y.: An LMI approach to H_∞ boundary control of semilinear parabolic and hyperbolic systems. Automatica 45(9), 2060–2066 (2009)
7. Filippov, A.F.: Differential Equations with Discontinuous Right-Hand Side. Kluwer, Dordrecht (1988); t control via variable structure and Lyapunov techniques. LNCIS. Springer, Berlin
8. Khalil, H.: Nonlinear systems, 3rd edn. Prentice Hall, Upper Saddle River (2002)
9. Krstic, M., Smyshlyaev, A.: Boundary control of PDEs: a course on backstepping designs. Advances in Design and Control Series. SIAM, Philadelphia (2008)
10. Krstic, M., Smyshlyaev, A.: Adaptive control of PDEs. Annual Reviews of Control 32, 149–160 (2008)
11. Krstic, M., Smyshlyaev, A.: Adaptive boundary control for unstable parabolic PDEs. Part I. Lyapunov design. IEEE Trans. Aut. Contr. 53, 1575–1591 (2008)
12. Levaggi, L.: Sliding modes in banach spaces. Diff. Integr. Equat. 15, 167–189 (2002)
13. Levant, A.: Sliding order and sliding accuracy in sliding mode control. Int. J. Contr. 58, 1247–1263 (1993)
14. Moreno, J., Osorio, M.: A Lyapunov approach to second-order sliding mode controllers and observers. In: Proc. 47 Conference on Decision and Control (CDC 2008), Cancun, Mexico, pp. 2856–2861 (December 2008)
15. Orlov, Y., Utkin, V.I.: Sliding mode control in infinite-dimensional systems. Automatica 23, 753–757 (1987)
16. Orlov, Y.: Discontinuous Systems Lyapunov Analysis and Robust Synthesis under Uncertainty Conditions. Communications and Control Engineering Series. Springer, Berlin (2009)
17. Orlov, Y.: Finite-time stability and robust control synthesis of uncertain switched systems. SIAM J. Contr. Opt. 43, 1253–1271 (2005)
18. Orlov, Y.: Discontinuous unit feedback control of uncertain infinite-dimensional systems. IEEE Trans. Aut. Contr. 45(12), 834–843 (2000)
19. Orlov, Y., Lou, Y., Christofides, P.D.: Robust stabilization of infinite-dimensional systems using sliding-mode output feedback control. Int. J. Contr. 77(12), 1115–1136 (2004)
20. Orlov, Y., Pisano, A., Usai, E.: Boundary second-order sliding-mode control of an uncertain heat process with spatially varying diffusivity. In: Proc. 50th IEEE Conference on Decision and Control and European Control Conference (CDC-ECC 2011), Orlando, FL, USA, December 12-15 (2011)

21. Orlov, Y., Pisano, A., Scodina, S., Usai, E.: On the Lyapunov-based second order SMC design for some classes of distributed-parameter systems. IMA Journal of Mathematical Control and Information (2011), doi:10.1093/imamci/dns003, Published online: 29 FEB 2012
22. Orlov, Y., Pisano, A., Usai, E.: Exponential stabilization of the uncertain wave equation via distributed dynamic input extension. IEEE Trans. Aut. Contr. 56(1), 212–217 (2011b)
23. Orlov, Y., Pisano, A., Usai, E.: Continuous state-feedback tracking of an uncertain heat diffusion process. Syst. Contr. Lett. 59(12), 754–759 (2010)
24. Pisano, A., Orlov, Y., Usai, E.: Tracking Control of the Uncertain Heat and Wave Equation via Power-Fractional and Sliding-Mode Techniques. SIAM J. Contr. Optimizat. 49, 363–382 (2011a)
25. Pisano, A., Orlov, Y.: Boundary second-order sliding-mode control of an uncertain heat process with unbounded matched perturbation. Automatica (2012) (in press), doi:10.1016/j.automatica.2012.05.041
26. Cheng, M.-B., Radisavljevic, V., Su, W.-C.: Sliding mode boundary control of a parabolic PDE system with parameter variations and boundary uncertainties. Automatica 47, 381–387
27. Utkin, V.I.: Sliding Modes In Control and Optimization. Springer, Berlin (1972)

Chapter 5
Practical Relative Degree Approach in Sliding-Mode Control

Arie Levant

Abstract. The high-order sliding-mode approach offers a robust way to solve numerous output-regulation problems when the system relative degree is known. Still the difficult cases remain when the relative degree does not exist, is very high, or the mathematical model is not reliable. The notion of practical relative degree is proposed, which generalizes the standard relative-degree notion for the cases of uncertain systems lacking certain mathematical model. Practical output regulation is ensured. Computer simulation and practical results confirm the theoretical approach.

5.1 Introduction

Control under heavy uncertainty conditions is one of the main subjects of the modern control theory, and the main idea to deal with such problems is to single out and keep properly chosen constraints successively lowering the system dimension. Sliding-mode (SM) or high-gain control are the corresponding methods ⟦12,20,42⟧. The simplest problem of such kind is to make the output σ of a single-input single-output (SISO) "black-box" system vanish.

Sliding mode is accurate and insensitive to disturbances ⟦12, 42⟧. The informal definition of the relative degree (RD) ⟦18⟧ is the least order of the total output derivative, which explicitly contains the control. While standard SMs are applicable to make a sliding variable vanish, if its RD is 1 ⟦12, 40, 42⟧, high-order sliding modes (HOSMs) ⟦2, 3, 8, 9, 11, 14, 22, 24, 36, 37, 39, 41, 44⟧ are capable of keeping constraints of higher RDs. One of the main reasons for their application is the possibility ⟦2, 22, 28⟧ to effectively attenuate the so-called chattering effect ⟦7, 15, 16⟧ caused by the high control-switching frequency.

Establishing the needed constraint $\sigma = 0$ requires the stabilization of the sliding variable σ at zero. The corresponding dynamics of σ is of the order of the RD and is

Arie Levant
Tel-Aviv University, 69978 Tel-Aviv, Israel
e-mail: levant@post.tau.ac.il

B. Bandyopadhyay et al. (Eds.): *Advances in Sliding Mode Control*, LNCIS 440, pp. 97–115.
DOI: 10.1007/978-3-642-36986-5_5

typically uncertain. Theoretically it also allows feedback linearization ⟦18⟧, though the uncertainty prevents its direct application. Finite-time stabilization is preferable, since it provides for higher robustness, simpler overall performance analysis, and higher accuracy in the presence of small sampling noises and delays. With the RD 1, such finite-time stabilization is easily obtained by means of the relay control, which is widely used in the standard sliding-mode control. With higher RDs the problem is much more complicated and corresponds to the HOSM approach ⟦22, 24⟧.

HOSM actually is a motion on the discontinuity set of a dynamic system understood in Filippov's sense ⟦13⟧. The sliding order characterizes the dynamics smoothness degree in the vicinity of the mode. Let the task be to make some smooth scalar function σ vanish, keeping it at zero afterwards. Then successively differentiating σ along trajectories, a discontinuity will be encountered sooner or later in the general case. Thus, a sliding mode $\sigma \equiv 0$ may be classified by the number r of the first successive total time derivative $\sigma^{(r)}$ which is not a continuous function of the state space variables or does not exist due to some reason, like trajectory nonuniqueness. That number is called the sliding order ⟦22, 24⟧, and the motion $\sigma \equiv 0$ is said to be in rth-order sliding (r-sliding) mode. If σ is a vector, also the sliding order is a vector.

Thus, with the RD r a discontinuous control providing for $\sigma \equiv 0$, inevitably generates an r-sliding mode. In order to attenuate the chattering, the control derivative is used as a new discontinuous control ⟦2, 22, 28⟧. The RD with respect to the new control turns out to be $r+1$ and an $(r+1)$-SM is to be established.

The main result of the HOSM control theory probably is the list of universal SISO controllers corresponding to each RD ⟦24, 26, 33⟧. Only a few parameters (usually the control magnitude and one differentiator parameter) are to be adjusted to make the "black-box" output exactly vanish in finite time. Thus, the RD turns out to be the main information needed about the system, and it is typically assumed to be known.

Recent results ⟦28, 31⟧ show that fast stable actuators ⟦31⟧ and sensors ⟦28⟧, which can be considered as singular perturbations ⟦20⟧, as well as any small system perturbations affecting the RD ⟦27⟧, only partially destroy the performance of homogeneous SM controllers. Thus, the system RD actually becomes a design parameter. When developing a mathematical model of a controlled process, one can deliberately neglect some dynamics, in order to simplify the model and the corresponding controller.

Unfortunately, the available model can be very complicated, and sometimes it is difficult to present it as a simple low-order system with negligible slow and stable singular disturbances. Moreover, the very existence of the system RD is rather restrictive. It requires the system to be described by a smooth ordinary differential equation which is linear in control. Hence, a designed controller critically depends on the chosen model form, even if the model is considered unreliable.

The question arises, whether it is possible to treat a system as a "black box", avoiding any dependence on the model. Some recent practical results ⟦17⟧ show that it is possible. An attempt is made in this paper to mathematically justify such approach. The corresponding notion, called practical relative degree (PRD), is formally introduced. It also admits experimental identification by simulation or real life tests.

The PRD is informally defined as the order of the output derivative, which is explicitly affected by control step-function. It can involve system delays and output noises, and it does not require a system mathematical description. In particular, if the system RD exists, it is also a system PRD, but the system can also have a few lower PRDs. A homogeneous SM control designed for a certain RD is shown to be applicable also with the same PRD. The accuracy of the corresponding output regulation is determined by the output reaction delay value.

Even if the RD exists and is known, while theoretically providing for the exact output regulation, a corresponding HOSM controller can be practically impossible to realize due to high information or energy demand. In such a case the PRD approach turns to be a real alternative. Moreover, one can try to apply a number of simple controllers corresponding to lower RDs, without real detection of a PRD. The natural application of such approach lies in the field of artificial-intelligence research.

Computer simulation and recent practical results of blood glucose control ⟦17⟧ demonstrate the feasibility of the suggested approach.

5.2 Homogeneous Sliding Mode Control

5.2.1 Standard SISO Regulation Problem

Definition 5.1. Consider a discontinuous differential equation $\dot{x} = f(x)$ (Filippov differential inclusion $\dot{x} \in F(x)$ ⟦13, 25⟧) with a smooth output function $\sigma = \sigma(x)$, and let it be understood in the Filippov sense. Then, provided that

1. successive total time derivatives σ, $\dot{\sigma}$, ..., $\sigma^{(r-1)}$ are continuous functions of x,
2. the set
$$\sigma = \dot{\sigma} = \ldots = \sigma^{(r-1)} = 0 \tag{5.1}$$
is a non-empty integral set,
3. the Filippov set of admissible velocities at the r-sliding points (5.1) contains more than one vector,

the motion on set (5.1) is said to exist in r-sliding (rth-order sliding) mode ⟦22, 24⟧. In the non-autonomous case the additional coordinate t is formally added, $\dot{t} = 1$.

Consider a dynamic system of the form

$$\dot{x} = a(t,x) + b(t,x)u, \ \sigma = \sigma(t,x), \tag{5.2}$$

where $x \in \boldsymbol{R}^n$, a, b and σ: $\boldsymbol{R}^{n+1} \to \boldsymbol{R}$ are unknown smooth functions, $u \in \boldsymbol{R}$, the dimension n might be also uncertain. Only measurements of σ are available in real time. The task is to provide in finite time for exactly keeping $\sigma \equiv 0$.

The relative degree r of the system is assumed to be constant and known. In other words, for the first time the control explicitly appears in the rth total time derivative of σ and

$$\sigma^{(r)} = h(t,x) + g(t,x)u, \tag{5.3}$$

where $h(t,x) = \sigma^{(r)}\Big|_{u=0}$, $g(t,x) = \frac{\partial}{\partial u}\sigma^{(r)} \neq 0$. It is supposed that

$$0 < K_m \leq \frac{\partial}{\partial u}\sigma^{(r)} \leq K_M, \ \left|\sigma^{(r)}\Big|_{u=0}\right| \leq C \tag{5.4}$$

holds for some $K_m, K_M, C > 0$. It is always true at least in compact operation regions. Trajectories of (5.2) are assumed infinitely extendible in time for any Lebesgue-measurable bounded control $u(t, x)$.

Finite-time stabilization of smooth systems at an equilibrium point by means of continuous control is considered in [[1,6]]. In our case any continuous control

$$u = \varphi(\sigma, \dot{\sigma}, \ldots, \sigma^{(r-1)}), \tag{5.5}$$

providing for $\sigma \equiv 0$, should satisfy the equality $\varphi(0,0,\ldots,0) = -h(t,x)/g(t,x)$, whenever (5.1) holds. Since the problem of uncertainty prevents it, *the control has to be discontinuous at least on the set* (5.1). Hence, the r-sliding mode $\sigma = 0$ is to be established.

As follows from (5.3), (5.4)

$$\sigma^{(r)} \in [-C, C] + [K_m, K_M]u. \tag{5.6}$$

The obtained inclusion does not "remember" anything on system (5.2) except the constants r, C, K_m, K_M. Thus, provided (5.4) holds, the finite-time stabilization of (5.6) at the origin simultaneously solves the stated problem for all systems (5.2).

Note that the realization of this plan requires real-time differentiation of the output. The controllers, which are further designed, are *r-sliding homogeneous* [[25]]. The corresponding notion is introduced below.

5.2.2 Homogeneous Sliding Modes

Definition 5.2. A function f: $\mathbf{R}^n \to \mathbf{R}$ (respectively a vector-set field $F(x) \subset \mathbf{R}^n$ (see [[25]]), $x \in \mathbf{R}^n$, or a vector field f: $\mathbf{R}^n \to \mathbf{R}^n$) is called *homogeneous of the degree $q \in \mathbf{R}$ with the dilation* [[1]]

$$d_\kappa : (x_1, x_2, ..., x_n) \mapsto (\kappa^{m_1}x_1, \kappa^{m_2}x_2, ..., \kappa^{m_n}x_n),$$

where m_1, ..., m_n are some positive numbers (*weights*), if for any $\kappa > 0$ the identity $f(x) = \kappa^{-q} f(d_\kappa x)$ holds (respectively $F(x) = \kappa^{-q} d_\kappa^{-1} F(d_\kappa x)$, or $f(x) = \kappa^{-q} d_\kappa^{-1} f(d_\kappa x)$). The non-zero homogeneity degree q of a vector field can always be scaled to ± 1 by an appropriate proportional change of the weights m_1, ..., m_n.

The homogeneity of a vector field $f(x)$ (a vector-set field $F(x)$) can equivalently be defined as the invariance of the differential equation $\dot{x}=f(x)$ (differential inclusion $\dot{x}\in F(x)$) with respect to the combined time-coordinate transformation

$$G_\kappa : (t,x) \mapsto (\kappa^p t, d_\kappa x),$$

where p = - q, might naturally be considered as the weight of t. Indeed, the homogeneity condition can be rewritten as

$$\dot{x} \in F(x) \Leftrightarrow \frac{d(d_\kappa x)}{d(\kappa^p t)} \in F(d_\kappa x).$$

It was proved in [[25]] that if $\dot{x} \in F(x)$ is a homogeneous Filippov inclusion with a negative homogeneous degree -p, then uniform finite-time stability, uniform asymptotic stability and the contractivity feature [[25]] are equivalent and the maximal settling time is a homogeneous function of the initial conditions of the degree p. Furthermore it was proved there that in the presence of variable delays of the order τ^p, and sampling noises of x_i of the order τ^{m_i} the trajectories converge in finite-time into a region featuring $x_i = O(\tau^{m_i})$. Finite-time stability of homogeneous discontinuous differential equations was also considered in [[38]].

Suppose that feedback (5.5) imparts homogeneity properties to the closed-loop inclusion (5.5), (5.6). Due to the term [-C, C], the right-hand side of (5.5) can only have the homogeneity degree 0 with $C \neq 0$. Thus, the homogeneity degree of $\sigma^{(r-1)}$ is to be opposite to the degree of the whole system, i.e. $\deg \sigma^{(r-1)} = \deg t = -q$.

Scaling the system homogeneity degree to -1, would achieve the homogeneity weights of t, σ , $\dot{\sigma}$, ..., $\sigma^{(r-1)}$ to be 1, r, r - 1, ..., 1 respectively. This homogeneity is further called the *r-sliding homogeneity*. The inclusion (5.5), (5.6) is called *r-sliding homogeneous* if for any $\kappa > 0$ the combined time-coordinate transformation

$$G_\kappa : (t,\sigma,\dot{\sigma},\dots,\sigma^{(r-1)}) \mapsto (\kappa t, \kappa^r \sigma, \kappa^{r-1}\dot{\sigma},\dots,\kappa\sigma^{(r-1)}) \tag{5.7}$$

preserves the closed-loop inclusion (5.5), (5.6).

Transformation (5.7) transfers (5.5), (5.6) into

$$\frac{d^r(\kappa^r\sigma)}{d(\kappa t)^r} \in [-C,C] + [K_m,K_M]\varphi(\kappa^r\sigma,\kappa^{r-1}\dot{\sigma},\dots,\kappa\sigma^{(r-1)}).$$

Obviously, (5.5), (5.6) is r-sliding homogeneous if $\deg \varphi = 0$, i.e.

$$\varphi(\kappa^r\sigma,\kappa^{r-1}\dot{\sigma},\dots,\kappa\sigma^{(r-1)}) \equiv \varphi(\sigma,\dot{\sigma},\dots,\sigma^{(r-1)}). \tag{5.8}$$

Definition 5.3. Controller (5.5) is called r-sliding homogeneous (rth order sliding homogeneous) if (5.8) holds for any $(\sigma,\dot{\sigma},...,\sigma^{(r-1)})$ and $\kappa > 0$. The corresponding sliding mode is also called homogeneous (if exists).

Such a homogeneous controller is inevitably discontinuous at the origin (0, ..., 0), unless φ is a constant function. Most known r-sliding controllers, $r \geq 2$, are based on

r-sliding homogeneous controllers. An important exception is the terminal 2-sliding controller maintaining 1-sliding mode $\dot{\sigma} + \beta\sigma^{\rho} \equiv 0$, where $\rho = (2k+1)/(2m+1)$, $\beta > 0$, $k < m$, and k, m are natural numbers ⟦36, 44⟧. Indeed, the homogeneity requires $\rho = 1/2$ and $\sigma \geq 0$.

5.2.3 Arbitrary Order Sliding Mode Controllers

Following is one of the most known r-sliding controller families ⟦24, 26, 33⟧ called quasi-continuous. The controllers of the form

$$u = -\alpha\Psi_{r-1,r}(\sigma, \dot{\sigma}, \ldots, \sigma^{(r-1)}), \tag{5.9}$$

are defined by recursive procedures, have the magnitude $\alpha > 0$, and solve the general output regulation problem from Section 5.2.1 with the relative degree r. Quasi-continuous r-sliding controller is a feedback function of σ, $\dot{\sigma}$, ..., $\sigma^{(r-1)}$ being continuous everywhere except on the manifold $\sigma = \dot{\sigma} = \ldots = \sigma^{(r-1)} = 0$ of the r-sliding mode. In the presence of errors in evaluation of σ and its derivatives, these equalities never take place simultaneously with $r > 1$. Therefore, control practically turns out to be a continuous function of time.

The parameters of the controllers can be chosen in advance for each r. Only the magnitude α is to be adjusted for any fixed C, K_m, K_M, most conveniently by computer simulation, avoiding complicated and redundantly large estimations. Obviously, α is to be taken negative with $(\partial/\partial u)\sigma^{(r)} < 0$. In the following $\beta_1, \ldots, \beta_{r-1} > 0$ are the controller parameters, and i = 1,..., r-1.The following procedure defines a family of such controllers ⟦26⟧:

$$\varphi_{0,r} = \sigma,\; N_{0,r} = |\sigma|,\; \Psi_{0,r} = \varphi_{0,r}/N_{0,r} = \operatorname{sign}\sigma,$$
$$\varphi_{i,r} = \sigma^{(i)} + \beta_i N_{i-1,r}^{(r-i)/(r-i+1)}\Psi_{i-1,r},$$
$$N_{i,r} = |\sigma^{(i)}| + \beta_i N_{i-1,r}^{(r-i)/(r-i+1)},\; \Psi_{i,r} = \varphi_{i,r}/N_{i,r}, u = -\alpha\Psi_{r-1,r}.$$

Note that while increasing α enlarges the class of systems (5.4), to which the controller is applicable, parameters β_i are tuned to provide for the needed convergence rate ⟦32⟧. Asymptotic accuracies of these controllers are readily obtained from their homogeneity properties. In particular $\sigma^{(i)} = O(\tau^{r-i})$, i = 0, 1, …, r-1, if the measurements are performed with the sampling interval τ.

A controller providing for the time-optimal stabilization of the inclusion (5.6) under the restriction $|u| \leq \alpha$ was proposed in ⟦8⟧. Such controllers are also r-sliding homogeneous providing for the same asymptotic accuracy. Unfortunately, in practice they are only available for $r \leq 3$.

Controller Adjustment. The magnitude of the controllers ⟦24, 26⟧ can be increased without loss of the convergence. The corresponding controller gets the form

$$u = -\alpha\Phi(t,x)\Psi_{r-1,r}(\sigma, \dot{\sigma}, \ldots, \sigma^{(r-1)}), \tag{5.10}$$

where $\alpha > 0$, and $\Psi_{r-1,r}$ were introduced above. Note that controller (5.10) is not homogeneous. While the function Φ can be chosen large to control exploding systems, it is also reasonable to make the function Φ decrease and even vanish, when approaching the system operational point, therefore reducing the chattering [[32]].

It follows from [[33]] that the parameters $\beta_1, \ldots, \beta_{r-1}$ can be chosen one-by-one by means of relatively simple simulation of concrete differential equations $\varphi_{i,r} = 0$, all of which are to be finite-time stable. The controller with the resulting parameters formally provides for the universal solution of the stated problem. Nevertheless, in practice one often needs to adjust the convergence rate, either to slow it down relaxing the requirements to actuators, or to accelerate it in order to meet some system requirements. In that context note that redundantly enlarging the magnitude $\alpha\Phi$ of controller (5.9) does not accelerate the convergence, but only increases the chattering, while its reduction may lead to the convergence loss.

The main procedure is to take the controller

$$u = -\gamma^r \alpha \Psi_{r-1,r}(\sigma, \dot{\sigma}/\gamma, \ldots, \sigma^{(r-1)}/\gamma^{r-1}),\ \gamma > 0,$$

instead of (5.9), providing for the approximately γ times reduction of the convergence time [[32]]. With $0 < \gamma < 1$ the convergence is slowed down.

In the case of quasi-continuous controllers the form of the controller is preserved. The new parameters $\tilde{\beta}_1, \ldots, \tilde{\beta}_{r-1}, \tilde{\alpha}$ are calculated according to the formulas $\tilde{\beta}_1 = \gamma\beta_1$, $\tilde{\beta}_2 = \gamma^{r/(r-1)}\beta_2$, $\ldots$, $\tilde{\beta}_{r-1} = \gamma^{r/2}\beta_{r-1}$, $\tilde{\alpha} = \gamma^r\alpha$. Larger the γ,faster the convergence. Following are the resulting quasi-continuous controllers with $r \le 4$, simulation-tested β_i and a general gain function Φ:

1. $u = -\gamma\alpha\Phi \operatorname{sign}\sigma$,
2. $u = -\gamma^2\alpha\Phi\,(\dot{\sigma} + \gamma|\sigma|^{1/2}\operatorname{sign}\sigma)/(|\dot{\sigma}| + \gamma|\sigma|^{1/2})$,
3. $u = -\gamma^3\alpha\Phi\,[\ddot{\sigma} + 2\gamma^{3/2}(|\dot{\sigma}| + \gamma|\sigma|^{2/3})^{-1/2}(\dot{\sigma} + \gamma|\sigma|^{2/3}\operatorname{sign}\sigma)]/$
$[|\ddot{\sigma}| + 2\gamma^{3/2}(|\dot{\sigma}| + \gamma|\sigma|^{2/3})^{1/2})]$,
4. $\varphi_{3,4} = \dddot{\sigma} + 3\gamma^2[\ddot{\sigma} + \gamma^{4/3}(|\dot{\sigma}| + 0.5\gamma|\sigma|^{3/4})^{-1/3}(\dot{\sigma} + 0.5\gamma)|\sigma|^{3/4}\operatorname{sign}\sigma)]$
$[|\ddot{\sigma}| + \gamma^{4/3}(|\dot{\sigma}| + 0.5\gamma|\sigma|^{3/4})^{2/3}]^{-1/2}$,
$N_{3,4} = |\dddot{\sigma}| + 3\gamma^2[|\ddot{\sigma}| + \gamma^{4/3}(|\dot{\sigma}| + 0.5\gamma|\sigma|^{3/4})^{2/3}]^{1/2}$,
$u = -\gamma^4\alpha\Phi\,\varphi_{3,4}/N_{3,4}$.

Chattering Attenuation. The standard chattering attenuation procedure is to consider the control derivative as a new control input, increasing the relative degree and the sliding order by one [[1, 22, 28]]. It was successfully applied in practice (for example see [[3]]) many times, though formally the convergence is only locally ensured in some vicinity of the $(r + 1)$-sliding mode $\sigma \equiv 0$. Global convergence can be easily obtained in the case of the transition from the relative degree 1 to 2 [[22]]; semi-global convergence can be assured with higher relative degrees using integral $(r + 1)$-sliding modes [[30]].

5.3 Arbitrary Order Robust Exact Differentiation

Any r-sliding homogeneous controller can be complemented by an (r-1)th order differentiator [4, 41, 43] producing an output-feedback controller. In order to preserve the demonstrated exactness, finite-time stability and the corresponding asymptotic properties, the natural way is to calculate $\dot{\sigma}$, ..., $\sigma^{(r-1)}$ in real time by means of a robust finite-time convergent exact *homogeneous* differentiator [23, 24]. Its application is possible due to the boundedness of $\sigma^{(r)}$ provided by the boundedness of the feedback function φ in (5.5).

5.3.1 Standard Arbitrary-Order Robust Exact Differentiator

Let the input signal $f(t)$ be a function defined on $[0, \infty)$ and consisting of a bounded Lebesgue-measurable noise with unknown features, and of an unknown base signal $f_0(t)$, whose kth derivative has a known Lipschitz constant $L > 0$. The problem of finding real-time robust estimations of $\dot{f}_0(t)$, $\ddot{f}_0(t)$, ... ,$f_0^{(k)}(t)$ being exact in the absence of measurement noises is solved by the differentiator [24]

$$\begin{aligned}
&\dot{z}_0 = v_0, v_0 = -\lambda_k L^{1/(k+1)}|z_0 - f(t)|^{k/(k+1)}\text{sign}(z_0 - f(t)) + z_1,\\
&\dot{z}_1 = v_1, v_1 = -\lambda_{k-1} L^{1/k}|z_1 - v_0|^{(k-1)/k}\text{sign}(z_1 - v_0) + z_2,\\
&\dots\\
&\dot{z}_{k-1} = v_{k-1}, v_{k-1} = -\lambda_1 L^{1/2}|z_{k-1} - v_{k-2}|^{1/2}\text{sign}(z_{k-1} - v_{k-2}) + z_k,\\
&\dot{z}_k = -\lambda_0 L\,\text{sign}(z_k - v_{k-1}).
\end{aligned} \tag{5.11}$$

The parameters $\lambda_0, \lambda_1, \dots, \lambda_k > 0$ being properly chosen, the following equalities are true in the absence of input noises after a finite time of the transient process:

$$z_0 = f_0(t); \quad z_i = v_{i-1} = f_0^{(i)}, \quad i = 1, \dots, k.$$

Note that the differentiator has a recursive structure. Once the parameters λ_0, λ_1, $\dots$, λ_{k-1} are properly chosen for the (k - 1)th order differentiator with the Lipschitz constant L, only one parameter λ_k is needed to be tuned for the kth order differentiator with the same Lipschitz constant. The parameter λ_k is just to be taken sufficiently large. Any $\lambda_0 > 1$ can be used to start this process. Such differentiator can be used in any output feedback.

Thus an infinite sequence of parameters λ_k can be built, valid for all k. In particular, one can choose $\lambda_0 = 1.1$, $\lambda_1 = 1.5$, $\lambda_2 = 2$, $\lambda_3 = 3$, $\lambda_4 = 5$, $\lambda_5 = 8$, which is enough for $k \leq 5$. Another possible choice of the differentiator parameters with $k \leq 5$ is $\lambda_0 = 1.1$, $\lambda_1 = 1.5$, $\lambda_2 = 3$, $\lambda_3 = 5$, $\lambda_4 = 8$, $\lambda_5 = 12$ [25].

The homogeneity features imply the asymptotic accuracy of the differentiator [25]. Let the measurement noise be any Lebesgue-measurable function with the magnitude not exceeding ε. Then the accuracy $|z_i(t) - f_0^{(i)}(t)| = O(\varepsilon^{(k+1-i)/(k+1)})$ is obtained. That accuracy is shown to be the best possible [21, 23]. Differentiators

(5.11) with constant and variable parameters L have been already proved useful in practical and theoretical observation ⟦3, 5⟧.

Due to the specific homogeneity features of the differentiator (5.11), its output-feedback combination with controller (5.9) produces an r-sliding homogeneous controller; which when used in the feedback closure of (5.6) an r-sliding homogeneous differential inclusion is produced. Thus the asymptotic accuracy of the output-feedback controller remains the same as of the controller (5.9) with *direct* measurements.

5.3.2 Homogeneous Tracking Differentiator

The following construction is called a *homogeneous tracking differentiator* ⟦29⟧ of the order k. As previously let the input be a function $f(t) = f_0(t) + \eta(t)$, $|f_0^{(k+1)}| \leq L$, $|\eta| \leq \varepsilon$. Construct an auxiliary dynamic system $w^{(k+1)} = v$ rewritten as

$$\begin{aligned} \dot{w}_0 &= w_1, \cdots, \dot{w}_{k-1} = w_k, \\ \dot{w}_k &= v \end{aligned} \tag{5.12}$$

with the input v, output w_0 and the measured signal $f(t)$ to be tracked. For the further use rewrite differentiator (5.11) symbolically by the formula $z = D_{k,\lambda,L}(f)$. Now, close system (5.12) by the feedback

$$\begin{aligned} v &= -\varpi L \Psi_{k,k+1}(z), \\ z &= D_{k,\lambda,\varpi_1 L}(w_0 - f) \end{aligned}. \tag{5.13}$$

where $\varpi_1 \geq \varpi > 1$. Here $\Psi_{k,k+1}$ is the quasicontinuous controller introduced in Section 5.2.3, but any other $(k+1)$-sliding homogeneous controller can also be used. With sufficiently large ϖ, a system which starts to track the function $f_0(t)$ in finite time is obtained. That implies the following simple theorem.

Theorem 5.1. *With sufficiently large $\varpi > 1$ and any $\varpi_1 \geq \varpi$, tracking differentiator (5.12), (5.13) provides for the finite-time convergence of w_i to $f_0^{(i)}$, $i = 0, 1, \ldots, k$. The asymptotic accuracy of the tracking differentiator (5.12), (5.13) is exactly the same as of the standard differentiator ⟦24⟧ (5.11). In particular, with continuous-time sampling the tracking accuracies $|w_i - f_0^{(i)}| \leq \mu_i L^{i/(k+1)} \varepsilon^{(k+1-i)/(k+1)}$ are obtained. The coefficients μ_i only depend on the tracking differentiator parameters.*

The parameter $\varpi > 1$ can be chosen once and forever. The value $\varpi_1 \geq \varpi$ is adjusted according to the circumstances, in particular, larger values are to be considered in the presence of significant noises. It can also be shown that the digital-implementation asymptotic accuracy of (5.12), (5.13) is exactly the same as of the standard differentiator (5.11) ⟦24⟧. Also its output-feedback combination with controller (5.9) produces an r-sliding homogeneous controller which can be used for the solution of the problem from Section 5.2.1. Moreover, the asymptotic accuracy of the output-feedback controller remains the same as of the controller (5.9) with *direct* measurements.

The main advantage of the tracking differentiator is that its estimations w_i of the input derivatives $f_0^{(i)}$, $i = 0, 1, \ldots, k-1$, are successive integrals of the kth-order derivative estimation w_k.

5.4 Practical Relative Degree Concept

5.4.1 Practical Relative Degree (PRD) Definition

Consider a SISO system with a scalar input $u \in \mathbf{R}$ (the control), and output $\tilde{\sigma} \in \mathbf{R}$. The output depends on the internal state of the system, which changes in time. The control influences the state in some way. The nature of the state remains unknown. The task is to keep the output $\tilde{\sigma}$ close to zero.

The input belongs to a certain class. For example, it should be Lebesgue-measurable, or continuous, etc. It is assumed in the following that the system accepts Lebesgue-measurable inputs, but the results do not change if inputs are required to have any predefined smoothness.

Definition 5.4. A natural number r is called a *practical relative degree* (PRD) of the SISO system with the input (control) $u \in \mathbf{R}$, and output $\tilde{\sigma} \in \mathbf{R}$, if there exist positive ε, δ_t, α_m, α_M, L, L_m, $\alpha_m \leq \alpha_M$, $L \leq L_m$, and $u_0 \in \mathbf{R}$, $\lambda_\sigma = \pm 1$ such that

1. The system accepts any bounded input $u(t)$, $|u - u_0| \leq \alpha_M$. The corresponding output can be always represented as a sum of two components, $\tilde{\sigma}(t) = \sigma(t) + \eta(t)$, where $|h| \leq \varepsilon$. With $r > 0$ the function σ is assumed r-1 times differentiable with $\sigma^{(r-1)}$ having a uniform Lipschitz constant L. Respectively $\sigma^{(r)}$ exists almost everywhere, $\sigma^{(r)} \leq L$.

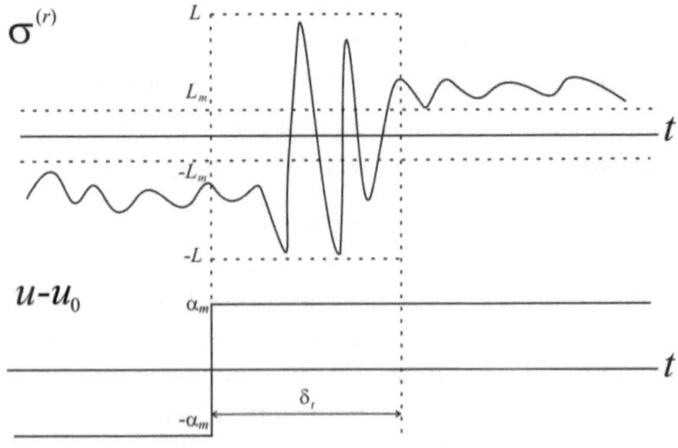

Fig. 5.1 Reaction of a system with PRD r to a step function

2. Let $w = \lambda_s \sigma$. For any time moment t_0, if starting from t_0 the inequality $\alpha_M \geq u(t) - u_0 \geq \alpha_m$ ($-\alpha_M \leq u(t) - u_0 \leq -\alpha_m$) is kept, then starting from the moment $t_0 + \delta_t$ the output satisfies $w^{(r)} \geq L_m$ (respectively $w^{(r)} \leq -L_m$).

Parameters u_0, λ_s, δ_t and ε are respectively called the *reference input*, the *input influence direction*, the *delay* and the *approximation* parameters.

In the following $u_0 = 0$, $\lambda_s = 1$ are assumed. The characteristic reaction of $\sigma^{(r)}$ to the input step function is shown in Fig. 5.1.

Definition 5.5. A natural number r is called a *local practical relative degree*, if there exist three time values t_1, t_2, T, $t_1 < t_2$, $\delta_t < T$, such that requirement 2 of Definition 5.4 is fulfilled for each $t_0 \in [t_1, t_2]$ over the time interval of the length T, and the first requirement is true over the time interval $[t_1, t_2 + T]$.

Obviously, in general one cannot hope to keep $\sigma = 0$ exactly. It should be stressed that the function σ does not necessarily has to have some real meaning. It can be simply an output of some smoothing filter, in particular, of a tracking differentiator. Though ε and δ_t are naturally assumed to be small with respect to T and t_2 - t_1 , it is not formally required. Local practical relative degree is a temporary feature of a system, usable for temporarily controlling its output.

The following Proposition is obvious.

Proposition 5.1. *If system (5.1) satisfies the assumptions of Section 5.2.1 it also has the practical relative degree r with* $\varepsilon = \delta_t = 0$, $u_0 = 0$, $\lambda_\sigma = 1$. *If condition (5.4) is not globally satisfied, but the relative degree is still r, then it has the local practical relative degree r over any compact region of the extended state space* t, x.

Indeed, one can choose any values $\alpha_M > \alpha_m > C/K_m$, $L > C + K_M \alpha_M$, $L_m = L - C - K_M \alpha_M$. If condition (5.4) is only locally satisfied $\lambda_\sigma = \text{sign}\, g$ is taken (recall that since the RD is r, the function $g = \frac{\partial}{\partial u}\sigma^{(r)}$ does never vanish).

Choose a controller for a system with the PRD r. The following construction is valid for any r-sliding homogeneous controller [24–26, 33] of the form (5.10). Let

$$u = u_0 - \alpha_m \lambda_s \Phi(z_0, z_1, \ldots, z_{r-1}) \Psi_{r-1,r}(z_0, z_1/\gamma, \ldots, z_{r-1}/\gamma^{r-1}), \tag{5.14}$$

$$z = D_{r-1,\lambda,\tilde{L}}(z_0 - \tilde{\sigma}). \tag{5.15}$$

Here functions Φ, $\Psi_{r-1,r}$ are described above, and meantime $\Phi \equiv 1$, $D_{r-1,\lambda,\tilde{L}}$ is the homogeneous (r-1)th order differentiator (5.11). Choose α so that control (5.10) provides for the finite-time stabilization of the simple system $\sigma^{(r)} = u$; and choose $\gamma > 0$ so that $\gamma^r \alpha \leq L_m$. The differentiator parameter $\tilde{L} > L$ is taken.

Theorem 5.2. *With the parameters chosen as above, controller (5.14), (5.15) provides in finite time for the accuracy* $\sigma^{(i)} \leq \eta_i \max[\delta_t^{r-i}, \varepsilon^{(r-i)/r}]$, $|\tilde{\sigma}| \leq \tilde{\eta}_0 \max[\delta_t^r, \varepsilon]$, $\tilde{\eta}_0$, η_i *being constants only depending on the parameters* L, α_m, L_m *of definition 5.4 and the choice of the controller parameters* $\tilde{L}$, γ.

Note that with $\varepsilon = \delta_t = 0$, exact output regulation is obtained, which, in particular, extends the known HOSM control results to systems of any nature, for example to possibly discontinuous systems nonlinear in control.

Idea of the Proof. After the differentiator converges, the arguments of the controller (5.12) are close to the corresponding derivatives of σ with the errors described in Section 5.3.1. Thus, in the additional time δ_t, the state space with coordinates $(\sigma, \dot{\sigma}, \ldots, \sigma^{(r-1)})$ is divided in the two regions. The first region corresponds to the points where respectively to the PRD definition, the trajectories satisfy the differential inclusion

$$\sigma^{(r)} \in -[L_m, L]\Psi_{r-1,r}(\sigma, \dot{\sigma}, \ldots, \sigma^{(r-1)}),$$

while in the second one any other values of $\sigma^{(r)}$, $\sigma^{(r)} \in -[L, L]$, are possible. Consider the homogeneity transformation

$$G_\kappa : (t, \delta_t, \varepsilon, \sigma, \dot{\sigma}, ..., \sigma^{(r-1)}) \mapsto (\kappa t, \kappa \delta_t, \kappa^r \varepsilon, \kappa^r \sigma, \kappa^{r-1} \dot{\sigma}, ..., \kappa \sigma^{(r-1)}). \quad (5.16)$$

The second region is proved to be described by homogeneous inequalities, which are preserved by transformation (5.16). As a result, the trajectories satisfy some homogeneous differential inclusion invariant with respect to (5.16). The further proof follows the standard homogeneity technique [[25]]. ■

Since under the conditions of the PRD definition, exact satisfaction of $\sigma \equiv 0$ is impossible with $\varepsilon > 0$, one can use continuous control without compromising the system accuracy. Indeed, it is enough to take the gain function Φ equal to a homogeneous norm of $(\sigma/\gamma_1^r, \dot{\sigma}/\gamma_1^{r-1}, ..., \sigma^{(r-1)}/\gamma_1)$ saturated at 1 with a sufficiently large $\gamma_1 > 0$ in (5.14).

5.4.2 PRD Identification

One can identify a PRD using analytical methods developed in the sequel. Another way is to experimentally identify a PRD by simulation or even by a real-life test. According to definition 5.4, the measured system output $\tilde{\sigma}(t)$ is to be the sum of a smooth component and a bounded (preferably small) additional term, $\tilde{\sigma}(t) = \sigma(t) + \eta(t)$. Let the function $\sigma^{(r)}$ be absolutely continuous with its derivative almost everywhere bounded by some number $\tilde{L}$. Such additional smoothness is usual due to the presence of sensors. Respectively call the PRD *strong*. Apply a tracking differentiator to single out the smooth component $\sigma(t)$:

$$\begin{aligned} &\dot{w}_0 = w_1, \cdots, \dot{w}_{r-1} = w_r, \\ &\dot{w}_r = -\varpi \tilde{L} \Psi_{r,r+1}(z), \\ &z = D_{r,\lambda,\varpi_1 \tilde{L}}(w_0 - \tilde{\sigma}) \end{aligned} \quad (5.17)$$

Here $\varpi > 1$ is chosen in advance, and any $\varpi_1 \geq \varpi$ fits. The following proposition is an easy consequence of Theorem 5.1.

Proposition 5.2. *Let the system have a strong PRD r, $|\sigma^{(r+1)}| \leq \tilde{L}$. Then observer (5.17) provides for the accuracies $|w_0^{(i)} - \sigma^{(i)}| \leq \mu_i \tilde{L}^{i/(r+1)} \varepsilon^{(r-i+1)/(r+1)}$, $i = 0, 1, \ldots, r$, established in uniformly bounded time with $\mu_i > 1$ being constants only depending on the observer parameters. Thus, with $L_m > \mu_r \tilde{L}^{r/(r+1)} \varepsilon^{1/(r+1)}$ the function $\sigma(t)$ can be redefined as w_0 in definition 5.4 with the corresponding change of other parameters. On the other hand, if after some transient, the output w_0 of the differentiator differs from the system output $\tilde{\sigma}$ by not more than some constant, and requirements 1, 2 of Definition 5.4 are satisfied for $\sigma = w_0$, then the system has the strong PRD r.*

Note that, as follows from Theorem 5.1, if the component σ is already known from the simulation context, the PRD itself can be identified by a bit simpler standard differentiator of the form (5.11).

5.4.3 PRD Features

In this subsection, general examples and properties of the practical relative degree are demonstrated. Consider a SISO system

$$\mu_z \dot{z} = f_z(z, u_z), v_z = v_{z0}(z) + \eta_z(t), \tag{5.18}$$

where $z \in \mathbf{R}^m$, $u_z \in \mathbf{R}$, $|u_z| \leq U_z$, is the control and the input of the actuator, v_{z0} is a continuous output function, $\mu_z > 0$ is a time-constant parameter, and $\eta_z(t)$ is some deterministic Lebesgue-measurable noise of the magnitude ε_z. The system is understood in Filippov's sense [[13]] and features the Bounded Input - Bounded State property. The initial values $z(0)$ are assumed belonging to a compact. Thus, z forever belongs to a larger compact W_z. The function $f_z(z, u_z)$ is assumed piece-wise continuous in the region $z \in W_z$, $|u_z| \leq U_z$ with a finite number of compact continuity regions and continuity components extendible up to the region boundaries.

System (5.6) is further called *a transmission unit*, if the above conditions are satisfied, and there is such $k \neq 0$ that with $\mu_z = 1$ and any $u_z = \text{const}$, the output v_{z0} converges to ku_z uniformly in u_z and initial values of z. That means that for any t_0, $\delta > 0$ there exists $T > 0$ such that with any $z(t_0)$, u_z, $u_z = \text{const}$, the inequality $|v_{z0} - ku_z| \leq \delta$ is kept, starting from the moment T.

Examples. Any LTI stable system with the transfer function $P(\mu_z w)/Q(\mu_z w)$ is a transmission unit, provided $\deg Q - \deg P > 0$, Q is a Hurwitz polynomial, $P(0)/Q(0) = k$. With infinitesimally small μ_z traditional models of actuators and sensors are produced. Another example: $\ddot{v}_z = -\alpha \,\text{sign}(v_z - ku_z) - \beta \dot{v}_z$, $\alpha, \beta > 0$, $z = (v_z, \dot{v}_z)$.

Proposition 5.3. *Provided the inputs are required to have a fixed Lipschitz constant, transmission units have PRD equal to 0.*

See [[28,31]] for the proof. Note that the time constants of the units are not required to be infinitesimal. Low-pass filters also have zero PRD. In the following propositions, the cascade connections are assumed to satisfy the obvious compliance conditions of input and output bounds.

Proposition 5.4. *A cascade connection of a SISO system of the PRD 0 at the input and another SISO system of the PRD r has the PRD equal to r with the delay parameter being the sum of two delay parameters. The new approximation parameter equals that parameter of the last system, i.e. of the system at the output.*

A cascade system considered in [[28]] with actuator and sensor also depending on the middle-system internal state, has the PRD equal to the RD of the system in the middle.

Proposition 5.5. *A cascade connection of a SISO system of the practical relative degree r_1 with the zero approximation parameter and a system of the form (5.1)-(5.3) of the relative degree r_2 with some bounded output noise forms a new SISO system of the practical relative degree $r_1 + r_2$.*

Note that putting a system with regular RD before a system with PRD may even lead to the loss of PRD. The above propositions allow constructing a lot of systems with PRD. Similar results are also true with respect to the local PRD.

Obviously a system can have a few PRDs. For example, consider a cascade system of successively connected smooth transmission units with RDs r_1, r_2, and a SISO system (5.1)-(5.3) with the RD r between them. Let all approximation parameters be zero. Then the resulting system has PRDs $r_1 + r + r_2$, $r_1 + r$, $r + r_2$, and r.

5.5 Simulation and Applications

5.5.1 Disturbed-Kinematic-Car-Model Control

Consider a simple kinematic model of car control

$$\dot{x} = V\cos\varphi,\ \dot{y} = V\sin\varphi,\ \dot{\varphi} = \frac{V}{\Delta}\tan\theta,\ \dot{\theta} = u,$$

where x and y are Cartesian coordinates of the rear-axle middle point, φ is the orientation angle, V is the longitudinal velocity, Δ is the length between the two axles and θ is the steering angle (i.e. the real input), u is the system input (control). The task is to steer the car from a given initial position to the trajectory $y = g(x)$, where $g(x)$ and y are assumed to be available in real time. The relative degree of the system is 3.

Now consider a disturbed system.

$$\dot{x} = V\cos\phi,\ \dot{y} = V\sin\phi,\ \ddot{\phi} = -4\,\mathrm{sign}(\phi - \varphi) - 6\dot{\phi},$$
$$\dot{\varphi} = \frac{V}{\Delta}\tan\theta,\ \dot{\theta} = \zeta_1.$$

Let $V = \text{const} = 10m/s$, $\Delta = 5m$, $x = y = \varphi = \theta = 0$ at $t = 0$, $g(x) = 10\sin(0.05x) + 5$. Introduce the actuator transmission unit

$$\ddot{\zeta}_1 = -100(2(\zeta_1 - u) + 0.01\dot{\zeta}_1)^3 - 100(\zeta_1 - u) - 2\dot{\zeta}_1,$$

and the sensor transmission unit

$$\dddot{\zeta}_2 = -100(\zeta_2 - y) - 2\dot{\zeta}_2 - 0.02\ddot{\zeta}_2, \;\; \sigma = \zeta_2 + 0.01\dot{\zeta}_2 - g(x), \;\; \tilde{\sigma} = \sigma + \eta(t),$$

which produces the noisy output $\tilde{\sigma}$ with σ being a smooth component. Here η is a noise, $|\eta| \leq 0.01m$; $\zeta_2 = -10$, $\dot{\zeta}_2 = 2000$, $\ddot{\zeta}_2 = -80000$, $\zeta_1 = \dot{\zeta}_1 = \phi = \dot{\phi} = 0$ at $t = 0$.

Note that the disturbed system does not have a relative degree, for it is not smooth. With $\phi \equiv \varphi$, $\eta = 0$ the RD would be equal to 8. Propositions 5.2-5.5 show that it has a local strong PRD equal to 3. The PRD identification results obtained by means of the third-order differentiator $z = D_{3,\{1.1,1.5,2,3\},100}(\tilde{\sigma})$

$$\begin{aligned}
\dot{z}_0 &= v_0,\; v_0 = -9.49|z_0 - \tilde{\sigma}(t)|^{3/4}\operatorname{sign}(z_0 - \tilde{\sigma}(t)) + z_1,\\
\dot{z}_1 &= v_1,\; v_1 = -9.28|z_1 - v_0|^{2/3}\operatorname{sign}(z_1 - v_0) + z_2,\\
\dot{z}_2 &= v_2,\; v_2 = -15|z_2 - v_1|^{1/2}\operatorname{sign}(z_2 - v_1) + z_3,\\
\dot{z}_3 &= -110\operatorname{sign}(z_3 - v_2).
\end{aligned}$$

of the form (5.11) are demonstrated in Fig. 5.2.

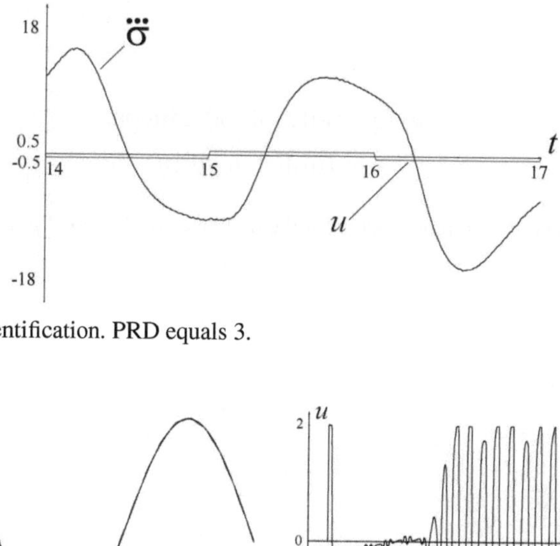

Fig. 5.2 PRD identification. PRD equals 3.

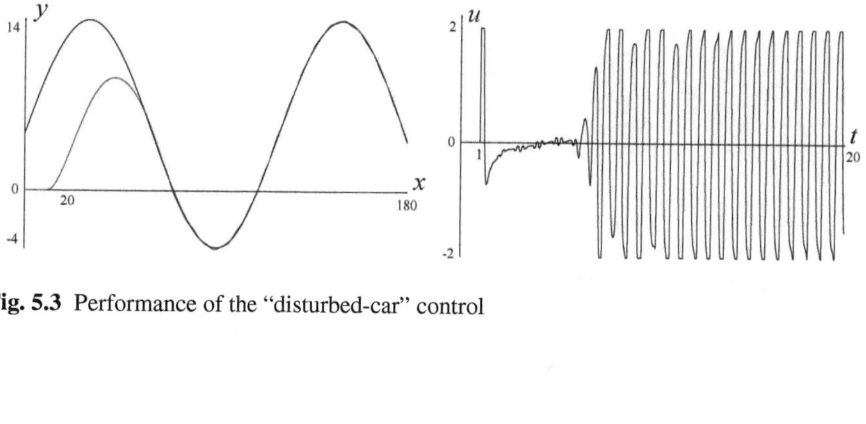

Fig. 5.3 Performance of the “disturbed-car” control

The applied control consists of the quasi-continuous 3-sliding controller

$$u = 0,\ \ 0 \le t < 1,$$
$$u = -2[s_2 + 2(|s_1| + |s_0|^{2/3})^{-1/2}(s_1 + |s_0|^{2/3}\,\mathrm{sign}\,s_0)]/[|s_2| + 2(|s_1| + |s_0|^{2/3})^{1/2}]$$
$$\text{with } t \ge 1;$$

and the second-order differentiator

$$\dot{s}_0 = v_1,\ v_1 = -9.28|s_0 - \tilde{\sigma}(t)|^{2/3}\,\mathrm{sign}(s_0 - \tilde{\sigma}(t)) + s_1,$$
$$\dot{s}_1 = v_1,\ v_{1,} = -15|s_1 - v_0|^{1/2}\,\mathrm{sign}(s_1 - v_0) + s_2,$$
$$\dot{s}_2 = -110\,\mathrm{sign}(s_2 - v_1).$$

The tracking results are shown in the coordinates x, y of the "car" in Fig. 5.3. The obtained accuracy is $|y - g(x)| \le 0.16m$.

Model	RD	No. States
Bergman	3	3
Candas-Radziuk	3	4
Cobelli	3	7
Hovorka	5	8
Dalla Man	5	8
Sorensen	5	18

- Output: blood glucose
- Input: insulin

Fig. 5.4 Summary of different models for the dynamics of the blood glucose concentration

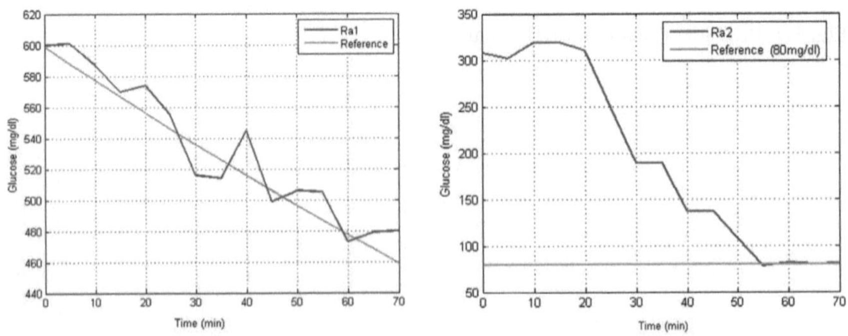

Fig. 5.5 Control of the glucose concentration in the blood of live rats. PRD equals 3.

5.5.2 Practical Control of Glucose Blood Concentration

Consider now the practical problem of controlling the glucose concentration in the human blood [[10, 17, 19, 35]]. The concentration is measured in real time once per minute, and a pump injects insulin, when it is needed. A number of models are available with the relative degrees changing from 3 to 5 and the number of variables changing from 3 to 18 (Fig. 5.4, courtesy to A.G. Gallardo-Hernandez, [[17]]).

Computer simulation shows that most models have the PRD 3 [[17]] (not all models were checked). A controller of the same form, as for the above "car" control, was applied. Negative values of the insulin injections were simply zeroed. The recent experimental results on **live rats** with the same control are shown in Fig. 5.5 (courtesy to A.G. Gallardo-Hernandez, [[17]]).

5.6 Conclusions

A new concept of practical relative degree is introduced, which generalizes the standard relative degree notion to systems of arbitrary nature, not necessarily described by ordinary differential equations. The results are new already for smooth dynamic systems nonlinear in control.

A new homogeneous tracking differentiator is proposed, featuring the same asymptotic accuracy as the standard homogeneous differentiator [[24]], but at the same time producing smooth estimations of the input derivatives with higher-order estimations being exact derivatives of the lower order ones. It has found natural application in identification of the practical relative degree.

Propositions 5.1-5.5 provide many examples of systems with practical relative degrees. One system can have a few practical relative degrees, which means that controllers developed for different relative degrees can prove to be efficient for the same system. The lowest practical relative degree is not always the best choice. A lot depends also on on the corresponding delay and approximation parameters. The new concept significantly generalizes the previous results showing that one can neglect fast stable actuators [[31]], fast stable sensors [[28]] and small perturbations changing the relative degree [[27]].

Thus, actually a new class of systems is singled out. The further theoretical research is to find various practical examples of systems with practical relative degrees and estimation of their delay and approximation parameters, which actually can be of no resemblance to real noises and delays. The notion can be probably extended to multi-input multi-output case.

Another natural application is in the artificial intelligence research. In such a case, one can successively try universal HOSM controllers corresponding to lower practical relative degrees 1, 2, 3, even without performing an attempt of the practical relative degree identification.

References

1. Bacciotti, A., Rosier, L.: Liapunov functions and stability in control theory. Springer, London (2005)
2. Bartolini, G., Ferrara, A., Usai, E.: Chattering avoidance by second-order sliding mode control. IEEE Trans. Automat. Control 43(2), 241–246 (1998)
3. Bartolini, G., Pisano, A., Punta, E., Usai, E.: A survey of applications of second-order sliding mode control to mechanical systems. International Journal of Control 76(9/10), 875–892 (2003)
4. Bartolini, G., Pisano, A., Usai, E.: First and second derivative estimation by sliding mode technique. Journal of Signal Processing 4(2), 167–176 (2000)
5. Bejarano, F.J., Fridman, L.: High order sliding mode observer for linear systems with unbounded unknown inputs. International Journal of Control 83(9), 1920–1929 (2010)
6. Bhat, S.P., Bernstein, D.S.: Finite time stability of continuous autonomous systems. SIAM J. Control Optim. 38(3), 751–766 (2000)
7. Boiko, I., Fridman, L.: Analysis of chattering in continuous sliding-mode controllers. IEEE Trans. Automatic Control 50(9), 1442–1446 (2005)
8. Dinuzzo, F., Ferrara, A.: Higher order sliding mode controllers with optimal reaching. IEEE Trans. Automatic Control 54(9), 2126–2136 (2009)
9. Defoort, M., Floquet, T., Kokosy, A., Perruquetti, W.: A novel higher order sliding mode control scheme. Systems & Control Letters 58, 102–108 (2009)
10. Dorel, L.: Glucose level regulation via integral high-order sliding modes. Mathematical Biosciences and Engineering 8(2), 549–560 (2011)
11. Evangelista, C., Puleston, P., Valenciaga, F.: Wind turbine efficiency optimization. Comparative study of controllers based on second order sliding modes. International Journal of Hydrogen Energy (2010), available online January 8, 2010
12. Edwards, C., Spurgeon, S.K.: Sliding mode control: theory and applications. Taylor & Francis (1998)
13. Filippov, A.F.: Differential equations with discontinuous right-hand side. Kluwer, Dordrecht (1988)
14. Floquet, T., Barbot, J.-P., Perruquetti, W.: Higher-order sliding mode stabilization for a class of nonholonomic perturbed systems. Automatica 39, 1077–1083 (2003)
15. Fridman, L.: Chattering analysis in sliding mode systems with inertial sensors. International Journal of Control 76(9/10), 906–912 (2003)
16. Furuta, K., Pan, Y.: Variable structure control with sliding sector. Automatica 36, 211–228 (2000)
17. Gallardo-Hernandez, A.G., Fridman, L., Levant, A., Shtessel, Y., Leder, R., Islas-Andrade, S., Revilla-Monsalve, C.: High-order sliding-mode control of blood glucose concentration via practical relative degree identification. In: Proc. IEEE CDC 2011, Orlando, FL, USA, December 12-15 (2011)
18. Isidori, A.: Nonlinear control systems, 2nd edn. Springer, New York (1989)
19. Kaveh, P., Shtessel, Y.B.: Blood glucose regulation using higher-order sliding mode control. International Journal of Robust and Nonlinear Control 18(4-5), 557–569 (2008)
20. Kokotovic, P.V., Khalil, H.K., O'Reilly, J.: Singular perturbation methods in control: analysis and design. SIAM (1999)
21. Kolmogoroff, A.N.: On inequalities between upper bounds of consecutive deriva-tives of an arbitrary function defined on an infinite interval. Amer. Math. Soc. Transl. 2, 233–242 (1962)
22. Levant, A., Levantovsky, L.V.: Sliding order and sliding accuracy in sliding mode control. International Journal of Control 58(6), 1247–1263 (1993)

23. Levant, A.: Robust exact differentiation via sliding mode technique. Automatica 34(3), 379–384 (1998)
24. Levant, A.: Higher-order sliding modes, differentiation and output-feedback control. International Journal of Control 76(9/10), 924–941 (2003)
25. Levant, A.: Homogeneity approach to high-order sliding mode design. Automatica 41(5), 823–830 (2005)
26. Levant, A.: Quasi-continuous high-order sliding-mode controllers. IEEE Trans. Automat. Control 50(11), 1812–1816 (2006)
27. Levant, A.: Robustness of homogeneous sliding modes to relative degree fluctuations. In: Proc. of 6th IFAC Symposium on Robust Control Design, Haifa, Israel, June 16-18 (2009)
28. Levant, A.: Chattering analysis. IEEE Transactions on Automatic Control 55(6), 1380–1389 (2010)
29. Levant, A.: Digital sliding-mode-based differentiation. In: Proc. 11th Scientific Workshop VSS 2012, Mumbay, India, January 12-14 (2012)
30. Levant, A., Alelishvili, L.: Integral high-order sliding modes. IEEE Trans. Automat. Control 52(7), 1278–1282 (2007)
31. Levant, A., Fridman, L.: Accuracy of homogeneous sliding modes in the presence of fast actuators. IEEE Transactions on Automatic Control 55(3), 810–814 (2010)
32. Levant, A., Michael, A.: Adjustment of high-order sliding-mode controllers. International Journal of Robust and Nonlinear Control 19(15), 1657–1672 (2009)
33. Levant, A., Pavlov, Y.: Generalized homogeneous quasi-continuous controllers. International Journal of Robust and Nonlinear Control 18(4-5), 385–398 (2008)
34. Levant, A., Pridor, A., Gitizadeh, R., Yaesh, I., Ben-Asher, J.Z.: Aircraft pitch control via second order sliding technique. J. of Guidance, Control and Dynamics 23(4), 586–594 (2000)
35. Kaveh, P., Shtessel, Y.B.: Blood glucose regulation using higher-order sliding mode control. International Journal of Robust and Nonlinear Control 18(4-5), 557–569 (2008)
36. Man, Z., Paplinski, A.P., Wu, H.R.: A robust MIMO terminal sliding mode control for rigid robotic manipulators. IEEE Trans. Automat. Control 39(12), 2464–2468 (1994)
37. Massey, T., Shtessel, Y.: Continuous traditional and high order sliding modes for satellite formation control. AIAA J. Guidance, Control, and Dynamics 28(4), 826–831 (2005)
38. Orlov, Y.: Finite time stability and robust control synthesis of uncertain switched systems. SIAM J. Cont. Optim. 43(4), 1253–1271 (2005)
39. Pisano, A., Davila, J., Fridman, L., Usai, E.: Cascade control of PM DC drives via second-order sliding-mode technique. IEEE Transactions on Industrial Electronics 55(11), 3846–3854 (2008)
40. Sira-Ramirez, H.: On the dynamical sliding mode control of nonlinear systems. International Journal of Control 57(5), 1039–1061 (1993)
41. Shtessel, Y.B., Shkolnikov, I.A.: Aeronautical and space vehicle control in dynamic sliding manifolds. International Journal of Control 76(9/10), 1000–1017 (2003)
42. Utkin, V.I.: Sliding modes in optimization and control problems. Springer, New York (1992)
43. Yu, X., Xu, J.X.: An adaptive signal derivative estimator. Electronic Letters 32(16) (1996)
44. Yu, S., Yu, X., Shirinzadeh, B., Man, Z.: Continuous finite-time control for robotic manipulators with terminal sliding mode. Automatica 41(11), 1957–1964 (2005)

Chapter 6
Higher Order Sliding Mode Based Accurate Tracking of Unmatched Perturbed Outputs

Leonid Fridman, Antonio Estrada, and Alejandra Ferreira de Loza

Abstract. Three approaches for higher-order sliding-mode based unmatched uncertainty compensation are summarized. Firstly, an algorithm is proposed based on the block control and quasi-continuous higher order sliding modes techniques. This method provides for the finite-time exact tracking of a smooth desired signal in spite of unmatched perturbations and allows the reduction of the controller gains in the case of partial knowledge of the system model. Thereafter, the combination of integral high-order sliding modes with the hierarchical quasi-continuous controller is proposed allowing finite-time convergence theoretically. Finally, high-order sliding mode observers are employed for exact state and uncertainties/perturbations reconstruction. A sliding mode control design is proposed which ensures theoretically exact compensation of the uncertainties/perturbations for the corresponding unmatched states based on the identified perturbation values. An inverted pendulum simulation example is considered for illustrating the feasibility of the proposed approach.

6.1 Introduction

It is a known issue that classical sliding mode (SM) control [37] is not able to compensate unmatched perturbations [11].

Leonid Fridman
Department of Control, Division of Electrical Engineering, Engineering Faculty, National Autonomous University of Mexico, UNAM, Ciudad Universitaria, 04510, Mexico, D.F.
e-mail: `lfridman@unam.mx`

Antonio Estrada
LUNAM Université, Ecole Centrale de Nantes, IRCCyN UMR CNRS 6597, Nantes, France
e-mail: `xheper@yahoo.com`

Alejandra Ferreira de Loza
University of Bordeaux, IMS-Lab, Automatic Control Group, 351 Cours de la Libération, 33405 Talence, France
e-mail: `da_ferreira@yahoo.com`

B. Bandyopadhyay et al. (Eds.): *Advances in Sliding Mode Control*, LNCIS 440, pp. 117–144.
DOI: 10.1007/978-3-642-36986-5_6

The combination of different robust techniques and SM has been applied to deal with systems with unmatched uncertainties [[9]]- [[5]]. In order to reduce the effects of the unmatched uncertainties [[6]] proposes a method that combines $\mathscr{H}_\infty$ and integral sliding mode control. The main idea is to choose such a projection matrix, ensuring that unmatched perturbations are not amplified and moreover minimized. For uncertain nonlinear systems in strict-feedback form [[23]], [[24]] developed the technique known as backstepping where a virtual control based on Lyapunov methods is constructed step by step. In a similar manner to backstepping, multiple surface sliding control is proposed in [[36]] to simplify the controller design of systems where model differentiation is difficult.

The combination of the backstepping design and sliding mode control is studied in [[3]] for systems in strict-feedback form with parameter uncertainties and it was extended to the multi-input case in [[16]]. The procedure proposed in [[3]], [[16]] reduces the computational load, as compared with the standard backstepping strategy, because only retains $(n-2)$ steps of the original backstepping technique, coupling them with an auxiliary second order subsystem to which a second order sliding mode control is applied. In [[34]] the combination of dynamical adaptive backstepping and first and second order sliding mode control is applied to both triangular and nontriangular uncertain observable minimum phase nonlinear systems.

Another approach to the problem of unmatched uncertainty compensation is based on the Nonlinear Block Controllable form (NBC-form) [[31]]. In [[31]] the conventional sliding mode technique is applied to compensate the matched perturbations. A high gain approach is used to achieve compensation of unmatched uncertainty and stabilization of the sliding mode dynamics. In [[20]] a sliding mode controller is designed using the combination of block control [[30]], a sigmoid approximation to the integral sliding mode control [[38]] and nested sliding mode control [[1]]. A coordinate transformation is applied to design a nonlinear sliding manifold. This transformation requires smoothness of each virtual control, that is why sigmoid, instead of signum functions are used. With the use of the high gain approach in [[31]] and sigmoid functions in [[20]], [[1]] they prove that asymptotic tracking is achieved.

In [[14]] a new design algorithm for systems in strict-feedback form, a special case of the BC-form, is proposed. This algorithm achieves finite-time **exact** tracking of the desired output in the presence of smooth unmatched perturbations. These features are accomplished via the use of quasi-continuous high-order sliding modes (HOSM) and a hierarchical design approach. In the first step the desired dynamic for the first state is defined by the desired tracking signal. After the first step, the desired dynamic for each state is defined by the previous one. Each virtual control is divided in two parts, the first one is intended to compensate the nominal nonlinear part of the system and the second one is aimed to achieve the desired dynamics in spite of perturbations.

Difficulties arise when initial conditions lead to big initial errors because then, the smoothness needed to achieve and keep the HOSM of each virtual control could be broken in some of them, leading to lost of control. One possible solution is to increase the gains of the HOSM term included in each virtual control, nevertheless it goes against a key motivation of the algorithm which is to reduce discontinuous control gain via the use of information on the known nominal part of the system. The solution proposed in the present paper is to apply the integral HOSM approach reported in [25] in which the desired reference is reached by means of a previously designed auxiliary smooth trajectory that depends on the initial conditions of the error and its derivatives up to the order of the HOSM control used. Thus each state starts in the proper auxiliary sliding motion and the whole internal dynamics remains controlled since the beginning.

On the other hand, the problem of unmatched uncertainties considering only output information has been tackled in [13], [8]. In [8] a linear matrix inequalities (LMI) based method for designing an output feedback variable structure control system is presented. The author proposes an LMI based sliding surface design considering H_2 performance. Another possible solution to overcome the full state requirement is to use an observer to estimate the state. In [13] an output robust stabilization problem for a class of systems with matched and mismatched uncertainties using sliding mode techniques is considered. The idea is to use an asymptotic nonlinear observer to estimate system states, then a variable structure controller is proposed to stabilize the system. Here, a HOSM observer is employed to reconstruct the state and identify the unknown inputs theoretically exact [4]. With these informations, a sliding manifold is designed such that the system's motion along the manifold meets the specified performance: the regulation of the non-actuated states and the theoretically exact unmatched uncertainties compensation. Thus, a discontinuous control law is designed such that the system's state is driven towards the manifold and stays there for all future time, regardless of disturbances and uncertainties.

The present chapter summarizes the results of papers [14], [15], [17]. In Section 6.2.1 the class of nonlinear systems to be treated and the problem formulation are described, the Quasi-continuous control is also introduced in this section. In 6.2.2 the hierarchical design algorithm is presented. The section 6.2.3 introduces the integral HOSM approach and the modification to the hierarchical design algorithm. At the final part of the both, section 6.2.2 and 6.2.3, the corresponding proposed algorithms are applied to an example and simulation results are presented. The HOSM based exact unmatched compensation control is introduced in Section 6.2.4. The state estimation as well as the perturbations identification methodologies based on HOSM techniques are explained in In 6.2.4.1. The output sliding mode controller rejecting the unmatched uncertainty is presented in 6.2.4.2. At the end of the section, a simulation example illustrates the performance of such controller. The note then concludes with a brief comment on the proposed algorithms.

6.2 HOSM Based Unmatched Uncertainties Compensation

It is a known issue that classical sliding mode (SM) control [37] is not able to compensate unmatched perturbations [11]. Nevertheless, controllers based on HOSM algorithms may be applied in order to reject the effect of unmatched perturbations. Next, some of these schemes are presented.

6.2.1 Black Box Control via HOSM

Consider a Single-Input-Single-Output system of the form

$$\begin{aligned} \dot{x}_1 &= f_1(x_1,t)+B_1(x_1,t)x_2+\omega_1(x_1,t) \\ \dot{x}_i &= f_i(\bar{x}_i,t)+B_i(\bar{x}_i,t)x_{i+1}+\omega_i(\bar{x}_i,t) \\ \dot{x}_n &= f_n(x,t)+B_n(x,t)u+\omega_n(x,t) \\ i &= 2,...,n-1 \\ \sigma &: (t,x) \mapsto \sigma(t,x) \in \mathrm{IR} \end{aligned} \tag{6.1}$$

where $x \in \mathrm{IR}^n$ is the state vector, $x_i \in \mathrm{IR}$, $\bar{x}_i = [x_1 \dots x_i]^T$, and $u \in \mathrm{IR}$ is the control. Moreover f_i and B_i are smooth scalar functions, ω_i is a bounded unknown perturbation term due to parameter variations and external disturbances with at least $n-i$ bounded derivatives w.r.t. system (6.1), $B_i \neq 0 \quad \forall x \in X \subset \mathrm{IR}^n, t \in [0,\infty)$ and σ is the measured output. The task is to achieve $\sigma \equiv 0$.

It is assumed that system (6.1) has a constant and known relative degree r. Then it follows that $\sigma^{(r)} = h(t,x)+g(t,x)u, \quad g(t,x) \neq 0$ holds, where $h(t,x) = \sigma^{(r)}|_{u=0}$, $g(t,x) = \frac{\partial}{\partial u}\sigma^{(r)}$ if the inequalities $0 < K_m \leq \frac{\partial}{\partial u}\sigma^{(r)} \leq K_M$, $|\sigma^{(r)}|_{u=0}| \leq C$ are fulfilled for some $K_m, K_M, C > 0$. The trajectories of (6.1) are assumed infinitely extendible in time for any Lebesgue-measurable bounded control $u(t,x)$. The next differential inclusion is implied

$$\sigma^{(r)} \in [-C,C]+[K_m,K_M]u \tag{6.2}$$

As it was described earlier, the above problem may be solved by the Quasi-Continuous Higuer-Order Sliding Mode (QC-HOSM) controllers [26], which is constructed according to (6.3), ensuring that $\sigma = \dot{\sigma} = \ldots = \sigma^{(r-1)} = 0$ is established in finite time.

$$\begin{aligned} \varphi_{0,r} &= \sigma,\ N_{0,r} = |\sigma|,\ \Psi_{0,r} = \varphi_{0,r}/N_{0,r} \\ \varphi_{i,r} &= \sigma^{(i)} + \beta_i N_{i-1,r}^{(r-i)/(r-i+1)} \Psi_{i-1,r} \\ N_{i,r} &= |\sigma^{(i)}| + \beta_i N_{i-1,r}^{(r-i)/(r-i+1)} \\ H_{i,r}(\cdot) &= \varphi_{i,r}/N_{i,r}; \qquad i = 0,\ldots,r-1. \end{aligned} \tag{6.3}$$

The above result is claimed in the next theorem [26]

Theorem 6.1. *[26] Provided that $\beta_1,\ldots,\beta_{r-1},\alpha > 0$ are chosen sufficiently large in the listed order, the above design result in the r-sliding homogeneous controller*

$$u = -\alpha H_{r-1,r}(\sigma,\dot{\sigma},\ldots,\sigma^{(r-1)}) \tag{6.4}$$

providing for the finite-time stability of (6.2),(6.4). The finite-time stable r-sliding mode $\sigma \equiv 0$ is established in system (6.1),(6.4).

In [32] compensation of unmatched perturbations is tackled using the block control approach combined with HOSM algorithms in order to consider unmodelled actuators in the controller design.

6.2.2 Model Based Application of HOSM

In ⟦14⟧, see also ⟦15⟧, a new design algorithm for systems in strict-feedback form, a special case of the BC-form, is proposed. This algorithm achieves finite-time **exact** tracking of the desired output in the presence of smooth unmatched perturbations. These features are accomplished via the use of QC-HOSM controllers and a hierarchical design approach. In the first step the desired dynamic for the first state is defined by the reference tracking signal. After the first step, the desired dynamic for each state is defined by the previous one. Each virtual control is divided into two parts, the first one is intended to compensate the nominal nonlinear part of the system and the second one is aimed at achieving the desired dynamics in spite of perturbations.

Consider the class of systems of equation (6.1), the control problem is to design a controller such that the output $y = x_1$ tracks a smooth desired reference y_d with bounded derivatives, in spite of the presence of unknown bounded perturbations. The whole state vector x is assumed to be known.

At each step i the constraint $\sigma_i = 0$ is established and kept by means of the virtual control $x_{i+1} = \phi_i$, which forms the constraint $\sigma_{i+1} = x_{i+1} - \phi_i$ for the next step.

Step 1. Defining $x_2 = \phi_1$, the next virtual controller is constructed

$$\begin{aligned} \phi_1(x_1,t,u_1) &= B_1(x_1,t)^{-1}\{-f_1(x_1,t)+u_1\} \\ u_1^{(n-1)} &= -\alpha_1 H_n(\sigma_1,\dot{\sigma}_1,\ldots,\sigma_1^{(n-1)}) \end{aligned} \tag{6.5}$$

where $\sigma_1 = x_1 - y_d$ and H_n is an n-th order sliding mode algorithm that is introduced in ϕ_1 through $n-1$ integrators. The derivatives $\sigma_1, \dot{\sigma}_1, \ldots, \sigma_1^{(n-1)}$ are calculated by means of robust differentiators with finite-time convergence ⟦27⟧.

Step i. The remaining virtual controls are constructed as follows.

$$\begin{aligned} \phi_i(\bar{x}_i,t,u_i) &= B_i(\bar{x}_i,t)^{-1}\{-f_i(\bar{x}_i,t)+u_i\} \\ u_i^{(n-i)} &= -\alpha_i H_{n-i+1}(\sigma_i,\dot{\sigma}_i,\ldots,\sigma_i^{(n-i)}) \\ \sigma_i &= x_i - \phi_{i-1}; \quad i = 2,\ldots,n. \end{aligned} \tag{6.6}$$

where H_{n-i+1} is an $n-i+1$-th order sliding algorithm. Notice that in *step n*, the real control is calculated i.e. $\phi_n = u$.

$$u = B_n(x,t)^{-1}\{-f_n(x,t)+u_n\} \tag{6.7}$$

$$u_n = -\alpha_n \operatorname{sign}(\sigma_n). \tag{6.8}$$

It is possible to smooth out the control signal by raising the order of the QC controller in each ϕ. If this is done, the super-twisting algorithm can also be used in u_n. The following theorem describes the result.

Theorem 6.2. *Provided that $\omega_i(\overline{x}_i,t)$ in system (6.1) and y_d are smooth functions with $n-i$ and n bounded derivatives respectively the above hierarchic design results in an ultimate controller u, providing for the finite time stability of $\sigma_1 = x_1 - y_d = \dot{\sigma}_1 = \ldots = \sigma_1^{(n-1)} = 0$ in system (6.1).*

Proof.
• Consider the state x_n

$$\begin{aligned} \dot{x}_n &= f_n(x,t) + B_n(x,t)u + \omega_n(x,t) \\ \text{with} \quad u &= B_n(x,t)^{-1}\{-f_n(x,t) - \alpha_n \text{sign}(\sigma_n)\}; \\ \sigma_n &= x_n - \phi_{n-1}; \quad \phi_{n-1} \text{ sufficiently smooth.} \end{aligned}$$

Thus $\dot{\sigma}_n = -\alpha_n \text{sign}(\sigma_n) + \omega_n(x,t) - \dot{\phi}_{n-1}$. Taking $\alpha_n \geq |\omega_n(x,t)| + |\dot{\phi}_{n-1}|$, provides for the appearance of a 1-sliding mode for the constraint σ_n.
• Now for the state x_{n-1}, we have

$$\begin{aligned} \dot{\sigma}_{n-1} &= \dot{x}_{n-1} - \dot{\phi}_{n-2} \\ &= f_{n-1}(\overline{x}_{n-1},t) + B_{n-1}(\overline{x}_{n-1},t)\phi_{n-1} \\ &\quad + \omega_{n-1}(\overline{x}_{n-1},t) - \dot{\phi}_{n-2} \\ &= u_{n-1} + \omega_{n-1}(\overline{x}_{n-1},t) - \dot{\phi}_{n-2} \\ \ddot{\sigma}_{n-1} &= \dot{u}_{n-1} + \dot{\omega}_{n-1}(\overline{x}_{n-1},t) - \ddot{\phi}_{n-2} \end{aligned} \tag{6.9}$$

and according to (6.6)

$$\dot{u}_{n-1} = -\alpha_{(n-1)} H_{1,2}(\sigma_{n-1}, \dot{\sigma}_{n-1}). \tag{6.10}$$

That is (6.9) takes the form:

$$\ddot{\sigma}_{n-1} = h_{n-1}(t,x) + g_{n-1}(t,x)\dot{u}_{n-1} \tag{6.11}$$

$$\begin{aligned} \text{with} \quad h_{n-1}(t,x) &= \ddot{\sigma}_{n-1}|_{\dot{u}_{n-1}=0} = \dot{\omega}_{n-1} - \ddot{\phi}_{n-2} \\ g_{n-1}(t,x) &= \partial \ddot{\sigma}_{n-1} / \partial \dot{u}_{n-1}. \end{aligned}$$

If for some $K_{m_{n-1}}, K_{M_{n-1}}, C_{n-1} > 0$ the inequalities $0 < K_{m_{n-1}} \leq g_{n-1} \leq K_{M_{n-1}}$, $|h_{n-1}| \leq C_{n-1}$ are fulfilled, then the next differential inclusion is implied

$$\ddot{\sigma}_{n-1} \in [-C_{n-1}, C_{n-1}] + [K_{m_{n-1}}, K_{M_{n-1}}]\dot{u}_{n-1} \tag{6.12}$$

and controller (6.10) provides for the finite time stability of ((6.10),6.12). The finite-time stable 2-sliding mode is established for the constraint σ_{n-1}.

• It is possible to obtain an analogous equation to (6.9) for each of the remaining states, thus for the state x_1

$$\sigma_1^{(n)} = h_1(t,x) + g_1(t,x)u_1^{(n-1)} \tag{6.13}$$

$$u_1^{(n-1)} = -\alpha_1 H_{n-1,n}(\sigma_1, \dot{\sigma}_1, \ldots, \sigma_1^{(n-1)}) \tag{6.14}$$

$$\sigma_1^{(n)} \in [-C_1, C_1] + [K_{m_1}, K_{M_1}]u_1^{(n-1)} \tag{6.15}$$

the differential inclusion (6.15) is implied for some constants K_{m_1}, K_{M_1} and C_1. The controller (6.14) provides for the finite time stability of (6.15). The finite time stable n-sliding mode is established for the constraint σ_1. □

Due to the dependence on states of functions in (6.13), the inclusion (6.15) may be ensured only locally. The same applies to the inclusion obtained for each virtual control.

6.2.2.1 Example

Consider the perturbed third order system

$$\begin{aligned} \dot{x}_1 &= 2\sin(x_1) + 1.5x_2 + \omega_1(x_1,t) \\ \dot{x}_2 &= 0.8x_1x_2 + x_3 + \omega_2(\bar{x}_2,t) \\ \dot{x}_3 &= -x_3^2 + 2u + \omega_3(x,t) \end{aligned} \tag{6.16}$$

functions ω_1, ω_2 are unmatched bounded perturbations and function ω_3 is a matched bounded perturbation; these functions were defined as follows

$$\begin{aligned} \omega_1(x_1,t) &= 0.2\sin(t) + 0.1x_1 + 0.12 \\ \omega_2(\bar{x}_2,t) &= 0.3\sin(2t) + 0.2x_1 + 0.2x_2 - 0.4 \\ \omega_3(x,t) &= 0.2\sin(2t) + 0.2x_1 + 0.3x_2 + 0.2x_3 + 0.3. \end{aligned}$$

Tracking of $y_d = 2\sin(0.15t) + 4\cos(0.1t) - 4$ by x_1 is desired.

• Step 1. According to the tracking objective, the first error signal is defined as $\sigma_1 = x_1 - y_d$ and the first virtual controller as follows

$$\begin{aligned} \phi_1 &= \frac{1}{1.5}\{-2\sin(x_1) + u_1\}; \\ \ddot{u}_1 &= -\alpha_1 H_{2,3}(\sigma_1, \dot{\sigma}_1, \ddot{\sigma}_1) \\ H_{2,3} &= \frac{\ddot{\sigma}_1 + 2(|\dot{\sigma}_1| + |\sigma_1|^{2/3})^{-1/2}(\dot{\sigma}_1 + |\sigma_1|^{2/3}\text{sign}(\sigma_1))}{|\ddot{\sigma}_1| + 2(|\dot{\sigma}_1| + |\sigma_1|^{2/3})^{1/2}} \end{aligned}$$

• Step 2. Defining $\sigma_2 = x_2 - \phi_1$, the next expression corresponds to ϕ_2

$$\begin{aligned} \phi_2 &= -0.8x_1x_2 + u_2 \\ \dot{u}_2 &= -\alpha_2 H_{1,2}(\sigma_2, \dot{\sigma}_2) \\ H_{1,2} &= \frac{\dot{\sigma}_2 + |\sigma_2|^{1/2}\text{sign}(\sigma_2)}{|\dot{\sigma}_2| + |\sigma_2|^{1/2}} \end{aligned}$$

• Step 3. With $\sigma_3 = x_3 - \phi_2$, the real control is

$$u = \frac{1}{2}\{x_3^2 + u_3\}$$
$$u_3 = -\alpha_3 \text{sign}(\sigma_3).$$

Results obtained in simulation are shown in figures (6.1)-(6.5), using $\alpha_1 = 6$, $\alpha_2 = 10, \alpha_3 = 16$ and the initial conditions $x_\circ = [0.1,\ 0,\ 0]^T$.

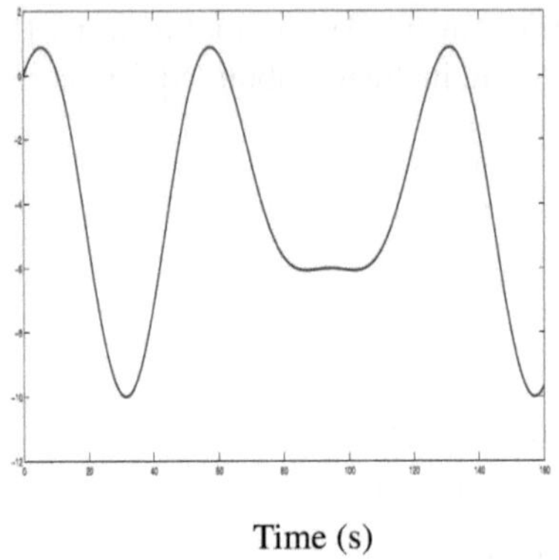

Fig. 6.1 Signals x_1 and y_d

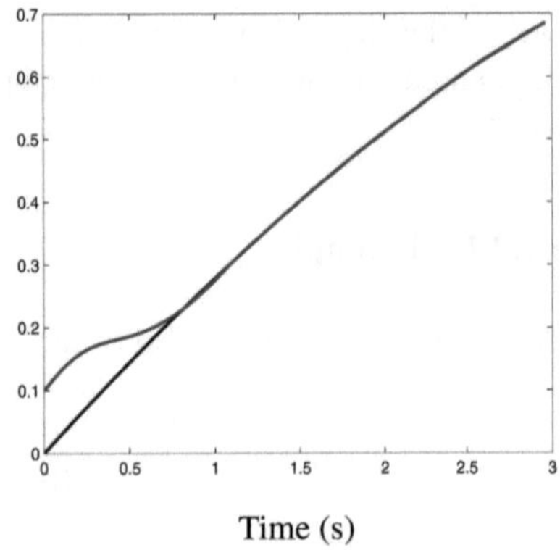

Fig. 6.2 Zoom in of the x_1 and y_d signals

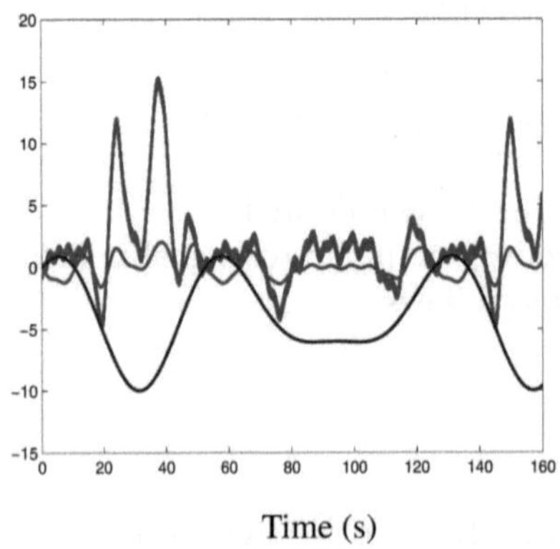

Fig. 6.3 States x_1,x_2 and x_3

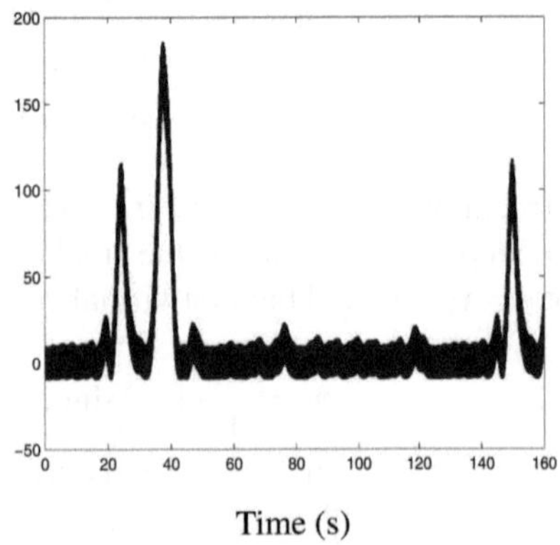

Fig. 6.4 Control signal

Since $\sigma_1 = x_1 - y_d$, straightforward algebra reveals that $u_1 = \dot{y}_d - \omega_1$ has to be accomplished in order to achieve that x_1 tracks y_d. Figure (6.5) depicts the fulfilling of the aforementioned equality.

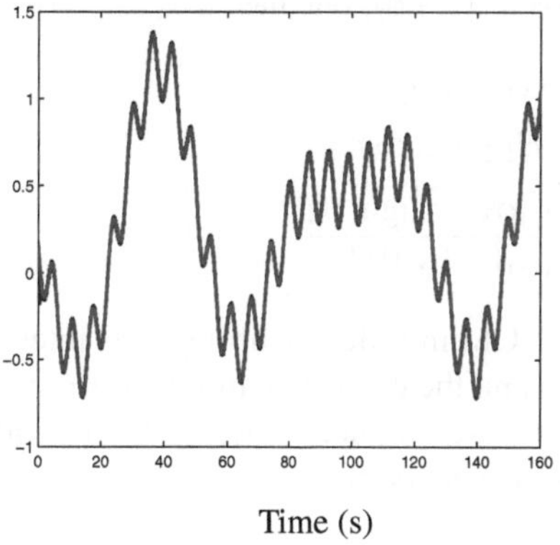

Fig. 6.5 Signals u_1 and $\dot{y}_d - \omega_1$

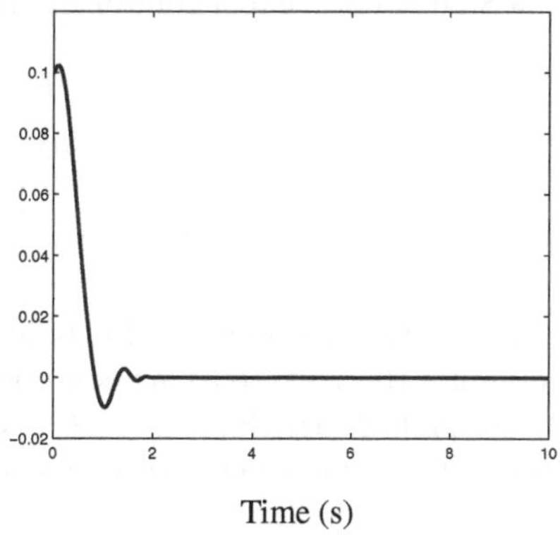

Fig. 6.6 Error signal σ_1

6.2.2.2 Example: Smoothness of the Control Signal

Consider the same nonlinear system (6.16) of the previous example. We apply the same control design procedure, but now, increasing by one the order of the QC-HOSM control present, on each virtual controller, with the aim of obtaining a smooth real control signal u.

• Step 2. The tracking error signal is $\sigma_1 = x_1 - y_d$ and next is the expression for the first virtual control

$$\phi_1(t, x_1, \sigma_1) = \frac{1}{1.5}\{-2\sin(x_1) + u_1\}$$
$$u_1^{(3)} = -\alpha_1 H_{3,4}(\sigma_1, \dot{\sigma}_1, \ddot{\sigma}_1, \dddot{\sigma}_1)$$

where

$$H_{3,4} = \Big\{\dddot{\sigma}_1 + 3[\ddot{\sigma}_1 + (|\dot{\sigma}_1| + 0.5|\sigma_1|^{3/4})^{-1/3}(\dot{\sigma}_1 + 0.5|\sigma|^{3/4}\mathrm{sign}(\sigma_1))] \times [|\ddot{\sigma}_1| + (|\dot{\sigma}_1| + 0.5|\sigma_1|^{3/4})^{2/3}]^{1/2}\Big\} \Big/ \Big\{|\dddot{\sigma}_1| + 3[|\ddot{\sigma}_1| + (|\dot{\sigma}_1| + 0.5|\sigma_1|^{3/4})^{2/3}]^{1/2}\Big\}$$

• Step 2. The new error signal $\sigma_2 = x_2 - \phi_1$, is used for the next virtual control

$$\phi_2 = -0.8x_1x_2 + u_2$$
$$\ddot{u}_2 = -\alpha_2 H_{2,3}(\sigma_2, \dot{\sigma}_2, \ddot{\sigma}_2)$$

where

$$H_{2,3} = \frac{\ddot{\sigma}_2 + 2(|\dot{\sigma}_2| + |\sigma_2|^{2/3})^{-1/2}(\dot{\sigma}_2 + |\sigma_1|^{2/3}\mathrm{sign}(\sigma_2))}{|\ddot{\sigma}_2| + 2(|\dot{\sigma}_2| + |\sigma_2|^{2/3})^{1/2}}$$

• Step 3. For the real control, defining $\sigma_3 = x_3 - \phi_2$, one has

$$u = -0.8x_1x_2 + u_3$$
$$\dot{u}_3 = -\alpha_3 H_{1,2}(\sigma_3, \dot{\sigma}_3)$$
$$H_{1,2}(\sigma_3, \dot{\sigma}_3) = \frac{\dot{\sigma}_3 + |\sigma_3|^{1/2}\text{sign}(\sigma_3)}{|\dot{\sigma}_3| + |\sigma_3|^{1/2}}$$

That is, the term u_3 in u, is a 2nd order QC-HOM introduced through one integrator. In Figures (6.7) and (6.8) the new control u and the corresponding new error signal σ_1 are plotted. The Figures (6.9) and (6.10) are zoomed views of the previously obtained non smooth control signal and the new smooth control.

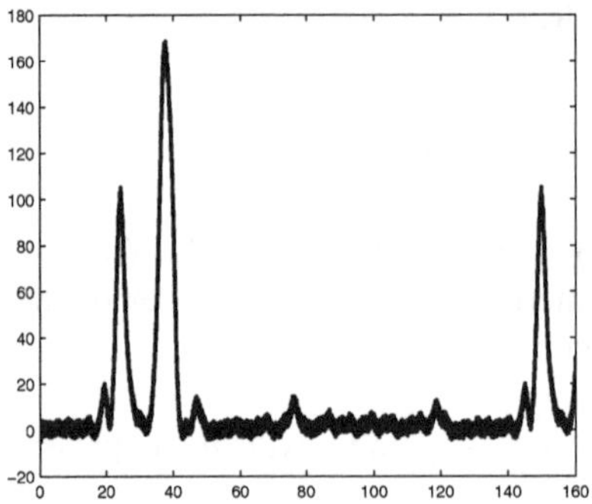

Fig. 6.7 Smooth control signal u

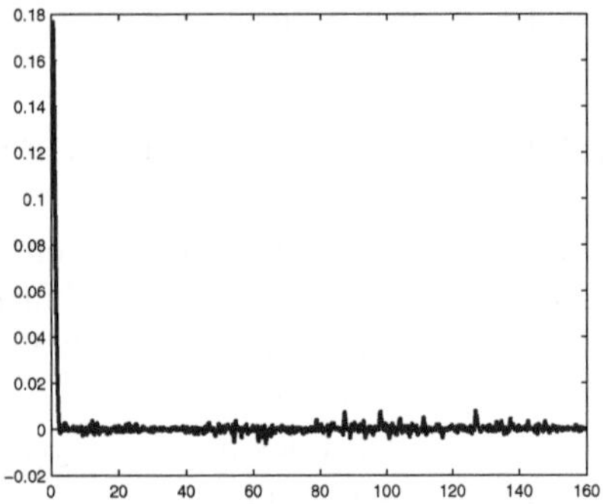

Fig. 6.8 Error signal σ_1

6.2.3 Hierachical Design Using Integral HOSM Approach

In order to illustrate the convenience of the integral HOSM approach in combination with the design algorithm presented in previous section, consider the state x_n of system (6.1) and recall the first step of the convergence proof for the control (6.7):

• For the state n

$$\dot{x}_n = f_n + B_n u + \omega_n(x,t)$$
$$\text{with} \quad u = B_n^{-1}\{-f_n + \alpha_n \text{sign}(\sigma_n)\}$$
$$\sigma_n = x_n - \phi_{n-1}; \quad \phi_{n-1} \text{ sufficiently smooth.}$$

Then $\dot{\sigma}_n = -\alpha_n \text{sign}(\sigma_n) + \omega_n(x,t) + \dot{\phi}_{n-1}$ and taking $\alpha_n \geq |\omega_n| + |\dot{\phi}_{n-1}|$ provides for the appearance of a 1-sliding mode for the constraint σ_n after a finite time T_n. Thus the subsystem

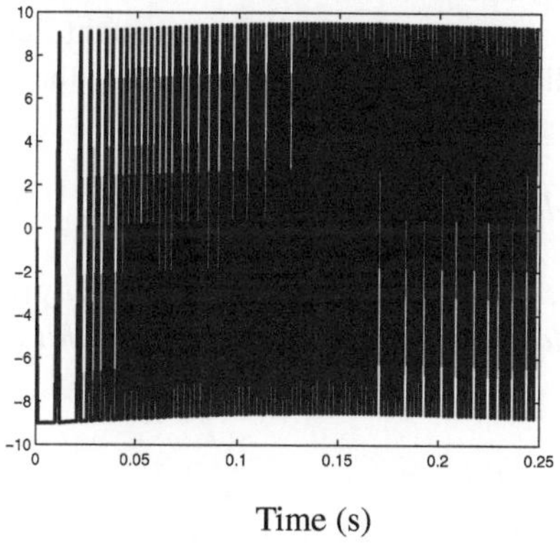

Fig. 6.9 Zoom in of non smooth control u

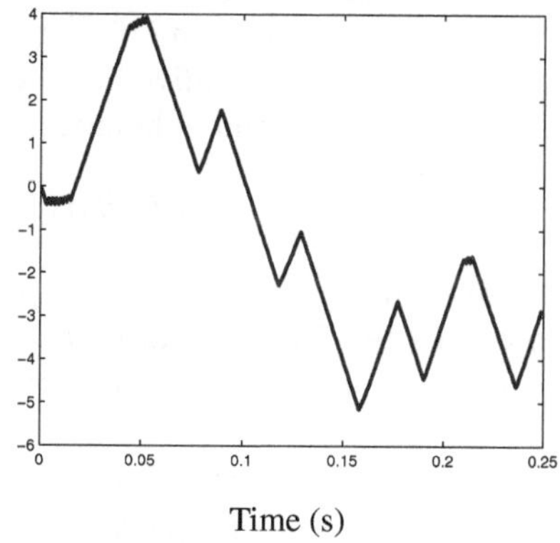

Fig. 6.10 Zoom in of smooth control u

$$\dot{x}_{n-1} = f_{n-1} + B_{n-1}x_{n-1} + \omega_{n-1}(\bar{x}_{n-1},t)$$

could be unstable in the transient, when $x_n \neq \phi_{n-1}$. The same can be said, unless the system is bounded-input-bounded-state (BIBS), for each of the remaining states used as virtual controls before they reach the desired dynamics.

In order to overcome the problem of transient dynamics, the application of the integral HOSM approach reported in [[25]] is proposed. The main idea is that each virtual control starts in the sliding mode of the proper order from the very beginning. The procedure is as follows:

Consider the r-sliding controller (6.3) and suppose a transient trajectory $\sigma(t,x(t)) = \rho(t)$, $t_0 \leq t \leq t_f$ such that:

$$\left.\begin{array}{c}\rho(t_0) = \sigma(t_0), \ldots, \rho^{(j)}(t_0) = \sigma^{(j)}(t_0) \\ j = 1, \ldots, r-1; \quad \rho(t) \equiv 0 \,\forall t \geq t_f\end{array}\right\} \tag{6.17}$$

Integral r-sliding mode. Let $\rho^{(r-1)}(t)$ be a Lipschitz function, then it has a globally bounded derivative $\rho^{(r)}(t)$ almost everywhere, and the new output $\Sigma(t,x) = \sigma(t,x) - \rho(t)$ satisfies

$$0 < K_m \leq \frac{\partial}{\partial u}\Sigma^{(r)} \leq K_M, |\Sigma^{(r)}|_{u=0}| \leq C$$

with some changed constants K_m, K_M, $C > 0$.
Let the $(r-1)-$smooth function $\rho(t)$ satisfying (6.17) have the form

$$\rho = (t - t_f)^r (c_0 + c_1(t - t_0) + \ldots + c_{r-1}(t - t_0)^{r-1}). \tag{6.18}$$

Parameters c_i are now to be found from conditions (6.17) after t_f is assigned. In order to avoid the necessity of very large control values to reach the $r-$sliding mode

$\vec{\sigma}(t_0) = 0$ (i.e., $\sigma = \dot{\sigma} = \ldots = \sigma^{(r-1)} = 0$) due to far distanced initial values, or a very low convergence rate if $\vec{\sigma}(t_0)$ is close to zero instead of a constant, let $t_f - t_0$ be a continuous positive-definite r-sliding homogeneous function of the initial conditions, $\vec{\sigma}(t_0)$, and of homogeneity degree 1, i.e.,

$$t_f - t_0 = T(\vec{\sigma}(t_0)), \quad T(d_\kappa \vec{\sigma}) = \kappa T(\vec{\sigma}) \, \forall \kappa > 0. \tag{6.19}$$

Theorem 6.3. *[[25]] The function $\rho(t-t_0, \vec{\sigma}(t_0))$ is uniquely determined by (6.17), (6.18), (6.19). Then with any sufficiently large α, independent of the initial conditions $\vec{\sigma}(t_0)$, the controller (6.20):*

$$u = \alpha H_{r-1,r}(\Sigma, \dot{\Sigma}, \ldots, \Sigma^{(r-1)}) \tag{6.20}$$

$$\Sigma(t,x) = \begin{cases} \sigma(t,x) - \rho(t-t_0, \vec{\sigma}(t_0)), & t_0 \le t \le t_0 + T(\vec{\sigma}(t_0)) \\ \sigma(t,x), & t \ge t_0 + T(\vec{\sigma}(t_0)) \end{cases}$$

establishes the finite-time-stable r-sliding mode $\sigma \equiv 0$ with the transient time (6.19). The equality $\sigma(t,x(t)) = \rho(t-t_0, \vec{\sigma}(t_0)))$ is kept during the transient process.

The function used in this paper for $T(\vec{\sigma}(t_0))$ is the one reported in [[25]], whose expression is

$$T(\vec{\sigma}(t_0)) = \lambda(|\sigma(t_0)|^{p/r} + |\dot{\sigma}(t_0)|^{p/(r-1)} + |\sigma^{(r-1)}(t_0)|^p)^{1/p}; \quad p, \lambda > 0. \tag{6.21}$$

As previously mentioned the stability advantages of integral HOSM, obtained through the use of the knowledge of the initial conditions of the system will be used in the hierarchical design; the details are explained next.

6.2.3.1 Modification of the Hierarchical QC-HOSM Controller

The modification consists in the substitution of each restriction σ_i for a new one, $\Sigma_i = \sigma_i - \rho_i$, as follows.

Step i. The i-th sliding surface is chosen as $\Sigma_i = \sigma_i - \rho_i$ where $\sigma_i = x_i - \phi_{i-1}$ (with the exception $\sigma_1 = x_1 - y_d$).

$$\begin{aligned} \phi_i(\bar{x}_i, t, u_i) &= B_i(\bar{x}_i, t)^{-1}\{-f_i(\bar{x}_i, t) + u_i\} \\ u_i^{(n-i)} &= -\alpha_i H_{n-i,n-i+1}(\Sigma_i, \dot{\Sigma}_i, \ldots, \Sigma_i^{(n-i)}). \end{aligned} \tag{6.22}$$

$H_{n-i,n-i+1}$ is defined as in (6.3), obviously using $\sigma = \Sigma_i$ in those equations and where ρ_i fulfills condition (6.17):

$$\left. \begin{array}{c} \rho_i(t_0) = \sigma(t_0), \ldots, \rho_i^{(n-i)}(t_0) = \sigma_i^{(n-i)}(t_0) \\ \rho_i(t) \equiv 0 \, \forall t \ge t_{fi} \end{array} \right\} \tag{6.23}$$

and is constructed according to (6.18) setting $r = n - i + 1$. The equation for (6.19), depending on the index i, is

$$\begin{aligned} T(\vec{\sigma_i}(t_0)) &= \lambda_i(|\sigma_i(t_0)|^{p/n-i+1} + |\dot{\sigma}_i(t_0)|^{p/(n-i)} + \\ &\dots + |\sigma_i^{(n-i)}(t_0)|^p)^{1/p}. \end{aligned} \tag{6.24}$$

Step n. $\Sigma_n = \sigma_n - \rho_n$ where $\sigma_n = x_n - \phi_{n-1}$.

$$\begin{aligned} u &= B_n(x,t)^{-1}\{-f_n(x,t) + u_n\} \\ \text{where } u_n &= -\alpha_n sign(\Sigma_n). \end{aligned} \tag{6.25}$$

Theorem 6.4. *If system (6.1) is BIBS then provided that ω_i and y_d are smooth functions with $n-i$ and n bounded derivatives respectively the above hierarchic design results in the controller (6.25) that assures the finite time stability of $\sigma_1 = x_1 - y_d = \dot{\sigma}_1 = \ldots = \sigma_1^{(n-1)} = 0$ in system (6.1) independently of their initial conditions $x(t_0)$.*

Remark 6.1. Observe that the BIBS condition is only a sufficient but not necessary condition as it can be seen in the convergence proof.

Proof.
• Consider the state x_n

$$\begin{aligned} \dot{x}_n &= f_n + B_n u + \omega_n(x,t) \\ \text{with} \quad u &= B_n^{-1}\{-f_n + \alpha_n sign(\Sigma_n)\} \\ \Sigma_n = \sigma_n - \rho_n,\ \sigma_n &= x_n - \phi_{n-1},\ \phi_{n-1} \text{ sufficiently smooth.} \end{aligned}$$

Then $\dot{\Sigma}_n = -\alpha_n sign(\sigma_n) + \omega_n(x,t) + \dot{\phi}_{n-1} - \dot{\rho}_n$ and taking $\alpha_n \geq |\omega_n| + |\dot{\phi}_{n-1}| + |\dot{\rho}_n|$ provides for the appearance of a 1-sliding mode for the constraint Σ_n after $t = t_0$, i.e. since the beginning, and for σ_n after $T_n = t_{fn} - t_0 = \lambda_n(|\sigma_n(t_0)|)$ choosing $p = 1$ for equation (6.24).

• Now for the state x_{n-1}, with ϕ_{n-1} defined according to (6.22) and $\Sigma_{n-1} = \sigma_{n-1} - \rho_{n-1}$, $\sigma_{n-1} = x_{n-1} - \phi_{n-2}$ then:

$$\begin{aligned} \dot{\Sigma}_{n-1} &= \dot{x}_{n-1} - \dot{\phi}_{n-2} - \dot{\rho}_{n-1} \\ &= f_{n-1} + B_{n-1}x_n + \omega_{n-1} - \dot{\rho}_{n-1} - \dot{\phi}_{n-2} \\ &= f_{n-1} + B_{n-1}(\phi_{n-1} + \rho_{n-1}) + \omega_{n-1} - \dot{\rho}_{n-1} - \dot{\phi}_{n-2}. \end{aligned}$$

The function f_{n-1}, may not be compensated right from $t = t_0$ because of the arbitrary initial condition of x_n. However due to the BIBS condition x_{n-1} remains bounded and after $t = t_0 + T_n$, when $x_n = \phi_{n-1}$:

$$\begin{aligned} \dot{\Sigma}_{n-1} &= u_{n-1} + \omega_{n-1}(\bar{x}_{n-1},t) - \dot{\rho}_n - \dot{\phi}_{n-2} \\ \ddot{\Sigma}_{n-1} &= \dot{u}_{n-1} + \dot{\omega}_{n-1}(\bar{x}_{n-1},t) - \ddot{\rho}_n - \ddot{\phi}_{n-2} \end{aligned} \tag{6.26}$$

that is (6.26) takes the form

$$\ddot{\Sigma}_{n-1} = h_{n-1}(t,x) + g_{n-1}(t,x)\dot{u}_{n-1} \tag{6.27}$$

$$\text{with} \quad h_{n-1}(t,x) = \ddot{\Sigma}_{n-1}|_{u_{n-1}=0};\ g_{n-1}(t,x) = \frac{\partial}{\partial \dot{u}_{n-1}}\ddot{\Sigma}_{n-1}$$

$$\dot{u}_{n-1} = -\alpha_{n-1}H_{1,2}(\Sigma_{n-1},\dot{\Sigma}_{n-1}). \tag{6.28}$$

If for some $K_{m_{n-1}}, K_{M_{n-1}}, C_{n-1} > 0$ the inequalities $0 < K_{m_{n-1}} \leq \frac{\partial}{\partial \dot{u}_{n-1}}\ddot{\Sigma}_{n-1} \leq K_{M_{n-1}}$ and $|\ddot{\Sigma}_{n-1}|_{\dot{u}_{n-1}=0}| \leq C_{n-1}$ holds, the next differential inclusion is implied

$$\ddot{\Sigma}_{n-1} \in [-C_{n-1}, C_{n-1}] + [K_{m_{n-1}}, K_{M_{n-1}}]\dot{u}_{n-1} \tag{6.29}$$

and controller (6.28) keeps (since it was established from t_0) stability of (6.29), (6.28). The finite-time stable 2-sliding mode is maintained for the constraint Σ_{n-1} from t_0 and for σ_{n-1} after $t_{fn-1} = t_0 + T_n$.

The same procedure can be applied to each one of the states of (6.1). □

Remark 6.2. As it was previously mentioned it becomes clear that the BIBS condition is not a necessary one, it will suffice that in each subsystem of (6.1)

$$\dot{x}_i = f_i(\bar{x}_i,t) + B_i(\bar{x}_i,t)x_{i+1} + \omega_i(\bar{x}_i,t)$$

x_i remains bounded with the input x_{i+1} bounded, at least during the time interval $t < t_{fi}$; because after that time f_i is compensated.

Notice that, with the use of integral HOSM in each virtual control, it is possible to introduce suitable dynamics on each of them. If direct application is used, in which only the input and the output is considered, this is not possible.

6.2.3.2 Example

Consider the perturbed nonlinear system (6.16), rewritten here along with the control problem statement for the reader convenience

$$\begin{aligned} \dot{x}_1 &= 2\sin(x_1) + 1.5x_2 + \omega_1(x_1,t) \\ \dot{x}_2 &= 0.8x_1x_2 + x_3 + \omega_2(\bar{x}_2,t) \\ \dot{x}_3 &= -1.5x_3^2 + 2u + \omega_3(x,t) \end{aligned} \tag{6.30}$$

where functions ω_1, ω_2 are the unmatched bounded perturbations and function ω_3 is the matched perturbation. These functions were defined as follows

$$\begin{aligned}\omega_1(x_1,t) &= 0.2\sin(t)+0.1x_1+0.12\\ \omega_2(\bar{x}_2,t) &= 0.3\sin(2t)+0.2x_1+0.2x_2-0.4\\ \omega_3(x,t) &= 0.2\sin(2t)+0.2x_1+0.3x_2+0.2x_3+0.3\end{aligned}$$

a controller that achieves tracking of $y_d = 2\sin(0.15t)+4\cos(0.1t)-4$ by x_1 is desired. In addition to the previous perturbations the nominal compensation term of the first two virtual controls is not exact.

The first sliding surface is $\Sigma_1 = \sigma_1 - \rho_1$, $\sigma_1 = x_1 - y_d$, and the virtual control for x_1:

$$\begin{aligned}\phi_1(x_1,t,u_1) &= \frac{1}{1.5}\{-1.8\sin(x_1)+u_1\}\\ \dot{u}_1^{(2)} &= -\alpha_1 H_{2,3}(\Sigma_1,\dot{\Sigma}_1,\ddot{\Sigma}_1);\end{aligned}$$

$$H_{2,3}(\Sigma_1,\dot{\Sigma}_1,\ddot{\Sigma}_1) = \frac{\ddot{\Sigma}_1+2(|\dot{\Sigma}_1|+|\Sigma_1|^{2/3})^{-1/2}(\dot{\Sigma}_1+|\Sigma_1|^{2/3}\mathrm{sign}(\Sigma_1))}{|\ddot{\Sigma}_1|+2(|\dot{\Sigma}_1|+|\Sigma_1|^{2/3})^{1/2}}$$

$$\begin{aligned}\rho_1 &= (t-t_{f1})^3(c_{10}+c_{11}(t-t_0)+c_{12}(t-t_0)^2)\\ T_1 &= \lambda_1(|\sigma_1(t_0)|^2+|\dot{\sigma}_1(t_0)|^3+|\ddot{\sigma}_1(t_0)|^6)^{1/6}.\end{aligned}$$

For the next state $\Sigma_2 = \sigma_2 - \rho_2$, $\sigma_2 = x_2 - \phi_1$ then

$$\begin{aligned}\phi_2(x_2,t,u_2) &= -0.7x_1x_2+u_2\\ \dot{u}_2 &= -\alpha_2 H_{1,2}(\Sigma_2,\dot{\Sigma}_2)\\ H_{1,2}(\Sigma_2,\dot{\Sigma}_2) &= \frac{\dot{\Sigma}_2+|\Sigma_2|^{1/2}\mathrm{sign}(\Sigma_2)}{|\dot{\Sigma}_2|+|\Sigma_2|^{1/2}}\\ \rho_2 &= (t-t_{f_2})^2(c_{20}+c_{21}(t-t_0))\\ T_2 &= \lambda_2(|\sigma_2(t_0)|+|\dot{\sigma}_2(t_0)|^2)^{1/2}.\end{aligned}$$

Finally for state x_3, $\Sigma_3 = \sigma_3 - \rho_3$, $\sigma_3 = x_3 - \phi_2$

$$\begin{aligned}u &= \frac{1}{2}\{1.5x_3^2+u_3\}\\ u_3 &= -\alpha_3\mathrm{sign}(\Sigma_3)\\ \rho_3 = (t-t_{f_3})(c_{30});\ & T_3 = \lambda_3(|\sigma_3(t_0)|).\end{aligned}$$

Results obtained in simulations are shown in figures (6.11)-(6.14), where the next parameters $\alpha_1 = 4$, $\alpha_2 = 10$, $\alpha_3 = 8$, $\lambda_1 = 6$, $\lambda_2 = 0.5$, $\lambda_3 = 1$ were used. In figures (6.11) and (6.12), the vector of initial conditions $x_\circ = [3,\ -2,\ 4]^T$ is used, whereas in figures (6.13) and (6.14), $x_\circ = [-3,\ 1.5,\ 2]^T$ was used and a phase lead of 30 seconds is introduced in y_d.

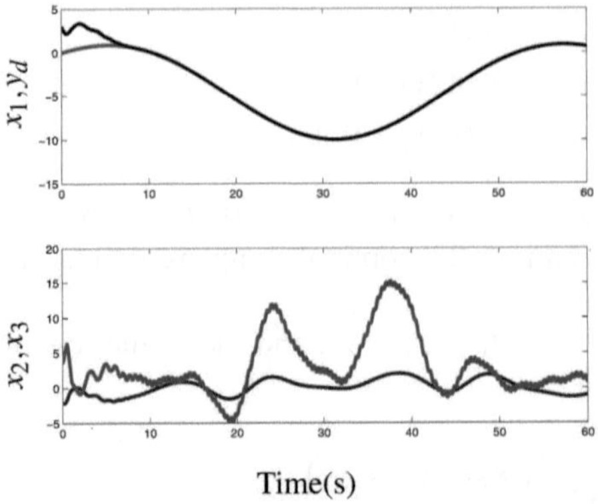

Fig. 6.11 States and reference evolution ($x(0) = [3,\ -2,\ 4]^T$)

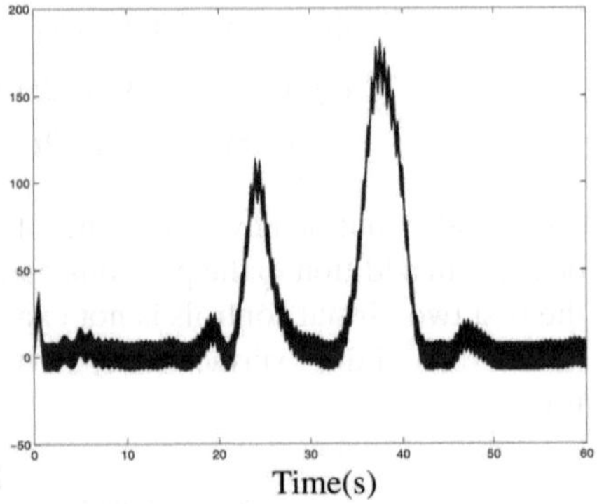

Fig. 6.12 Control signal u

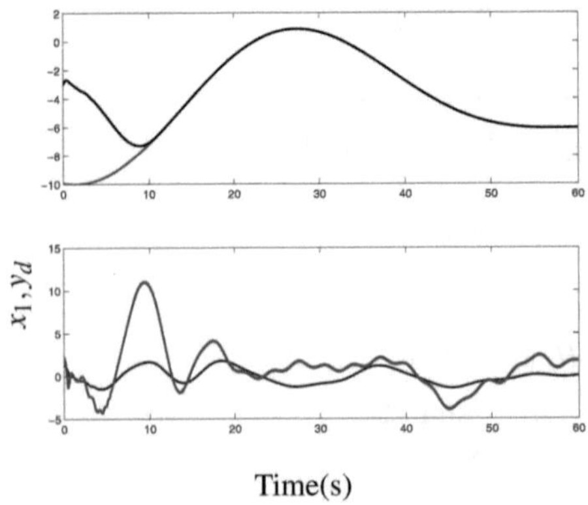

Fig. 6.13 States and reference evolution ($x(0) = [-3,\ 1.5,\ 2]^T$)

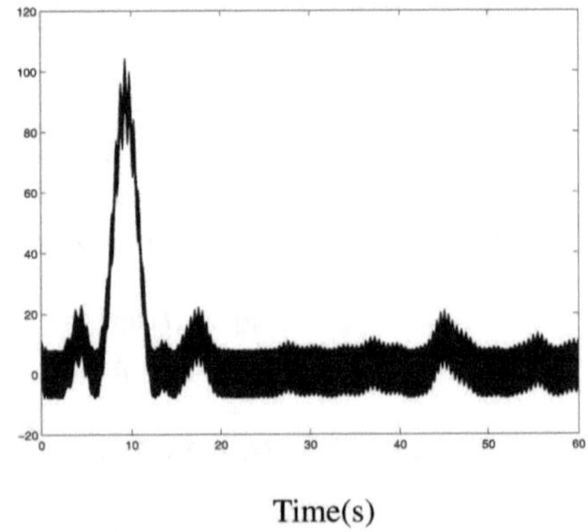

Fig. 6.14 Control signal u

6.2.4 *Exact Unmatched Uncertainties Compensation Based on HOSM Observation*

Let us consider a linear time invariant system with unknown inputs

$$\begin{aligned} \dot{x}(t) &= Ax(t) + Bu(t) + D\omega(t), \\ y(t) &= Cx(t), \end{aligned}$$

where $x \in \mathbb{R}^n$, $u \in \mathbb{R}^m$, $y \in \mathbb{R}^p$ and $\omega \in \mathbb{R}^q$ are the state vector, the control, the measured output of the system and the unknown input (or disturbance), respectively. In addition, and without loss of generality, let us assume that $\operatorname{rank} C = p$, $\operatorname{rank} B = m$, $\operatorname{rank} D = q$ and that the triplet (A, D, C) is strongly observable, such that the state $x(t)$ may be recovered in finite-time using only the output and its derivatives (through the use of the HOSM differentiator). Under the additional assumption on the smoothness of the unknown input w, i.e. $\|\dot{\omega}(t)\| \leq \omega^{+}$, an extra derivative of the estimated state can be computed, thus obtaining an estimate for $\dot{x}$. Under these considerations, an estimate for the unknown input may be obtained as

$$\hat{\omega} = D^{+}\left[\dot{x}(t) - Ax(t) - Bu(t)\right],$$

With this estimate of the unknown input, it is natural to try to compensate the effect of the unknown input in the system as much as possible. Direct compensation of the part of $\omega(t)$ that is matched to u is possible. To see this apply the state transformation $\mathcal{T} = \begin{bmatrix} B^{\perp} \\ B^{+} \end{bmatrix}$ which allows to rewrite the system as

$$\begin{aligned} \dot{x}_1(t) &= A_{11}x_1(t) + A_{12}x_2(t) + D_1\omega(t), \\ \dot{x}_2(t) &= A_{21}x_1(t) + A_{22}x_2(t) + D_2\omega(t) + u(t) \end{aligned} \tag{6.31}$$

with $x_1(t) \in \mathbb{R}^{n-m}$, $x_2(t) \in \mathbb{R}^{m}$. Then taking

$$u(t) = -D_2\hat{\omega}(t) + v(t),$$

the effect of matched disturbances can be reduced and, if all the derivatives are exact, completely removed without the direct application of a discontinuous control signal. Hence, $v(t) \in \mathbb{R}^{m}$, which may be designed following any control strategy, is a nominal control.

Another option is to consider the estimate of the disturbance into the sliding surface design in order to compensate the effects of the unmatched inputs. Thus, if

$$\operatorname{span} D_1 \in \operatorname{span} A_{12}, \tag{6.32}$$

the sliding surface is designed as $\sigma = x_2 + Kx_1 + G\hat{\omega}$, where matrices K and G are to be designed to provide for both stability and performance and the control is constructed as an unitary control

$$u(t) = -\varphi(x)\frac{\sigma}{\|\sigma\|}.$$

Now, to realize the reconstruction of the state, let us introduce a HOSM observer. The HOSM observer provides the theoretically exact value of the state vector and the unknown inputs identification in a finite time.

6.2.4.1 HOSM Observation and Identification Process

Basically, the HOSM observer design consists of two stages: firstly a Luenberger observer is used to maintain the norm of the estimation error bounded; then, by means of a differentiation scheme, the state vector is reconstructed. For further details see [4].

Before introducing the observer, let us define the following notation: let $f(t)$ be a vector function, $f^{[i]}(t)$ represents the i-th anti-differentiator of $f(t)$, i.e. $f^{[i]}(t) = \int_0^t \int_0^{\tau_1} \cdots \int_0^{\tau_{i-1}} f(\tau_i)\, d\tau_i \cdots d\tau_2 d\tau_1$, $f^{[0]}(t) = f(t)$.

Stage 1. In order to realize the differentiation process we need to be sure that the observation error will be bounded. Firstly, design an auxiliary dynamic system

$$\dot{\tilde{x}}(t) = A\tilde{x}(t) + Bu(t) + L(y(t) - \tilde{y},(t)),$$

where $\tilde{x} \in \mathbb{R}^n$ is an auxiliary state vector and $\tilde{y}(t) = C\tilde{x}(t)$ and the gain L must be designed such that the matrix $\tilde{A} := (A - LC)$ is Hurwitz (notice that strongly observable assumption implies that (A,C) pair is observable). Let $e(t) := x(t) - \tilde{x}(t)$, whose dynamic equations are

$$\dot{e}(t) = \tilde{A}e(t) + Dw(t). \tag{6.33}$$

Thus, in view of the boundness of the unknown input $\omega(t)$, $e(t)$ has a bounded norm, i.e., there exists a known constant e^+ and a finite time t_e, such that

$$\|e(t)\| \leq e^+, \text{ for all } t > t_e. \tag{6.34}$$

Stage 2. This part of the state reconstruction is based on an algorithm that allows decoupling the unknown inputs from the successive derivatives of the output of the linear estimation error system $y_{\mathrm{e}}(t) := y(t) - C\tilde{x}(t)$.

0. Define $M_1 := C$.
1. Derive a linear combination of the output $y_{\mathrm{e}}(t)$, ensuring that the derivative of this combination is unaffected by the uncertainties, i.e., $\frac{d}{dt}(M_1D)^{\perp} y_{\mathrm{e}}(t) = (M_1D)^{\perp} C\tilde{A}e(t)$. Thus, form the extended vector

$$\begin{bmatrix} \frac{d}{dt}(M_1D)^{\perp} y_{\mathrm{e}}(t) \\ y_{\mathrm{e}}(t) \end{bmatrix} = \underbrace{\begin{bmatrix} (M_1D)^{\perp} C\tilde{A} \\ C \end{bmatrix}}_{M_2} e(t). \tag{6.35}$$

   Then, moving the differentiation operator outside the parenthesis and defining $J_1 = (M_1D)^{\perp}$, the following equation is obtained

$$\frac{d}{dt}\begin{bmatrix} J_1 & 0 \\ 0 & I_p \end{bmatrix}\begin{bmatrix} y_{\mathrm{e}}(t) \\ y_{\mathrm{e}}^{[1]}(t) \end{bmatrix} = M_2 e(t), \tag{6.36}$$

   where $I_p \in \mathbb{R}^{p \times p}$ is an identity matrix.
2. Derive a linear combination of $M_2 e(t)$, ensuring that the derivative of this combination is unaffected by uncertainties, i.e. $\frac{d}{dt}(M_2D)^{\perp} M_2 e(t)$. Then form the extended vector

$$\begin{bmatrix} \frac{d}{dt}(M_2D)^{\perp}M_2e(t) \\ y_e(t) \end{bmatrix} = \underbrace{\begin{bmatrix} (M_2D)^{\perp}C\tilde{A} \\ C \end{bmatrix}}_{M_3} e(t). \tag{6.37}$$

Moving the differentiation operator outside the parenthesis from (6.37) we have that

$$\frac{d}{dt}\begin{bmatrix} (M_2D)^{\perp}M_2e(t) \\ y_e^{[1]}(t) \end{bmatrix} = M_3e(t).$$

From the above expression and from (6.36), and by moving the differentiation operator outside the parenthesis, it yields to

$$\frac{d^2}{dt^2}\begin{bmatrix} J_2 & 0 \\ 0 & I_p \end{bmatrix}\begin{bmatrix} y_e(t) \\ y_e^{[1]}(t) \\ y_e^{[2]}(t) \end{bmatrix} = M_3e(t), \tag{6.38}$$

where $J_2 = (M_2D)^{\perp}\begin{bmatrix} J_1 & 0 \\ 0 & I_p \end{bmatrix}$.

j. A general step j $(j \geq 1)$ can be summarized as follows. Derive $\left(M_{j-1}D\right)^{\perp}M_{j-1}e(t)$. Then, from the identity

$$\begin{bmatrix} \frac{d}{dt}\left(M_{j-1}D\right)^{\perp}M_{j-1}e(t) \\ y_e(t) \end{bmatrix} = \underbrace{\begin{bmatrix} \left(M_{j-1}D\right)^{\perp}M_{j-1}\tilde{A} \\ C \end{bmatrix}}_{M_j} e(t), \tag{6.39}$$

the next expression is obtained

$$\frac{d^{j-1}}{dt^{j-1}}\begin{bmatrix} J_{j-1} & 0 \\ 0 & I_p \end{bmatrix}\begin{bmatrix} y_e(t) \\ \vdots \\ y_e^{[j-1]}(t) \end{bmatrix} = M_je(t), \tag{6.40}$$

where $J_{j-1} = \left(M_{j-1}D\right)^{\perp}\begin{bmatrix} J_{j-2} & 0 \\ 0 & I_p \end{bmatrix}$.

Due to the strong observability assumption, there exists a unique positive integer k such that after k steps of the algorithm $(0 \leq k \leq n)$, the matrix M_k generated recursively by (6.40), satisfies the conditions $\texttt{rank}M_i < n$ for all $i < k$ and $\texttt{rank}M_i = n$ for all $i \geq k$ (see, e.g., [33]). This means that the algebraic equation

$$M_ke(t) = \frac{d^{k-1}}{dt^{k-1}}\begin{bmatrix} J_{k-1} & 0 \\ 0 & I_p \end{bmatrix}\begin{bmatrix} y_e(t) \\ \vdots \\ y_e^{[k-1]}(t) \end{bmatrix}$$

has a unique solution for $e(t)$. Such solution may be found by pre-multiplying both sides of the previous equation by $M_k^+ := \left(M_k^T M_k\right)^{-1} M_k^T$. That is

$$e(t) = \frac{d^{k-1}}{dt^{k-1}} M_k^+ \begin{bmatrix} J_{k-1} & 0 \\ 0 & I_p \end{bmatrix} \begin{bmatrix} y_e(t) \\ \vdots \\ y_e^{[k-1]}(t) \end{bmatrix} \tag{6.41}$$

Thus, the term $e(t)$ can be reconstructed in just one step using a high order differentiation; meanwhile the matrices M_k and J_{k-1} should be obtained in an iterative manner using (6.39) with $M_1 = C$.

From (6.41), the reconstruction of $x(t)$ is equivalent to the reconstruction of $e(t)$, which can be carried out by a linear combination of the output $y_e(t)$ and its $(k-1)$-th derivatives. Hence, a real time high order sliding mode differentiator will be used in order to provide the theoretically exact observation and unknown inputs identification.

The bondness of $\dot{\omega}(t)$ allows realizing a $k-th$ order sliding mode differentiator, such that we recover not only the state $x(t)$ but also the disturbance $\omega(t)$. Beforehand, let us define

$$\Theta(t) := M_k^+ \begin{bmatrix} J_{k-1} & 0 \\ 0 & I_p \end{bmatrix} \begin{bmatrix} y_e(t) \\ \vdots \\ y_e^{[k-1]}(t) \end{bmatrix}. \tag{6.42}$$

That is, from (6.41) and (6.42)

$$e(t) = \frac{d^{k-1}}{dt^{k-1}} \Theta(t). \tag{6.43}$$

The HOSM differentiator is given by

$$\begin{aligned}
\dot{z}_0(t) &= -\lambda_0 \Gamma^{\frac{1}{i+1}} \Psi^{\frac{i}{i+1}} \left(z_0(t) - \Theta(t)\right) + z_1(t) \\
\dot{z}_1(t) &= -\lambda_1 \Gamma^{\frac{1}{i}} \Psi^{\frac{i-1}{i}} \left(z_1(t) - \dot{z}_0(t)\right) + z_2(t) \\
&\ \vdots \\
\dot{z}_{k-1}(t) &= -\lambda_{k-1}(t) \Gamma^{\frac{1}{2}} \Psi^{\frac{1}{2}} \left(z_{k-1}(t) - \dot{z}_{k-2}(t)\right) + z_k(t) \\
\dot{z}_k(t) &= -\lambda_k \Gamma \Psi^0 \left(z_k(t) - \dot{z}_{k-1}(t)\right),
\end{aligned} \tag{6.44}$$

where $z_i(t), \Theta(t) \in \mathbb{R}^n$, $\lambda_i, \Gamma \in \mathbb{R}$. Consider $\vartheta = \left[\vartheta_1 \ \dots \ \vartheta_n\right]^T$, $\beta \in \mathbb{R}$, the function vector $\Psi^\beta(\sigma) \in \mathbb{R}^n$ is defined as $\Psi^\beta(\vartheta) = \left[|\vartheta_1|^\beta sign(\vartheta_1) \ \dots \ |\vartheta_n|^\beta sign(\vartheta_n)\right]^T$.

In [[27]] there was shown that there is a finite time T such that the identity

$$z_j(t) = \frac{d^j}{dt^j} \Theta(t) \tag{6.45}$$

is achieved for every $j = 0, ..., k$.

The values of the λ's can be calculated as it is shown in [[27]], Γ is a Lipschitz constant of $\Theta^{(k+1)}(t)$, which for our case can be calculated in the following way: from (6.34) and (6.43) $\left\|\Theta^{(k-1)}(t)\right\| \leq e^{+}$, the next derivative $\Theta^{(k)}(t) = \dot{e}(t)$ will be also bounded $\left\|\Theta^{(k)}(t)\right\| \leq \left\|\tilde{A}\right\| e^{+} + \|B\| \omega^{+}$. Finally, it can be verified that

$$\Gamma \geq \left\|\tilde{A}\right\|^{2} e^{+} + \left\|\tilde{A}\right\| \|D\| \omega^{+} + \|D\| \omega^{+}. \tag{6.46}$$

The vector $e(t)$ can be reconstructed from the $(k-1)$-th order sliding dynamics. Thus from (6.45), we achieve the identity $z_{k-1}(t) = e(t)$, and consequently

$$\hat{x}(t) := z_{k-1}(t) + \tilde{x}(t) \text{ for all } t \geq T, \tag{6.47}$$

where $\hat{x}$ represents the estimated value of x. Therefore, the identity

$$\hat{x}(t) \equiv x(t) \tag{6.48}$$

is achieved, for all $t \geq T$.

Now, from error dynamics (6.33), we can recover $\dot{e}(t)$ using the HOSM differentiator (6.44). From (6.45), the equality $z_k(t) = \dot{e}(t)$ is achieved for all $t \geq T$ and the next equation holds

$$\hat{\omega}(t) = D^{+}\left[z_k(t) - \tilde{A} z_{k-1}(t)\right], \tag{6.49}$$

where $\hat{\omega}(t)$ is the identified exact value of the unknown input $\omega(t)$. Thus, after a finite time T, when the HOSM differentiator converges (see [[27]]), the identity $\hat{\omega}(t) \equiv \omega(t)$ holds.

6.2.4.2 Control Design

The sliding surface is designed considering the estimated values of the state and the identified unknown input signal, $\left[\hat{x}_1 \ \hat{x}_2\right] \leftrightarrow \mathscr{T}\hat{x}$, as follows

$$s(t) = K\hat{x}_1(t) + \hat{x}_2(t) + G\hat{\omega}(t). \tag{6.50}$$

The matrix $K \in \mathrm{IR}^{m\times(n-m)}$ could be designed to prescribe the required performance of the reduced-order system. The term $G\hat{w}(t)$ is added to compensate unmatched uncertainties. The control law given in (6.2.4) is considered. First, it is necessary guarantee that the above control law induces a sliding motion despite the presence of uncertainties.

Due to (6.48), the identities $\hat{x}_1 = x_1$, $\hat{x}_2 = x_2$ are certainly obtained. Then, the time derivative of $\sigma(t)$ is given by

$$\dot{\sigma}(t) = \Phi x(t) + (KD_1 + D_2)\,\omega(t) + G\dot{\hat{\omega}}(t) + u(t), \tag{6.51}$$

where matrix $\Phi \in \mathrm{IR}^{m\times n}$ is defined as $\Phi := \left[KA_{11} + A_{21} \ KA_{12} + A_{22}\right]\mathscr{T}$.

Choosing a Lyapunov candidate function $V(\sigma) = \frac{\sigma^T(t)\sigma(t)}{2}$ and taking its derivative along time yields

$$\begin{aligned}\dot{V}(\sigma) &= \sigma^T(t)\left(\Phi x(t) + (KD_1 + D_2)\,\omega(t) + G\dot{\hat{\omega}}(t) - \varphi(x)\frac{\sigma(t)}{\|\sigma(t)\|}\right) \\ &\leq -\|\sigma(t)\|\left(\varphi(x) - \|\Phi\|\,\|x(t)\| - \theta\right) \end{aligned} \tag{6.52}$$

where $\theta := \|(KD_1 + D_2)\|\,\omega^+ + \|G\|\,\omega^+$. The scalar gain $\varphi(x)$ satisfies the condition

$$\varphi(x) - \|\Phi\|\,\|x(t)\| - \theta \geq \zeta > 0,$$

where ζ is a constant.

$$\varphi(x) > \|\Phi\|\,\|x(t)\| + \theta + \zeta. \tag{6.53}$$

Combining inequalities (6.52) and (6.53), it follows that the derivative of the Lyapunov function satisfies $\dot{V}(\sigma) \leq -\zeta V^{\frac{1}{2}}$ and, consequently, gain $\varphi(x)$ will induce the sliding motion.

When the system reaches the sliding surface $\sigma = 0$, we have

$$x_2(t) = -Kx_1(t) - G\hat{\omega}(t) \tag{6.54}$$

$$\dot{x}_1(t) = (A_{11} - A_{12}K)x_1(t) - A_{12}G\hat{\omega}(t) + D_1\omega(t). \tag{6.55}$$

It is well known that the (A_{11}, A_{12}) pair will be controllable since the (A, B) pair is controllable also [[12]]. Hence, there exists a matrix K such that matrix $A_s \triangleq (A_{11} - A_{12}K)$ has stable eigenvalues. The G gain matrix should be selected in order to compensate the unmatched uncertainties. In order to compensate $\omega(t)$ from $x_1(t)$, matrix D_1 must be matched with respect to A_{12}; therefore, im $(D_1) \subset$ im (A_{12}).

Thus, matrix $G \in \mathbb{R}^{m\times p}$ may be chosen so that

$$A_{12}G = D_1. \tag{6.56}$$

Then equation (6.55) yields

$$\dot{x}_1(t) = (A_{11} - A_{12}K)x_1(t) + D_1(\omega(t) - \hat{\omega}(t)), \tag{6.57}$$

so, in the ideal case after a finite time T, $w(t) \equiv \hat{\omega}(t)$, and therefore,

$$\dot{x}_1(t) = A_s x_1(t). \tag{6.58}$$

In particular, when $\texttt{rank}(A_{12}) = n - m$, matrix $G = A_{12}^{+}D_1$.

Since the eigenvalues of A_s have negative real part, equation (6.58) is exponentially stable. Hence, the unmatched uncertainties are compensated and coordinate x_1

is stabilized. The trajectories of the state x_1 will converge to a bounded region, i.e. there exist some constants $a_1, a_2 > 0$ such that

$$\|x_1(t)\| \leq a_1 \|x_1(0)\| \exp^{-a_2 t} \quad \forall t > T_\sigma,$$

where T_σ is the time taken to reach the sliding surface. Furthermore, $x_2(t)$ is bounded as well indeed during sliding motion. Taking the norm of equation (6.54) we have

$$\|x_2(t)\| \leq \|K\| \|x_1(t)\| + \|G\| w^+ \ \forall t > T_\sigma. \tag{6.59}$$

From the above equation, it is clear that the trajectories of $x_2(t)$ are bounded.

6.2.4.3 Simulations

Consider the inverted pendulum system shown in Fig. 6.15. The system consists of a cart (e) moving along a metal guiding bar (d). A cylindrical weight (f) is fixed to the cart by an axis (g). The cart is connected by a transmission belt (c) to a drive wheel (b). The wheel is driven by a current controlled direct current motor (a) which transforms the voltage u in torque such that the cart is accelerated. The state equations, considering the actuator dynamics are

$$\dot{x} = \begin{bmatrix} 0 & 0 & 1 & 0 & 0 \\ 0 & 0 & 0 & 1 & 0 \\ 0 & -\frac{m_p \ell^2 g}{(m_p+m_c)I+m_p m_c \ell^2} & 0 & 0 & \frac{K_t(I+m_p\ell^2)}{(m_p+m_c)I+m_p m_c \ell^2} \\ 0 & \frac{(m_p+m_c) m_p g \ell}{(m_p+m_c)I+m_p m_c \ell^2} & 0 & 0 & -\frac{m_p \ell K_t}{(m_p+m_c)I+m_p m_c \ell^2} \\ 0 & 0 & \frac{2\pi K_m}{I_m} & 0 & -\frac{R}{I_m} \end{bmatrix} x + \begin{bmatrix} 0 \\ 0 \\ 0 \\ 0 \\ \frac{1}{I_m} \end{bmatrix} u$$

$$+ \begin{bmatrix} 0 \\ 0 \\ \frac{I+m_p\ell^2}{(m_p+m_c)I+m_p m_c \ell^2} \\ -\frac{m_p \ell}{(m_p+m_c)I+m_p m_c \ell^2} \\ 0 \end{bmatrix} w \tag{6.60}$$

$$y = \begin{bmatrix} 1\,0\,0\,0\,0 \\ 0\,1\,0\,0\,0 \\ 0\,0\,0\,0\,1 \end{bmatrix} x, \tag{6.61}$$

The state vector is given by $x = \begin{bmatrix} r \ \phi \ \dot{r} \ \dot{\phi} \ di/dt \end{bmatrix}^T$, where r, ϕ, i, represent the longitudinal position of the cart, the angular position of the pendulum and the motor current respectively. The unknown input ω is a perturbing force acting on the cart. The variable description and their corresponding values are given in the next table. For this example, we are considering an unknown input $\omega = 2.5\sin(2.2t) + 1.5$.

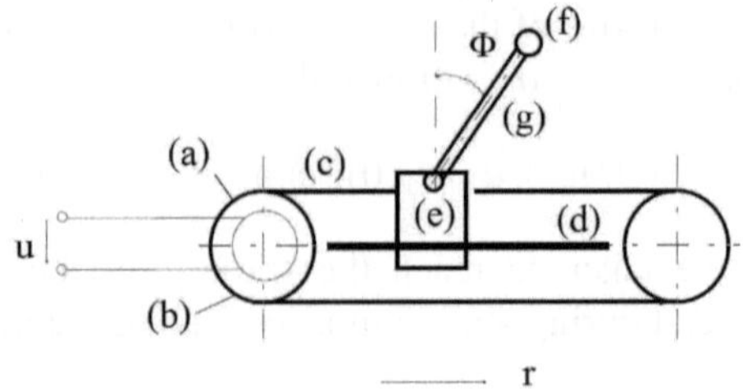

Fig. 6.15 Inverted cart pendulum system

Table 6.1 Inverted-Cart Pendulum System description

Symbol	Description	Value	Units
m_p	mass of the pendulum	0.36	kg
m_c	mass of the cart	4	kg
I	pendulum moment of inertia	0.084	$kg \cdot m^2$
ℓ	longitude	0.5	m
g	gravitational acceleration constant	9.81	m/s^2
K_t	motor torque constant	0.0295	$N \cdot m$
I_m	motor inductance	0.0087	H
K_m	motor back electromotive constant	0.212	$V/(rad/s)$
R	motor armature resistance	3.12	Ω

Observer design. First, a Luenberger-type observer is designed such that matrix $A - LC$ has a set of eigenvalues given by $\{-780, -9, -2.6, -2\}$. System (6.62)-(6.63) is strongly observable. A way to check the system's strong observability property is to apply the unknown inputs decoupling algorithm introduced in Observer Section. Thus, if the system is strongly observable, k iterations ($k \leq n$) are needed to find a full column rank matrix M_k. For system (6.62)-(6.63) $k = 2$. From (6.41) we need to differentiate once ($i.e.$ $(k-1) - times$) in order to reconstruct the state. Additionally, for recovering the unknown inputs a second differentiation is needed. The total order of the differentiator (6.44) is determined by the smoothness of the unknown input, we select a HOSM differentiator of $2^{nd} - order$. The input of the HOSM differentiator (6.42) is

$$\Theta(t) = \begin{bmatrix} -0.07 & 0 & 0 & 0.45 & 0.05 & 0.48 \\ 0 & 0 & 0 & 0.05 & 0.99 & -0.48 \\ -0.32 & 0 & 0 & -2.4 & 0.18 & -2.76 \\ 0.76 & 1 & 0 & 4.22 & 9.16 & 3.83 \\ 0.06 & 0 & 0 & 0.48 & -0.04 & 0.56 \end{bmatrix} \begin{bmatrix} y_e(t) \\ y_e^{[1]}(t) \end{bmatrix}.$$

Following [27] we select $\lambda_0 = 1.1$, $\lambda_1 = 1.5$, $\lambda_2 = 2$.The observer gain is $\Gamma = 2.8e6$. The sampling step is $\delta = 10(\mu s)$.

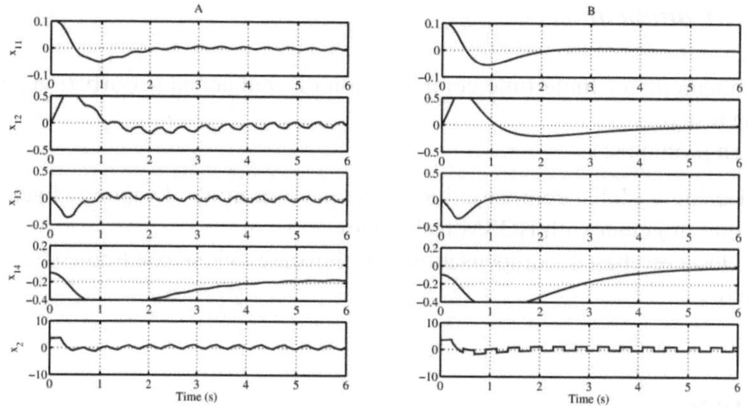

Fig. 6.16 Column (B) shows the compensated system and in column (A) the system without compensation. The underactuated states are $x_{11},\ldots,x_{14}$, while the completely actuated state is x_2

Control design. Regularizing the system

$$\dot{x}_1 = \begin{bmatrix} 0 & 0 & 1 & 0 \\ -0.4377 & 0 & 0 & 0 \\ 10.6 & 10 & 0 & 0 \\ 0 & -1 & 0 & 0 \end{bmatrix} x_1 + \underbrace{\begin{bmatrix} 0 \\ 0.8345 \\ -0.8632 \\ 0 \end{bmatrix}}_{A_{12}} x_2 + \underbrace{\begin{bmatrix} 0 \\ 0.2396 \\ -0.2479 \\ 0 \end{bmatrix}}_{D_1} \omega \tag{6.62}$$

$$\dot{x}_2 = \begin{bmatrix} 0 & -1.332 & 0 & 0 \end{bmatrix} x_1 - 386.35u. \tag{6.63}$$

From the above equation can be seen that condition im $(D_1) \subset \text{im}(A_{12})$ is satisfied. The compensator gain is selected as $G = 0.28$. The gain K was designed by eigenvalues assignment, such that the reduced order system has an eigenvalues set given by $\{-1 - 1.3 - 4.53 - 3.9\}$. From (6.53) the sliding mode gain is selected as $\varphi(x) = 1.2e3\,(\|x(t)\| + 1)$.

The simulation was carried comparing two approaches for sliding surface design. The approach in (A) was carried designing a conventional sliding mode surface (see [37]) i.e. $\sigma_A(t) = x_1(t) + Kx_2(t)$, while in (B), the surface was designed to cope with the unmatched perturbations (6.50) i.e. $\sigma_B(t) = x_1(t) + Kx_2(t) + G\hat{\omega}(t)$. Fig. (6.16) shows the states of the regularized system, column (A) shows the results when no compensation is carried: the perturbation effects are present in all the states. Column (B) shows the states when the compensation of unmatched uncertainties is done through the sliding surface, here the stabilization of state x_1 is achieved, while the trajectories of state x_2 remain bounded.

6.2.5 Conclusions

Three methods for finite time compensation of unmatched perturbations of inputs are suggested.

Open problems:

1. Design a global compensation technique joining the Lyapunov backstepping techniques together with HOSM techniques.
2. Observer based compensation for the case when condition (6.32) is not satisfied.

References

1. Adhami-Mirhosseini, A., Yazdanpanah, M.J.: Robust Tracking of perturbed systems by nested sliding mode control. In: Proc. of the International conference on Control and Automation, Budapest, Hungary, pp. 44–48 (2005)
2. Angulo, M., Fridman, L., Levant, A.: Exact Finite-Time Output Based Control using High-Order Sliding Modes. International Journal of Systems Science 42(11), 1847–1857 (2011)
3. Bartolini, G., Ferrara, A., Giacomini, L., Usai, E.: A combined backstepping/second order sliding mode approach to control a class of non-linear systems. In: Proceedings of the IEEE International Workshop on Variable Structure Systems, Tokyo, Japan, pp. 205–210 (1996)
4. Bejarano, F.J., Fridman, L.: High order sliding mode observer for linear systems with unbounded unknown inputs. International Journal of Control 83(9), 1920–1929 (2010)
5. Cao, W., Xu, J.: Nonlinear integral-type sliding surface for both matched and unmatched uncertain systems. IEEE Trans. Automat. Contr. 49, 1355–1360 (2004)
6. Castaños, F., Fridman, L.: Analysis and Design of Integral Sliding manifolds for Systems With Unmatched Perturbations. IEEE Trans. Automat. Contr. 51(5), 853–858 (2006)
7. Choi, H.H.: An LMI-based switching surface design method for a class of mismatched uncertain systems. IEEE Trans. Automat. Contr. 48, 1634–1638 (2003)
8. Choi, H.H.: Output feedback variable structure control design with an H_2 performance bound constraint. Automatica 48(9), 2403–2408 (2008)
9. Davis, R., Spurgeon, S.K.: A nonlinear control strategy for robust sliding mode performance in the presence of unmatched uncertainty. International Journal of Control 57, 1107–1123 (1993)
10. Davis, R., Spurgeon, S.K.: Robust implementation of sliding mode control schemes. International Journal of System Science 24, 733–743 (1993)
11. Drazenovic, B.: The Invariance Conditions in Variable Structure Systems. Automatica 5(3), 287–295
12. Edwards, C., Spurgeon, S.K.: Sliding Mode Control. Taylor and Francis, London (1998)
13. Yan, X.G., Spurgeon, S.K., Edwards, C.: Dynamic sliding mode control for a class of systems with mismatched uncertainty. European Journal of Control 11(1), 1–10 (2005)

14. Estrada, A., Fridman, L.: Integral HOSM Semiglobal Controller for Finite-Time Exact Compensation of Unmatched Perturbations. IEEE Trans. Automat. Contr. 55(11), 2644–2649 (2010)
15. Estrada, A., Fridman, L.: Quasi-continuous HOSM control for systems with unmatched perturbations. Automatica 46(11), 1916–1919 (2010)
16. Ferrara, A., Giacomini, L.: On multi-input backstepping design with second order sliding modes for a class of uncertain nonlinear systems. International Journal of Control 71(5), 767–788 (1998)
17. Ferreira de Loza, A., Bejarano, F.J., Fridman, L.: Unmatched uncertainties compensation based on high-order sliding mode observation. International Journal on Robust and Nonlinear Control (2012), doi:10.1002/rnc.2795
18. Filippov, A.F.: Differential Equations with Discontinuous Right-Hand Side. Kluwer, Dordrecht (1988)
19. Fridman, L.: Singularly perturbed analysis of chattering in relay control systems. IEEE Transactions on Automatic Control 47(12), 2079–2084 (2079)
20. Huerta-Avila, H., Loukianov, A.G., Cañedo, J.M.: Nested Integral Sliding Modes of Large Sacale Power Systems. In: Proc. of the 46th IEEE Conference on Decision and Control, New Orleans, LA, USA, pp. 1993–1998 (December 2007)
21. Freeman, R.A., Kokotovic, P.V.: Robust Nonlinear Control Design. Birkhauser, Boston (1996)
22. Isidori, A.: Nonlinear Control Systems, 2nd edn. Springer, New York (1989)
23. Kanellakopoulos, I., Kokotović, P.V., Morse, A.S.: Systematic design of adaptive controllers for feedback linearizable system. IEEE Trans. on Automatic Control 36(11), 1241–1253 (1991)
24. Krstic, M., Kanellakopoulous, I., Kokotovic, P.: Nonlinear and Adaptive Control Design. Wiley Interscience, NY (1995)
25. Levant, A., Alelishvili, L.: Integral High-Order Sliding Modes. IEEE Trans. Automat. Contr. 52(7), 1278–1282 (2007)
26. Levant, A.: Quasi-Continuous High-Order Sliding-Mode Controllers. IEEE Trans. Automat. Contr. 50(11), 1812–18162 (2005)
27. Levant, A.: High-order sliding modes: differentiation and output-feedback control. International Journal of Control 76, 924–941 (2003)
28. Levant, A., Michael, A.: Adjustment of high-order sliding-mode controllers. International Journal of Robust and Nonlinear Control 19(15), 1657–1672 (2009)
29. Loukyanov, A., Utkin, V.: Methods of reducing equations for dynamic systems to a regular form. Automatic and Remote Control 42(4), 413–420 (1993)
30. Loukianov, A.G.: Nonlinear Block Control with Sliding Mode. Automation and Remote Control 59(7), 916–933 (1998)
31. Loukianov, A.G.: Robust Block Decomposition Sliding Mode Control Design. Mathematical Problems in Engineering 8(4-5), 349–365 (2002)
32. Loukianov, A., Caedo, J., Fridman, L., Soto Cota, A.: High order block sliding mode controller for a synchronous generator with an exciter system. IEEE Transactions on Industrial Electronics 58(1), 337–347 (2011)
33. Molinari, B.P.: A strong controllability and observability in linear multivariable control. IEEE Transactions on Automatic Control 21(5), 761–764 (1976)
34. Scarratt, J.C., Zinober, A.S.I., Mills, R.E., Rios-Bolivar, M., Ferrara, A., Giacomini, L.: Dynamical adaptive first and second-order sliding backstepping control of nonlinear non-triangular uncertain systems. ASME Journal of Dynamic Systems, Measurement and Control 122(4), 746–752 (2000)

35. Shtessel, Y.B., Buffington, J., Banda, S.: Multiple Time Scale Flight Control Using Reconfigurable Sliding Modes. In: Proceedings of the 37th IEEE Conference on Decision and Control, Tampa, Florida, USA, pp. 4196–4201 (December 1998)
36. Won, M.C., Hedrick, J.K.: Multiple surface sliding control of a class of uncertain nonlinear systems. International Journal of Control 64(4), 693–706 (1996)
37. Utkin, V.I.: Sliding modes in control and optimization. Springer, Berlin (1992)
38. Utkin, V.I., Guldner, J., Shi, J.: Sliding Mode Control in Electromechanical Systems. Taylor & Francis, London (1999)

Chapter 7
Higher Order Sliding Mode Control by Keeping a 2-Sliding Constraint

Prasiddh Trivedi and Bijnan Bandyopadhyay

Abstract. This article investigates a new algorithm for higher order sliding mode control. The proposed control law keeps a constraint in 2-sliding mode such that the finite time stabilization of the chain of integrators is achieved. The proposed switching function has relative degree two with respect to the input and a second order sliding controller is used. The twisting controller is used for achieving finite time convergent 2-sliding mode to the switching manifold. The switching manifold is designed to provide finite time convergence of the integrator chain. The fractional powers in the switching function are carefully designed to prevent the unboundedness or singularity arising because of the switching constraint being kept at zero.

7.1 Introduction

Finite time stability and Sliding Mode Control (SMC) are closely related areas of active research. Since sliding mode control methods require finite time reaching to the sliding manifold, finite time stabilization methods can naturally be applied in SMC. However, SMC has more emphasis on robustness and sliding mode controllers should be robust. The Higher Order Sliding Mode Control (HOSM) is a necessity for achieving robust stabilization or tracking in the systems with outputs having relative degree more that one with respect to the input. The sliding order by definition gives better accuracy [[11]], [[13]].

The literature on second order sliding modes is abundant. However, few controllers exist for achieving finite time convergence for an arbitrary order dynamical system. In general this may not be possible but given boundedness of the states

Prasiddh Trivedi
Research Scholar, Systems and Control Engineering, IIT Bombay
e-mail: prasiddh@sc.iitb.ac.in

B. Bandyopadhyay
Professor, Systems and Control Engineering, IIT Bombay
e-mail: bijnan@ee.iitb.ac.in

B. Bandyopadhyay et al. (Eds.): *Advances in Sliding Mode Control*, LNCIS 440, pp. 145–163.
DOI: 10.1007/978-3-642-36986-5_7

finite time robust stabilization of an arbitrary chain of integrators is an interesting problem. Finite time stability with non-lipschitzian right hand sides is well documented [2], [8], [15]. Homogeneity properties of the vector field are also explored and shown to be quite useful in proving the finite time nature [1], [10]. The generalization to systems with dynamics of order higher than 2 has been established in different forms [3], [9]. These methods employ fractional powers and provide a continuous control law.

Terminal sliding mode (TSM) was first introduced in [7] as a robust finite time controller for a second order two-link robotic manipulator. The terminal sliding mode has an extraordinary feature of providing finite time convergence of all the system states through a non linear switching function. The more general formulation was presented in [17], where nested non linear switching functions are designed. All switching functions converge to zero in finite time sequentially and as a consequence ultimately all system states converge to zero. The terminal sliding mode can easily be used to achieve finite time stabilization of chain of integrators and thus providing for a higher order sliding mode. However, as mentioned by Levant [12], this form leads to unbounded control for systems of order three and higher. Thus, Filippov solutions are not well defined for this formulation.This has led to development of non-singular terminal sliding mode control. A slightly different switching function has been shown [6], [5] to provide terminal sliding without causing unbounded control. Non singular terminal sliding mode control has been used as second order sliding mode control of uncertain multivariable systems [5].

The discontinuous control i.e., the sliding mode control has also been developed for achieving finite time stabilization of higher order systems [12], [16]. The discontinuous feedback has the apparent advantage of robustness over the continuous feedback laws. The most popular HOSM controller is detailed in [12]. The idea in [12] for achieving finite time convergence is to keep a properly designed switching constraint in 1-sliding. The novel method proposed in [16] utilizes feedback control with matrix exponentials, but needs the knowledge of initial conditions to compute the gain matrix. The advantage is the reaching time can be easily specified and the designed control law is able to steer the trajectory to the desired sliding set in the specified time. The proposed controller in this article is based on the idea of holding a 2-sliding constraint. The organisation is as follows. The first section provides a basic introduction to HOSM. The second section describes the main proposal and detailed proof that keeping a properly designed constraint in 2-sliding can achieve finite time convergence of a triple integrator. Extension of the idea to the chain of integrators for higher order sliding mode control is described in the third section followed by numerical simulations and an example.

7.1.1 Higher Order Sliding Mode Control

The HOSM is introduced and well defined in terms of a family of real sliding trajectories [11]. This article follows the same definitions and assumptions. Let the dynamical system be described by,

$$\dot{x} = f(t,x) + g(t,x)u \tag{7.1}$$

with $x \in \mathbb{R}^n, u \in \mathbb{R}$. Let $\sigma(x) \in \mathbb{R}$ be some output of the system with stable zero dynamics and the problem is to steer it to zero in finite time. The classical sliding mode approach makes $\dot{\sigma}$ discontinuous using a relay feedback and provides for finite time stabilization. This approach would require $\sigma(x)$ to have relative degree 1 with respect to the input and it makes the input discontinuous. Apparently when the output has relative degree more than one, this approach would not work and HOSM concepts have to be employed. In some situations, discontinuous control might be unacceptable. Then the relative degree can be increased artificially by cascading integrators and shifting the discontinuity in to higher derivatives of input. This approach increases the smoothness of control as well as $\sigma(x)$.
Suppose $\sigma(x)$ has relative degree r with respect to the input (it may have been artificially increased as mentioned). Then the control appears in the r-th derivative of $\sigma(x)$ with the direction derivatives of system functions. Thus,

$$\sigma^{(r)} = \phi(t,x) + \gamma(t,x)u \tag{7.2}$$

where, $\phi(t,x)$ and $\gamma(t,x)$ contains directional derivative(Lie derivatives) of the system functions $f(t,x)$ and $g(t,x)$. It is assumed that

$$|\phi(t,x)| < \Phi,\ 0 < \Gamma_m \leq \gamma(t,x) \leq \Gamma_M \tag{7.3}$$

The problem is to find a control input u which stabilizes (7.2) in finite time.

It can also be stated as finite time stabilization of an integrator chain in the presence of uncertainties in the form of $\phi(t,x)$ and $\gamma(t,x)$. Also, finite time stabilization of an integrator chain without uncertainties is a good starting point for analysis of new controller. The next section describes a new controller for triple integrator without uncertainties and then extends it to the triple integrator with uncertainties.

7.2 Third Order Sliding via Non-singular Terminal Switching Function

This section introduces the idea of keeping a constraint in 2-sliding to obtain third order sliding. At first we consider the triple integrator without uncertainties. We show that finite time convergence of the triple integrator is obtained by holding a switching function in 2-sliding mode. The switching function is the one used in non-singluar terminal sliding mode control. The triple integrator input has relative degree two with respect to this switching function. The twisting controller is the obvious choice to confine the trajectory on the manifold defined by this constraint function.

7.2.1 Triple Integrator without Uncertainties

Let the triple integrator be $\sigma^{(3)} = u$. The constraint function, as mentioned earlier, has to have relative degree two with respect to the input. Thus, it must be a function of variables σ and $\dot{\sigma}$ such that the input appears in the second time derivative of the function. Consider the following function,

$$\psi(\sigma,\dot{\sigma}) = \sigma + \dot{\sigma}^{\alpha} \tag{7.4}$$

where, α is a ratio of positive odd integers. Specifically $\alpha = p/q$, with $p,q \in \mathbb{N}$ and odd. This function also defines a 2-sliding manifold in $\sigma,\dot{\sigma}$ and $\ddot{\sigma}$ coordinates as $\mathscr{S} = \{(\sigma,\dot{\sigma},\ddot{\sigma}) | \psi(\sigma,\dot{\sigma}) = \dot{\psi}(\dot{\sigma},\ddot{\sigma}) = 0\}$. We have the following proposition to show that keeping the constraint function (7.4) in a 2-sliding mode achieves finite time stabilization of the triple integrator.

Proposition 7.1. *The control law,*

$$u = -k_1 sign(\sigma + \dot{\sigma}^{\alpha}) - k_2 sign(\dot{\sigma} + \alpha\dot{\sigma}^{\alpha-1}\ddot{\sigma}) \tag{7.5}$$

stabilizes $\sigma^{(3)} = u$ in finite time with the following sufficient conditions.

- *α is a ratio of two odd integers with $1 < \alpha < 1.5$.*
- *$k_1 > k_2$ and $\Xi_3 + \alpha'\Xi_2\Xi_3^2 + \Xi_2(k_1 - k_2) < -\Xi_3 - \alpha'\Xi_2\Xi_3^2 + \Xi_2(k_1 + k_2)$.*

where, $|\dot{\sigma}| < \frac{1}{\alpha}\Xi_2^{1-\alpha}$, $|\ddot{\sigma}| < \Xi_3$ and $\Xi_2, \Xi_3 \in \mathbb{R}$ are known.

Proof. Let us define the integrator states as, $\xi_1 = \sigma, \xi_2 = \dot{\sigma}$ and $\xi_3 = \ddot{\sigma}$, then the state equations are

$$\begin{aligned} \dot{\xi}_1 &= \xi_2 \\ \dot{\xi}_2 &= \xi_3 \\ \dot{\xi}_3 &= u \end{aligned} \tag{7.6}$$

Consider the switching function $\psi(\xi) = \xi_1 + \xi_2^{\alpha}$. The proof is divided in two major parts. First we show that if the constraint $\psi(\xi)$ is held in 2-sliding, i.e., $\psi(\xi) = \dot{\psi}(\xi) = 0$ is kept then the reduced order dynamics of (7.6) is finite time stable. It is equivalent to say that the zero dynamics of the (7.6) with function $\psi(\xi)$ as output is finite time stable. The second part proposes a controller which keeps this constraint in 2-sliding i.e, makes the trajectory reach the 2-sliding manifold in finite time.

The zero dynamics can be easily obtained by equating $\psi(\xi)$ and $\dot{\psi}(\xi)$ to zero. That is,

$$\psi(\xi) = \xi_1 + \xi_2^{\alpha} = 0 \tag{7.7}$$

$$\dot{\psi}(\xi) = \xi_2 + \alpha\xi_2^{\alpha-1}\xi_3 = 0 \tag{7.8}$$

The algebraic relationships are obtained as,

$$\xi_2 = -\xi_1^{\frac{1}{\alpha}} \tag{7.9}$$

$$\xi_3 = -\frac{1}{\alpha}\xi_2^{2-\alpha} \tag{7.10}$$

Note that, (7.8) dictates two algebraic relations namely $\xi_2 = 0$ or $\xi_3 = -\frac{1}{\alpha}\xi_2^{2-\alpha}$. However, $\xi_2 = 0$ is not to be considered for reduced order dynamics for reasons explain later in the proof. Thus, when the constraint ψ is kelp in 2-sliding, the dynamics of (7.6) reduces to one differential equation $\dot{\xi}_1 = -\xi_1^{1/\alpha}$ which is finite time stable with $1 < \alpha < 2$, and the other two states are algebraically related with ξ_1. Here, it is obvious that for finite time stability $\alpha > 1$ is necessary but $\alpha < 2$ is apparent from the algebraic relation (7.10), where the $\xi_2^{2-\alpha}$ converges to zero only with $\alpha < 2$.

We have established that if the constraint $\psi(\xi) = \xi_1 + \xi_2^{\alpha}$ is kept in 2-sliding, then finite time convergence of (7.6) is achieved. Next we show that the proposed controller achieves finite time convergence to the 2-sliding manifold $\mathscr{S}$. To this end, consider the second time derivative of $\psi(\xi)$,

$$\ddot{\psi} = \xi_3 + \alpha(\alpha-1)\xi_2^{\alpha-2}\xi_3^2 + \alpha\xi_2^{\alpha-1}\dot{\xi}_3 \tag{7.11}$$

$$= \phi'(\xi) + \gamma'(\xi)u \tag{7.12}$$

where,

$$\phi'(\xi) = \xi_3 + \alpha(\alpha-1)\xi_2^{\alpha-2}\xi_3^2 \tag{7.13}$$

$$\gamma'(\xi) = \alpha\xi_2^{\alpha-1} \tag{7.14}$$

The functions $\phi'(\xi)$ and $\gamma'(\xi)$ are of utmost importance in the subsequent analysis. Some remarks about these functions, follow, which are useful in the proof.

Remark 7.1. The function $\phi'(\xi)$ is not globally bounded so the question can be raised about the existence of Filippov's solution. Note that the equation (7.11) is only a tool to understand the behaviour of ψ and $\dot{\psi}$. The solution of the system is dictated by (7.6). It is obvious that Filippov solutions for ξ are very well defined with the proposed control. The functions ψ and $\dot{\psi}$ are algebraically related with ξ so these are also well defined.

Remark 7.2. The set $\mathscr{S} = \{\xi \in \mathbb{R}^3 | \psi(\xi) = \dot{\psi}(\xi) = 0\}$ is intended to be a positively invariant set, hence, it is necessary to examine the dynamics when the trajectory is within $\mathscr{S}$ i.e., during the sliding mode. Consider (7.11) when $\psi(\xi) \equiv \dot{\psi}(\xi) \equiv 0$,i.e., substituting algebraic relations (7.9)-(7.10) into (7.11),

$$\ddot{\psi} = \alpha\xi_2^{\alpha-1}\left(\frac{\alpha-2}{\alpha^2}\xi_2^{2-3\alpha} + u\right) \tag{7.15}$$

Note that it contains the term $\xi^{2-3\alpha}$ and $2-3\alpha>0$ is necessary for the right hand side to be bounded while sliding. Thus, we obtain the condition $1<\alpha<1.5$.

Remark 7.3. $\gamma'(\xi)$ is a sign definite positive function. The α is ratio of two odd integers, so $\alpha-1$ has an even number in the numerator. Thus $\xi_2^{\alpha-1}$ is always positive. However, it does not have a lower bound. It is seen that $\gamma'(\xi)\to 0$ as $\xi_2\to 0$. Also as $\xi_2\to 0$ the function $\phi'(\xi)\to\infty$.

Remark 7.4. Also, it is crucial to note that on the manifold $\mathscr{S}$ the function $\phi'(\xi)$ is bounded. It is evident from the following limit.

$$\lim_{(\xi_2\to 0)|\mathscr{S}} \xi_3 + \frac{\alpha(\alpha-1)\xi_2^{\alpha}\xi_3^2}{\xi_2^2} = \frac{\alpha-2}{\alpha}\xi_2^{2-\alpha} = 0 \tag{7.16}$$

Thus, once the trajectories converge to the 2-sliding set $\mathscr{S}$, $\phi'(\xi)$ is bounded.

Consider some identities relating the functions ϕ' and γ'. These identities are useful in the proof.

$$\dot{\gamma}'(\xi) = \alpha(\alpha-1)\xi_2^{\alpha-2}\xi_3 = \alpha'\gamma'(\xi)\xi_2^{-1}\xi_3 \tag{7.17}$$
$$\phi'(\xi) = \xi_3 + \dot{\gamma}'(\xi)\xi_3 \tag{7.18}$$
$$\dot{\psi}(\xi) = \frac{\xi_2}{\alpha'}(\alpha'+\dot{\gamma}'),\quad \alpha'=\alpha-1 \tag{7.19}$$

The proof of finite convergence will consist of showing that the real trajectory is "majored" by a known fixed trajectory (majorant curve). This majorant curve converges to zero in finite time and thus the real trajectory converges to zero. To this end, consider the $\ddot{\psi}$ again in a more convenient form,

$$\ddot{\psi} = \xi_3 + \dot{\gamma}'(\xi)\xi_3 + \gamma'(\xi)u \tag{7.20}$$
$$= \xi_3 + \gamma'(\xi)\left(\alpha'\xi_2^{-1}\xi_3^2 + u\right) \tag{7.21}$$

Next, consider the projection of the trajectory on ψ-$\dot{\psi}$ coordinates. Without loss of generality one can consider the trajectory starting from a point $(0,\dot{\psi}_0)$. The trajectory enters the quadrant $\psi>0,\dot{\psi}>0$. We shall show that the real trajectory of the system with proposed control in this quadrant is confined by the axis $\psi=0$, $\dot{\psi}=0$ and the trajectory of the equation

$$\ddot{\psi} = \Xi_3 + \alpha'\Xi_2\Xi_3^2 - \Xi_2 k_M \tag{7.22}$$

where, $k_M = k_1+k_2$,Ξ_2 and Ξ_3 are bounds considered as $|\xi_3|<\Xi_3,\gamma'(\xi)<\Xi_2$.

For the justification of this fact let us write the differential inclusion from (7.20) as,

$$\ddot{\psi} \in [-\Xi_3,\Xi_3] + [0,\Xi_2]\left(\frac{\alpha'}{\xi_2}\left[-\Xi_3^2,\Xi_3^2\right]+u\right) \tag{7.23}$$

Table 7.1 Table gain conditions in state-space

Region	ξ_1 ξ_2 ξ_3	Condition for $\dot{\psi} > 0$	Condition for $\phi' < 0$
$\mathcal{O}_1$	+,+,+	Always	Never
$\mathcal{O}_2$	−,+,+	Always	Never
$\mathcal{O}_3$	+,−,+	$\dot{\gamma}' < -\alpha'$	$\dot{\gamma}' < -1$
$\mathcal{O}_4$	+,+,−	$\dot{\gamma}' > -\alpha'$	$\dot{\gamma}' > -1$
$\mathcal{O}_5$	−,+,−	$\dot{\gamma}' > -\alpha'$	$\dot{\gamma}' > -1$

Now, refer to the Table-7.1. The table details the regions in ξ-space where $\dot{\psi}$ can be positive. The third column lists the condition for $\dot{\psi} > 0$ and fourth column lists the condition for $\phi'(\xi)$ as in (7.18) to be negative. Interestingly the condition for $\dot{\psi} > 0$ implies the condition for $\phi'(\xi) < 0$ in the regions $\mathcal{O}_4$ and $\mathcal{O}_5$. For example, consider $\dot{\psi}_0 \in \mathcal{O}_4$. Then, from (7.19) $\dot{\gamma}' > -\alpha'$ is necessary for $\dot{\psi} > 0$. Moreover, in this region $\dot{\gamma}' > -1$ implies $\phi' < 0$. It is trivial that when $\phi'(\xi) < 0$ the real trajectory is obviously confined by trajectory of (7.22). Thus, the regions where $\phi'(\xi) > 0$ are of concern which are $\mathcal{O}_1$, $\mathcal{O}_2$ and $\mathcal{O}_3$. Observe that in these regions $\xi_3 > 0$ and in the quadrant $\psi > 0$, $\dot{\psi} > 0$ we have $\dot{\xi}_3 = -k_M$. Thus, ξ_3 becomes negative in finite time and enters regions $\mathcal{O}_4$ or $\mathcal{O}_5$. In these regions as we have already seen the real trajectory is bounded by the trajectory of (7.22).

However, there exist initial points on $\psi = 0$ axis such that $\ddot{\psi} > 0$. In this case the trajectory is not bounded by (7.22). Note that such initial points exist only in the regions $\mathcal{O}_1$, $\mathcal{O}_2$ and $\mathcal{O}_3$. We have already seen that the trajectory leaves this regions never to enter again. Let the majorant curve intersection point on the axis $\dot{\psi} = 0$ be $(\psi_M, 0)$, then

$$-2(\Xi_3 + \alpha'\Xi_2\Xi_3^2 - \Xi_2 k_M) = \dot{\psi}_0^2 \tag{7.24}$$

Similar analysis can be done to see that the trajectory in the quadrant $\psi > 0, \dot{\psi} < 0$ is confined by the trajectory of

$$\ddot{\psi} = -\Xi_3 - \alpha'\Xi_2\Xi_3^2 - \Xi_2 k_m \tag{7.25}$$

where, $k_m = k_1 - k_2$. Assume that the trajectory intersects the axis $\psi = 0$ at a point $(0, \dot{\psi}_1)$ then,

$$2(\Xi_3 + \alpha'\Xi_2\Xi_3^2 + \Xi_2 k_m) = \dot{\psi}_1^2 \tag{7.26}$$

For the convergence of the majorant trajectory it is sufficient to have $|\dot{\xi}_1|/|\dot{\xi}_0| < 1$. Thus, a sufficient condition can be written as

$$\Xi_3 + \alpha'\Xi_2\Xi_3^2 + \Xi_2 k_m < -\Xi_3 - \alpha'\Xi_2\Xi_3^2 + \Xi_2 k_M \tag{7.27}$$

This completes the proof. ∎

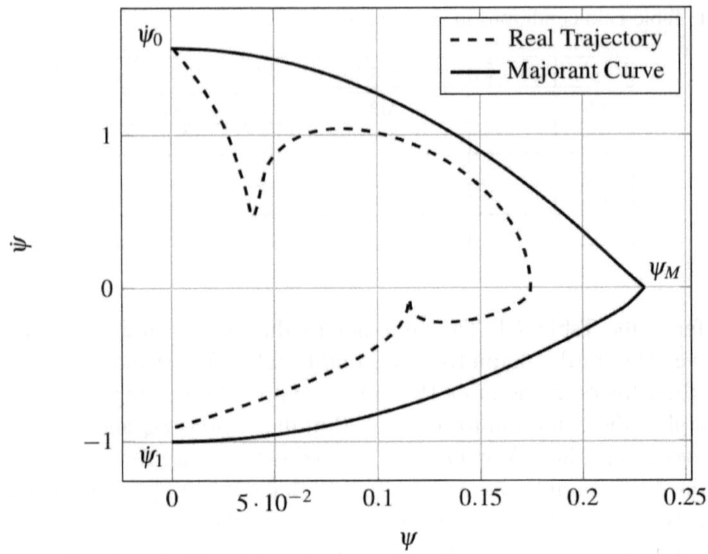

Fig. 7.1 Real trajectory and the Majorant Curve

Thus, it is established that finite time convergence to a 3-sliding set is possible by keeping a switching constraint in 2-sliding mode. Unlike the usual terminal sliding mode, the existence of Filippov solutions is also seen with a condition on the fractional power. It is important to note that the derived gain conditions are too conservative and in practice gains cannot be assigned using these inequalities.

7.2.2 Uncertain Triple Integrator

This section considers the triple integrator with uncertainties. The practical application of the 3-sliding algorithm requires stabilization in the presence of uncertain bounded system functions. Thus, it is necessary for the proposed controller to be able to achieve finite time stabilization of,

$$\sigma^{(3)} = \phi(t,x) + \gamma(t,x)u \tag{7.28}$$

with bounds given as,

$$|\phi(t,x)| < \Phi, 0 < \Gamma_m < \gamma(t,x) \leq \Gamma_M \tag{7.29}$$

Proposition 7.2. *The control* (7.5)*, stabilizes the uncertain integrator* (7.28) *in finite time, if the following sufficient conditions hold.*

- *α is a ratio of two odd integers with $1 < \alpha < 1.5$.*

- $\Gamma_m \Xi_3 k_M - \Xi_2 \Xi_3^2 - \Phi > \Gamma_M \Xi_3 k_m + \Xi_3 + \Xi_2 \Xi_3^2 + \Phi$.

Proof. The equation for $\ddot{\psi}$ is changed incorporating these changes as,

$$\ddot{\psi} = \xi_3 + \alpha\alpha' \xi_2^{\alpha-2} \xi_3^2 + \alpha \xi_2^{\alpha-1} \phi(t,x) + \alpha \xi_2^{\alpha-1} \gamma(t,x) u \tag{7.30}$$

Let $\phi'(t,x)$ and $\gamma'(t,x)$ be,

$$\phi'(t,x) = \xi_3 + \alpha\alpha' \xi_2^{\alpha-2} \xi_3^2 + \alpha \xi_2^{\alpha-1} \phi(t,x) \tag{7.31}$$
$$\gamma'(t,x) = \alpha \xi_2^{\alpha-1} \gamma(t,x) \tag{7.32}$$

Exactly the same analysis as in the proof of Proposition 7.1 can be repeated to determine the majorant curve in the ψ-$\dot{\psi}$ plane. The majorant curve in the first quadrant($\psi > 0, \dot{\psi} > 0$) is determined as

$$\ddot{\psi} = \Xi_3 + \Xi_2 \Xi_3^2 + \Phi - \Gamma_m k_M \tag{7.33}$$

In the second quadrant($\psi > 0, \dot{\psi} < 0$) the majorant curve is determined by

$$\ddot{\psi} = -\Xi_3 - \Xi_2 \Xi_3^2 - \Phi - \Gamma_M k_m \tag{7.34}$$

Thus, the sufficient condition in this case can be easily written as,

$$\Gamma_m \Xi_3 k_M - \Xi_2 \Xi_3^2 - \Phi > \Gamma_M \Xi_3 k_m + \Xi_3 + \Xi_2 \Xi_3^2 + \Phi \tag{7.35}$$

Thus, the proposed control produces a 2-sliding mode on the constraint $\psi(\xi)$ and in turn finite time stabilization of the triple integrator. ■

7.3 Extension to Higher Order Sliding

The idea presented in the previous section can also be used to achieve higher order sliding modes. The key idea is to hold a 2-sliding constraint to obtain higher order sliding. Consider the chain of r integrators with uncertainties which might be of the form (7.2).

$$\sigma^{(r)}(x) = \phi(t,x) + \gamma(t,x) u \tag{7.36}$$

Assume that the r-sliding set $\mathscr{S} = \{x \in \mathbb{R}^n | \sigma = \dot{\sigma} = \ldots = \sigma^{(r-1)} = 0\}$ is non empty. The objective is to find an input u such that the trajectory of (7.36) is finite time convergent to $\mathscr{S}$. To this end, a switching function $\psi(\sigma, \dot{\sigma}, \cdots, \sigma^{(r-2)})$ is designed such that if ψ and $\dot{\psi}$ are forced to zero the reduced dynamics of (7.36) is finite time convergent.

Theorem 7.1. *Let $\psi_0 = \sigma$ and further the switching functions defined as,*

$$\psi_i = \psi_{i-1} + \dot{\psi}_{i-1}^{\alpha_i}, \; i = 1, \ldots, r-2 \tag{7.37}$$

where, $r \geq 4$ and each α_i is a ratio of odd integers with $1 < \alpha_i < \frac{r-i+1}{r-i}$.

$$u = -k_1 sgn(\psi_{r-2}) - k_2 sgn(\dot{\psi}_{r-2}),\ k_1 > k_2 \tag{7.38}$$

The controller with sufficiently large gains k_1 and k_2 makes the trajectories of (7.36) *reach the r-sliding set $\mathscr{S}$ in finite time.*

Proof. The control law (7.5) first establishes a 2-sliding mode on ψ_{r-2}, i.e., the trajectory reaches the set $\{\psi_{r-2} = \dot{\psi}_{r-2} = 0\}$. This provides finite time convergence of ψ_{r-3} and $\dot{\psi}_{r-3}$. In turn the trajectory is successively transferred to the set $\mathscr{N} = \{\psi_0 = \psi_1 = \cdots = \psi_{r-2} = 0\}$. As seen earlier it is clear that if the trajectory reaches within $\mathscr{N}$ and stays in $\mathscr{N}$ then the reduced order dynamics are such that $\sigma, \dot{\sigma}, \cdots, \sigma^{(r-1)}$ converges to the origin in finite time.

The second time derivative of ψ_i can be obtained exactly as (7.14). A similar argument holds for existence of Filippov solutions and the evaluation of $\ddot{\psi}_i$ on the constraint manifolds is required which is obtained as (details are given in the appendix),

$$\begin{aligned}\ddot{\psi}_i = \prod_{j=1}^{r-2} \alpha_j \psi_{i-1}^{\alpha_i} \Big(\beta_i \dot{\psi}_{i-1}^{3-2\alpha_i} + \dot{\psi}_{i-2}^{\alpha_{i-1}-1} \Big(\beta_{i-1} \dot{\psi}_{i-2}^{4-3\alpha_{i-1}} \\ + \dot{\psi}_{i-3}^{\alpha_{i-2}-1} \Big(\beta_{i-2} \dot{\psi}_{i-3}^{5-4\alpha_{i-2}} + \ldots + u\Big)\Big)\Big)\end{aligned} \tag{7.39}$$

It is easy to see that for the right hand side of (7.39) to be well defined, the following inequalities are necessary.

$$\alpha_i < \frac{3}{2},\ \alpha_{i-1} < \frac{4}{3},\ \ldots,\ \alpha_1 < \frac{r}{r-1} \tag{7.40}$$

The (7.40) can be written collectively as

$$\alpha_i < \frac{r-i+1}{r-i} \tag{7.41}$$

As we have noted in previous sections with k_1, k_2 sufficiently large, ψ_{r-2} is attractive in finite time. Since, (7.37) define nested surfaces, ultimately the r-sliding set $\mathscr{S}$ is reached in finite time. ■

This proves to be another algorithm for stabilizing an uncertain chain of integrators in finite time. It can be noted that as this algorithm involves keeping the constraint in 2-sliding, it provides for extra accuracy compared to algorithms keeping a 1-sliding constraint. The following section considers some simulation examples to illustrate the proposed algorithm.

7.4 Simulation Examples

All simulations are carried out using MATLAB®'s ODE45 program with all tolerances set to 10^{-4}.

Example 7.1. This numerical simulation shows stabilization of an uncertain triple integrator with the proposed controller. Consider the perturbed triple integrator

$$\sigma^{(3)} = 1.5sin(2t) + (1.5 + 0.5sin\,(t) + 0.5cos\,(3t))\,u \tag{7.42}$$

The switching function $\psi(\sigma,\dot{\sigma})$ is taken as,

$$\psi(\sigma,\dot{\sigma}) = \sigma + \dot{\sigma}^{\frac{7}{5}} \tag{7.43}$$

According to the condition derived in Proposition 7.1, we have $1 < \frac{7}{5} < 1.5$ and the controller used for simulation is

$$u = -8sgn(\psi) - 6sgn(\dot{\psi}) \tag{7.44}$$

The Figure-7.3 shows that the switching function and its derivative reaches zero at about 1 unit of time. Figure-7.2 shows σ,$\dot{\sigma}$ and $\ddot{\sigma}$ reaching zero in finite time as desired.

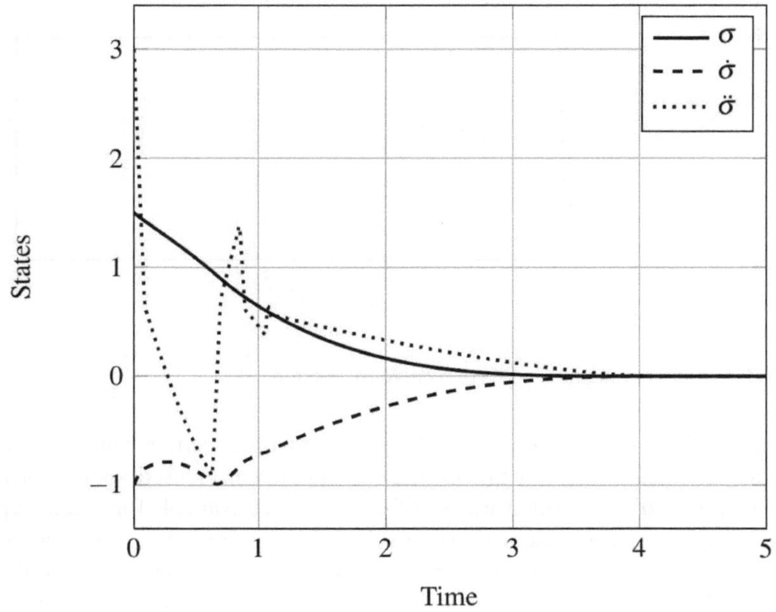

Fig. 7.2 Switching function and its derivatives in 3-sliding in Example-1

Example 7.2. This numerical simulation shows stabilization of pure fourth order integrator with the proposed control. Consider the fourth order integrator chain as $\sigma^{(4)} = u$. The switching functions are defined as

$$\psi_0 = \sigma \tag{7.45}$$

$$\psi_1 = \psi_0 + \dot{\psi}_0^{\frac{9}{7}} \tag{7.46}$$

$$\psi_2 = \psi_1 + \dot{\psi}_1^{\frac{7}{5}} \tag{7.47}$$

Since $\frac{7}{5} < \frac{3}{2}$, and $\frac{9}{7} < \frac{4}{3}$ these are proper choice of fraction powers according to the Proposition 7.2. The control law is determined as $u = -13sign(\psi_2) - 11sign(\dot{\psi}_2)$ and the initial conditions are $(-0.1\ 0.7\ 0.8\ 0.3)^T$. Figure-7.4 shows the 4-sliding trajectories of σ,$\dot{\sigma}$,$\ddot{\sigma}$, $\sigma^{(3)}$.

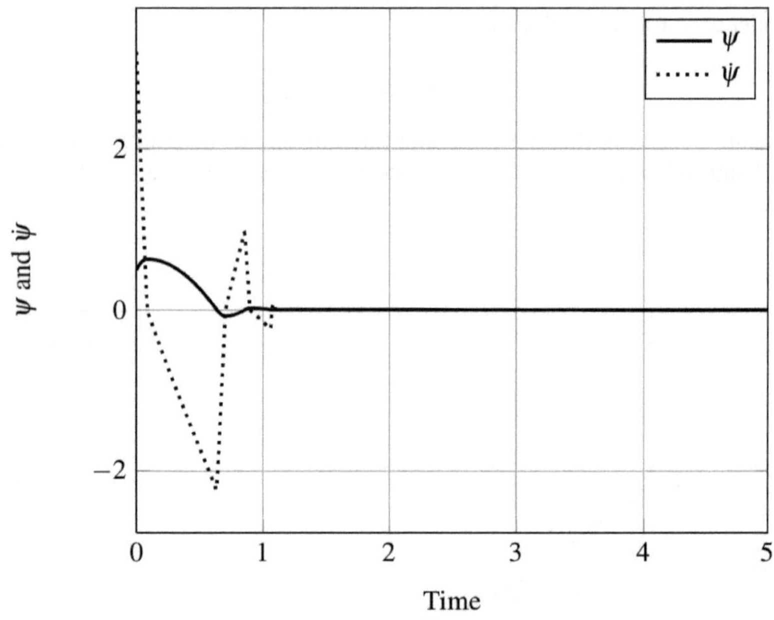

Fig. 7.3 Switching function and its derivative for Example-1

Example 7.3. To realize the proposed algorithm with uncertain functions, we consider a non linear DC motor model. DC motors have been used widely for motion control in industries. A wide variety of DC motors are available for specific applications. This example considers a series excited DC motor model. Motors are usually non linear devices but frequently linearized for control design. However, there are several applications where a linear model that ignores the friction nonlinearity is not adequate. For example, for orientation angle control of large telescopes. Since speeds are very small in these applications the friction force is dominant and cannot be ignored. One such non linear model is considered here.

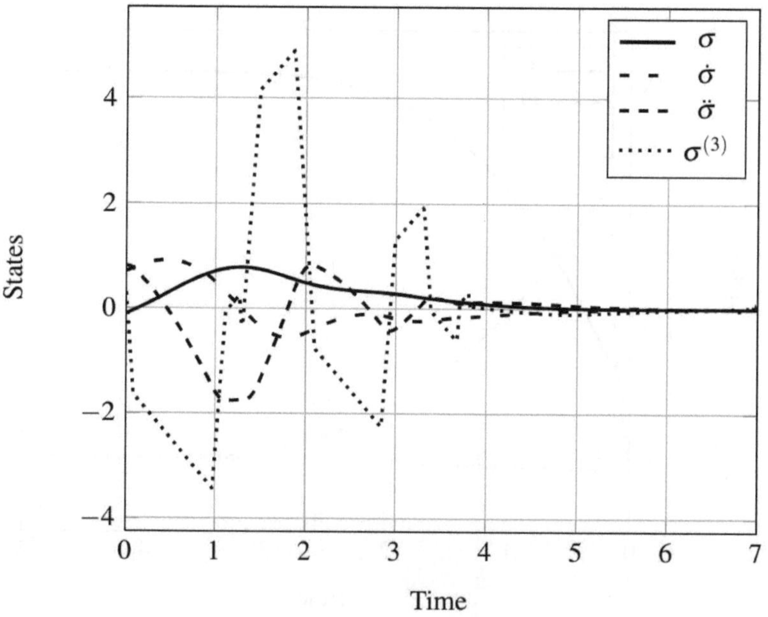

Fig. 7.4 Switching function and its derivatives in 4-sliding

The model is described by [4],

$$\begin{aligned}\dot{x}_1 &= x_2\\ \dot{x}_2 &= -m_1x_2 + m_2x_3 - m_6 sgn(x_2)\\ &\quad - m_7 e^{m_8|x_3|} sgn(x_2)\\ \dot{x}_3 &= -m_4x_2 - m_3x_3 + m_5u\end{aligned} \tag{7.48}$$

where, $x_1 = \theta$ (angular position), $x_2 = \omega$ (angular speed), $x_3 = i_a$ (motor current), $u = V_a$ (armature voltage). k_i are motor parameters given as $m_1 = 0.0110, m_2 = 16.16, m_3 = 50.66, m_4 = 1.31, m_5 = 15.8, m_6 = 19.27, m_7 = 10.11, m_8 = 0.0051$.

The problem is to design a control to track a given position i.e., $\theta_d(t)$ is given and the motor angle should track the desired angle at all times. Defining $\sigma = x_1 - x_{1d}$, it is clear that it has relative degree three with the input. Thus, the proposed method in Section 7.2 is applicable here. Please note that the *signum* function is $\dot{x}_2$ equation is just a notational aid to represent the friction force reversal when the direction of the rotation reverses. Thus, it does not affect the relative degree in any way.

The switching function is designed as $\psi = \sigma + \dot{\sigma}^{\frac{7}{5}}$ and $\dot{\psi} = \dot{\sigma} + \frac{7}{5}\dot{\sigma}^{2/5}\ddot{\sigma}$ is obtained. The gains for the control law are determined as $k_1 = 17$ and $k_2 = 13$. Figure-7.5,7.6 show the simulations with $\theta(0) = 0.35^R$ and all other initial conditions zero. The desired position is given as $\theta_d = \pi/2^R$.

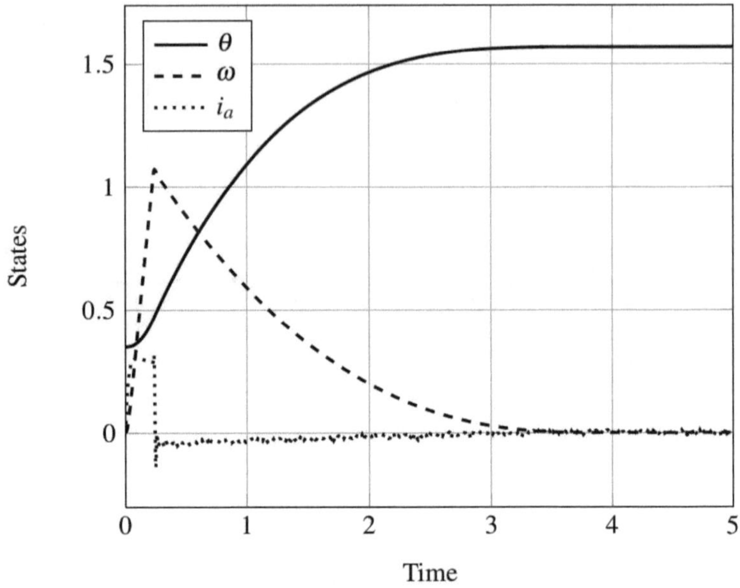

Fig. 7.5 Angle, speed and armature current with 3-sliding

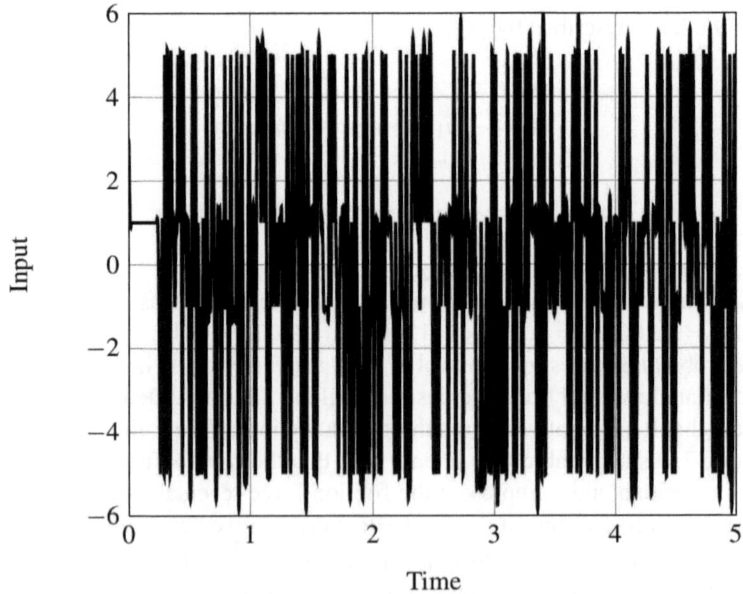

Fig. 7.6 Armature Voltage with 3-sliding

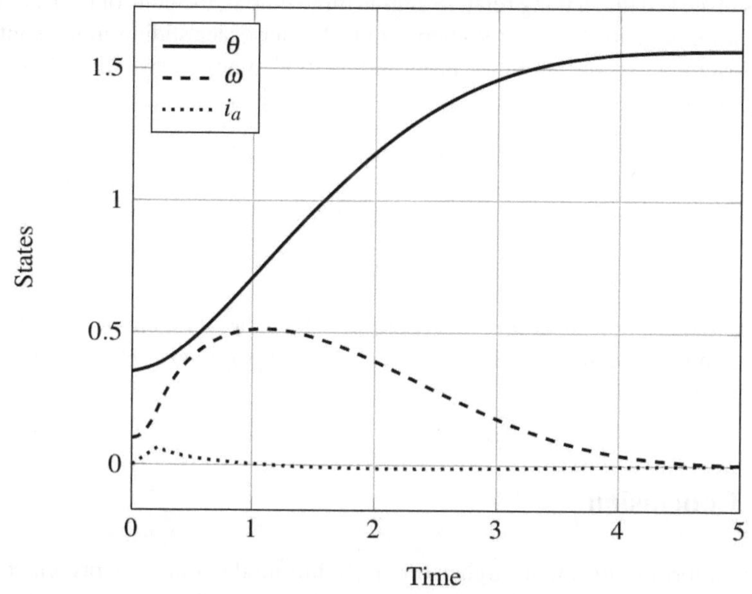

Fig. 7.7 State Trajectories with 4-sliding control

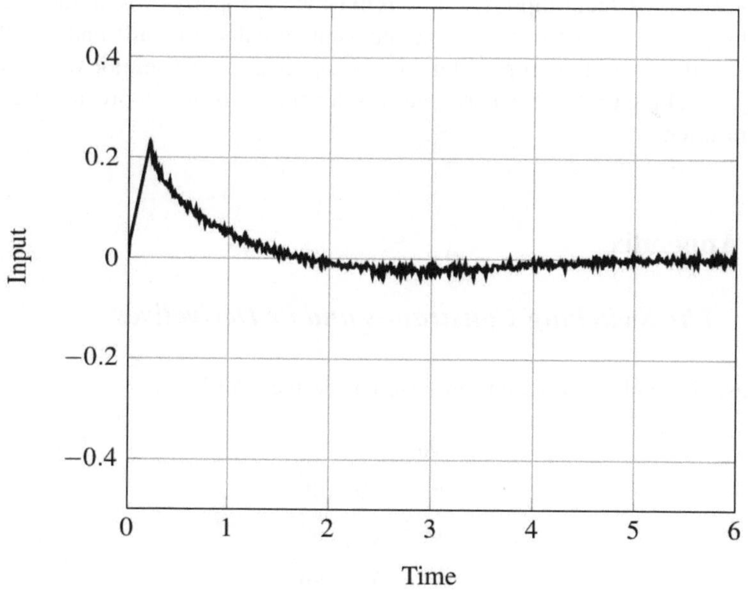

Fig. 7.8 Armature Voltage with 4-sliding control

It is observed that having relative degree three output, the control is discontinuous and if smooth control is desired then 4-th or higher order sliding mode control has to be designed. According to the proposed method, a nested switching function can be designed as

$$\psi_0 = x_1 - \theta_d \tag{7.49}$$

$$\psi_1 = \psi_0 + \dot{\psi}_0^{\frac{9}{7}} \tag{7.50}$$

$$\psi_2 = \psi_1 + \dot{\psi}_1^{\frac{7}{5}} \tag{7.51}$$

The controller is given as $u = -3sign(\psi_2) - 2sign(\dot{\psi}_2)$. Figure-7.7,7.8 shows trajectories and input with 4-sliding controller. Comparatively smooth input can be recognised.

7.5 Conclusion

A new algorithm to obtain higher order sliding mode control is presented in this paper. The proposed algorithm keeps a properly designed switching constraint in 2-sliding. The constraint design utilizes a form of non-singular terminal sliding mode switching function, and keeping it in 2-sliding is achieved using twisting controller. It has been shown that Filippov solutions exist with properly chosen fractional powers. The proposed controller is simulated numerically for third and fourth order sliding modes. Also the proposed method is applied to a DC motor with friction in an angle tracking problem, where fourth order sliding is used to avoid chattering in the controller.

7.6 Appendix

7.6.1 The Switching Constraints and its Derivatives

The nested switching functions are listed as defined in (7.37),

$$\begin{aligned}
\psi_0 &= \sigma \\
\psi_1 &= \psi_0 + \dot{\psi}_0^{\alpha_1} \\
\psi_2 &= \psi_1 + \dot{\psi}_1^{\alpha_2} \\
\psi_3 &= \psi_2 + \dot{\alpha}_2^{\alpha_3} \\
&\vdots
\end{aligned} \tag{7.52}$$

The time derivatives of the functions (7.52) can also be listed accordingly

$$\begin{aligned}
\dot{\psi}_1 &= \dot{\psi}_0 + \alpha_1 \dot{\psi}^{\alpha_1 - 1} \ddot{\psi}_0 \\
\dot{\psi}_2 &= \dot{\psi}_1 + \alpha_2 \dot{\psi}^{\alpha_2 - 1} \ddot{\psi}_1 \\
\dot{\psi}_3 &= \dot{\psi}_2 + \alpha_3 \dot{\psi}^{\alpha_3 - 1} \ddot{\psi}_2 \\
&\vdots
\end{aligned} \tag{7.53}$$

Since we are interested in the form of $\ddot{\psi}_i$, while $\psi_i = \dot{\psi}_i = 0$, it is necessary to evaluate the identities provided by (7.52) and (7.53).

$$\dot{\psi}_0 = -\psi_0^{\frac{1}{\alpha_1}} \tag{7.54}$$

$$\dot{\psi}_1 = -\psi_1^{\frac{1}{\alpha_1}} \tag{7.55}$$

$$\dot{\psi}_2 = -\psi_2^{\frac{1}{\alpha_1}} \tag{7.56}$$

$$\vdots \tag{7.57}$$

and from first time derivatives,

$$\ddot{\psi}_0 = -\frac{1}{\alpha_1} \dot{\psi}_0^{2-\alpha_1} \tag{7.58}$$

$$\ddot{\psi}_1 = -\frac{1}{\alpha_2} \dot{\psi}_1^{2-\alpha_2} \tag{7.59}$$

$$\ddot{\psi}_2 = -\frac{1}{\alpha_3} \dot{\psi}_2^{2-\alpha_3} \tag{7.60}$$

$$\vdots \tag{7.61}$$

Now, if desired sliding order is $r = 3$ then using (7.54) and (7.58),

$$\ddot{\psi}_1 = \alpha_1 \psi_0^{\alpha_1 - 1} \left(\frac{\alpha_1 - 2}{\alpha_1^2} \psi_0^{3 - 2\alpha_1} + \psi_0^{(3)} \right) \tag{7.62}$$

$$= \alpha_1 \psi_0^{\alpha_1 - 1} \left(\beta_1 \psi_0^{3 - 2\alpha_1} + \psi_0^{(3)} \right) \tag{7.63}$$

For $r = 4$, using (7.55) and (7.59)

$$\begin{aligned}
\ddot{\psi}_2 = \alpha_1 \alpha_2 \dot{\psi}_1^{\alpha_2 - 1} \Bigg(& \frac{\alpha_2 - 2}{\alpha_1 \alpha_2^2} \dot{\psi}_1^{3 - 2\alpha_2} \\
& + \dot{\psi}_0^{\alpha_1 - 1} \left(\left(\frac{(\alpha_1 - 2)^2}{\alpha_1^2} - \alpha_1 + 1 \right) \psi_0^{4 - 3\alpha_1} + \psi_0^{(4)} \right) \Bigg)
\end{aligned} \tag{7.64}$$

Denoting real number terms of α_1 and α_2 by β_1 and β_2 respectively

$$\ddot{\psi}_2 = \alpha_1 \alpha_2 \dot{\psi}_1^{\alpha_2 - 1} \left(\beta_2 \dot{\psi}_1^{3-2\alpha_2} + \dot{\psi}_0^{\alpha_1 - 1} \left(\beta_1 \dot{\psi}_0^{4-3\alpha_1} + \psi_0^{(4)} \right) \right) \tag{7.65}$$

For $r = 5$, using (7.56) and (7.60),

$$\ddot{\psi}_3 = \alpha_1 \alpha_2 \alpha_3 \dot{\psi}_2^{\alpha_3 - 1} \left(\beta_3 \dot{\psi}_2^{3-2\alpha_3} + \dot{\psi}_1^{\alpha_2 - 1} \left(\beta_2 \dot{\psi}_1^{4-3\alpha_2} + \dot{\psi}_0^{\alpha_1 - 1} \left(\beta_1 \dot{\psi}^{5-4\alpha_1} + \psi_0^{(5)} \right) \right) \right) \tag{7.66}$$

Thus, this recursion leads to the rth-order equation as,

$$\begin{aligned} \ddot{\psi}_i = \prod_{j=1}^{r-2} \alpha_j \dot{\psi}_{i-1}^{\alpha_i} \Big(\beta_i \dot{\psi}_{i-1}^{3-2\alpha_i} + \dot{\psi}_{i-2}^{\alpha_{i-1}-1} \Big(\beta_{i-1} \dot{\psi}_{i-2}^{4-3\alpha_{i-1}} \\ + \dot{\psi}_{i-3}^{\alpha_{i-2}-1} \Big(\beta_{i-2} \dot{\psi}_{i-3}^{5-4\alpha_{i-2}} + \ldots + u \Big) \Big) \Big) \end{aligned} \tag{7.67}$$

References

1. Bhat, S.P., Bernstein, D.S.: Finite-Time Stability of Homogeneous Systems. In: Proceedings of American Control Conference, pp. 2513–2514 (1997)
2. Bhat, S.P., Bernstein, D.S.: Continuous, Bounded, Finite-Time Stabilization of the Translational and Rotational Double Integrators. IEEE Transactions on Automatic Control 43(5), 678–682 (1998)
3. Bhat, S.P., Bernstein, D.S.: Geometric homogeneity with applications to finite-time stability. Mathematics of Control, Signals, and Systems 17(2), 101–127 (2005)
4. Cong, S., Li, G., Feng, X.: Parameters Identification of Nonlinear DC Motor Model Using Compound Evolution Algorithms. In: Proceedings of World Congress on Engineering, London (June 2010)
5. Feng, Y., Han, X., Wang, Y., Yu, X.: Second-order terminal sliding mode control of uncertain multivariable systems. International Journal of Control 80(6), 856–862 (2007)
6. Feng, Y., Yu, X., Man, Z.: Non-singular terminal sliding mode control of rigid manipulators. Automatica 38(12), 2159–2167 (2002)
7. Gulati, S., Venkataraman, S.T.: Control of nonlinear systems using terminal sliding modes. In: American Control Conferece, pp. 891–893 (1992)
8. Haimo, V.T.: Finite time controllers. SIAM Journal on Control and Optimization 24, 760–770 (1986)
9. Hong, Y.: Finite-time stabilization and stabilizability of a class of controllable systems. Systems & Control Letters 46(4), 231–236 (2002)
10. Kawski, M.: Geometric Homogeneity and Stabilization. In: Proc. IFAC Nonlinear Control Symposium, Lake Tahoe, CA, pp. 164–169 (1995)
11. Levant, A.: Sliding order and sliding accuracy in sliding mode control. International Journal of Control 58, 1247–1263 (1993)
12. Levant, A.: Universal single-input-single-output (SISO) sliding-mode controllers with finite-time convergence. IEEE Transactions on Automatic Control 46(9), 1447–1451 (2001)

13. Levant, A.: Construction Principles of Output Feedback 2-Sliding Mode Design. In: 41st IEEE Conference on Decision and Control, pp. 317–322. IEEE (2002)
14. Levant, A.: Homogeneity approach to high-order sliding mode design. Automatica 41(5), 823–830 (2005)
15. Mouley, E., Perruquetti, W.: Finite-Time Stability and Stabilization:State of the art. Springer (2006)
16. Plestan, F., Glumineau, A., Laghrouche, S.: A new algorithm for high-order sliding mode control. International Journal of Robust and Nonlinear Control 18, 441–453 (2008)
17. Yu, X., Man, Z.: Multi-input uncertain linear systems with terminal sliding-mode control. Automatica 34(3), 389–392 (1998)
18. Zhihong, M., Yu, X.H.: Terminal sliding mode control of MIMO linear systems. IEEE Transactions on Circuits and Systems I: Fundamental Theory and Applications 44, 1065–1070 (1997)

Chapter 8
Applying Sliding Mode Technique to Filter and Controller Design for Nonlinear Polynomial Stochastic Systems*

Michael Basin and Pablo Rodriguez-Ramirez

Abstract. This chapter addresses the mean-square and mean-module filtering problems for stochastic polynomial systems with Gaussian white noises. The obtained solution contains a sliding mode term, signum of the innovations process. It is shown that the designed sliding mode mean-square filter generates the mean-square estimate, which has the same minimum estimation error variance as the best estimate given by the conventional mean-square polynomial filter, although the gain matrices of both filters are different. The designed sliding mode mean-module filter generates the mean-module estimate, which yields a better value of the mean-module criterion in comparison to the conventional polynomial mean-square filter. The theoretical results are complemented with illustrative examples verifying performance of the designed filters. It is demonstrated that the estimates produced by the designed sliding mode mean-square filter and the conventional polynomial mean-square filter yield the same estimation error variance, and there is an advantage in favor of the designed sliding mode mean-module filter. The chapter then presents the solution to the optimal controller problems for a polynomial system over linear observations with respect to a Bolza-Meyer criterion, where the integral control and state energy terms are quadratic and the non-integral term is of the first degree. The simulation results confirm an advantage in favor of the designed sliding mode controller.

8.1 Introduction

Since the sliding mode control was invented in the beginning of 1970s (see a historical review in [[1–3]]), it has been applied to solve several classes of problems. For

Michael Basin · Pablo Rodriguez-Ramirez
Department of Physical and Mathematical Sciences,
Autonomous University of Nuevo Leon, San Nicolas de los Garza, Nuevo Leon, Mexico
e-mail: mbasin@fcfm.uanl.mx, mbasin2007@gmail.com,
pablo.rodriguezrm@uanl.edu.mx

* The author thanks the Mexican National Science and Technology Council (CONACyT) for financial support under Grants 55584, 129081 and joint Mexico-EU FONCICyT Grant 93302.

B. Bandyopadhyay et al. (Eds.): *Advances in Sliding Mode Control*, LNCIS 440, pp. 165–185.
DOI: 10.1007/978-3-642-36986-5_8

instance, the sliding mode control methodology has been used in stabilization ⟦4,5⟧, tracking ⟦6, 7⟧, observer design ⟦8, 9⟧, frequency domain analysis ⟦10⟧, and other control problems. Promising modifications of the original sliding mode concept, such as integral sliding mode ⟦11⟧ and higher order sliding modes ⟦3, 12⟧, have been developed. The sliding mode optimal regulator has been recently designed for a linear system with a non-quadratic Bolza-Meyer criterion ⟦13⟧. Application of the sliding mode method is extended even to stochastic systems ⟦14–16⟧ and stochastic filtering problems ⟦17, 18⟧. However, to the best of the authors' knowledge, no sliding mode filtering algorithms solving the mean-square filtering problem have been designed for stochastic polynomial systems. The mean-square and mean-module filtering problems, as well as the corresponding mean-square controller problem, are considered in this chapter.

This chapter presents the solutions to the mean-square and mean-module filtering problems for stochastic polynomial systems, which contain a sliding mode term, signum of the innovations process. It is shown that the designed sliding mode mean-square filter generates the mean-square estimate, which has the same minimum estimation error variance as the best estimate given by the the mean-square polynomial filter ⟦19⟧, although the gain matrices of both filters are different. This is the first designed sliding mode filter that is optimal with respect to the mean-square criterion and yields an estimate with the same structural properties as the conventional mean-square optimal filter. On the other hand, the designed sliding mode filter generates the mean-module estimate, which yields a better value of the mean-module criterion in comparison to the mean-square polynomial filter ⟦19⟧. Again, this is the first designed sliding mode filter that is optimal with respect to the mean-module criterion. The theoretical result is complemented with an illustrative example verifying the performance of the designed filters. It is demonstrated that the estimates produced by the designed sliding mean-square filter and conventional mean-square optimal filter yield the estimate with the same minimum estimation error variance, whereas there is an advantage in favor of the designed sliding mode mean-module filter.

This chapter then presents the solutions to the optimal controller problem for a polynomial system over linear observations with respect to a Bolza-Meyer criterion, where the integral control and state energy terms are quadratic and the non-integral term is of the first degree. The simulation results confirm an advantage in favor of the designed sliding mode controller.

The chapter is organized as follows. Section 2 states the mean-square and mean-module filtering problems for stochastic polynomial systems with Gaussian white noises. The sliding mode solutions to the mean-square and mean-module filtering problems are given in Sections 3 and 4, respectively, which also contain illustrative examples. Section 5 addresses the controller design problem for a polynomial system over linear observations with respect to the Bolza-Meyer criteria with first degree terms.

8.2 Filtering Problem Statement

Let (Ω, F, P) be a complete probability space with an increasing right-continuous family of σ-algebras $F_t, t \geq t_0$, and let $(W_1(t), F_t, t \geq t_0)$ and $(W_2(t), F_t, t \geq t_0)$ be independent standard Wiener processes. The F_t-measurable random process $(x(t), y(t))$ is described by a polynomial differential equation for the system state

$$dx(t) = f(x,t)dt + b(t)dW_1(t), \quad x(t_0) = x_0, \tag{8.1}$$

and a linear differential equation for the observation process

$$dY(t) = A(t)x(t)dt + B(t)dW_2(t). \tag{8.2}$$

Here, $x(t) \in R^n$ is the state vector and $y(t) \in R^m$, $m \leq n$, is the observation process. The initial condition $x_0 \in R^n$ is a Gaussian vector such that x_0, $W_1(t)$, and $W_2(t)$ are independent. It is assumed that $B(t)B^T(t)$ is a positive definite matrix. All coefficients in (8.1),(8.2) are deterministic functions of time of appropriate dimensions.

The nonlinear function $f(x,t)$ is considered polynomial of n variables, components of the state vector $x(t) \in R^n$, with time-dependent coefficients. Since $x(t) \in R^n$ is a vector, this requires a special definition of the polynomial for $n > 1$. A p-degree polynomial of a vector $x(t) \in R^n$ is regarded as a p-linear form of n components of $x(t)$

$$f(x,t) = a_0(t) + a_1(t)x(t) + a_2(t)xx^T + \ldots + a_p(t)x_{\ldots p \; times \ldots}x,$$

where $a_0(t)$ is a vector of dimension n, a_1 is a matrix of dimension $n \times n$, a_2 is a 3D tensor of dimension $n \times n \times n$, a_p is a $(p+1)$D tensor of dimension $n \times \ldots_{(p+1) \; times} \ldots \times n$, and $x \times \ldots_{p \; times} \ldots \times x$ is a pD tensor of dimension $n \times \ldots_{p \; times} \ldots \times n$ obtained by p times spatial multiplication of the vector $x(t)$ by itself. Such a polynomial can also be expressed in the summation form

$$f_k(x,t) = a_{0\,k}(t) + \sum_i a_{1\,ki}(t)x_i(t) + \sum_{ij} a_{2\,kij}(t)x_i(t)x_j(t)$$

$$+ \ldots + \sum_{i_1 \ldots i_p} a_{p\,ki_1 \ldots i_p}(t)x_{i_1}(t) \ldots x_{i_p}(t), \quad k, i, j, i_1 \ldots i_p = 1, \ldots, n.$$

The state and observation equations can also be written in an alternative form

$$\dot{x}(t) = f(x,t) + b(t)\psi_1(t), \quad x(t_0) = x_0, \tag{8.1*}$$

$$y(t) = A(t)x(t) + B(t)\psi_2(t), \tag{8.2*}$$

where $y(t) = \dot{Y}(t)$, and $\psi_1(t)$ and $\psi_2(t)$ are white Gaussian noises, which are the weak mean square derivatives of standard Wiener process $W_1(t)$, and $W_2(t)$ (see [[20]]). The representations (8.1),(8.2) and (8.1*),(8.2*) are equivalent ([[21]]). The equations (1*),(2*) present the conventional form for the equations (8.1),(8.2), which is actually used in practice.

The estimation problem is to find the estimate $\hat{x}(t)$ of the system state $x(t)$, based on the observation process $Y(t) = \{y(s), t_0 \leq s \leq t\}$, that minimizes the mean-square norm

$$J = E[(x(t) - \hat{x}(t))^T (x(t) - \hat{x}(t)) \mid F_t^Y] \tag{8.3}$$

at every time moment t. Here, $E[z(t) \mid F_t^Y]$ means the conditional expectation of a stochastic process $z(t) = (x(t) - \hat{x}(t))^T (x(t) - \hat{x}(t))$ with respect to the σ - algebra F_t^Y generated by the observation process $Y(t)$ in the interval $[t_0, t]$. As known [20], this estimate is given by the conditional expectation

$$\hat{x}(t) = m(t) = E(x(t) \mid F_t^Y)$$

of the system state $x(t)$ with respect to the σ - algebra F_t^Y generated by the observation process $Y(t)$ in the interval $[t_0, t]$. As usual, the matrix function

$$P(t) = E[(x(t) - m(t))(x(t) - m(t))^T \mid F_t^Y]$$

is the estimation error variance.

The solution to this filtering problem for stochastic polynomial systems is given by the mean-square polynomial filter [19] generalizing the optimal Kalman-Bucy filter [22] for linear systems. An alternative solution involving the sliding mode term is given in the next section and proved in Subsection 3.2. As demonstrated, the obtained sliding mode filter is optimal with respect to the criterion (8.3).

This chapter also addresses the mean-module filtering problem to find the estimate $\hat{x}(t)$ of the system state $x(t)$, based on the observation process $Y(t) = \{y(s), t_0 \leq s \leq t\}$, that minimizes the mean-module norm

$$J = E[(\mid x(t) - \hat{x}(t) \mid) \mid F_t^Y] \tag{8.4}$$

at every time moment t. Here, $\mid x \mid = [\mid x_1 \mid, \ldots, \mid x_n \mid] \in R^n$ is defined as the vector of absolute values of the components of the vector $x \in R^n$

The solution to the stated filtering problem, involving the sliding mode term, is given in Section 4 and proved in Subsection 4.2. As demonstrated, the obtained sliding mode filter is optimal with respect to the criterion (8.4).

8.3 Sliding Mode Mean-Square Filter Design

The solution to the mean-square filtering problem for the polynomial stochastic system (8.1) and the criterion (8.3) is given as follows. The mean-square estimate satisfies the differential equation with the sliding mode term

$$\dot{m}(t) = E(f(x,t)|F_t^Y) + Q(t)A^T(t)(B(t)B^T(t))^{-1}\times \tag{8.5}$$

$$A(t)Sign[A^T(t)(A(t)A^T(t))^{-1}y(t) - m(t)].$$

with the initial condition $m(t_0) = E(x(t_0) \mid F_{t_0}^Y)$.

Here, the Signum function of a vector $x = [x_1, \ldots, x_n] \in R^n$ is defined as $Sign[x] = [sign(x_1), \ldots, sign(x_n)] \in R^n$, and the signum function of a scalar x is defined as $sign(x) = 1$, if $x > 0$, $sign(x) = 0$, if $x = 0$, and $sign(x) = -1$, if $x < 0$ ([23]).

The matrix function $Q(t)$ satisfies the matrix equation with time-varying coefficients

$$\dot{Q}(t) = E(f(x,t)(x(t) - m(t))^T | F_t^Y) + (b(t)b^T(t)) * | A^T(t)(A(t)A^T(t))^{-1}y(t) - m(t) |, \tag{8.6}$$

with the initial condition

$$Q(t_0) = E[(x(t_0) - m(t_0)(x(t_0) - m(t_0)^T \mid F_{t_0}^Y] * | A^T(t_0)(A(t_0)A^T(t_0))^{-1}y(t_0) - m(t_0) | .$$

Here, $| x | = [| x_1 |, \ldots, | x_n |] \in R^n$ is defined as the vector of absolute values of the components of the vector $x \in R^n$, and $A * b$ denotes a product between a matrix $A \in R^{n \times n}$ and a vector $b \in R^n$, that results in the matrix defined as follows: all entries of the j-th column of the matrix A are multiplied by the j-th component of the vector b, $j = 1, \ldots, n$.

Note that the equations (8.5) and (8.6) do not form a closed system of equations due to the presence of polynomial terms depending on x, $E(f(x,t) \mid F_t^Y)$, and $E(f(x,t)(x(t) - m(t))^T) \mid F_t^Y)$, which are not expressed yet as functions of the filter variables, $m(t)$ and $Q(t)$ (or $P(t)$). However, as shown in [24], the closed system of the mean-square filtering equations can be obtained for any polynomial state (8.1) over linear observations (8.2). In the next section, the sliding mode mean-square filter is obtained in a particular case of a third-order polynomial state. Consider the case where $f(x,t)$ is,

$$f(x,t) = a_0 + a_1(t)x + a_2(t)xx^T + a_3(t)xxx^T \tag{8.7}$$

be a third-order polynomial, where x is an n-dimensional vector, $a_0(t)$ is a vector of dimension n, a_1 is a matrix of dimension $n \times n$, a_2 is a 3D tensor of dimension $n \times n \times n$, a_3 is a 4D tensor of dimension $n \times n \times n \times n$. In this case, the following filtering equations for the optimal estimate $m(t)$ and the filter gain matrix $Q(t)$ are obtained

$$\dot{m}(t) = a_0 + a_1 m(t) + a_2 m(t)m(t)^T + a_2 P(t) + 3a_3 P(t)m(t) + a_3 m(t)m(t)m(t)^T + Q(t)A^T(t)(B(t)B^T(t))^{-1}A(t)Sign[A^T(t)(A(t)A^T(t))^{-1}y(t) - m(t)], \tag{8.8}$$

with the initial condition $m(t_0) = E(x(t_0) \mid F_{t_0}^Y)$,

$$\dot{Q}(t) = a_1 Q(t) + 2a_2 m(t)Q(t) + (a_2 m(t)Q(t))^T \tag{8.9}$$
$$+3(a_3(P(t)Q(t) + m(t)m(t)^T Q(t)))$$
$$+(a_3(3P(t)Q(t) + 2m(t)m(t)^T Q(t)))^T$$
$$+(b(t)b^T(t)) * \mid A^T(t)(A(t)A^T(t))^{-1}y(t) - m(t) \mid,$$

with the initial condition

$$Q(t_0) = E[(x(t_0) - m(t_0)(x(t_0) - m(t_0)^T \mid F_{t_0}^Y] * \mid A^T(t_0)(A(t_0)A^T(t_0))^{-1}y(t_0) - m(t_0) \mid,$$

$$\dot{P}(t) = a(t)P(t) + P(t)a^T(t) + 2a_2(t)m(t)P(t) \tag{8.10}$$
$$+2(a_2(t)m(t)P(t))^T + 3(a_3(P(t)P(t) + m(t)m^T(t)P(t)))$$
$$+3(a_3(P(t)P(t) + m(t)m^T(t)P(t)))^T + b(t)b^T(t)$$
$$-P(t)A^T(t)(B(t)B^T(t))^{-1}A(t)P(t),$$

with the initial condition

$$P(t_0) = E[(x(t_0) - m(t_0)(x(t_0) - m(t_0)^T \mid F_{t_0}^Y].$$

Consequently, the main result is formulated in the following theorem.

Theorem 1. The mean-square filter for the third degree polynomial stochastic system state (8.7) over the linear observations (8.2) is given by the equation (8.8) for the estimate $m(t) = E(x(t) \mid F_t^Y)$ and the equations (8.9) and (8.10) for the filter gain matrix $Q(t)$ and the error variance matrix $P(t)$.

8.3.1 Example 1

This section presents an illustrative example of designing the mean-square sliding mode filter for a second degree polynomial state (8.7) over linear observations (8.2), using the filtering equations (8.5)–(8.10).

Consider a scalar polynomial unmeasured state

$$\dot{x}(t) = 0.1x^2(t) + \psi_1(t), \quad x(0) = x_0, \tag{8.11}$$

and the scalar linear observation process

$$y(t) = x(t) + \psi_2(t), \tag{8.12}$$

where $\psi_1(t)$ and $\psi_2(t)$ are white Gaussian noises, which are the weak mean square derivatives of standard Wiener processes (see [20]). The equations (8.11),(8.12) correspond to the alternative conventional form (1*),(2*) for the equations (8.1),(8.2).

The filtering problem is to find the mean-square estimate for the second degree polynomial state (8.11), using linear observations (8.12) confused with independent and identically distributed disturbances modeled as white Gaussian noises.

The filtering equations (8.8),(8.9) and (8.10) take the following particular form for the system (8.11),(8.12)

$$\dot{m}(t) = 0.1m^2(t) + 0.1P(t) + Q(t)sign[y(t) - m(t)], \tag{8.13}$$

with the initial condition $m(0) = E(x(0) \mid y(0)) = m_0$,

$$\dot{Q}(t) = 0.3Q(t) + \mid y(t) - m(t) \mid, \tag{8.14}$$

with the initial condition $Q(0) = E((x(0) - m(0))(x(0) - m(0))^T \mid y(0)) * \mid y(0) - m(0) \mid$.

$$\dot{P}(t) = 1 + 0.4m(t)P(t) - P^2(t), \tag{8.15}$$

with the initial condition $P(0) = E((x(0) - m(0))(x(0) - m(0))^T \mid y(0))$. The estimates obtained upon solving the equations (8.13)–(8.15) are also compared to the estimates satisfying the mean-square filtering equations [19] for the second degree polynomial system (8.11),(8.12)

$$\dot{m}_K(t) = 0.1m_K^2(t) + 0.1P(t) + P(t)[y(t) - m_K(t)], \tag{8.16}$$

with the initial condition $m(0) = E(x(0) \mid y(0)) = m_0$,

$$\dot{P}(t) = 1 + 0.4m(t)P(t) - P^2(t), \tag{8.17}$$

with the initial condition $P(0) = E((x(0) - m(0))(x(0) - m(0))^T \mid y(0))$.

Numerical simulation results are obtained solving the systems of filtering equations (8.13)–(8.15) and (8.16),(8.17). The obtained values of the estimates $m(t)$ and $m_K(t)$ satisfying the equations (8.13) and (8.16), respectively, are compared to the real values of the state variables $x(t)$ in (8.11).

For each of the two filters (8.13)–(8.15) and (8.16),(8.17) and the reference system (8.11),(8.12), involved in simulation, the following initial values are assigned: $x_0 = 1$, $m_0 = 10$, $P(0) = 100$, $Q(0) = 943.74$. The filtering horizon is set to $T = 0.4$. Gaussian disturbances $\psi_1(t)$ and $\psi_2(t)$ in (8.7),(8.8) are realized using the built-in MATLAB white noise function.

The following graphs are obtained: graphs of the reference state $x(t)$, satisfying the equation (8.11), the mean-square sliding mode filter estimate $m(t)$, satisfying the equations (8.13), and the mean-square polynomial filter estimate $m_K(t)$, satisfying the equation (8.16), are shown in the entire simulation interval $[0, 0.4]$ in Fig. 8.1. In addition, the graph of the mean-square sliding mode filter estimate $m(t)$ averaged by a Butterworth filter and all the variables of Fig. 8.1 are shown in detail in the interval $[0.2, 0.4]$ in Fig. 8.2.

It can be observed that the estimates given by both filters generate the same minimum estimation error variance, although the gain matrices $Q(t)$ and $P(t)$ are different.

8.4 Sliding Mode Mean-Module Filter Design

The solution to the mean-module filtering problem for the polynomial system (8.1) and the criterion (8.4) is given as follows. The mean-module estimate satisfies the differential equation with the sliding mode term

$$\dot{m}(t) = E(f(x,t) \mid F_t^Y)dt + Q(t)A^T(t)(B(t)B^T(t))^{-1} \times \qquad (8.19)$$
$$A(t)Sign[A^T(t)(A(t)A^T(t))^{-1}y(t) - m(t)].$$

with the initial condition $m(t_0) = E(x(t_0) \mid F_{t_0}^Y)$.

The matrix function $Q(t)$ satisfies the matrix equation with time-varying coefficients

$$\dot{Q}(t) = b(t)b^T(t) + E(f(x,t)(x(t) - m(t))^T) \mid F_t^Y), \qquad (8.20)$$

with the initial condition
$Q(t_0) = E[(x(t_0) - m(t_0))(Sign(A^T(t_0)(A(t_0)A^T(t_0))^{-1}A(t_0)x(t_0) - m(t_0)))^T \mid F_{t_0}^Y]$.

Note that the equations (8.4) and (8.5) do not form a closed system of equations due to the presence of polynomial terms depending on x, $E(f(x,t) \mid F_t^Y)$, and $E(f(x,t)(x(t) - m(t))^T) \mid F_t^Y)$, which are not expressed yet as functions of the filter variables, $m(t)$ and $Q(t)$ (or $P(t)$). Similarly to the mean-square case [24], the closed system of the filtering equations can be obtained for any polynomial state (8.1) over linear observations (8.2), using the technique of representing of superior moments of the conditionally Gaussian random variable $x(t) - m(t)$ as functions of only two of its lower conditional moments, $m(t)$ and $P(t)$ (see [24] for more details of this technique). Apparently, the polynomial dependence of $f(x,t)$ and $f(x,t)(x(t) - m(t))^T$ on x is the key point making this representation possible.

Next, a closed form of the filtering equations is obtained from (8.19) and (20) for a third-order function $f(x,t)$ in the equation (1), as follows. It should be noted, however, that application of the same procedure would result in designing a closed system of the filtering equations for any polynomial function $f(x,t)$ in (8.1).

Let the function

$$f(x,t) = a_0(t) + a_1(t)x + a_2(t)xx^T + a_3(t)xxx^T \qquad (8.21)$$

be a third-order polynomial, where x is an n-dimensional vector, $a_0(t)$ is an n-dimensional vector, $a_1(t)$ is a $n \times n$-dimensional matrix, $a_2(t)$ is a 3D tensor of dimension $n \times n \times n$, $a_3(t)$ is a 4D tensor of dimension $n \times n \times n \times n$. In this case, the following filtering equations for the optimal estimate $m(t)$ and the filter gain matrix $Q(t)$ are obtained

$$\dot{m}(t) = a_0(t) + a_1(t)m(t) + a_2(t)m(t)m^T(t) + \tag{8.22}$$
$$a_2(t)Q(t) * \mid A^T(t)(A(t)A^T(t))^{-1}y(t) - m(t) \mid +$$
$$3a_3(t)m(t)Q(t) * \mid A^T(t)(A(t)A^T(t))^{-1}y(t) - m(t) \mid +$$
$$a_3(t)m(t)m(t)m^T(t) + Q(t)A^T(t)(B(t)B^T(t))^{-1} \times$$
$$A(t)Sign[A^T(t)(A(t)A^T(t))^{-1}y(t) - m(t)],$$
$$m(t_0) = E(x(t_0) \mid F_t^Y)),$$
$$\dot{Q}(t) = a_1(t)Q(t) + 2a_2(t)m(t)Q(t) + \tag{8.23}$$
$$a_3(t)[Q(t)Q(t) * \mid A^T(t)(A(t)A^T(t))^{-1}y(t) - m(t) \mid +$$
$$3m(t)m^T(t)Q(t)]) + b(t)b^T(t),$$
$$Q(t_0) = E[(x(t_0) - m(t_0))(Sign(A^T(t_0)(A(t_0)A^T(t_0))^{-1}A(t_0)x(t_0) - m(t_0)))^T \mid F_{t_0}^Y].$$

Consequently, this result is formulated in the following theorem.

Theorem 2. The mean-module filter for the third degree polynomial system state (8.21) over the linear observations (8.2) is given by the equation (8.22) for the estimate $m(t)$ and the equation (8.23) for the filter gain matrix $Q(t)$.

8.4.1 Example 2

This section presents an illustrative example of designing the mean-module sliding mode filter for a second degree polynomial state (8.21) over linear observations (8.2), using the filtering equations (8.22),(8.23).

Consider a scalar linear unmeasured state

$$\dot{x}(t) = 0.1x^2(t) + \psi_1(t), \quad x(0) = x_0, \tag{8.24}$$

and the scalar linear observation process

$$y(t) = x(t) + \psi_2(t), \tag{8.25}$$

where $\psi_1(t)$ and $\psi_2(t)$ are white Gaussian noises.

The filtering problem is to find the mean-module estimate for the second degree polynomial state (8.24), using linear observations (8.25) confused with independent and identically distributed disturbances modeled as white Gaussian noises.

The filtering equations (8.22),(8.23) take the following particular form for the system (8.24),(8.25)

$$\dot{m}(t) = 0.1m^2(t) + \tag{8.26}$$
$$0.1Q(t) \mid y(t) - m(t) \mid + Q(t)sign[y(t) - m(t)],$$

with the initial condition $m(0) = E(x(0) \mid y(0)) = m_0$,

$$\dot{Q}(t) = 0.2m(t)Q(t) + 1, \tag{8.27}$$

with the initial condition $Q(0) = E((x(0) - m(0))(Sign(x(0) - m(0)))^T \mid y(0))$.

The estimates obtained upon solving the equations (8.26),(8.27) are also compared to the estimates satisfying the mean-square filtering equations [19] for the second degree polynomial system (8.24),(8.25)

$$\dot{m}_P(t) = 0.1m_P^2(t) + 0.1P(t) + P(t)[y(t) - m_P(t)], \tag{8.28}$$

with the initial condition $m(0) = E(x(0) \mid y(0)) = m_0$,

$$\dot{P}(t) = 1 + 0.4m_P(t)P(t) - P^2(t), \tag{8.29}$$

with the initial condition $P(0) = E((x(0) - m(0))(x(0) - m(0))^T \mid y(0))$.

Numerical simulation results are obtained solving the systems of filtering equations (8.26),(8.27) and (8.28),(8.29). The obtained values of the estimates $m(t)$ and $m_P(t)$ satisfying the equations (8.26) and (8.28), respectively, are compared to the real values of the state variables $x(t)$ in (8.24).

For each of the two filters (8.26),(8.27) and (8.28),(8.29) and the reference system (8.24),(8.25), involved in simulation, the following initial values are assigned: $x_0 = 1$, $m_0 = 4$, $P(0) = Q(0) = 100$. The filtering horizon is set to $T = 0.4$. Gaussian disturbances $\psi_1(t)$ and $\psi_2(t)$ in (8.24),(8.25) are realized using the built-in MatLab white noise function.

Note that the initial conditions $P(0)$ and $Q(0)$ are assigned equal for simulation purposes, since the results should be compared with respect to the mean-module criterion (8.4). If the initial value for Q is assigned as $Q(0) = 10$, the mean-square

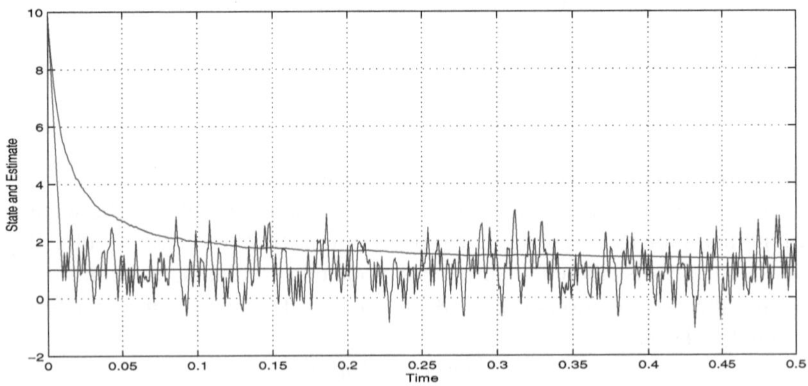

Fig. 8.1 Graphs of the unmeasured state (11) $x(t)$ (blue), the mean-square sliding mode estimate (13) $m(t)$ (green), and the mean-square polynomial filter estimate (16) $m_K(t)$ (red) in the interval $[0, 0.4]$.

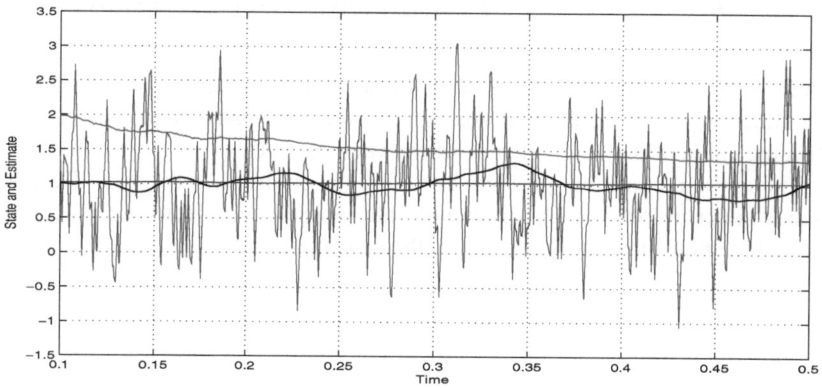

Fig. 8.2 Graphs of the unmeasured state (11) $x(t)$ (blue), the mean-square sliding mode estimate (13) $m(t)$ (green), the mean-square sliding mode estimate (13) $m(t)$ averaged by a Butterworth filter (light blue), and the mean-square polynomial estimate (16) $m_K(t)$ (red) in the interval $[0.1, 0.4]$.

polynomial filter of [19] would yield a better result as the mean-square polynomial filter.

The following graphs are obtained: graphs of the reference state $x(t)$, satisfying the equation (8.24), the mean-module sliding mode filter estimate $m(t)$, satisfying the equations (8.26), and the mean-square polynomial filter estimate $m_P(t)$, satisfying the equation (8.28), are shown in the entire simulation interval $[0, 0.4]$ in Fig. 8.1.

It can be observed that the mean-module sliding mode filter (8.26),(8.27) yields a certainly better value of the mean-module criterion (8.4) in comparison to the mean-square polynomial filter (8.28),(8.29).

Note that the comparison of the designed mean-module sliding mode filter (8.26),(8.27) to the mean-square polynomial filter (8.28),(8.29) with respect to the criterion (8.4) is conducted for illustration purposes, since the filter (8.26),(8.27) should theoretically yield a better result, as follows from Theorem 1.

In the classical linear optimal controller problem [25, 26], the criterion to be minimized is defined as a quadratic Bolza-Meyer functional:

$$J_3 = E[\frac{1}{2}[x(T)]^T \psi[x(T)] + \frac{1}{2}\int_{t_0}^{T}(u^T(s)R(s)u(s) + x^T(s)L(s)x(s))ds], \qquad (8.30)$$

where $R(t)$ is positive and ψ, $L(t)$ are nonnegative definite symmetric matrix functions, and $T > t_0$ is a certain time moment. The symbol $E[f(x)]$ means the expectation (mean) of a function f of a random variable x, and a^T denotes transpose to a vector (matrix) a. The solution to this problem was recently obtained in [31].

In this chapter, the criterion to be minimized includes a non-quadratic terminal term is defined as follows:

$$J_1 = E[\sum_{i=1}^{n} \psi_{ii} \mid x_i(T) \mid + \frac{1}{2}\int_{t_0}^{T} (u^T(s)R(s)u(s) + x^T(s)L(s)x(s))ds], \qquad (8.31)$$

where $R(s)$ is positive and $L(s)$ is a nonnegative definite continuous symmetric matrix functions, ψ is a diagonal nonnegative definite matrix, and $\mid x_i \mid$ denotes the absolute value of the component x_i of the vector $x \in R^n$.

The optimal controller problem is to find the control $u^*(t)$, $t \in [t_0, T]$, that minimizes the criterion J_1 or J_2 along with the unobserved trajectory $x^*(t)$, $t \in [t_0, T]$, generated upon substituting $u^*(t)$ into the state equation (8.1).

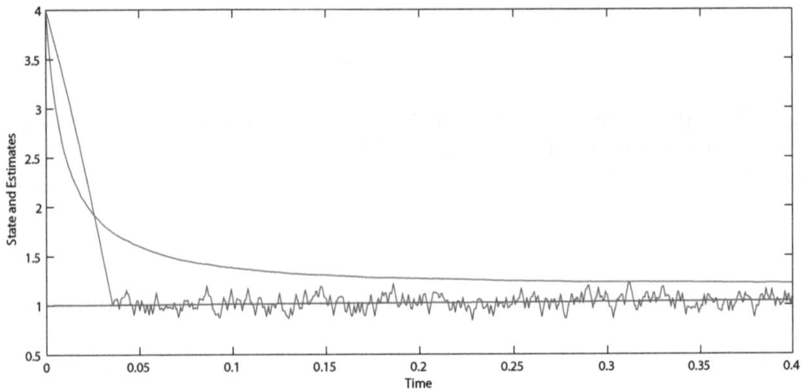

Fig. 8.3 Graphs of the unmeasured state (24) $x(t)$ (blue), the mean-module sliding mode estimate (26) $m(t)$ (green), and the mean-square polynomial filter estimate (28) $m_P(t)$ (red) in the interval $[0, 0.4]$.

8.5 Mean-Square Controller Design

8.5.1 Separation Principle

Solving the problem corresponding to the criterion (8.31), the unmeasured polynomial state $x(t)$, satisfying (8.1), is replaced with its mean-square estimate $m(t)$ over linear observations $y(t)$ (8.2), which is obtained using the mean-square sliding mode filter for polynomial systems (see [27] for the corresponding filtering problem statement and solution)

$$\dot{m}(t) = E(f(x,t) \mid F_t^Y) + B(t)u(t) + K(t)A^T(t)(G(t)G^T(t))^{-1} \times \qquad (8.32)$$
$$A(t)Sign[A^T(t)(A(t)A^T(t))^{-1}y(t) - m(t)].$$

with the initial condition $m(t_0) = E(x(t_0) \mid F_{t_0}^Y)$. Here, the Signum function of a vector $x = [x_1, \ldots, x_n] \in R^n$ is defined as $Sign[x] = [sign(x_1), \ldots, sign(x_n)] \in R^n$, and the signum function of a scalar x is defined as $sign(x) = 1$, if $x > 0$, $sign(x) = 0$, if $x = 0$, and $sign(x) = -1$, if $x < 0$ ([[23]]).

The matrix function $K(t)$ satisfies the matrix equation with time-varying coefficients

$$\dot{K}(t) = E(f(x,t)(x(t) - m(t))^T | F_t^Y) + \\ (b(t)b^T(t)) * | A^T(t)(A(t)A^T(t))^{-1}y(t) - m(t) |, \tag{8.33}$$

with the initial condition

$$K(t_0) = E[(x(t_0) - m(t_0)(x(t_0) - m(t_0)^T \mid F_{t_0}^Y] * \\ | A^T(t_0)(A(t_0)A^T(t_0))^{-1}y(t_0) - m(t_0) | .$$

Here, $| x |= [| x_1 |, \ldots, | x_n |] \in R^n$ is defined as the vector of absolute values of the components of the vector $x \in R^n$, and $A * b$ denotes a product between a matrix $A \in R^{n \times n}$ and a vector $b \in R^n$, that results in the matrix defined as follows: all entries of the j-th column of the matrix A are multiplied by the j-th component of the vector b, $j = 1, \ldots, n$.

Recall that $m(t)$ is the mean-square estimate for the state vector $x(t)$, based on the observation process $Y(t) = \{y(s), t_0 \leq s \leq t\}$, that minimizes the mean-square norm

$$H = E[(x(t) - m(t))^T (x(t) - m(t)) \mid F_t^Y]$$

at every time moment t. Here, $E[\xi(t) \mid F_t^Y]$ means the conditional expectation of a stochastic process $\xi(t) = (x(t) - m(t))^T (x(t) - m(t))$ with respect to the σ - algebra F_t^Y generated by the observation process $Y(t)$ in the interval $[t_0, t]$. As known [[20]], this optimal estimate is given by the conditional expectation

$$m(t) = E(x(t) \mid F_t^Y)$$

of the system state $x(t)$ with respect to the σ - algebra F_t^Y generated by the observation process $Y(t)$ in the interval $[t_0, t]$. As usual, the matrix function

$$P(t) = E[(x(t) - m(t))(x(t) - m(t))^T \mid F_t^Y]$$

is the estimation error variance.

Remark 1. Note that the equations (8.32) and (8.33) do not form a closed system of equations due to the presence of polynomial terms depending on x, $E(f(x,t) \mid F_t^Y)$, and $E(f(x,t)(x(t) - m(t))^T) \mid F_t^Y)$, which are not expressed yet as functions of the filter variables, $m(t)$ and $K(t)$ (or $P(t)$). However, as shown in [[24]], the closed system of the mean-square filtering equations can be obtained for any polynomial state (8.1) over linear observations (8.2).

It is readily verified that the optimal control problem for the system state (8.1) and cost function (8.31) is equivalent to the optimal control problem for the estimate (8.32) and the cost function J_1 represented as

$$J_1 = \sum_{i=1}^{n} \psi_{ii} \mid m_i(T) \mid + \frac{1}{2} \int_{t_0}^{T} (u^T(s)R(s)u(s) + m^T(s)L(s)m(s))ds + \tag{8.34}$$

$$\frac{1}{2} \int_{t_0}^{T} tr[P(s)L(s)]ds + tr[P^{-1}(T)K(T)\psi],$$

where $tr[A]$ denotes trace of a matrix A. Since the latter part of J_1 does not depend on control $u(t)$ or state $x(t)$, the reduced effective cost function M_1 to be minimized takes the form

$$M_1 = \sum_{i=1}^{n} \psi_{ii} \mid m_i(T) \mid + \frac{1}{2} \int_{t_0}^{T} (u^T(s)R(s)u(s) + m^T(s)L(s)m(s))ds. \tag{8.35}$$

Thus, the solution for the optimal control problem specified by (8.1),(8.31) can be found solving the optimal control problem given by (8.32),(8.35). Finally, the minimal value of the criterion J_1 should be determined using (8.34). This conclusion presents the separation principle for polynomial systems with a non-quadratic criterion (8.31).

8.5.2 Controller Design

The optimal solution to the control problem defined by (32),(35) is given in [[28]]. Applying the separation principle from the previous subsection to the sliding mode mean-square filter in [[27]] and the sliding mode optimal regulator in [[28]], the optimal controller solving the original problem (1),(2),(31) is given by the following theorem.

Theorem 3. The mean-square controller for a polynomial system (8.1) over linear observations (8.2) with respect to a non-quadratic criterion (8.31) is given by the control law

$$u(t) = R^{-1}(t)B^T(t)Q(t)Sign[m(t)], \tag{8.36}$$

where the matrix function $Q(t)$ is the solution of the matrix equation

$$\dot{Q}(t) = L(t) * \mid m(t) \mid - [a_1(t) + 2a_2(t)m(t) + \tag{8.37}$$

$$3a_3(t)m(t)m^T(t) + \ldots + pa_p(t)m(t)\ldots_{p-1 \text{ times}}\ldots m(t)]^T Q(t).$$

The terminal condition for the equation (8.37) is defined as $Q(T) = -\psi$, if the state $m(t)$ does not reach the sliding manifold $m(t) = 0$ within the time interval $[t_0, T]$, $m(t) \neq 0, t \in [t_0, T]$. Otherwise, if the state $m(t)$ reaches the sliding manifold $m(t) = 0$ within the time interval $[t_0, T]$, $m(t) = 0$ for some $t \in [t_0, T]$, then $Q(t)$ is set equal to a matrix function $Q_0(t)$ that is such a solution of (10) that $m(t)$ reaches the sliding manifold $m(t) = 0$ under the control law (36) with the matrix $Q_0(t)$ exactly at the final time moment $t = T$, $m(T) = 0$, but $m(t) \neq 0, t < T$.

Upon substituting the optimal control (8.36) into the equation (8.32), the following optimally controlled state estimate equation is obtained

$$\dot{m}(t) = E(f(x,t) \mid F_t^Y) + \tag{8.38}$$

$$B(t)R^{-1}(t)B^T(t)Q(t)Sign[m(t)] + K(t)A^T(t)(G(t)G^T(t))^{-1}\times$$

$$A(t)Sign[A^T(t)(A(t)A^T(t))^{-1}y(t) - m(t)],$$

with the initial condition $m(t_0) = E(x(t_0) \mid F_t^Y)$.

As commented in Remark 1, the obtained controller equations (8.32),(8.33), (8.36)–(8.38), do not present an explicit closed system of equations, if a specific form of the polynomial $f(x,t)$ in (1) is not provided. Therefore, in the next subsection, a closed form of the filtering equations is obtained from (32) and (33) for a third-order function $f(x,t)$ in the equation (8.1), as follows. It should be noted, however, that application of the same procedure would result in designing a closed system of the filtering equations for any polynomial function $f(x,t)$ in (8.1).

8.5.2.1 Controller Design for a Third-Order Polynomial State

Let the function

$$f(x,t) = a_0 + a_1(t)x + a_2(t)xx^T + a_3(t)xxx^T \tag{8.39}$$

be a third-order polynomial, where x is an n-dimensional vector, $a_0(t)$ is a vector of dimension n, a_1 is a matrix of dimension $n\times n$, a_2 is a 3D tensor of dimension $n\times n\times n$, a_3 is a 4D tensor of dimension $n\times n\times n\times n$. In this case, the following filtering equations for the optimal estimate $m(t)$ and the filter gain matrix $Q(t)$ are explicitly obtained

$$\dot{m}(t) = a_0 + a_1 m(t) + a_2(t)m(t)m(t)^T + a_2(t)P(t) \tag{8.40}$$

$$+3a_3(t)P(t)m(t) + a_3(t)m(t)m(t)m(t)^T$$

$$+K(t)A^T(t)(G(t)G^T(t))^{-1}\times$$

$$A(t)Sign[A^T(t)(A(t)A^T(t))^{-1}y(t) - m(t)].$$

with the initial condition $m(t_0) = E(x(t_0) \mid F_{t_0}^Y)$,

$$\dot{K}(t) = a_1(t)K(t) + 2a_2(t)m(t)K(t) + (a_2(t)m(t)K(t))^T \tag{8.41}$$

$$+3(a_3(t)(P(t)K(t) + m(t)m(t)^T K(t)))$$

$$+(a_3(t)(3P(t)K(t) + 2m(t)m(t)^T K(t)))^T$$

$$+(b(t)b^T(t))* \mid A^T(t)(A(t)A^T(t))^{-1}y(t) - m(t) \mid,$$

with the initial condition

$$K(t_0) = E[(x(t_0) - m(t_0)(x(t_0) - m(t_0)^T \mid F_{t_0}^Y]*$$

$$\mid A^T(t_0)(A(t_0)A^T(t_0))^{-1}y(t_0) - m(t_0) \mid,$$
$$\dot{P}(t) = a(t)P(t) + P(t)a^T(t) + 2a_2(t)m(t)P(t) \tag{8.42}$$
$$+2(a_2(t)m(t)P(t))^T + 3(a_3(t)(P(t)P(t) + m(t)m^T(t)P(t)))$$
$$+3(a_3(t)(P(t)P(t) + m(t)m^T(t)P(t)))^T + b(t)b^T(t)$$
$$-P(t)A^T(t)(B(t)B^T(t))^{-1}A(t)P(t),$$

with the initial condition

$P(t_0) = E[(x(t_0) - m(t_0)(x(t_0) - m(t_0)^T \mid F_{t0}^Y]$.

Consequently, the obtained result is formulated in the following theorem.

Theorem 4. The mean-square controller for a third degree polynomial stochastic system state (8.39) over linear observations (8.2) with respect to a non-quadratic criterion (8.31) is given by the control law (8.36), the equation (8.40) for the estimate $m(t) = E(x(t) \mid F_t^Y)$, the equations (8.41) and (8.42) for the filter gain matrix $K(t)$ and the error variance matrix $P(t)$, and the equation (8.10) for the control gain matrix $Q(t)$.

8.5.3 Example

This section presents an example of designing the optimal sliding mode controller for a polynomial system (8.1) over linear observations (8.2) with a non-quadratic criterion (8.31), using the controller (8.36),(8.37),(8.40)–(8.42), and comparing it to the mean-square third-order polynomial controller [31].

Consider a quadratic state equation

$$\dot{x}(t) = 0.1x^2(t) + u(t) + \psi_1(t), \quad x(0) = 1, \tag{8.43}$$

and a linear observation process

$$y(t) = x(t) + \psi_2(t), \tag{8.44}$$

where $\psi_1(t)$ and $\psi_2(t)$ are white Gaussian noises, which are the weak mean square derivatives of standard Wiener processes (see [20]). The equations (8.43),(8.44) correspond to the alternative conventional form (1*),(2*) for the equations (8.1),(8.2).

The controller problem is to find the control $u(t)$, $t \in [0,T]$, $T = 0.4$, that minimizes the criterion

$$J_1 = 50 \mid x(T) \mid + \frac{1}{2}\int_0^T (u^2(t) + x^2(t))dt. \tag{8.45}$$

Applying the designed sliding-mode controller (8.36),(8.37),(8.40)–(8.42), the control law (8.36) is given by

$$u(t) = Q(t)sign[m(t)], \tag{8.46}$$

where $m(t)$ satisfies the equation

$$\dot{m}(t) = 0.1m^2(t) + 0.1P(t) + K(t)sign[y(t) - m(t)], \tag{8.47}$$

with the initial condition $m(0) = E(x(0) \mid y(0)) = m_0$,

$$\dot{K}(t) = 0.3K(t) + \mid y(t) - m(t) \mid, \tag{8.48}$$

with the initial condition $K(0) = E((x(0) - m(0))(x(0) - m(0))^T \mid y(0)) * \mid y(0) - m(0) \mid$,

$$\dot{P}(t) = 1 + 0.4m(t)P(t) - P^2(t), \tag{8.49}$$

with the initial condition $P(0) = E((x(0) - m(0))(x(0) - m(0))^T \mid y(0))$, and

$$\dot{Q}^*(t) = \mid m(t) \mid - 0.2m(t)Q^*(t), \tag{8.50}$$

with the terminal condition $Q(0.4) = -50$, if $m(t) \neq 0$ for any $t < 5$, and $Q^*(t^*) = 0$, where t^* is the time that the estimate $m(t)$ reaches the sliding manifold $m = 0$ at the final moment $t = T$, otherwise.

Upon substituting the control (8.46) into (8.47), the optimally controlled state estimate equation takes the form

$$\dot{m}(t) = 0.1m^2(t) + 0.1P(t) + \tag{8.51}$$

$$Q(t)sign[m(t)] + K(t)sign[y(t) - m(t)],$$

with the initial condition $m(0) = E(x(0) \mid y(0)) = m_0$. The obtained system (8.47)–(8.51) can be solved using simple numerical methods, such as "shooting". This method consists in varying initial conditions of (8.50) until the given terminal condition is satisfied.

For numerical simulation of the system (8.43),(8.44) and the controller (8.47)–(8.51), the initial values $x(0) = 1$, $m(0) = 10$, and $P(0) = 866.25$ are assigned. The final time is set to $T = 0.4$. The disturbances $\psi_1(t)$ in (8.43) and $\psi_2(t)$ in (8.44) are realized using the built-in MatLab white noise function.

The system (8.47)–(8.51) is first simulated with the terminal condition $Q^*(0.4) = -50$. As the simulation shows, the state $m(t)$ reaches zero before the final moment $T = 0.4$. Accordingly, the terminal condition for the equation (50) is reset to $Q^*(0.4) = -\psi_0$ such that $m(0.4) = 0$, and the system (47)–(51) is simulated again. The value of the criterion (8.45) at the final moment $T = 0.4$ is $J_1(0.4) = 6.8286$.

The designed sliding mode controller (8.36),(8.37),(8.40)–(8.42) is compared to the best polynomial controller for the criterion J_3 with the quadratic non-integral term

$$J_3 = 25x^2(T) + \frac{1}{2}\int_0^T (u^2(t) + x^2(t))dt. \tag{8.52}$$

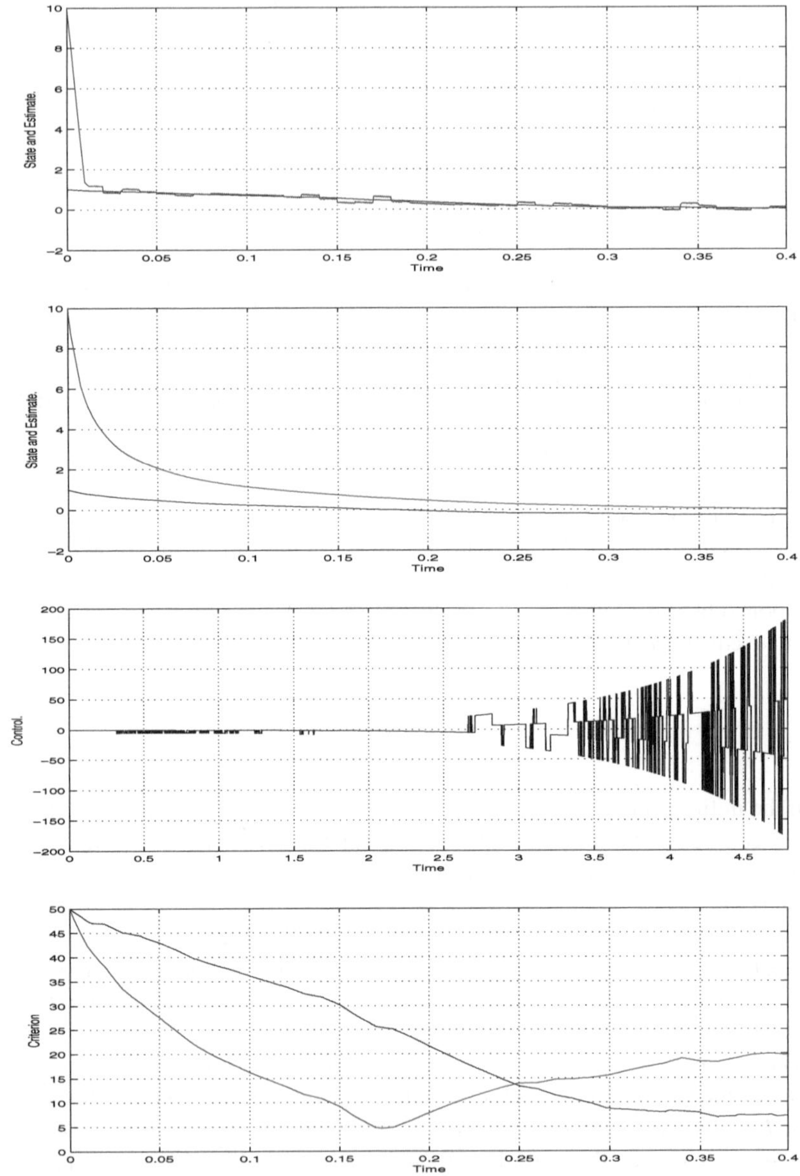

Fig. 8.4 Sliding mode controller optimal with respect to criterion J_1 vs. feedback polynomial-quadratic controller in the entire simulation interval $[0, 0.4]$. **1. Sliding mode controller.** Graphs of the controlled state (43) $x(t)$ (blue) and the controlled estimate (51) $m(t)$ (green); **2. Feedback polynomial-quadratic controller.** Graphs of the controlled state (43) $x(t)$ (blue) and the controlled estimate (57) $m(t)$ (green); **3. Control.** Graphs of the sliding mode control (46) $u^*(t)$ (blue) and the feedback polynomial-quadratic control (53) $u(t)$ (green); **4. Criterion.** Graphs of the criterion (45) J_1 produced by the sliding mode controller (blue) and by the feedback polynomial-quadratic controller (green).

As follows from the polynomial-quadratic control theory [31], the control law is given by

$$u(t) = Q(t)m(t), \tag{8.53}$$

where $m(t)$ satisfies the equation

$$\dot{m}(t) = 0.1m^2(t) + 0.1P(t) + u(t) + P(t)[y(t) - m(t)], \tag{8.54}$$

with the initial condition $m(0) = E(x(0) \mid y(0)) = m_0$,

$$\dot{P}(t) = 1 + 0.4m(t)P(t) - P^2(t), \tag{8.55}$$

with the initial condition $P(0) = E((x(0) - m(0))(x(0) - m(0))^T \mid y(0))$, and

$$\dot{Q}(t) = 1 - 0.3m(t)Q(t) - Q^2(t), \quad Q(0.4) = -50. \tag{8.56}$$

Upon substituting the control (8.53) into (8.54), the controlled state estimate equation takes the form

$$\dot{m}(t) = 0.1m^2(t) + 0.1P(t) + Q(t)m(t) + P(t)[y(t) - m(t)], \tag{8.57}$$

with the initial condition $m(0) = E(x(0) \mid y(0)) = m_0$.

Note that the comparison of the designed sliding mode controller (8.36),(8.37), (8.40)–(8.42) to the polynomial-quadratic controller (8.53)–(8.57) with respect to the criterion (8.45) is conducted for illustration purposes, since the controller (8.36),(8.37),(8.40)–(8.42) should theoretically yield a better result, as follows from Theorem 3. The value of the criterion (8.45) at the final moment $T = 0.4$ is $J_1(0.4) = 19.7499$.

It can be observed that the sliding mode controller (8.36),(8.37),(8.40)–(8.42) yields a certainly better value of the criterion (8.45) in comparison to the feedback polynomial-quadratic controller (8.53)–(8.57). Note that the feedback polynomial-quadratic controller fails to provide a causal control for the criterion (8.45).

References

1. Utkin, V.I.: Sliding Modes in Control and Optimization. Springer (1992)
2. Edwards, C., Spurgeon, S.K.: Sliding Mode Control: Theory and Applications. Taylor and Francis, London (1998)
3. Fridman, L., Levant, A.: Higher order sliding modes. In: Perruquetti, W., Barbot, J.P. (eds.) Sliding Mode Control in Engineering, pp. 53–101. Marcel Dekker, Inc., New York (2002)
4. Utkin, V.I., Guldner, J., Shi, J.: Sliding Mode Control in Electromechanical Systems. Taylor and Francis, London (1999)
5. Suzuki, S., Pan, Y., Furuta, K., Hatakeyama, S.: VS-control with time-varying sliding sector: Design and application to pendulum. Asian Journal of Control 6, 307–316 (2004)
6. Castaños, F., Fridman, L.: Analysis and design of integral sliding manifolds for systems with unmatched perturbations. IEEE Trans. Automatic Control 51, 853–858 (2006)

7. Baev, S., Shtessel, Y.B., Edwards, C., Spurgeon, S.K.: Output feedback tracking in causal nonminimum-phase nonlinear systems using HOSM techniques. In: Proc. 10th International Workshop on Variable Structure Systems, pp. 209–214 (2008)
8. Azemi, A., Yaz, E.: Sliding mode adaptive observer approach to chaotic synchronization. ASME Transactions. J. Dynamic Systems, Measurements and Control 122, 758–765 (2000)
9. Spurgeon, S.K.: Sliding mode observers: A survey. Intern. Journal of Systems Science 39, 751–764 (2008)
10. Boiko, I., Fridman, L., Pisano, A., Usai, E.: Analysis of chattering in systems with second order sliding modes. IEEE Trans. Automatic Control 52, 2085–2102 (2007)
11. Utkin, V.I., Shi, L.: Integral sliding mode in systems operating under uncertainty conditions. In: Proc. 35th Conference on Decision and Control, Kobe, Japan, pp. 4591–4596 (1996)
12. Bartolini, G., Ferrara, A., Levant, A., Usai, E.: On Second Order Sliding Mode Controllers. In: Young, K.D., Ozguner, U. (eds.) Variable Structure Systems, Sliding Mode and Nonlinear Control. LNCIS, vol. 247, pp. 329–350. Springer, Heidelberg (1999)
13. Basin, M.V., Ferrara, A., Calderon-Alvarez, D.: Sliding mode regulator as solution to optimal control problem. In: Proc. 47th Conference on Decision and Control, Cancun, Mexico, pp. 2184–2189 (2008)
14. Xia, Y., Jia, Y.: Robust sliding mode control for uncertain stochastic time-delay systems. IEEE Trans. Automatic Control 48, 1086–1092 (2003)
15. Niu, Y., Ho, D.W.C., Lam, J.: Robust integral sliding mode control for uncertain stochastic systems with time-varying delay. Automatica 41, 873–880 (2005)
16. Shi, P., Xia, Y., Liu, G.P., Rees, D.: On designing of sliding mode control for stochastic jump systems. IEEE Trans. Automatic Control 51, 97–103 (2006)
17. Basin, M.V., Fridman, L., Skliar, M.: Optimal and robust sliding mode filter for systems with continuous and delayed measurements. In: Proc. 41st Conference on Decision and Control, Las Vegas, NV, pp. 2594–2599 (2002)
18. Basin, M.V., Fridman, L., Rodriguez-Gonzalez, J.G., Acosta, P.: Integral sliding mode design for robust filtering and control of linear stochastic time-delay systems. Intern. J. Robust Nonlinear Control 15, 407–421 (2005)
19. Basin, M.V., Calderon-Alvarez, D.A., Skliar, M.: Optimal filtering for incompletely measured polynomial states over linear observations. International J. Adaptive Control and Signal Processing 22, 482–494 (2008)
20. Pugachev, V.S., Sinitsyn, I.N.: Stochastic Systems: Theory and Applications. World Scientific (2001)
21. Åström, K.J.: Introduction to Stochastic Control Theory. Academic Press, New York (1970)
22. Kalman, R.E., Bucy, R.S.: New results in linear filtering and prediction theory. ASME Trans., Part D (J. of Basic Engineering) 83, 95–108 (1961)
23. Filippov, A.F.: Differential Equations with Discontinuous Righthand Sides. Kluwer (1988)
24. Basin, M.V.: New Trends in Optimal Filtering and Control for Polynomial and Time-Delay Systems. Springer, Heidelberg (2008)
25. Kwakernaak, H., Sivan, R.: Linear Optimal Control Systems. Wiley-Interscience, New York (1972)
26. Fleming, W.H., Rishel, R.W.: Deterministic and Stochastic Optimal Control. Springer (1975)
27. Basin, M.V., Rodriguez-Ramirez, P.: Sliding mode filtering for polynomial systems. In: Proc. 12th IEEE Workshop on Variable Structure Systems, pp. 365–360. Bombay, India (2012)

28. Basin, M.V., Calderon-Alvarez, D.: Sliding mode regulator as solution to optimal control problem for non-linear polynomial systems. Journal of The Franklin Institute 347, 910–922 (2010)
29. Basin, M.V., Rodriguez-Ramirez, P.: Sliding mode mean-module filter design for polynomial systems. In: Proc. 2011 American Control Conference, pp. 632–636 (2011)
30. Basin, M.V., Loukianov, A., Hernandez-Fabian, R.: Sliding mode regulator as solution to optimal control problem with a non-quadratic criterion. In: Proc. 11th IEEE Workshop on Variable Structure Systems, Mexico City, Mexico, pp. 106–111 (2010)
31. Basin, M.V., Calderon-Alvarez, D.: Optimal controller for uncertain stochastic polynomial systems. Journal of The Franklin Institute 346, 206–222 (2009)

Chapter 9
A Review on Self-oscillating Relay Feedback Systems and Its Application to Underactuated Systems with Degree of Underactuation One

Luis T. Aguilar, Igor Boiko, Leonid Fridman, and Rafael Iriarte

Abstract. A tool for the design of a periodic motion in underactuated systems via generating a self-excited oscillation of a desired amplitude and frequency driven by a variable structure control is reviewed. In this chapter, we overview the capabilities of the two-relay controller to induce oscillations in dynamical systems. In this chapter, we will focus on underactuated mechanical systems with degree of underactuation one, that is, n degrees-of-freedom and $n-1$ actuators only. Three methods to set the frequency and amplitude of oscillation and its application to one-degree of underactuation systems are reviewed: describing function method, Locus of the perturbed relay system design (LPRS), and Poincaré map based design. Theoretical and practical open problems are also discussed.

9.1 Introduction

Researchers have been investigating and applying limit cycle behaviour to many different engineering fields. We can find several research works on this subject (see, e.g., [26]) but in the present survey we will focus on limit cycles induced by relay feedback systems only. In this chapter, we review the control of one of the simplest

Luis T. Aguilar · Rafael Iriarte
Instituto Politécnico Nacional, Ave. del parque 1310 Mesa de Otay,
Tijuana Baja California 22510
e-mail: {laguilar,ririarte}@citedi.mx

Igor Boiko
Electrical Engineering Department, The Petroleum Institute,
P.O. BOX 2533, Abu Dhabi, U.A.E.
e-mail: i.boiko@ieee.org

Leonid Fridman
Universidad Nacional Autónoma de México Departamento de Ingeniería de Control,
UNAM, México, D.F., 04510
e-mail: lfridman@unam.mx

B. Bandyopadhyay et al. (Eds.): *Advances in Sliding Mode Control*, LNCIS 440, pp. 187–205.
DOI: 10.1007/978-3-642-36986-5_9

types of functional motion: generation of a periodic motion in underactuated systems which could be non-minimum-phase. Current representative works on periodic motions and orbital stabilization of underactuated systems involve finding and using a reference model as a generator of limit cycles (see, e.g., [9, 25]), thus considering the problem of obtaining a periodic motion as a servo problem. Orbital stabilization of underactuated systems finds applications in electrical and mechanical systems.

Electrical systems. In power electronics applications, the idea of using self-oscillating switching has been explored in dc–dc inverters [22] because zero sensitivity to load changes and high performance have been demonstrated. Such inverters are attractive to operate in dc–ac converters since two buck-boost dc–dc inverters are commonly used. Several topologies have been proposed to design these converters, for example, Sanchis *et al.* [27] design a buck-boost dc–ac inverter using a double-loop control for both buck-boost dc–dc converter. Youssef and Jain [32] present a self-sustained oscillating controller for power factor correction circuits. Several circuit topologies for nonconventional dc–ac inverters are illustrated in J. Lai [19]. Under the framework of the present chapter, the dc–ac converter will consists of a two relay controller and a second order filter. For example, in an uninterruptible power supply (UPS) units, the desired voltage and frequency will be taken as 120 V_{rms} and 60 Hz, respectively according to North America standard. Thus, the purpose of the two relay controller is to induce a periodic voltage at the output of the filter with desired frequency and amplitude avoiding fast switching.

Underactuated mechanical systems. In particular, one of the most interesting applications of self-oscillation is to develop motion-planning algorithms which allow an underactuated robot to execute reliable maneuvers which is still a challenge for this class of systems for example in the coordinated motion of biped robots [12]. The formulation is different from the typical formulation of the tracking control problem for fully actuated mechanical systems [30] where the reference trajectory can be arbitrarily given, because underactuated systems are not feedback linearizables due to the insufficient number of actuators. Therefore, special attention is required in the selection of the desired trajectory for the systems under study. Different approaches for orbital stabilization have been proposed. For example, Shiriaev *et al.* [29] introduces a constructive tool for generation and orbital stabilization of periodic motion in underactuated nonlinear system through virtual constraint approach. Grizzle *et al.* [18] demonstrate asymptotic tracking for an unactuated link by finding conditions for the existence of a set of outputs that yields a system with a one-dimensional exponentially stable zero dynamics. In Orlov *et al.* [24] and Santiesteban *et al.* [28] an asymptotic harmonic generator was introduced through a modified Van der Pol equation tested on a friction pendulum to solve the swing up problem for an inverted pendulum. Berkemeier and Fearing [9] derive a set of exact trajectories of the nonlinear equations of motion of Acrobot, which involve inverted periodic motions. Martínez *et al.* [21] made a study of motion planning and oscillatory control for underactuated systems under geometric control theory.

In this chapter, underactuated systems are considered as the systems with internal (unactuated) dynamics with respect to the actuated variables. It allows us to propose a method of generating a periodic motion in an underactuated system where the same behavior can be seen via second order sliding mode (SOSM) algorithms, that is, generating self-excited oscillations using the same mechanism as the one that produces chattering. However, the generalization of the SOSM algorithms and the treatment of the unactuated part of the plant as additional dynamics result in the oscillations that may not necessarily be fast and of small amplitude.

There exist two approaches to analysis of periodic motions in sliding mode systems due to the presence of additional dynamics: the time-domain approach, which is based on the state space representation, and the frequency-domain approach. The Poincaré maps [31] are successfully used to ensure the existence and stability of periodic motions in the relay control systems (see, e.g., [16]). We can find application of Poincaré maps in several tasks such as biped locomotion [23] and switched converter systems [13]. The describing function (DF) method [15] allows approximate values of the frequency and the amplitude of periodic motions to be found in systems with linear plants driven by sliding mode controllers. The locus of perturbed relay system (LPRS) method [10] provides an exact solution of the periodic problem in discontinuous control systems, including finding exact values of the amplitude and the frequency of the self-excited oscillation.

Biped robots, gymnastic robot, and mechanical systems that evolve coordinated motion in order to emulate human walking patterns and climbing are examples of slow motion systems while switching power supplies are examples where fast oscillations are required. In this chapter, we will focus in underactuated mechanical systems with degree of underactuation one, that is, n degrees-of-freedom and $n-1$ actuators. Examples of such systems are inertia wheel pendulum, acrobot, pendubot, among others.

9.2 Problem Statement

Let the underactuated mechanical system, which is a plant in the system where a periodic motion is supposed to occur, be given by the Lagrange equation:

$$M(q)\ddot{q}+H(q,\dot{q}) = Su \tag{9.1}$$

where $q(t) \in \mathbb{R}^m$ is the vector of joint positions; $u(t) \in \mathbb{R}$ is the vector of applied joint torques where $m < n$; $S = [0_{(m-1)}, 1]^T$ is the input that maps the torque input to the joint coordinates space; $M(q) \in \mathbb{R}^{m\times m}$ is the symmetric positive-definite inertia matrix; and $H(q,\dot{q}) \in \mathbb{R}^m$ is the vector that contains the Coriolis, centrifugal, gravity, and friction torques.

The following two-relay controller is proposed for the purpose of exciting a periodic motion:

$$u = -c_1 \operatorname{sign}(y) - c_2 \operatorname{sign}(\dot{y}), \tag{9.2}$$

where c_1 and c_2 are parameters designed such that the scalar output of the system (the position of a selected link of the plant)

$$y = h(q) \tag{9.3}$$

has a steady periodic motion with the desired frequency and amplitude.

The *analysis and design objectives* are formulated as follows: Find the parameter values c_1 and c_2 in (9.2) such that the system (9.1) has a periodic motion with the desired frequency Ω and desired amplitude of the output signal A_1. Therefore, the main objective of this research is to find mapping G to be able to tune c_1 and c_2 values.

9.3 Methodologies Review

In Aguilar *et al.* ⟦5, 6⟧ was demonstrated the capabilities of the two relay controller to induce oscillations in nonlinear dynamical systems. Indeed, due to the simplicity of the controller to solve such problem, it looks attractive for its physical implementation. The TRC were successfully tested in academic underactuated systems (inverted pendulums). However, we can found in literature many underactuated systems with certain degree of complexity where motion control is required. Analysis of periodic motions in variable structure system were studied by Fridman ⟦17⟧. The introductory works ⟦2–4⟧ were motivated by the original works of Boiko and Fridman compiled in ⟦11⟧ where the analysis of chattering in linear systems was pursued using frequency domain tools such as describing function and locus of a perturbed relay systems (LPRS) methods.

9.3.1 Describing Function

Let firstly, the linearized plant be given by:

$$\begin{aligned} \dot{x} &= Ax + Bu \\ y &= Cx \end{aligned}, \qquad x \in \mathbb{R}^n, \quad y \in \mathbb{R}, \quad n = 2m \tag{9.4}$$

which can be represented in the transfer function form as follows:

$$W(s) = C(sI - A)^{-1}B.$$

Let us assume that matrix A has no eigenvalues at the imaginary axis and the relative degree of (9.4) is greater than 1.

The describing function (DF), N, of the variable structure controller (9.2) is the first harmonic of the periodic control signal divided by the amplitude of $y(t)$ ⟦8⟧:

$$N = \frac{\omega}{\pi A_1} \int_0^{2\pi/\omega} u(t)\sin(\omega t)dt + j\frac{\omega}{\pi A_1} \int_0^{2\pi/\omega} u(t)\cos(\omega t)dt \tag{9.5}$$

where A_1 is the amplitude of the input to the nonlinearity (of $y(t)$ in our case) and ω is the frequency of $y(t)$. However, the algorithm (9.2) can be analyzed as the parallel connection of two ideal relay where the input to the first relay is the output variable and the input to the second relay is the derivative of the output variable. For the first relay the DF is:

$$N_1 = \frac{4c_1}{\pi A_1},$$

and for the second relay it is [8]:

$$N_2 = \frac{4c_2}{\pi A_2},$$

where A_2 is the amplitude of dy/dt. Also, take into account the relationship between y and dy/dt in the Laplace domain, which gives the relationship between the amplitudes A_1 and A_2: $A_2 = A_1\Omega$, where Ω is the frequency of the oscillation. Using the notation of the algorithm (9.2) we can rewrite this equation as follows:

$$N = N_1 + sN_2 = \frac{4c_1}{\pi A_1} + j\Omega\frac{4c_2}{\pi A_2} = \frac{4}{\pi A_1}(c_1 + jc_2), \tag{9.6}$$

where $s = j\Omega$. Let us note that the DF of the algorithm (9.2) depends on the amplitude value only. This suggests the technique of finding the parameters of the limit cycle via the solution of the harmonic balance equation [8]:

$$W(j\Omega)N(a) = -1, \tag{9.7}$$

where a is the generic amplitude of the oscillation at the input to the nonlinearity, and $W(j\omega)$ is the complex frequency response characteristic (Nyquist plot) of the plant. Using the notation of the algorithm (9.2) and replacing the generic amplitude with the amplitude of the oscillation of the input to the first relay this equation can be rewritten as follows:

$$W(j\Omega) = -\frac{1}{N(A_1)}, \tag{9.8}$$

where the function at the right-hand side is given by:

$$-\frac{1}{N(A_1)} = \pi A_1 \frac{-c_1 + jc_2}{4(c_1^2 + c_2^2)}.$$

Equation (9.7) is equivalent to the condition of the complex frequency response characteristic of the open-loop system intersecting the real axis in the point $(-1, j0)$. The function $-1/N$ is a straight line the slope of which depends on c_2/c_1 ratio. The point of intersection of this function and of the Nyquist plot $W(j\omega)$ provides the solution of the periodic problem.

Here, we summarize the steps to tune c_1 and c_2:

a) Identify the quadrant in the Nyquist plot where the desired frequency Ω is located, which falls into one of the following categories (sets):

$$\begin{aligned}
Q_1 &= \{\omega \in \mathbb{R} : \mathrm{Re}\{W(j\omega)\} > 0, \mathrm{Im}\{W(j\omega)\} \geq 0\} \\
Q_2 &= \{\omega \in \mathbb{R} : \mathrm{Re}\{W(j\omega)\} \leq 0, \mathrm{Im}\{W(j\omega)\} \geq 0\} \\
Q_3 &= \{\omega \in \mathbb{R} : \mathrm{Re}\{W(j\omega)\} \leq 0, \mathrm{Im}\{W(j\omega)\} < 0\} \\
Q_4 &= \{\omega \in \mathbb{R} : \mathrm{Re}\{W(j\omega)\} > 0, \mathrm{Im}\{W(j\omega)\} < 0\}.
\end{aligned}$$

b) The frequency of the oscillations depends only on the c_2/c_1 ratio, and it is possible to obtain the desired frequency Ω by tuning the $\xi = c_2/c_1$ ratio:

$$\xi = \frac{c_2}{c_1} = -\frac{\mathrm{Im}\{W(j\Omega)\}}{\mathrm{Re}\{W(j\Omega)\}}. \tag{9.9}$$

Since the amplitude of the oscillations is given by

$$A_1 = \frac{4}{\pi}|W(j\Omega)|\sqrt{c_1^2 + c_2^2}, \tag{9.10}$$

then the c_1 and c_2 values can be computed as follows

$$c_1 = \begin{cases} \frac{\pi}{4}\frac{A_1}{|W(j\Omega)|}\left(\sqrt{1+\xi^2}\right)^{-1} & \text{if } \Omega \in Q_2 \cup Q_3 \\ -\frac{\pi}{4}\frac{A_1}{|W(j\Omega)|}\left(\sqrt{1+\xi^2}\right)^{-1} & \text{elsewhere} \end{cases} \tag{9.11}$$

$$c_2 = \xi \cdot c_1. \tag{9.12}$$

9.3.2 Locus of a Perturbed Relay System Design (LPRS)

The LPRS proposed in [10] provides an exact solution of the periodic problem in a relay feedback system having a plant (9.4) and the control given by the hysteretic relay. The LPRS is defined as a characteristic of the response of a linear part to an unequally spaced pulse control of variable frequency in a closed-loop system [10]. This method requires a computational effort but will provide an exact solution. The LPRS can be computed as follows:

$$J(\omega) = \sum_{k=1}^{\infty}(-1)^{k+1}\mathrm{Re}\{W(k\omega)\} + j\sum_{k=1}^{\infty}\frac{1}{2k-1}\mathrm{Im}\{W[(2k-1)\omega]\}. \tag{9.13}$$

The frequency of the periodic motion for the algorithm (9.2) can be found from the following equation [10]:

$$\mathrm{Im}J(\Omega) = 0.$$

In fact, we are going to consider the plant being nonlinear, with the second relay transposed to the feedback in this equivalent plant. Introduce the following function, which will be instrumental in finding a response of the nonlinear plant to the periodic square-wave pulse control.

$$L(\omega,\theta)=\sum_{k=1}^{\infty}\frac{1}{2k-1}(\sin[(2k-1)2\pi\theta]\mathrm{Re}\{W[(2k-1)\omega]\}$$
$$+\cos[(2k-1)2\pi\theta]\mathrm{Im}\{W[(2k-1)\omega]\}). \tag{9.14}$$

The function $L(\omega,\theta)$ denotes a linear plant output (with a coefficient) at the instant $t=\theta T$ (with T being the period: $T=2\pi/\omega$) if a periodic square-wave pulse signal of unity amplitude is applied to the plant:

$$L(\omega,\theta)=\left.\frac{\pi y(t)}{4c}\right|_{t=2\pi\theta/\omega}$$

with $\theta\in[-0.5,0.5]$ and $\omega\in[0,\infty]$, where $t=0$ corresponds to the control switch from -1 to $+1$.

With $L(\omega,\theta)$ available, we obtain the following expression for $\mathrm{Im}\{J(\omega)\}$ of the equivalent plant:

$$\mathrm{Im}\{J(\omega)\}=L(\omega,0)+\frac{c_2}{c_1}L(\omega,\theta). \tag{9.15}$$

The value of the time shift θ between the switching of the first and second relay can be found from the following equation

$$\dot{y}(\theta)=0.$$

As a result, the set of equations for finding the frequency Ω and the time shift θ is as follows:

$$c_1L(\Omega,0)+c_2L(\Omega,\theta)=0,\quad c_1L_1(\Omega,-\theta)+c_2L_1(\Omega,0)=0. \tag{9.16}$$

The amplitude of the oscillations can be found as follows. The output of the system is:

$$y(t)=\frac{4}{\pi}\sum_{i=1}^{\infty}\{c_1\sin[(2k-1)\Omega+\varphi_L((2k-1)\Omega)]$$
$$+c_2\sin[(2k-1)\Omega t+\varphi_L((2k-1)\Omega)+(2k-1)2\pi\theta]\}A_L((2k-1)\Omega) \tag{9.17}$$

where $\varphi_L(\omega)=argW(\omega)$, which is a response of the plant to the two square pulse-wave signals shifted with respect to each other by the angle $2\pi\theta$. Therefore, the amplitude is

$$A_1=\max_{t\in[0;2\pi/\omega]}y(t). \tag{9.18}$$

Yet, instead of the true amplitude we can use the amplitude of the fundamental frequency component (first harmonic) as a relatively precise estimate. In this case, we can represent the input as the sum of two rotating vectors having amplitudes $4c_1/\pi$ and $4c_2/\pi$, with the angle between the vectors $2\pi\theta$. Therefore, the amplitude of the control signal (first harmonic) is

$$A_u = \frac{4}{\pi}\sqrt{c_1^2 + c_2^2 + 2c_1c_2\cos(2\pi\theta)}, \tag{9.19}$$

and the amplitude of the output (first harmonic) is

$$A_1 = \frac{4}{\pi}\sqrt{c_1^2 + c_2^2 + 2c_1c_2\cos(2\pi\theta)}A_L(\Omega), \tag{9.20}$$

where $A_L(\omega) = |W(j\omega)|$. We should note that despite using approximate value for the amplitude in (9.20), the value of the frequency is exact. Expressions (9.16), (9.20) if considered as equations for Ω and A_1 provide one with mapping F. From (9.16) one can see that the frequency of the oscillations depends only on the ratio $c_2/c_1 = \xi$. Therefore, Ω is invariant with respect to c_2/c_1: $\Omega(\lambda c_1, \lambda c_2) = \Omega(c_1, c_2)$. It also follows from (9.20) that there is the following invariance for the amplitude: $A_1(\lambda c_1, \lambda c_2) = \lambda A_1(c_1, c_2)$. Therefore, Ω and A_1 can be manipulated independently in accordance with mapping G considered below.

Mapping G (inverse of F) can be derived from (9.16), (9.20) if c_1, c_2 and θ are considered unknown parameters in those equations. For any given Ω, from equation (9.16) the ratio $c_2/c_1 = \xi$ can be found (as well as θ). Therefore, we can find first $\xi = c_2/c_1 = h(\Omega)$, where $h(\Omega)$ is an implicit function that corresponds to (9.16). After that c_1 and c_2 can be computed as per the following formulas:

$$c_1 = \frac{\pi}{4}\frac{A_1}{A_L(\Omega)}\frac{1}{\sqrt{1 + 2\xi\cos(2\pi\theta) + \xi^2}} \tag{9.21}$$

$$c_2 = \frac{\pi}{4}\frac{A_1}{A_L(\Omega)}\frac{\xi}{\sqrt{1 + 2\xi\cos(2\pi\theta) + \xi^2}}. \tag{9.22}$$

9.3.3 Poincaré-Map-Based Design

To construct the Poincaré map, one has to choose a surface of section S in the state space $\mathbb{R}^4$ and consider the points of successive intersections of a given trajectory with this surface. Switching occur on the level surfaces defined by

$$\begin{aligned} S_1 &= \{x : y = 0, \dot{y} < 0\}, \qquad & S_2 &= \{x : y < 0, \dot{y} = 0\}, \\ S_3 &= \{x : y = 0, \dot{y} > 0\}, \qquad & S_4 &= \{x : y > 0, \dot{y} = 0\}. \end{aligned} \tag{9.23}$$

The space $\mathbb{R}^4$ is divided into four regions by S_i, $i = 1, \ldots, 4$:

$$\begin{aligned} R_1 &= \{x : y < 0, \dot{y} < 0\}, \qquad & R_2 &= \{x : y < 0, \dot{y} > 0\}, \\ R_3 &= \{x : y > 0, \dot{y} > 0\}, \qquad & R_4 &= \{x : y > 0, \dot{y} < 0\}. \end{aligned} \tag{9.24}$$

Depending on the state, the system is governed by one of the four models defined by

$$\begin{aligned}M_1 &: \dot{x} = Ax + B(c_1 + c_2),\\ M_2 &: \dot{x} = Ax + B(c_1 - c_2),\\ M_3 &: \dot{x} = Ax - B(c_1 + c_2),\\ M_4 &: \dot{x} = Ax + B(-c_1 + c_2).\end{aligned}$$

The solution of M_1 on the time interval $[0;t_1]$, where t_1 is the transition time from S_1 to S_2, subject to the initial conditions $x(0) = \rho_p$, where "$(\cdot)_p$" stands for "periodic", such that (without loss of generality)

$$\begin{aligned}y(0) &= Cx(0) = C\rho_p = 0,\\ \dot{y}(0) &= C(Ax(0) + Bu) = CA\rho_p < 0\end{aligned} \tag{9.25}$$

is given by

$$x(t) = e^{At}x(0) + \int_0^t e^{A\tau}d\tau Bu,$$

where

$$\int_0^t e^{A\tau}d\tau = \sum_{i=1}^{\infty} A^{i-1}t^i/i! = A^{-1}(e^{At} - I)$$

and $u = c_1 + c_2$. The transition to S_2 and switching to $u = c_1 - c_2$ is ensured under the technical transversality condition

$$\ddot{y}(t_1) = CA^2\eta_k > 0. \tag{9.26}$$

Under this condition, the trajectory will enter the region R_2, and since the matrix A is Hurwitz it will reach either S_3 or return back to S_2. We will assume for now that the latter does not happen.

Analogously, for the case of the twisting projection of the motion onto the $(y, \dot{y})$-plane, the four state transitions initiated at $\rho_k = \rho_p$ are given by

$$\begin{aligned}\eta_k &= e^{At_1}\rho_k + A^{-1}(e^{At_1} - I)B(c_1 + c_2),\\ \rho_k^- &= e^{At_2}\eta_k + A^{-1}(e^{At_2} - I)B(c_1 - c_2),\\ \eta_k^- &= e^{At_3}\rho_k^- - A^{-1}(e^{At_3} - I)B(c_1 + c_2),\\ \rho_{k+1} &= e^{At_4}\eta_k^- - A^{-1}(e^{At_4} - I)B(c_1 - c_2),\end{aligned} \tag{9.27}$$

where t_2 is the time interval between S_2 and S_3, t_3 is the time interval between S_3 and S_4, and t_4 is the time interval between S_4 and S_1.

The fixed point of the Poincaré map, corresponding to an isolated periodic solution of system (9.4) driven by the two-relay controller, is determined by equation $\rho_{k+1} = \rho_k = \rho_p$. Skipping the sequential numbers of switching in (9.27) and using the principle of symmetry one can write the following: $\rho_p^- = -\rho_p$. For the T-periodic (symmetric) solution we will use the following notation: $t_1 = t_3 = \theta_1,\ t_2 = t_4 = \theta_2 = T/2 - \theta_1$.

The equation for the fixed point together with the switching conditions can be rewritten as follows:

$$-\rho_p = e^{A\theta_2}\eta_p + A^{-1}(e^{A\theta_2} - I)B(c_1 - c_2) \tag{9.28}$$

and, with the help of $y(0) = \dot{y}(\theta_1) = 0$ and $CB = 0$,

$$\begin{aligned} \eta_p &= e^{A\theta_1}\rho_p + A^{-1}(e^{A\theta_1} - I)B(c_1 + c_2) \\ C\rho_p &= 0, \quad CA\eta_p = 0, \quad CA\rho_p < 0, \quad CA^2\eta_p > 0. \end{aligned} \tag{9.29}$$

We assume in (9.28) and (9.29) that there are no additional switches on intervals $t \in (0;t_1)$ and $t \in (t_1;t_2)$, respectively since $\dot{y} < 0$ initially and y monotonically decreases from zero and cannot cross zero before $\dot{y}$ changes sign at $t = t_1$. This condition can easily be verified after parameters θ_1 and θ_2 are determined.

It is left to formalize the condition ensuring transition from S_2 to S_3 without leaving R_2. Defining two hypothetical (for the fixed control input $u = c_1 - c_2$) boundary crossing times as $\bar{t}_2$ and t_2, we have

$$t_2 = \min\left\{t > 0 : C\left(e^{At}\eta_p + A^{-1}(e^{At} - I)B(c_1 - c_2)\right) = 0\right\}$$

and

$$\bar{t}_2 = \min\left\{t > 0 : CA\left(e^{At}\eta_p + A^{-1}(e^{At} - I)B(c_1 - c_2)\right) = 0\right\}.$$

Hence, we require

$$t_2 < \bar{t}_2 \tag{9.30}$$

to ensure that our analysis of the limit cycle with exactly four switches is correct. In the case when the transition time is sufficiently small, dropping smaller-order terms in the definitions of t_2 and $\bar{t}_2$, one can derive the following simplified approximate algebraic assumption[1]

$$0 < t_2 \approx -\frac{2CA^2\eta_p}{CA^3\eta_p + c_2 - c_1} < \sqrt{\frac{2C\eta_p}{-CA^2\eta_p}} \approx \bar{t}_2.$$

Let us move onto defining the amplitude and frequency of the oscillations.

The system (9.28) and (9.29) can be considered as a system of algebraic equations for design of the two-relay controller providing for the system (9.4) the desired periodic solution with a given frequency Ω and amplitude A_1. Taking into account that

$$y(\theta_1) = C\eta_p = A_1, \quad \theta_1 + \theta_2 = \pi/\Omega = T/2 \tag{9.31}$$

(9.28), (9.29), and (9.31) can be reduced to a system of five nonlinear algebraic equations with respect to five variables: c_1, c_2, θ_1 and the first and the second coordinates of the vector ρ_p. Once the resolving set of parameters is found, the two-relay controller gains that provide the periodic solution of the system (9.4) with the desired amplitude and frequency are designed. This can be summarized as follows.

[1] Here we have used the identities $CA\eta_p = 0$, $CB = CAB = 0$, and $CA^2B = 1$, and dropped the third-order terms in the series expansions for the matrix exponents.

Theorem 9.1. *Suppose the system (9.28), (9.29), and (9.31) possesses a solution satisfying (9.30). If the desired amplitude A_1 is sufficiently small to avoid singularity in the matrix $M(q)$ defined in (9.1) and assuming that there are no additional switches on intervals $t \in (0;t_1)$ and $t \in (t_1;t_2)$, then the closed-loop system (9.1)–(9.2) has the desired periodic solution.*

Note however that (9.28), (9.29), and (9.31) is a system of nonlinear algebraic equations and might be hard to solve and even have no solutions for a particular values of A_1 and Ω.

It turns out that the linearity of the (transformed) plant and the fact that the control in the periodic motion can be represented as a sum of two relay controls, with finding the response of the plant as a linear combination (sum) of the two periodic relay controls of amplitudes c_1 and c_2, allows for a reduction of complexity of the original problem. Let us develop an approach that might simplify finding fixed points of the Poincaré map utilizing the concepts of the locus of a perturbed relay system method.

9.4 Linearized-Poincaré-Map-Based Analysis of Orbital Stability

Let us use (9.27) to analyze the deviation of a trajectory initiated on the surface S_1 at $x(0) = \rho_k = \rho_p + \delta_p$ from a periodic trajectory initiated from some ρ_p for sufficiently small initial deviations δ_p. Using the equation in (9.27) for η_k, the equation in (9.29) for η_p, and the Taylor expansion $e^{At_1} = e^{A\theta_1} + e^{A\theta_1}A\Delta t + O(\Delta t^2)$, $\Delta t = t_1 - \theta_1$ one can proceed as follows:

$$\begin{aligned}\eta_k &= e^{At_1}(\rho_p + \delta_p) + A^{-1}(e^{At_1} - I)B(c_1 + c_2)\\ &= \left(e^{A\theta_1} + e^{A\theta_1}A\Delta t\right)(\rho_p + \delta_p) + O\left(\Delta t^2\right)\\ &\quad + A^{-1}\left(e^{A\theta_1} + (e^{A\theta_1} - I + I)A\Delta t - I\right)B(c_1 + c_2)\end{aligned}$$

so that

$$\eta_k = e^{A\theta_1}\left(\delta_p + A\delta_p\Delta t\right) + (I + A\Delta t)\eta_p + B(c_1 + c_2)\Delta t + O\left(\Delta t^2\right).$$

Now, since $CA\eta_k = CA\eta_p = 0$, premultiplying this equation by CA which results in

$$CAe^{A\theta_1}\left(\delta_p + A\delta_p\Delta t\right) + CA(A\eta_p + B(c_1 + c_2))\Delta t = O\left(\Delta t^2\right),$$

one immediately concludes that $\Delta t = O(\delta_p)$ and obtains an estimate for $t_1 = \theta_1 + \Delta t$, that can be substituted back:

$$\eta_k = \eta_p + \delta_\eta = \eta_p + \varphi_1\delta_p + O(\delta_p^2),$$

where

$$\varphi_1 = \left(I - \frac{v_1 CA}{CAv_1}\right) e^{A\theta_1}, \qquad v_1 = A\eta_p + B(c_1 + c_2). \tag{9.32}$$

Following the second equation in (9.27) and computing t_2 using $C\rho_k^- = C\rho_p = 0$, one, in a similar way, obtains

$$\rho_k^- = -\rho_p + \delta_{\rho-} = -\rho_p + \varphi_2\delta_\eta + O(\delta_\rho^2),$$

where

$$\varphi_2 = \left(I - \frac{v_2 C}{Cv_2}\right) e^{A\theta_2}, \qquad v_2 = A\rho_p + B(c_1 - c_2). \tag{9.33}$$

Following the third equation in (9.27) and computing t_3 using $CA\eta_k^- = CA\eta_p = 0$, one obtains

$$\eta_k^- = -\eta_p + \delta_{\eta-} = -\eta_p + \varphi_3\delta_{\rho-} + O(\delta_\rho^2),$$

where $\varphi_3 = \varphi_1$.

Following the last equation in (9.27) and computing t_4 using $C\rho_{k+1} = C\rho_p = 0$, one obtains

$$\rho_{k+1} = \rho_p + \varphi_4\delta_{\rho-} + O(\delta_\rho^2),$$

where $\varphi_4 = \varphi_2$.

Finally, we have for small $\delta_\rho = \rho_k - \rho_p$: $\rho_{k+1} - \rho_p = \Phi\cdot(\rho_k - \rho_p) + O(\delta_\rho^2)$, with

$$\Phi = (\varphi_2\cdot\varphi_1)^2. \tag{9.34}$$

Since we have just computed a linearization for the Poincaré map, we conclude with the following.

Theorem 9.2. *Suppose that the parameters c_1 and c_2 induce a periodic trajectory for the closed-loop system controlled by the two-relay algorithm, that is, (9.1)–(9.2). This solution is orbitally exponentially stable if and only if all the eigenvalues of the matrix Φ, defined by (9.32), (9.33), and (9.34), are inside the unit circle.*

9.5 Robust Control Design

The system (9.1)–(9.2) is not a free-disturbance system except if there exist a desired frequency and amplitude such that the coefficients of (9.2) satisfy the twisting condition $c_1 > c_2 > w_{\max} > 0$, where $w_{\max}$ is the upper bound of disturbance $w(t)$, that is $\|w(t)\| \le w_{\max}$. Moreover, the system (9.1) is nonlinear and the design was done considering the linearization of the model, which might result in a closed-loop that is not robust in the presence of unknown inputs. In particular, in Aguilar *et al.* [[7]], imperfections of oscillation characteristics due to Coulomb friction forces, dead zone, mechanical vibrations, etc., were reported. Motivated by the need of a robust closed-loop scheme, we present a non-autonomous scheme to deal with this problem.

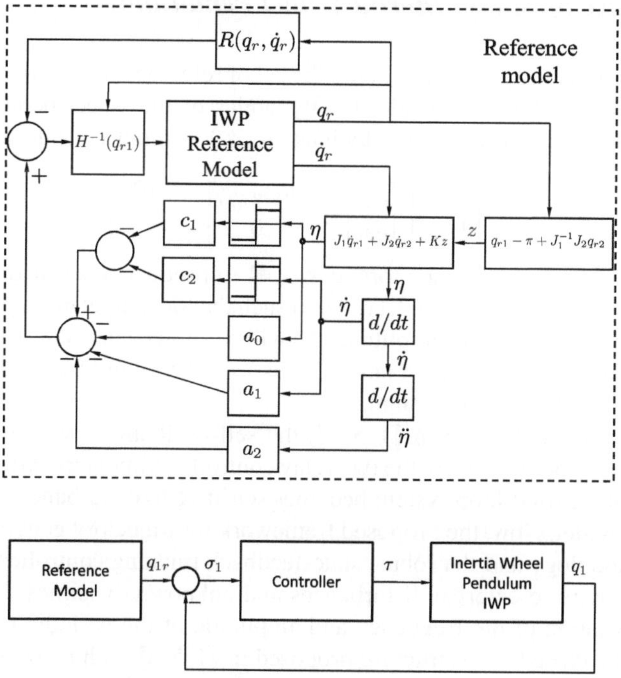

Fig. 9.1 Block diagram of the two-relay controller for real-time trajectory generation for orbital stabilization of inertia wheel pendulum

Let us start by explaining how to find a set of desired oscillations around its upright position. To begin with, let us rewrite (9.1) in terms of the reference positions and velocities $(q_r, \dot{q}_r)$

$$M(q_r)\ddot{q}_r + H(q_r, \dot{q}_r) = Su(y_r, \dot{y}_r). \tag{9.35}$$

We need to find $u(y_r, \dot{y}_r) \in \mathbb{R}$ to produce a set of desired periodic motion of the reference model (y_r) such that its output has a periodic motion with desired frequency and amplitude. Figure 9.1 shows a block diagram of the control applied to the inertia wheel pendulum with the real time trajectory generator.

Finally, a feedback law must be designed to ensure

$$\lim_{t\to\infty} \|y_r(t) - y(t)\| = \lim_{t\to\infty} \|\sigma(t)\| = 0 \tag{9.36}$$

where $\sigma(t) \in \mathbb{R}$ stands for the output error, and $y_r(t)$ is the desired output while providing boundedness of $\dot{q}_2$ and attenuating external disturbances where the reference output $y_r(t)$ is computed online from (9.35).

9.5.1 Case of Study: Inertia Wheel Pendulum

We will illustrate the procedure to find the set of reference trajectories for the inertia wheel pendulum. Let us consider the dynamics of the wheel pendulum in terms of the reference positions and velocities $(q_r, \dot{q}_r)$ without considering the viscous friction force

$$\begin{bmatrix} J_1 & J_2 \\ J_2 & J_2 \end{bmatrix} \begin{bmatrix} \ddot{q}_{1r} \\ \ddot{q}_{2r} \end{bmatrix} + \begin{bmatrix} h\sin q_{1r} \\ 0 \end{bmatrix} = \begin{bmatrix} 0 \\ 1 \end{bmatrix} \tau_r. \tag{9.37}$$

We need to find the reference torque $\tau_r \in \mathbb{R}$ to produce a set of desired periodic motion of the underactuated link ($y = q_{1r}$) such that the output has a periodic motion with desired frequency and amplitude. As will be shown later, viscous friction is not required in the above equation since it acts as a damping force thus ensuring the stability of the closed-loop system.

In the results published in [1, 5, 6], the self-oscillations were generated in an inertia wheel pendulum using the two relay controller without tracking control. Consequently, the closed-loop system becomes sensitive to disturbances and uncertainties of the model. Now, the proposed framework for trajectory generation under the same methodology and the robust state-feedback tracking controller contributes to avoid sensitivity to external disturbances and unknown dynamics. Here, we mean that the deviation of the frequency and amplitude of the periodic trajectory at the output of the closed-loop structure proposed in [1, 5, 6] with respect to the desired ones, depended on the uncertainties of the parameters of the model because formulas to compute the two values of the two-relay controller (c_1 and c_2) depends of the values of the inertia, length of the link, and masses, only; while viscous friction level was not considered as part of the formulas however exists in the system. Now, the proposed scheme is robust with respect to effect of viscous friction and external disturbances which will be rejected using a second-order sliding mode tracking controller.

The inertia wheel pendulum has underactuation degree one and satisfies certain structural property noted in [18]. As a result, it is possible to make exact linearization thus achieving local stability of zero dynamics. Following [18], let us take

$$\begin{aligned} p_1 &= q_{1r} - \pi + J_1^{-1} J_2 q_{2r} \\ \eta &= J_1 \dot{q}_{1r} + J_2 \dot{q}_{2r} + K p_1 \end{aligned}$$

where $K > 0$ is a constant. It is easy to verify that

$$J_1 \dot{p}_1 = \eta - K p_1$$

while

$$\begin{aligned} \dot{\eta} &= K J_1^{-1} J_2 \dot{q}_{2r} - h\sin(q_{1r}) + K\dot{q}_{1r}, \\ \ddot{\eta} &= -h\cos(q_{1r})\dot{q}_{1r} - K J_1^{-1} h \sin(q_{1r}), \\ \dddot{\eta} &= R(q_{1r}, \dot{q}_{1r}) + H(q_{1r})\tau_r \end{aligned}$$

where

$$\begin{aligned} H(q_{1r}) &= \frac{h\cos(q_{1r})}{J_1 - J_2}, \\ R(q_{1r},\dot{q}_{1r}) &= h\left(\dot{q}_{1r}^2 + H(q_{1r})\right)\sin(q_{1r}) - \frac{hK}{J_1}\dot{q}_{1r}\cos(q_{1r}). \end{aligned} \tag{9.38}$$

Hence, we can take

$$\tau_r = H^{-1}(q_{1r})\left(u_r - a_0\eta - a_1\dot{\eta} - a_2\ddot{\eta} - R(q_{1r},\dot{q}_{1r})\right), \tag{9.39}$$

where $H(q_r)$ is nonsingular around the equilibrium point $(q_{1r}^\star, \dot{q}_{1r}^\star) = (\pi, 0)$, a_0, a_1, and a_2 are positive constants. Introducing the new state coordinates $x = (x_1, x_2, x_3) = (\eta, \dot{\eta}, \ddot{\eta})$, we obtain

$$\begin{bmatrix} \dot{x}_1 \\ \dot{x}_2 \\ \dot{x}_3 \end{bmatrix} = \begin{bmatrix} 0 & 1 & 0 \\ 0 & 0 & 1 \\ -a_0 & -a_1 & -a_2 \end{bmatrix} \begin{bmatrix} x_1 \\ x_2 \\ x_3 \end{bmatrix} + \begin{bmatrix} 0 \\ 0 \\ 1 \end{bmatrix} u_r, \tag{9.40}$$

$$\dot{p}_1 = -\frac{K}{J_1}p_1 + \frac{1}{J_1}y_r, \qquad y_r = \begin{bmatrix} 1 & 0 & 0 \end{bmatrix} x. \tag{9.41}$$

The following *two-relay controller* is proposed for the purpose of exciting a periodic motion in (9.40):

$$u_r = -c_1\,\mathrm{sign}(y_r) - c_2\,\mathrm{sign}(\dot{y}_r) \tag{9.42}$$

where c_1 and c_2 are scalars parameters designed such that the scalar-valued function output $y_r(t)$ has a periodic motion with the desired frequency Ω and amplitude A_1. Let us recall that the difference between (9.42) and the second order sliding mode controller given, for example, in [20] is that c_1 is not constrained to be positive and greater than c_2.

Let us define the tracking error as:

$$\begin{aligned} \sigma(t) &= q_{1r}(t) - q_1(t) \\ \dot{\tilde{q}}_2(t) &= \dot{q}_{2r}(t) - \dot{q}_2(t). \end{aligned} \tag{9.43}$$

The second-order sliding mode controller can be straightforwardly synthesized from (9.43) obtaining:

$$\begin{aligned} \tau = &-\frac{\Delta}{J_2}\left(\alpha_1\,\mathrm{sign}(\sigma) + \alpha_2\,\mathrm{sign}(\dot{\sigma}) + \beta_1\sigma + \beta_2\dot{\sigma} - \gamma\dot{\tilde{q}}_2\right) \\ &- h\sin(q_{1r} - \sigma) + f_s(\dot{q}_{2r} - \dot{\tilde{q}}_2) - \frac{\Delta}{J_2}\ddot{q}_{1r} \end{aligned} \tag{9.44}$$

where $\Delta = (J_1 - J_2)J_2$. The role of $\gamma\dot{\tilde{q}}_2$ is to avoid that the wheel velocity saturates after a while because the pendulum is influenced by the acceleration of the wheel.

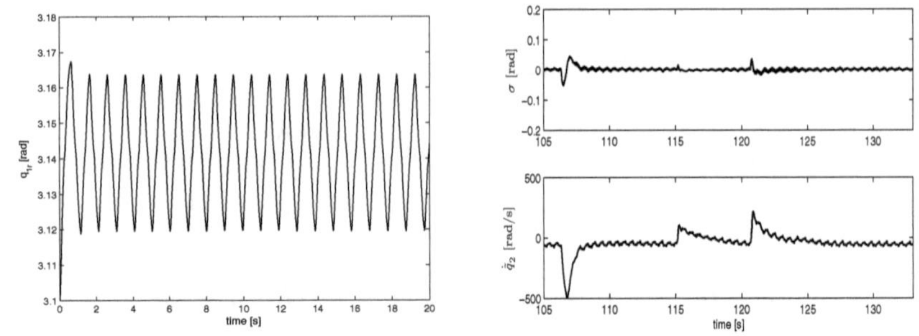

Fig. 9.2 Periodic reference signal at $\Omega = 2\pi$ [rad/s] and $A_1 = 0.07$ generated by the two-relay controller reference model under the parameters $c_1 = 2$, $c_2 = -0.1$, $K = 1 \times 10^{-4}$, $a_0 = 350$, $a_1 = 155$, and $a_2 = 22$ (left). Tracking error of the underactuated link σ under disturbances and perturbed velocity of the disk (right).

Fig. 9.2 shows the trajectory of the reference signal for the reaction wheel pendulum used as reference model. Figure 9.2 also shows the experimental trajectories of the pendulum for the unperturbed and perturbed case, respectively. Details in using quasi-continuous high-order sliding modes is presented by Estrada *et al.* [[14]].

9.6 Comments

Of course, the considered two-relay control algorithm is not the only one that can be used for the purpose of inducing oscillations in dynamical systems. Development of efficient algorithms for purpose of inducing oscillations in mechanical systems, methods of their analysis and design is still an open problem mainly for systems with degree of underactuation higher than one. In fact, the described method just presents a new approach, which is promising in the authors' opinion. In particular, this concerns the use of such simple and efficient methods as the describing function and LPRS methods. In the describing function analysis, sacrificing of exactness of amplitude/frequency computing in favor of simplicity is quite justified by the availability of qualitative results and important conclusions. The LPRS method provides the same functionality even without sacrifice of exactness.

The key feature of the proposed method is that the underactuated system can be considered as a system with unactuated dynamics with respect to the actuated variables. For generation of the self-excited oscillations with desired output amplitude and frequencies, a two relay controller is proposed. A systematic approach for two-relay controller parameter adjustment is proposed. The DF method provides approximate values of the controller parameters for the plants with the low-pass filtering properties. The LPRS gives exact values of the controller parameters for linear plants. The Poincaré maps provides the values of the controller parameters ensuring the existence of the locally orbitally stable periodic motions for an arbitrary mechanical plant. The effectiveness of the proposed design procedures is supported by experiments carried out on the Furuta pendulum and inertia wheel pendulum from Quanser.

Open questions and problems arising from the above presented results include (a) define conditions of orbital stability for nonlinear plants, (b) generalization of the results for systems that are exactly linearizable, (c) analysis and formulas for bounded control input and bounded outputs systems, (d) define the two-relay controller formulation for underactuated systems with degree of underactuation higher than one.

Acknowledgements. The authors gratefully acknowledge the financial support from CONACYT (Consejo Nacional de Ciencia y Tecnología), grants 132125 and 127575. Programa de Apoyo a Proyectos de Investigación e Innovación Tecnológica (PAPIIT) UNAM, grant 117211.

References

1. Aguilar, L., Boiko, I., Fridman, L., Freidovich, L.: Generating oscillations in inertia wheel pendulum via two relay controller. International Journal of Robust and Nonlinear Control 22, 318–330 (2012)
2. Aguilar, L., Boiko, I., Fridman, L., Iriarte, R.: Generation of periodic motions for underactuated mechanical system via second-order sliding-modes. In: Proc. of the American Control Conference, Minnesota, USA, pp. 5396–5400 (2006)
3. Aguilar, L., Boiko, I., Fridman, L., Iriarte, R.: Output excitation via continuous sliding-modes to generate periodic motion in underactuated systems. In: Proc. of the IEEE Conference on Decision and Control, San Diego, USA, pp. 1629–1634 (2006)
4. Aguilar, L., Boiko, I., Fridman, L., Iriarte, R.: Output excitation via second-order sliding-modes to generate periodic motion for underactuated systems. In: Proc. of the 9th International Workshop on Variable Structure Systems, Alghero, Italy, pp. 359–364 (2006)
5. Aguilar, L., Boiko, I., Fridman, L., Iriarte, R.: Generating self-excited oscillations for underactuated mechanical systems via two relay controller. International Journal of Control 82(9), 1678–1691 (2009)
6. Aguilar, L., Boiko, I., Fridman, L., Iriarte, R.: Generating self-excited oscillations via two-relay controller. IEEE Trans. on Automatic Control 54(2), 416–420 (2009)
7. Aguilar, L., Boiko, I., Iriarte, R., Fridman, L.: Periodic motion of underactuated mechanical systems self-generated by variable structure controllers: design and experiments. In: 2007 European Control Conference, Kos, Greece, pp. 3796–3801 (2007)
8. Atherton, D.: Nonlinear control engineering–Describing Function Analysis and Design. Van Nostrand, Workingham (1975)
9. Berkemeier, M., Fearing, R.: Tracking fast inverted trajectories of the underactuated acrobot. IEEE Transactions on Robotics and Automation 15(4), 740–750 (1999)
10. Boiko, I.: Oscillations and transfer properties of relay servo systems – the locus of a perturbed relay system approach. Automatica 41, 677–683 (2005)
11. Boiko, I.: Discontinuous control systems: Frequency-domain analysis and design. Birkhäuser, Boston (2009)

12. Chevallereau, C., Abba, G., Aoustin, Y., Plestan, E., Canudas-de-Wit, C., Grizzle, J.: Rabbit: A testbed for advanced control theory. IEEE Control Systems Magazine 23(5), 57–79 (2003)
13. Dranga, O., Navy, I.: Stability analysis of feedback controlled resonant DC-DC converter using Poincaré map function. In: IEEE Int. Symposium on Industrial Electronics, Pusan, Korea, pp. 2142–2147 (2001)
14. Estrada, A., Fridman, L.: Exact compensation of unmatched perturbation via quasi-continuous HOSM. In: 47th IEEE Conference on Decision and Control, Cancún, México, pp. 2202–2207 (2008)
15. Fendrich, O.: Describing functions in limit cycles. IEEE Transactions on Automatic Control 37(4), 486–488 (1992)
16. Fridman, L.: An averaging approach to chattering. IEEE Transactions on Automatic Control 46(8), 1260–1265 (2001)
17. Fridman, L.: Slow periodic motion in variable structure systems. International Journal of Systems Science 33(14), 1145–1155 (2002)
18. Grizzle, J., Moog, C., Chevallereau, C.: Nonlinear control of mechanical systems with an unactuated cyclic variable. IEEE Transactions on Automatic Control 50(5), 559–576 (2005)
19. Lai, J.: Power conditioning circuit topologies: Power conversion from low-voltage dc to high-voltage ac for single-phase grid-tie applications. IEEE Ind. Electronics Mag. 3(2), 24–34 (2009)
20. Levant, A.: High-order sliding modes: differentiation and output-feedback control. International Journal of Control 76, 924–941 (2003)
21. Martínez, S., Cortés, J., Bullo, F.: Motion Planning and Control Problems for Underactuated Robot. In: Bicchi, A., Christensen, H.I., Prattichizzo, D. (eds.) Control Problems in Robotics. STAR, vol. 4, pp. 59–74. Springer, Heidelberg (2003)
22. Martínez-Salamero, L., Valderrama-Blavi, H., Giral, R., Alonso, C., Estibals, B., Cid-Pastor, A.: Self-oscillating DC-to-DC switching converters with transformer characteristics. IEEE Transactions on Aerospace and Electronic Systems 41(2), 710–716 (2005)
23. Nakamura, Y., Suzuki, T., Koinuma, M.: Nonlinear behavior and control of a nonholonomic free-joint manipulator. IEEE Transactions on Robotics and Automation 13(6), 853–862 (1997)
24. Orlov, Y., Aguilar, L., Acho, L., Ortiz, A.: Asymptotic harmonic generator and its application to finite time orbital stabilization of a friction pendulum with experimental verification. International Journal of Control 81(2), 227–234 (2008)
25. Orlov, Y., Riachy, S., Floquet, T., Richard, J.: Stabilization of the cart-pendulum system via quasi-homogeneous switched control. In: Proc. of the 2006 Int. Workshop on Variable Structure Systems, Alghero, Italy, pp. 139–142 (2006)
26. Robinett, I.R., Wilson, D.: What is a limit cycle? International Journal of Control 81(12), 1886–1900 (2008)
27. Sanchis, P., Ursua, A., Gubia, E., Marroyo, L.: Buck-boost DC-AC inverter for a new control strategy. In: 35th Annual IEEE Power Electronics Specialist Conference, Aachen, Germany, pp. 3994–3998 (2004)
28. Santiesteban, R., Floquet, T., Orlov, Y., Riachy, S., Richard, J.: Second order sliding mode control for underactuated mechanical system II: orbital stabilization of an inverted pendulum with application to swing up/balancing control. International Journal of Robust Nonlinear Control 18(4–5), 544–556 (2008)

29. Shiriaev, A., Perram, J., Canudas-de-Wit, C.: Constructive tool for orbital stabilization of underactuated nonlinear systems: virtual contraint approach. IEEE Transactions on Automatic Control 50(8), 1164–1176 (2005)
30. Utkin, V., Guldner, J., Shi, J.: Sliding Mode Control in Electromechanical Systems. CRC Press, Boca Raton (1999)
31. Varigonda, S., Georgiou, T.: Dynamics of relay relaxation oscillators. IEEE Transactions on Automatic Control 46(1), 65–77 (2001)
32. Youssef, M., Jain, P.: A novel single stage AC–DC self-oscillating series-parallel resonant converter. IEEE Transactions on Power Electronics 21(6), 1735–1744 (2006)

Chapter 10
Design of Sliding Mode Controller with Actuator Saturation

Deepak Fulwani and Bijnan Bandyopadhyay

Abstract. This chapter discusses two methods of designing a sliding surface in the face of an actuator saturation constraint for a class of nonlinear uncertain systems. The first approach uses an ARE based approach to design the sliding surface and the second approach uses the parametric Lyapunov equation to design the surface. These methods are based on the low gain approach proposed by Lin et al. The design methods give a surface matrix as a function of the designed parameter. This parameter can be modulated to reduce the control amplitude which ensures that the control limits are respected in a region of the state space. This region can be made sufficiently large by choosing appropriate values of the design parameter.

10.1 Introduction

Beginning in the late 1970s and continuing today, sliding mode control has received plenty of attention due to its insensitivity to disturbances and parameter variations. The well known *sliding mode control* is a particular type of Variable Structure Control System (VSCS). Recently many successful practical applications of sliding mode control (SMC) have established the importance of sliding mode theory which has mainly been developed in the last three decades. This fact is also witnessed by many special issues of leading journals focusing on sliding mode control [2,4]. The research in this field was initiated by Emel'yanov and his colleagues [6,7], and the design paradigm now forms a mature and an established approach for robust control and estimation. The idea of sliding mode control (SMC) was not known to the control community at large until an article published by Utkin [16] and a book by Itkis [11].

Deepak Fulwani
Indian Institute of Technology Jodhpur, India
e-mail: fulwani@gmail.com

Bijnan Bandyopadhyay
Indian Institute of Technology Bombay, India
e-mail: bijnan@ee.iitb.ac.in

B. Bandyopadhyay et al. (Eds.): *Advances in Sliding Mode Control*, LNCIS 440, pp. 207–219.
DOI: 10.1007/978-3-642-36986-5_10

SMC is an established method to deal with uncertainty- inevitable in most practical systems. However, for any practical systems, the input is always limited in magnitude. Therefore it is necessary to consider this limitation *a priori* while designing the SMC. Design of a first order sliding mode is done in two steps viz. design of a stable sliding surface and a control law which produces a sliding mode in finite time. To ensure that the actuator does not saturate for a given set of initial conditions, the sliding surface (switching function) design should incorporate this limitation. Some authors have contributed in this regard, Corradini and Orlando [5] proposed a nonlinear surface to handle actuator saturation. Bartoszewicz and Nowacka [3] proposed an optimal sliding surface to handle input constraints. Ferrara and Rubagotti [8] proposed an effective algorithm to handle saturation in the higher order sliding mode framework.

In this chapter, we present two methods to design a sliding surface by which the control magnitude can be made arbitrarily small by choosing an appropriate surface matrix. Our method is based on the low gain approach proposed in [12–14, 18]. To enforce sliding motion, the required control input has two components, one linear and the other, discontinuous (for first order sliding mode). The discontinuous component is decided by the maximum amplitude of uncertainty therefore it does not provide any flexibility to reduce the control input to avoid actuator saturation. The linear component (linearly) depends on the sliding function matrix; flexibility to design the sliding surface matrix can be explored to avoid actuator saturation. We present two methods, based on the low gain approach as mentioned earlier, to design the surface. In the first method, the sliding surface matrix is parameterized by the parameter ε and in the second method, it is parameterized by γ. These parameters can be altered to reduce the control amplitude. The rest of the chapter is organized as follows. The work presented in this chapter is based on our work in [9]

Section 10.2 discusses the system description and problem statement. The surface design is discussed in Section 10.3. The effect of actuator saturation is discussed in Section 10.4. Section 10.5 discusses another method to design the sliding surface based on a parameterized Lyapunov equation. To verify the design methodology a numerical example is simulated in Section 10.6 followed by concluding Section 10.7.

10.2 System Description and Problem Statement

Consider the following class of nonlinear uncertain systems

$$\begin{aligned} \dot{x}_1 &= x_2 \\ \dot{x}_2 &= x_3 \\ &\vdots \\ \dot{x}_n &= f(x,t) + b_2 sat(u(t)) \end{aligned} \tag{10.1}$$

The function $f(x,t)$ satisfies the classical condition for the existence and uniqueness. $sat(u(t))$ is a saturation function and is defined as follows

$$sat(u(t)) = sign(u(t)) \times min(u_{max}, |u|)$$

In the above equation, u_{max} is the maximum value of $u(t)$. We make the following assumptions for the above system:

Assumption 1. b_2 is a non-zero scalar.
Assumption 2. Uncertain nonlinear function $f(x,t)$ satisfies
$\|f(x,t)\| \leq R_1\|x\| + R_2 \ \forall x \times t \in R^n \times R$ here R_1 and R_2 are positive constants.
Assumption 3. In a region Σ in the state space, there exists a constant Q such that $\forall x \in \Sigma$, $Q \geq R_1\|x\| + R_2 + \beta$ where $\beta > 0$ is a small positive constant which satisfies

$$\forall x \in \Sigma, \quad Q \leq \delta u_{max} \tag{10.2}$$

where $0 < \delta < 1$.

Assumption 3 ensures that the maximum amplitude of disturbance/uncertainty is smaller than the available control amplitude $\forall x \in \Sigma$. This assumption is necessary to enforce sliding mode.

Let the switching function for the above system be

$$s := c^T(\varepsilon)x(t) \tag{10.3}$$

Here ε is a design parameter which will be discussed later. Control input to ensure sliding mode ($s = 0$) in finite time can be defined as

$$u(t) := -(c(\varepsilon)^T B)^{-1}\{c^T(\varepsilon)Ax + Qsgn(s)\} \tag{10.4}$$

It should be noted that the region Σ in the state space is the region where stability with saturated actuator is ensured. Furthermore, we define the following matrices:

$$B := \begin{bmatrix} 0 \\ 0 \\ 0 \\ b_2 \end{bmatrix}, \quad A := \begin{bmatrix} 0 & 1 & 0 & \cdots & 0 \\ 0 & 0 & 1 & \cdots & 0 \\ \vdots & \vdots & \vdots & \ddots & \vdots \\ 0 & 0 & 0 & & 0 \end{bmatrix}$$

The objective is to design a sliding surface matrix $c(\varepsilon)^T$ such that $\forall x \in \Sigma$, the control law (10.4) respects the saturation limit and resulting closed loop system remains stable.

10.3 Design of Switching Function

Consider the following representation of the system defined in (10.1)

$$\dot{z}_1 = A_{11}z_1 + A_{12}z_2 \tag{10.5a}$$

$$\dot{z}_2 = A_{21}z_1 + A_{22}z_2 + f(x,t) + b_2u(t) \tag{10.5b}$$

Here

$$z_1 := [x_1\ x_2\ ...\ x_{n-1}]^T$$

$$z_2 := x_n$$

$$A_{11} = \begin{bmatrix} 0 & I_{n-2} \\ 0 & 0 \end{bmatrix}, \quad A_{12} = \begin{bmatrix} 0 \\ 0 \\ \vdots \\ 1 \end{bmatrix}, \quad A_{21} = [0, \cdots, 0], \quad A_{22} = 0.$$

It is convenient to design the switching function in $z-$ coordinates. The switching function in $z-$ coordinates is defined as

$$s := c^T(\varepsilon)x := \begin{bmatrix} c_1(\varepsilon) & 1 \end{bmatrix} \begin{bmatrix} z_1 \\ z_2 \end{bmatrix} \tag{10.6}$$

Here $c_1(\varepsilon) \in R^{1\times(n-1)}$ is to be designed.

Design of $c_1(\varepsilon)$
The control law (10.4) ensures that sliding mode $s = 0$ occurs in finite time. This leads to

$$c_1(\varepsilon)z_1 + z_2 = 0$$

$$\Rightarrow z_2 = -c_1(\varepsilon)z_1 \tag{10.7}$$

Using (10.5a) and (10.7)

$$\dot{z}_1 = (A_{11} - A_{12}c_1(\varepsilon))z_1 \tag{10.8}$$

$c_1(\varepsilon)$ should be designed such that the above closed loop system remains stable. $c_1(\varepsilon)$ is defined as follows

$$c_1(\varepsilon) := A_{12}^T P_1(\varepsilon) \tag{10.9}$$

$P_1(\varepsilon) > 0$ is a symmetric matrix and obtained by solving the ARE

$$A_{11}^T P_1(\varepsilon) + P_1(\varepsilon)A_{11} - P_1(\varepsilon)A_{12}A_{12}^T P_1(\varepsilon) + Q_1(\varepsilon) = 0 \tag{10.10}$$

$Q_1(\varepsilon)$ can be chosen as $Q_1(\varepsilon) = \varepsilon I$ as proposed in low gain design approach [13, 14].

Theorem 10.1. *For some $\varepsilon \in (0,1]$ there exists a $P_1(\varepsilon)$ which solves ARE in* (10.10) *and it satisfies*

$$\lim_{\varepsilon \to 0} c_1(\varepsilon) = 0.$$

Proof. The ARE in (10.10) is the result of minimization of the following cost function

$$J(x,u) = \frac{1}{2}\int_0^\infty [\varepsilon z_1(t)^T z_1(t) + z_2^T(t) z_2(t)]dt \tag{10.11}$$

This is a standard LQR problem and the existence of a unique positive definite $P_1(\varepsilon)$ for $\forall\varepsilon > 0$ was proved by Willems (1971) [[17]] for LTI system of the form

$$\dot{x}(t) = Ax(t) + Bu(t)$$

The pair (A,B) should be stabilizable. Replacing A by A_{11} and B by A_{12}, the existence of $P_1(\varepsilon)$ can be proved in a similar way as it is proved in Willems (1971) [[17]]. The continuity of solution i.e. $P_1(\varepsilon) \to 0$ as $\varepsilon \to 0$ was proved in [[15]]. Using (10.9) it is straightforward to infer $\lim_{\varepsilon\to 0} c_1(\varepsilon) = 0$.

Remark 10.1. *With this condition ($\lim_{\varepsilon\to 0} c_1(\varepsilon) = 0$), we can find some ε to ensure arbitrarily small norm of $c_1(\varepsilon)$.*
We need to find a region Σ such that $\forall x \in \Sigma$ implies $|u| \leq u_{max}$. Define

$$c_1(\varepsilon) := A_{12}^T P_1(\varepsilon) := \left[\bar{c}_1(\varepsilon)\ \bar{c}_2(\varepsilon) \cdot\cdot\ \bar{c}_{n-1}(\varepsilon)\right] \tag{10.12}$$

Here $\bar{c}_i(\varepsilon),\quad i = 1\cdots(n-1)$ are constants which depend on ε, moreover, $\lim_{\varepsilon\to 0} c_i(\varepsilon)$ $= 0$.

Remark 10.2. *It should be noted that all eigenvalues of A_{11} are at origin. Any nonzero $\varepsilon \in (0,1]$ ensures that closed loop system (10.8) is stable which also ensures stability of sliding surface.*

10.4 Effect of Actuator Saturation

For most of the physically realizable systems, the actuator capacity (amplitude) is limited. This requires that for a given set of initial conditions, the control should respects its boundaries. Consider the control law (10.4)

$$\begin{aligned}
u(t) &= -(b_2)^{-1}\{c(\varepsilon)^T Ax + Q sgn(s)\} \\
&= -(b_2)^{-1}\left[\bar{c}_1(\varepsilon)\ \bar{c}_2(\varepsilon) \cdot\cdot\ \bar{c}_{n-1}(\varepsilon)\ 1\right] \times \\
&\quad \begin{bmatrix} 0\,1\,0 \cdot\cdot\, 0 \\ 0\,0\,1 \cdot\cdot\, 0 \\ \vdots\ \vdots\ \vdots \cdot\cdot \vdots \\ 0\,0\,0\quad 0 \end{bmatrix} \begin{bmatrix} x_1 \\ x_2 \\ \vdots \\ x_n \end{bmatrix} - (b_2)^{-1} Q sgn(s) \\
&= -(b_2)^{-1}\left[\bar{c}_1(\varepsilon)\ \bar{c}_2(\varepsilon) \cdot\cdot\ \bar{c}_{n-1}(\varepsilon)\right] \begin{bmatrix} x_2 \\ x_3 \\ \vdots \\ x_n \end{bmatrix} \\
&\quad -(b_2)^{-1} Q sgn(s) \\
&= -(b_2)^{-1} c_1(\varepsilon)\bar{x} - (b_2)^{-1} Q sgn(s)
\end{aligned} \tag{10.13}$$

Where $\bar{x} = \begin{bmatrix} x_2 \\ x_3 \\ \vdots \\ x_n \end{bmatrix}$

The control law in (10.13) has two parts linear and nonlinear. The nonlinear component of the control law is decided by the maximum value of uncertainty. However, norm of linear component $c_1(\varepsilon)$ can be made arbitrarily small by choosing a small value of ε which allows to reduce the contribution of linear component to the required proportion. Considering the equation (10.2), the linear part of the control law (10.13) is limited by

$$|(b_2)^{-1}c(\varepsilon)^T\bar{x}| \leq (1-\delta)u_{max} \tag{10.14}$$

Suppose Σ is the region in state space, such that $\bar{x} \in \Sigma$ implies $|u| \leq (1-\delta)u_{max}$. This region can be obtained as follows [10].

Find

$$g := \text{Max } \bar{x}^T P_1(\varepsilon)\bar{x} \Rightarrow |(b_2)^{-1}c^T(\varepsilon)\bar{x}| \leq (1-\delta)u_{max} \tag{10.15}$$

This actually has an analytic solution which can be obtained using [10].

$$\begin{aligned} g &= \frac{(1-\delta)^2 u_{max}^2}{(b_2)^{-1}c(\varepsilon)^T P_1(\varepsilon)^{-1}c(\varepsilon)(b_2)^{-1}} \\ &= \frac{b_2^2(1-\delta)^2 u_{max}^2}{A_{12}^T P_1(\varepsilon)A_{12}} \end{aligned} \tag{10.16}$$

Remark 10.3. *We need only initial condition of $\bar{x} \in \Sigma$ to respect saturation limits. Recall that $\bar{x} = [x_2\ x_3\ \cdots\ x_n]^T$, this implies that the state x_1 can take any value without affecting the control input. Therefore initial condition of state x_1 does not influence control much. This is verified through simulation example.*

Remark 10.4. *It is desirable that the region Σ should include all possible initial conditions. This region can be made arbitrarily large by choosing sufficiently small value of ε. Consider* (10.16)*, as it is discussed earlier, $\lim_{\varepsilon \to 0} P_1(\varepsilon) = 0$ therefore a small value of ε results in a large value of parameter 'g' and thus region Σ also becomes large.*

Remark 10.5. *Using control law* (10.13)*, existence of sliding mode can be easily proved using* (10.2) *in the region Σ.*

10.5 Parametric Lyapunov Based Approach to Design Switching Function

In this section, we will study another method to design switching function. This method is based on low gain approach proposed in [18]. Consider the minimization of cost function for the system defined in (10.5)

$$J_1(z(t)) = \int_0^\infty e^{\gamma t}(z_1^T M z_1 + z_2^T N z_2)dt \tag{10.17}$$

where $M = E^T E \geq 0$, $N > 0$ and γ is a positive scalar. Using [18], we have the following proposition

Proposition 10.1. *Consider the system equation (10.5a) and the cost function (10.17). Assuming pair (A_{11}, E) is detectable and the pair (A_{11}, A_{12}) stabilizable. Stabilizability of pair (A,B) ensures the stabilizability of pair (A_{11}, A_{12}). With this assumptions, the value of z_2 which minimizes the cost function $J_1(z(t))$ in (10.17)*

$$z_2^* = -N^{-1}A_{12}^T P_2(\gamma) z_1(t) \tag{10.18}$$

$P_2(\gamma)$ is the unique positive-definite solution of the following ARE

$$(A_{11} + \frac{\gamma}{2}I)^T P_2(\gamma) + P_2(\gamma)(A_{11} + \frac{\gamma}{2}I) - P_2(\gamma)A_{12}N^{-1}A_{12}^T P_2(\gamma) = -M \tag{10.19}$$

Closed loop system (10.5) and (10.18) is exponentially stable with convergence rate faster than $e^{-\gamma/2\, t}$.
With $M = 0$, (10.19) becomes

$$A_{11}^T P_2(\gamma) + P_2(\gamma)A_{11} - P_2(\gamma)A_{12}N^{-1}A_{12}^T P_2(\gamma) = -\gamma P_2(\gamma) \tag{10.20}$$

and corresponding cost function becomes

$$J_1(z_2) = \int_0^\infty e^{\gamma t} z_2^T N z_2 dt \tag{10.21}$$

It should be noted that during the sliding mode, the state z_2 behaves as an 'input' to the system and with the above cost function we are not penalizing z_1. However, the convergence rate is controlled by the parameter γ.

The ARE (10.20) has a unique positive definite solution

$$P_2(\gamma) = W_2^{-1}(\gamma) \tag{10.22}$$

$W_2 > 0$ is obtained by solving the following Lyapunov equation

$$W_2(A_{11} + \frac{\gamma}{2}I)^T + (A_{11} + \frac{\gamma}{2}I)W_2 = A_{12}N^{-1}A_{12}^T \tag{10.23}$$

the above equation can be easily obtained by rearranging (10.20).
z_2 which minimizes (10.21)

$$z_2 = -N^{-1}A_{12}^T P_2(\gamma) z_1(t) \tag{10.24}$$

By using (10.24), sliding surface matrix can be obtained as

$$c^T(\gamma) = \left[N^{-1}A_{12}^T P_2(\gamma)\ 1\right] \tag{10.25}$$

During the sliding mode $s = 0$, and the resulting closed loop system becomes

$$\dot{z}_1 = (A_{11} - A_{12}N^{-1}A_{12}^T P_2(\gamma))z_1 \tag{10.26}$$

To prove stability of sliding surface, we need to prove stability of the above closed loop system for $\gamma > 0$.

Theorem 10.2. *All the eigenvalues of the closed loop system in* (10.26) *have negative real parts, thus the system is stable.*

Proof. Consider (10.20)

$$\begin{aligned} A_{11}^T P_2(\gamma) + P_2(\gamma)A_{11} - P_2(\gamma)A_{12}N^{-1}A_{12}^T P_2(\gamma) &= -\gamma P_2(\gamma) \\ \Rightarrow P_2^{-1}(\gamma)A_{11}^T P_2(\gamma) + A_{11} - A_{12}N^{-1}A_{12}^T P_2(\gamma) &= -\gamma I \\ \Rightarrow A_{11} - A_{12}N^{-1}A_{12}^T P_2(\gamma) = P_2^{-1}(\gamma)(-A_{11}^T - \gamma I)P_2(\gamma) \end{aligned}$$

It is clear from the above equation that the matrices $A_{11} - A_{12}N^{-1}A_{12}^T P_2(\gamma)$ and $-A_{11}^T - \gamma I$ are similar matrices and so have the same eigenvalues. It should be noted that all $n-1$ eigenvalues of A_{11} are located at the origin. Any nonzero positive value of γ ensures that eigenvalues of $-A_{11}^T - \gamma I$ have negative real part and thus stability of sliding surface is proved.

Remark 10.6. *When eigenvalues of A_{11} are located anywhere in the s-plane, the scalar γ should be selected such that* [[1, 18]]

$$\gamma > -2min\{Re(\lambda(A_{11}))\} \tag{10.27}$$

where $Re(\lambda(A_{11}))$ denotes real part of eigenvalue of A_{11}.

Remark 10.7. *The matrix $P_2(\gamma)$ is differentiable and monotonically increasing with respect to γ,* [[1, 18]]

$$\frac{dP_2(\gamma)}{d\gamma} > 0 \tag{10.28}$$

By choosing appropriate value of γ, norm of matrix $P_2(\gamma)$ can be chosen sufficiently small and, as we discussed in the previous section, we can limit linear part of the control. This property is needed to prove existence of sliding mode with saturated actuator in a region of state space.

With this method we can use parameter γ in a similar way as we used the parameter ε with the ARE based method. However, it should be noted that $\varepsilon \in (0, 1]$ while γ can be any positive value for the given system.

10.6 Simulation Studies

In this section, we will simulate a fourth order nonlinear uncertain system. We will design surface only by one method i.e. ARE based method. The following parame-

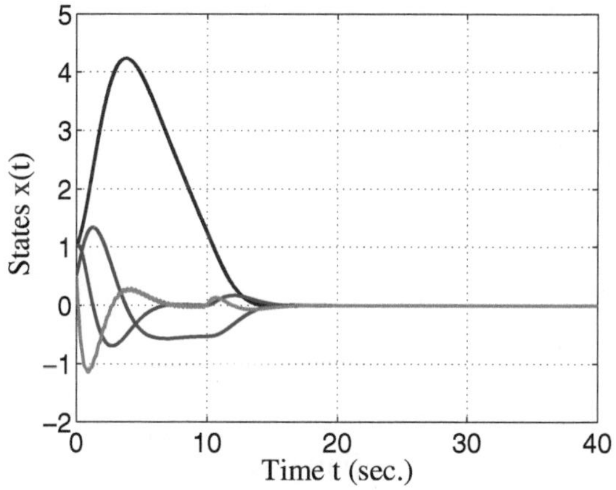

Fig. 10.1 Plot of states for Case-I

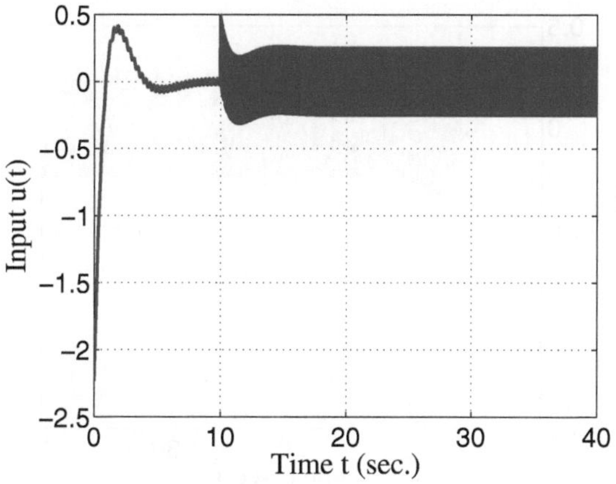

Fig. 10.2 Input for Case-I

ters are taken for the system in (10.1) $f(x,t) = 0.4sint(t)$, $b_2 = 2$, $u_{max} = 2.5$. The maximum value of $f(x,t)$ is 0.4, therefore parameter Q in control law is chosen as $Q = 0.5$ which satisfies (10.2). Linear part of the control depends on sliding surface matrix $c(\varepsilon)^T$. Linear part of the control can be reduced by choosing appropriate value of ε. The maximum possible value of linear part becomes 2. We design and simulate the system with two different values of ε parameter.

Case I. $\varepsilon = 0.9$
Solving ARE in (10.10) gives

$$P_1 = \begin{bmatrix} 2.1969 & 2.2312 & 0.9487 \\ 2.2312 & 4.4975 & 2.3157 \\ 0.9487 & 2.3157 & 2.3519 \end{bmatrix} \tag{10.29}$$

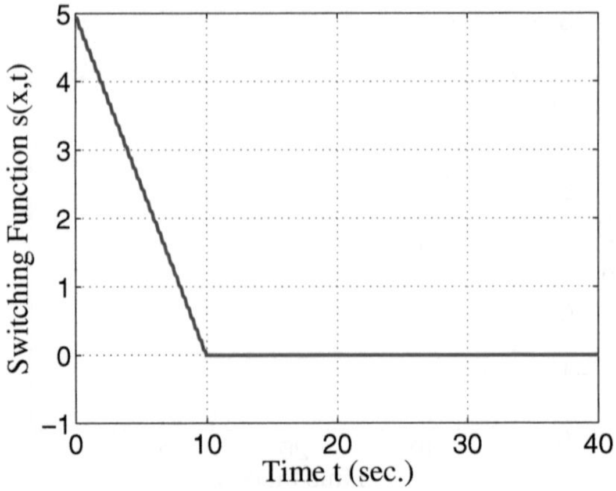

Fig. 10.3 Plot of switching function for Case-I

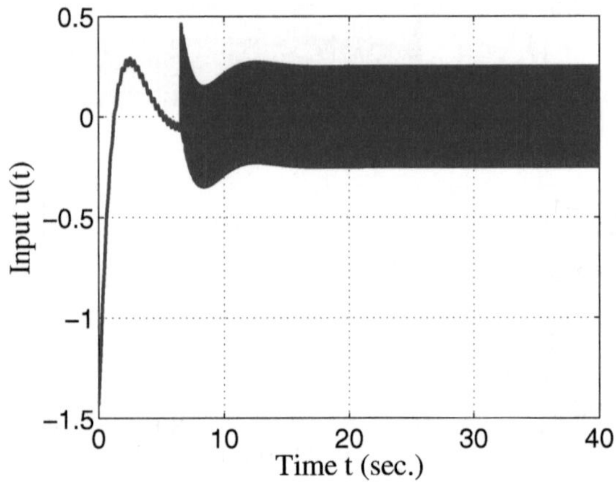

Fig. 10.4 Input for Case-II

Thus surface matrix becomes

$$c^T(\varepsilon) = \begin{bmatrix} 0.9487 \ 2.3157 \ 2.3519 \ 1.0000 \end{bmatrix} \tag{10.30}$$

With initial condition $x(0) = \begin{bmatrix} 1 \ 0.5 \ 1 \ 0.5 \end{bmatrix}^T$ the system is simulated with Runge-Kutta 4th order algorithm with maximum sampling time 0.001 sec. Fig. 10.2 shows input. Fig 10.1 shows evolution of states and Fig. 10.3 shows evolution of switching function with time. It is evident that system is stable and sliding mode establishes at $t = 10sec.$ and thereafter system remains in sliding mode.

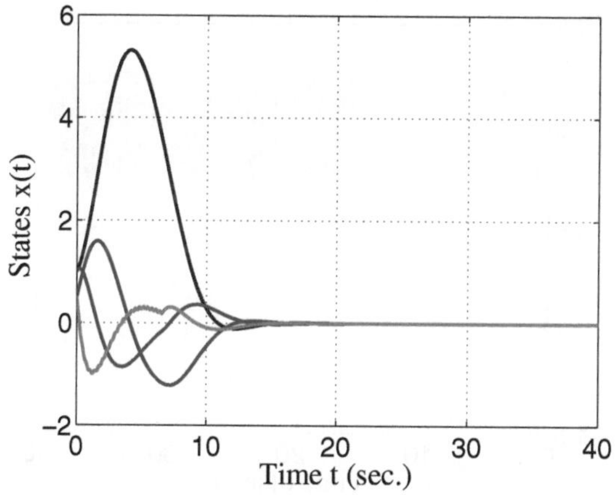

Fig. 10.5 Plot of states for Case-II

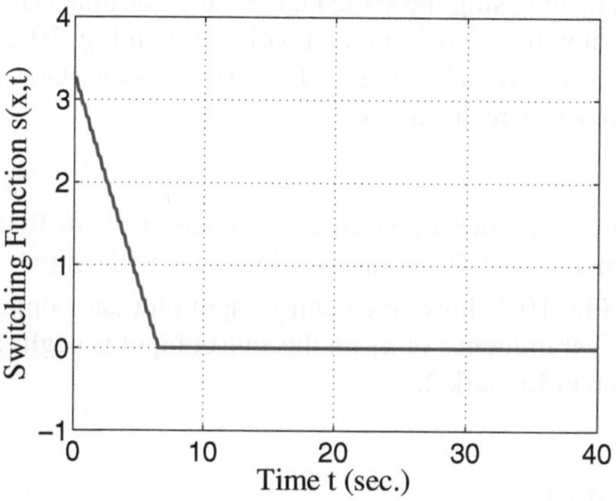

Fig. 10.6 Plot of switching function for Case-II

Case II. $\varepsilon = 0.2$

By proceeding in a similar way as we did in Case I, we obtain

$$P_1 = \begin{bmatrix} 0.5828 & 0.7492 & 0.4472 \\ 0.7492 & 1.7360 & 1.3032 \\ 0.4472 & 1.3032 & 1.6752 \end{bmatrix} \tag{10.31}$$

Thus surface matrix becomes

$$c^T(\varepsilon) = \begin{bmatrix} 0.4472 & 1.3032 & 1.6752 & 1.0000 \end{bmatrix} \tag{10.32}$$

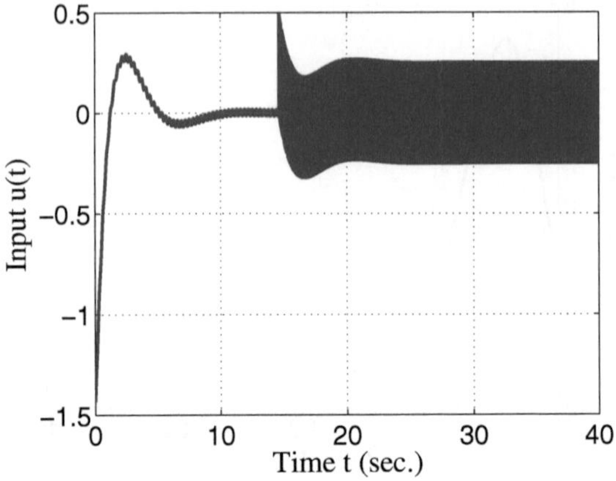

Fig. 10.7 Input for Case-III

The system is simulated using the same numerical algorithm as in Case I. Fig. 10.4 shows control input with $\varepsilon = 0.2$ and comparing it with Fig. 10.2 verifies that control input reduces as ε is reduced. Figs. 10.5, 10.6 show the states and the evolution of the switching function respectively.

Case III

The value of initial condition of x_1 state is increased by 10 fold and by keeping the same value of $\varepsilon = 0.2$. This case is simulated with initial condition $x(0) = \begin{bmatrix} 10\ 0.5\ 1\ 0.5 \end{bmatrix}^T$ Fig. 10.7 shows the control input plot and comparing it with Fig. 10.4 it is evident that influence of x_1 on thecontrol input is negligible which agrees with the discussion in Remark 2.

10.7 Conclusion

The sliding surface design to handle actuator saturation has been presented for a class of nonlinear system. The control amplitude can be controlled by a parameter ε. It has been observed that the control amplitude is not affected significantly by the value of the state x_1. Simulation studies verify the theoretical claims.

References

1. Amin, M.H.: Optimal pole shifting for continuous multivariable linear systems. International Journal of Control 41(3), 701–707 (1985)
2. Bartoszewicz, A., Kaynak, O., Utkin, V.I.: Special section on sliding mode control in industrial applications. IEEE Transactions on Industrial Electronics 55(11), 3806–4074 (2008)

3. Bartoszewicz, A., Nowacka-Leverton, A.: Itae optimal sliding modes for third-order systems with input signal and state constraints. IEEE Transactions on Automatic Control 55(8), 1928–1932 (2010)
4. Bartoszewicz, A., Patton, R.J.: Sliding mode control: Special issue. International Journal of Adaptive Control and Signal Processing 21, 635–822 (2007)
5. Corradini, M.L., Orlando, G.: Linear unstable plants with saturating actuators: Robust stabilization by a time varying sliding surface. Automatica 43, 88–94 (2007)
6. Emel'yanov, S.V.: Method of designing complex control algorithm using an error and its first time derivative only. Automation and Remote Control 18(10) (1957) (in Russian)
7. Emel'yanov, S.V., Burovoi, I.A., et al.: Mathematical models of process in technology and development of variable structure control system. Metallurgy (Moscow) 18(07) (1964) (in Russian)
8. Ferrara, A., Rubagotti, M.: A sub-optimal second order sliding mode controller for systems with saturating actuators. IEEE Transactions on Automatic Control 54(5), 1082–1087 (2009)
9. Fulwani, D., Bandyopadhyay, B.: Sliding surface design with saturated actuator. In: 2012 12th International Workshop on Variable Structure Systems (VSS), pp. 485–490 (2012)
10. Henrion, D., Garcia, G., Tarbouriech, S.: Piecewise linear robust control of systems with input saturation. European Journal of Control 5, 157–166 (1999)
11. Itkis, U.: Control Systems of Variable Structure. Wiley, New York (1976)
12. Lin, Z., Saberi, A.: Semi-global exponential stabilization of linear discrete-time systems subject to input saturation via linear feedbacks. Systems and Control Letters 24(2), 125–132 (1995)
13. Lin, Z., Stoorvogels, A.A., Saberi, A.: Output regulation for linear systems subject to input saturation. Automatica 32(1), 29–47 (1996)
14. Teel, A.: Semi-global stabilizability of linear null controllable systems with input nonlinearities. IEEE Transactions on Automatic Control 40(1), 96–100 (1995)
15. Trentelman, H.: Families of linear-quadratic problems: Continuity properties. IEEE Transactions on Automatic Control 32(4), 323–329 (1987)
16. Utkin, V.I.: Variable structure system with sliding mode. IEEE Transaction on Automatic Control 22(2), 212–221 (1977)
17. Willems, J.: Least squares stationary optimal control and the algebraic riccati equation. IEEE Transactions on Automatic Control 16(6), 621–634 (1971)
18. Zhou, B., Duan, G., Lin, Z.: A parametric lyapunov equation approach to the design of low gain feedback. IEEE Transaction on Automatic Control 53, 1548–1554 (2008)

Chapter 11
Quantization Behaviors in Equivalent-Control Based Sliding-Mode Control Systems

Yan Yan and Xinghuo Yu

Abstract. In this chapter, we study the quantization behaviors of the equivalent-control based second order single-input sliding-mode control (SMC) systems in the presence of quantized state feedback. We show that the class of SMC systems with both uniform and logarithmic quantizers can make the system states converge into a band, which relates to the quantization parameters. Properties of the system trajectories with quantized state feedback are characterized by using the concepts of quantized sliding mode (QSM) and quantized sliding-mode control (QSMC) system. We show that the QSM is piecewise constant. Various simulations illustrate the behaviors of the equivalent-control based SMC system under uniform and logarithmic quantized state feedback.

11.1 Introduction

Sliding-mode control systems have been extensively researched and applied to solve many practical control problems due to their robustness to external disturbance and parametric variations with a rather simple control structure. Major research efforts in SMC have been devoted to the development of various controllers using specific guiding principles [[4]], [[6]]- [[10]], [[13]], [[23]], [[24]], [[26]]- [[28]]. With the widespread implementation of SMC in practical systems, the SMC in digital systems has been widely studied these years. It is generally divided into the following two directions.

Yan Yan
School of Information Science and Technology, Dalian Maritime University,
116026, Dalian, Liaoning, China
e-mail: yy_spring@163.com

Xinghuo Yu
School of Electrical and Computer Engineering, RMIT University, Melbourne,
VIC 3000, Australia, and School of Automation, Southeast University, Nanjing, China
e-mail: x.yu@rmit.edu.au

B. Bandyopadhyay et al. (Eds.): *Advances in Sliding Mode Control*, LNCIS 440, pp. 221–241.
DOI: 10.1007/978-3-642-36986-5_11

One direction is that of discrete-time SMC systems. The other direction is that of SMC systems with quantization feedback.

Nowadays, SMC strategies are widely implemented in discrete-time and the switching frequency of the control variables is limited by h^{-1}, where h is the sampling time. The boundary layer of the chattering phenomenon in the vicinity of the sliding manifold was studied in [16], [18], [24]. The stable sliding mode designed for continuous-time systems may also become unstable after discretization [18]. Various SMC designs for discrete-time systems with relatively low switching frequency were proposed in [8], [6], [13]. The complex dynamical behaviors such as periodic trajectories and strange attractors due to zero-order-holder (ZOH) and Euler discretization were fully studied in [7], [19], [20], [26], [27]. A comprehensive survey was provided in [28].

The aforementioned papers assume that state feedback of SMC systems with infinite precision, which is not possible in some real practical systems. For example, in order to improve the performance of the networked control systems (NCSs), the effect of data quantization must be taken into account when designing a controller [2]. With the fast development of networked control and signal processing, the control systems with quantization feedback were widely studied in [2], [3], [5], [11], [12], [14], [29].

To the best of our knowledge, there have been only a few papers examining the quantization problem in the context of SMC [1], [15], [21], [22], [25], [30]. In [15], the level quantization effect in SMC systems with sliding modes of first and second orders were studied. A sliding-mode quantizer and feedback compensator were designed for a class-D audio power amplifier in [25]. The effect of quantization error on quasi sliding mode control was investigated in [1]. A novel switching-based sliding mode quantised feedback control design method was proposed in [30]. The recent papers [21]- [22] investigated the quantization effect on the second-order system based on the original SMC method. The definition and the occurrence conditions of 'quantised sliding mode' (QSM) were proposed. Discretization behavior and the characteristics of the QSM were analyzed, which showed that the effect of quantization on SMC systems was similar to discrete-time SMC systems.

Following the work of [21]- [22], in this chapter we explore the quantization behaviors on the most popular SMC method – the equivalent-control based SMC system. Particular attention is placed on the second-order systems so that insightful details of quantization effects can be explored. This chapter is organized as follows. It starts with the description of the equivalent-control based SMC systems. Furthermore, two typical quantization methods, the uniform quantizer and logarithmic quantizer, are introduced in Section 11.2. The second-order QSMC systems with both the uniform and logarithmic quantizers are analyzed in Section 11.3. Since the system steady bounds depend on the control parameter and quantization dense of the quantizer, the upper boundaries to guarantee the stability of the systems are established in Section 11.3. Various simulations are shown in Section 11.4 to verify the theoretical investigations. The chapter ends with the concluding remarks in Section 11.5.

11.2 Problem Statement

11.2.1 System Description

Consider the following single-input linear SMC system with

$$\dot{x}(t) = Ax(t) + bu \tag{11.1}$$

where $x(t) \in R^2$ is a state vector, $u \in R$ is the control input and $b = [0, 1]^T$. Assume $A \neq 0$ and A is a 2×2 constant matrix as $A = \begin{bmatrix} 0, & 1 \\ -a_1, & -a_2 \end{bmatrix}$. The switching manifold which represents asymptotically stable dynamics is defined as

$$s(x(t)) = cx(t) \tag{11.2}$$

where the constant matrix $c = [c_1, 1]$ with the slope of sliding mode $c_1 > 0$. The equivalent-control based SMC is [17]

$$u = u_{eq} + u_s \tag{11.3}$$

where the equivalent control u_{eq} is derived by solving $\dot{s} = 0$ subject to (11.1) and $u_{eq} = -(cb)^{-1}cAx(t)$, $u_s = -(cb)^{-1}\alpha \text{sgn}(s(x(t)))$, $\alpha > 0$ is the control parameter. The signum function presented by sgn() is defined as $\text{sgn}(x) = +1$ for $x > 0$, $\text{sgn}(x) = -1$ for $x < 0$ and $\text{sgn}(x) = 0$ for $x = 0$.

11.2.2 The Quantization Schemes and the Effect of Quantization to System State

In some systems such as NCSs, control signal data are quantized before transmission. Hence the dynamical system evolves in continuous time, but the state variables are quantized by a quantizer as shown in Fig. 11.1. Quantization from a mathematical viewpoint is the process of mapping a large set of input values to a small set C such as rounding values to some unit of precision. A device or algorithmic function that performs quantization is called a quantizer. The error introduced by quantization is referred to as quantization error. We choose two kinds of most popular quantizers here, the simple and commonly used uniform quantizer and logarithmic quantizer. They are introduced as follows.

11.2.2.1 Uniform Quantizer

The mapping of the uniform quantizer with quantizing level q is defined as

$$Q_{uni}(x) = q \cdot \text{round}(\frac{x}{q}) \tag{11.4}$$

where q is called quantizing level and $Q_{uni}(\cdot)$ is the quantization operator defined by a function round$(\cdot)$ that rounds a number to the nearest integer. The quantizer maps a real value input $x \in R$ to a countable set of output with discrete equidistant real values separated by distant q. The input-output characteristic of the quantizer is shown in Fig. 11.2(a). For all input values $x \in R$, the quantization error is $e = Q_{uni}(x) - x$ and $|e| \leq \frac{q}{2}$. The error characteristic of the quantizer is presented in Fig. 11.2(b).

To study the quantization effect, first let us look at the situation when the state variable is 1-D, as presented in Fig. 11.3(a), the axis of x is divided into many equal quantized bands by the uniform quantizer. In each quantized band $x_1 \in (\frac{2i^1-1}{2}q, \frac{2i^1+1}{2}q)$, the quantized value of x_1 is $\hat{x}_1 = i^1 q$, where $i^1 \in N$. Similarly, in 2-D, the $x_1 - x_2$ phase plane of (11.1) is divided into many quantized cells with the same size as in Fig. 11.3(b). In each quantized cell, $x_1 \in (\frac{2i^1-1}{2}q, \frac{2i^1+1}{2}q)$, $x_2 \in (\frac{2i^2-1}{2}q, \frac{2i^2+1}{2}q)$ and $i^1, i^2 \in N$, the quantized value of state is $Q(x) = (i^1 q, i^2 q)$, and it is located in the center of the cell.

11.2.2.2 Logarithmic Quantizer

The quantization levels of a logarithmic quantizer are described by

$$v = \{\mu_i = \rho^i \mu_0 : i = 0, \pm 1, \pm 2, ...\} \bigcup \{0\}, \tag{11.5}$$

where $\mu_0 > 0$ is a scaling parameter and $\rho \in (0,1)$ is quantization density. A small ρ implies coarse quantization whereas a large ρ implies dense quantization. $i < r$ and $r \in N^+$ is the smallest quantization level except zero. The quantization operator $Q_{log}(\cdot)$ is defined as:

$$Q_{log}(x) = \begin{cases} \rho^i \mu_0, & \text{if } \frac{1}{1+\delta}\rho^i \mu_0 < x \leq \frac{1}{1-\delta}\rho^i \mu_0, \\ 0, & \text{if } -\frac{1}{1+\delta}\rho^r \mu_0 \leq x \leq \frac{1}{1+\delta}\rho^r \mu_0, \\ -Q_{log}(-x), & \text{if } x < -\frac{1}{1+\delta}\rho^r \mu_0. \end{cases} \tag{11.6}$$

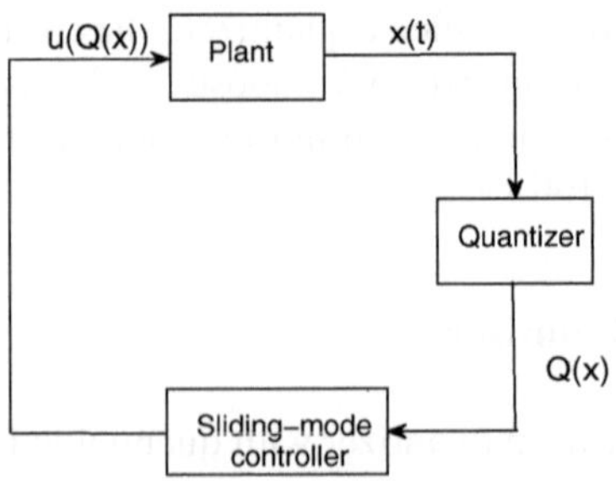

Fig. 11.1 The SMC system with quantized state feedback

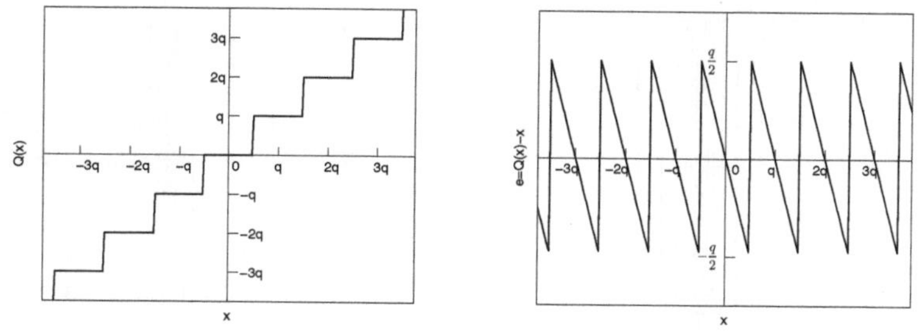

(a) The input-output characteristic.

(b) The error characteristic.

Fig. 11.2 The uniform quantizer

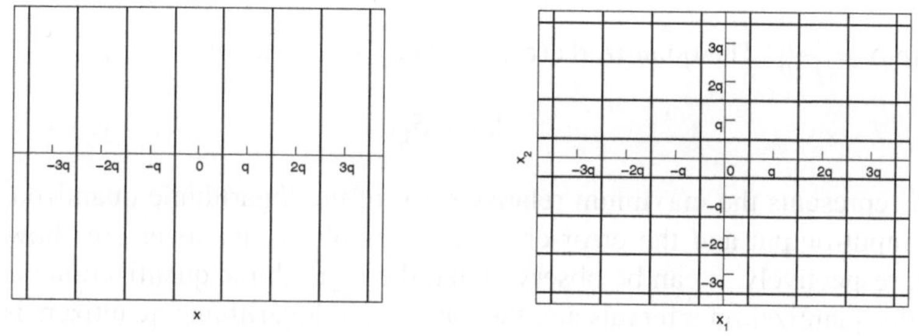

(a) One dimension.

(b) Two dimension.

Fig. 11.3 The uniform partitions

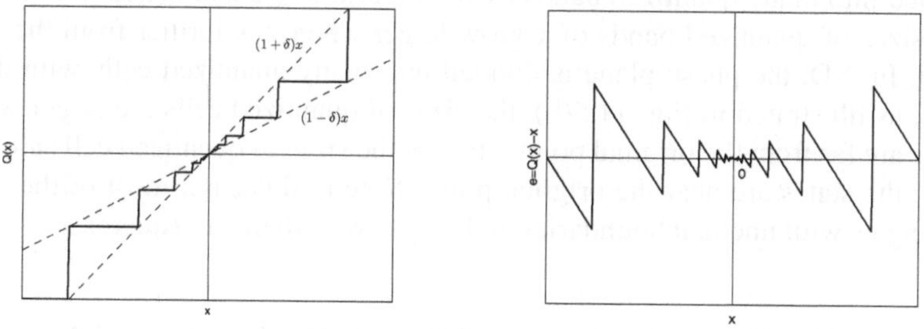

(a) The input-output characteristic.

(b) The error characteristic.

Fig. 11.4 The logarithmic quantizer

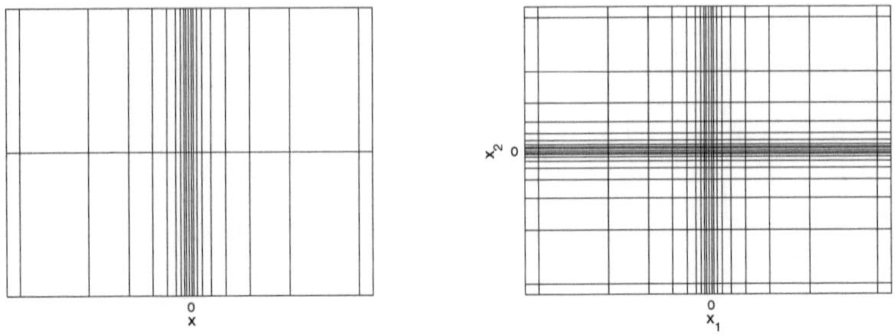

(a) One dimension. (b) Two dimension.

Fig. 11.5 The logarithmic partitions

where $\delta = \frac{1-\rho}{1+\rho}$. The quantized error $e = Q_{log}(x) - x$ satisfies

$$|e| \leq \delta |x| \tag{11.7}$$

So δ represents the maximum relative error of the logarithmic quantized variable. The input-output and the error characteristics of the quantizer are shown in Fig. 11.4, respectively. It can be observed that the logarithmic quantization has varying lengths quantization intervals and the partition in logarithmic quantizers is efficient for stabilization purpose. The magnitudes of the quantization error caused by the logarithmic quantizer are multiplicative, and reduce as the input signal becomes small.

For the system space divided by the logarithmic quantizer, in 1-D, the axis of x is divided into many quantized bands by the logarithmic quantizer as in Fig. 11.5(a). The sizes of quantized bands of x grow larger when x is further from the original point. In 2-D, the phase plane is divided into many quantized cells with different sizes, as illustrated in Fig. 11.5(b), the sizes of quantized cells are larger when the states are far from the original point whereas the sizes of quantized cells are smaller when the states are near the original point. Note in this case, most of the cells are rectangles with unequal boundaries and only few of them are squares.

11.2.3 The Equivalent-Control Based SMC System with Quantized State Feedback

Systems with quantization can be naturally viewed as hybrid systems [[11]]. By letting the states pass through a quantizer, the plant states are quantized as $\hat{x}_i(t) = Q(x_i(t))$, where $\hat{x}_i$ denotes the quantized values of x_i, $i = 1, 2$. The signals $\hat{u}$ and $\hat{s}$ are applied to denote the controller and switching manifold with quantized state feedback, respectively. Hence, the equivalent-control based SMC system with quantized state feedback can be rewritten as

$$\begin{aligned} \dot{x}_1 &= x_2 \\ \dot{x}_2 &= -a_1 x_1 - a_2 x_2 + \hat{u} \end{aligned} \tag{11.8}$$

$$\hat{s} = c_1 \hat{x}_1 + \hat{x}_2 \tag{11.9}$$

$$\hat{u} = \hat{u}_{eq} + \hat{u}_s \tag{11.10}$$

where $\hat{u}_{eq} = a_1\hat{x}_1 + (a_2 - c_1)\hat{x}_2$, $\hat{u}_s = -\alpha \mathrm{sgn}(\hat{s})$ and $\hat{x}$ is piecewise continuous and represents a discrete state.

Here two definitions in [[22]] are given for the SMC system with quantized state feedback.

Definition 11.1. The SMC system with quantized state feedback is called a quantized sliding-mode control (QSMC) system.

Definition 11.2. The sliding mode in a QSMC system is called a quantized sliding mode (QSM).

According to Definitions 11.1-11.2, (11.8)-(11.10) is a QSMC system and $\hat{s}$ is a QSM. Note $\hat{s}$ is nonlinear and piecewise constant due to $\hat{s}$ is the function of the piecewise constant variable $\hat{x}$, which is different with the linear sliding manifold in (11.2). In the following section, we will introduce the characteristics of the QSM. The questions to be addressed in this chapter are:

- What is the quantization effect on equivalent-control based SMC systems?
- What kinds of properties does the QSM exhibit with different quantizers and quantization parameters?

11.3 Quantization Behavior Analysis

System (11.8)-(11.10) can be considered as a dynamical system switching between two continuous-time systems containing discrete state $\hat{x}$. The dynamic behavior becomes much more complex due to the discrete state involved in the control input.

For the conventional SMC (11.3), u_{eq} stands for the linear part of controller and u_s stands for the nonlinear part, whereas for the QSMC system (11.8)-(11.10), both $\hat{u}_{eq}$ and $\hat{u}_s$ are nonlinear. Hence the switching happens for two reasons. One is due to the sign function in $\hat{u}_s$, the other is due to the fact that $\hat{u}_{eq}$ is function of $\hat{x}$ which is piecewise constant variable. The switching only happens in the boundaries of quantized cells. $\hat{u}_{eq}$ and $\hat{u}_s$ change according to the change of $\hat{x}$ and $\mathrm{sgn}(\hat{s})$, respectively, i.e., both $\hat{u}_{eq}$ and $\hat{u}_s$ will not change when the state stays inside quantized cells.

In the following, two typical quantization methods, the uniform quantization and logarithmic quantization are used in the QSMC system. The system dynamics in the two cases will be analyzed in the phase plane and the performance boundaries would be given, respectively.

11.3.1 The Equivalent-Control Based SMC System with Uniform Quantized State Feedback

As introduced in Section 11.2.2, the uniform quantized values of the system states are $\hat{x}_1 = q \cdot \text{round}(\frac{x_1}{q})$ and $\hat{x}_2 = q \cdot \text{round}(\frac{x_2}{q})$. Comparing to the conventional equivalent-control based SMC system, the closed loop dynamics of the QSMC system, with the uniform quantizer, is affected by quantization errors. The errors are as dynamic uncertainty in the system and come from two parts. One is the terms $\hat{x}$ as shown in the controller and the quantizing error $|e| = |\hat{x} - x| \leq \frac{q}{2}$ is as a kind of boundary disturbance in each quantized cells in state space. The other is the difference between $\text{sgn}(\hat{s})$ and $\text{sgn}(s)$. We divide the phase plane into regions according to the relation of $\text{sgn}(\hat{s})$ and $\text{sgn}(s)$. As shown in Fig. 11.6, the dash-dot line is the designed sliding manifold $s = cx$, the piecewise-constant line is the QSM, two parallel solid lines $s \pm \frac{(1+c_1)q}{2} = 0$ restrict the region of QSM, due to $\hat{s} = c\hat{x} = c(x+e) = s + ce$, and $|\hat{s} - s| \leq \frac{(1+c_1)q}{2}$, $\hat{s} \in [s - \frac{(1+c_1)q}{2}, s + \frac{(1+c_1)q}{2}]$. Note that $\text{sgn}(s) = \text{sgn}(\hat{s}) = +1$ in region I whereas $\text{sgn}(s) = \text{sgn}(\hat{s}) = -1$ in region III. However, it is possible that $\text{sgn}(s) \neq \text{sgn}(\hat{s})$ in regions II and IV, which causes the error due to the term $-\alpha \text{sgn}(\hat{s})$ in the controller. In region II $\text{sgn}(\hat{s}) = +1$ and in region IV $\text{sgn}(\hat{s}) \leq 0$ are shown in Fig. 11.6. The region V is the deadzone area, i.e., $\hat{s} = \hat{u} = 0$ due to $\hat{x} = 0$. The uniform QSM is located in the regions II and IV in Fig. 11.6 and is piecewise continuous. The response of a SMC system in general consists of three phases, namely, the reaching phase, sliding phase and steady-state phase [[29]]. When a quantizer is applied to the SMC system, the state response of the system can also be separated into the reaching, sliding and steady-state modes. The QSMC system (11.8)-(11.10) will be analyzed in the three phases, respectively.

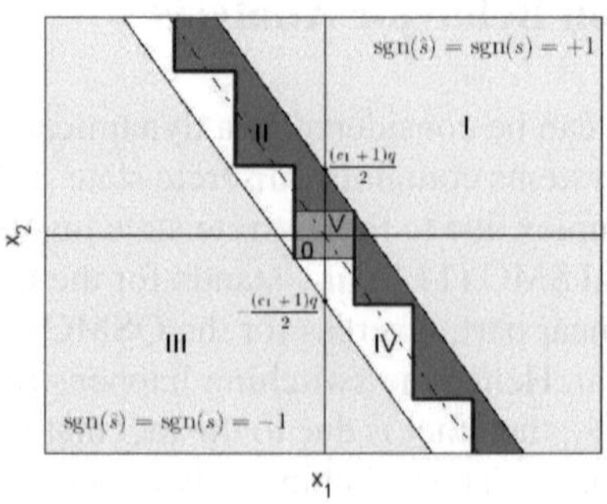

Fig. 11.6 Phase plane of the QSMC system with uniform quantizer

11.3.1.1 Reaching Phase

The reaching phase is the phase in which the trajectory moves into the QSM from any initial state. The situation is divided into the following two cases: if the state is in regions I and III in Fig. 11.6, $\hat{u}_s = u_s$ and $\hat{u}_{eq} \neq u_{eq}$; if the state is in regions II and IV, $\hat{u}_s \neq u_s$ and $\hat{u}_{eq} \neq u_{eq}$. In both cases u_{eq} and u_s stand for the controller without quantized state feedback as shown in (11.3). The switching in this phase has a low-frequency, and happens when the state crosses the boundaries of quantized cells.

Note the dynamics of s in regions I and III can be described as

$$\dot{s} = a_1 e_1 + (a_2 - c_1) e_2 - \alpha sgn(s) \tag{11.11}$$

where $|e_1| \leq \frac{q}{2}$ and $|e_2| \leq \frac{q}{2}$ are the quantizing errors of x_1 and x_2, respectively. The term $a_1 e_1 + (a_2 - c_1) e_2$ can be viewed as a kind of boundary disturbance due to e_1 and e_2 are bounded as shown in Fig. 11.2(b). In this situation, the system has linear dynamics in each quantized cell and piecewise-linear dynamics in phase plane.

11.3.1.2 Sliding Phase

The sliding phase is the phase in which the trajectory slides along the designed SM. It begins from the moment the trajectory reaches and starts to slide along the designed SM. Here $s = c_1 x_1 + x_2$ under the quantized state feedback turns into $\hat{s} = c_1 \hat{x}_1 + \hat{x}_2$. The variable $\hat{x}_1$ in $\hat{u}_s$ makes the QSM change only in the boundaries of x_1 in every quantized cell.

The sliding phase emerges when there exist two adjacent quantized state cells whose central values are $Q_1(\hat{x}_{1Q_1}, \hat{x}_{2Q_1})$ and $Q_1'(\hat{x}_{1Q_1'}, \hat{x}_{2Q_1'})$ with the properties that $\hat{x}_{1Q_1} = \hat{x}_{1Q_1'}$ and $\hat{x}_{2Q_1} = \hat{x}_{2Q_1'} \pm q$. The value of $\hat{s}$ has the property that $\hat{s}_{Q_1} > 0$ and $\hat{s}_{Q_1'} \leq 0$. Then the QSM in the band of $\hat{x}_1 = \hat{x}_{1Q_1}$ is $\hat{s}^1 : x_2 = \frac{\hat{x}_{2Q_1} + \hat{x}_{2Q_1'}}{2}$. As shown in Fig. 11.7(a), the dash-line cell is the quantized cell and the point G stands for the beginning of the sliding phase. The motion along $\hat{s}^1$ changes when the trajectory runs out of the quantized band $\hat{x}_1 = \hat{x}_{1Q_1}$, as the point F in the figure. Assuming in each quantized band of x_1, the quantized state value is $\hat{x}_1 = \hat{x}_{1Q_k}$, $k = 1, 2, ..., m$, m being the number of quantized bands of x_1 in which the QSM goes through in the sliding phase. $\hat{x}_1 = \hat{x}_{1Q_1}$ is the quantized band of x_1 where the first local QSM happens. Then the last local QSM happens in the band $\hat{x}_{1Q_m} = \pm q$. Hence, in each band of $\hat{x}_{1Q_k}$, there exist quantized cells with the centers $Q_k(\hat{x}_{1Q_k}, \hat{x}_{2Q_k})$ such that the absolute value of the QSM $\hat{s}$ is the smallest in the band. The switching happens between Q_k and its adjacent quantized state point $Q_k'(\hat{x}_{1Q_k'}, \hat{x}_{2Q_k'})$, where $\hat{x}_{1Q_k} = \hat{x}_{1Q_k'}$ and $\hat{x}_{2Q_k'} = \hat{x}_{2Q_k} \pm q$. The local QSM locates in the common boundary of the quantized cells whose centers are Q_k and Q_k', i.e., the local sliding mode in the quantized band $\hat{x}_1 = \hat{x}_{1Q_k}$ is $\hat{s}^k$: $x_2 = \frac{\hat{x}_{2Q_k} + \hat{x}_{2Q_k'}}{2}$. Moreover, it is obvious that the length of the

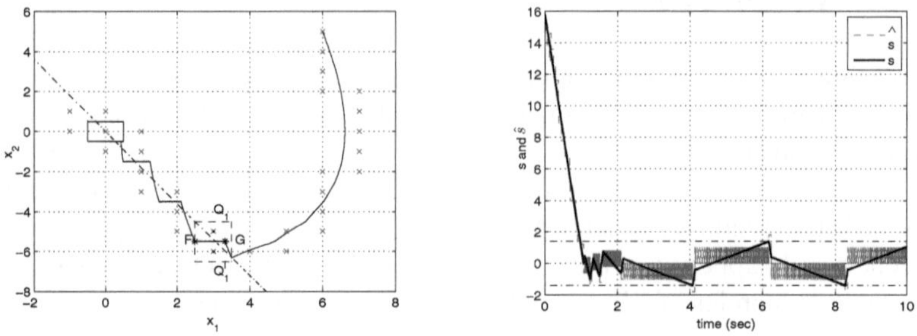

(a) The phase portrait. (b) Time responses of s and $\hat{s}$.

Fig. 11.7 The effect of the uniform quantizer to the SMC system

local uniform QSM equals to the quantizing level q. In summary, the switching function of the QSM is defined as

$$\hat{S} = \bigcup_{k=1}^{m} \hat{s}^k \tag{11.12}$$

where $\hat{s}^k : x_2 = \frac{\hat{x}_{2Q_k} + \hat{x}_{2Q'_k}}{2}$ is the local QSM in the band of $\hat{x}_1 = \hat{x}_{1Q_k}$. The characteristics of the QSM are summarized as follows:

1. The QSM is piecewise continuous and parallel to the axis of x_1 in phase plane of the system. After the trajectory departs the quantized band including the first local QSM, it will reach the next local QSM and slide along it. The similar motion is repeated until the trajectory reaches the dead zone of the uniform quantizer.
2. In each quantized band of x_1, under the action of the SMC, the local QSM is continuous and located in the $\hat{x}_2$ boundary of quantized cells whose absolute value of $\hat{s}$ is the smallest in the band.

11.3.1.3 Steady-State Phase

The uniform quantizer contains the quantizing level $Q_{uni}(x) = 0$, i.e., $\hat{x} = 0$ in the condition $|x| < \frac{q}{2}$. Note that when the states stay in the region $|x| < \frac{q}{2}$, $\hat{u}_{eq} = \hat{u}_s = 0$, hence there exists a dead-zone area for the controller. If system (11.8) is unstable without the control, its state will converge to a stable limit cycle as shown in Fig. 11.7(a). Trajectories both inside and outside the limit cycle move towards it.

One particular system behavior caused by quantization is the deterioration of dynamical performance near the equilibrium, as the difference between the current and the desired values of the state becomes small, higher precision is required, and so in the presence of quantization errors asymptotic convergence is impossible [[11]]. For the uniform QSMC system, the system is no longer asymptotic stable, and will not be stable if the control parameter α and quantizing level q are not chosen suitably.

The following theorem provides a sufficient condition under which all trajectories of the QSMC system (11.8)-(11.10) converge to the dead-zone area of the uniform quantizer in the steady-state phase.

Theorem 11.1. *Consider the second-order equivalent-control based SMC system (11.8)-(11.10), with the application of the uniform quantizer between the system state feedback and control input. If the condition*

$$\alpha \geq \frac{(|a_1|+|a_2-c_1|)q}{2} \tag{11.13}$$

is satisfied, then the following statements hold:

1. *The sliding mode will converge to the region $|s| \leq \frac{(1+c_1)q}{2}$ monotonically.*
2. *The closed-loop system is globally bounded.*
3. *In the steady state, the trajectory is chattering around the region $|x_1| \leq \frac{q}{2}$, $|x_2| \leq \frac{q}{2}$, and the quantized state is bounded by $|\hat{x}_1| \leq q$, $|\hat{x}_2| \leq q$.*

Proof. For the reaching phase, in regions I and III in Fig. 11.6, $\text{sgn}(s) = \text{sgn}(\hat{s})$, so the dynamic equation of the designed switching manifold is shown in (11.11). Consider the Lyapunov function

$$V = \frac{1}{2}s^2 \tag{11.14}$$

Substituting (11.11) into the time derivative of the Lyapunov function yields

$$\dot{V} = s\dot{s} = (a_1e_1 + (a_2-c_1)e_2)s - \alpha|s| \tag{11.15}$$

where the term $a_1e_1 + (a_2-c_1)e_2$ is considered as a disturbance in each quantized cell and

$$|a_1e_1 + (a_2-c_1)e_2| \leq \frac{(|a_1|+|a_2-c_1|)q}{2}$$

If the condition in (11.13) is satisfied then $\dot{V} < 0$ for $s \neq 0$ which indicates that s will converge to the region $|s| \leq \frac{(1+c_1)q}{2}$. The sliding mode s will decrease monotonically until the trajectory enters regions II and IV in Fig. 11.6. The boundaries of the region is presented by the two dash-dot lines in Fig. 11.7(b)

Substituting the controller (11.10) and $\hat{x}_2 = x_2 + e_2$ into (11.8), the dynamic behavior of x_2 can be rewritten as

$$\dot{x}_2 = a_1e_1 + (a_2-c_1)e_2 - c_1x_2 - \alpha sgn(\hat{s}) \tag{11.16}$$

In regions II and IV, before the trajectory reaches and slides on the QSM, the dynamical behaviors heavily depend on the location of $\hat{x}$ on two sides of the QSM. When the trajectory is in region $\text{sgn}(\hat{s}) = \text{sgn}(s)$, the dynamic is the same as in regions I and III as presented in the last paragraph. Once the trajectory enters the region $|s| \leq \frac{(1+c_1)q}{2}$, the value of s chatters until the trajectory arrives at the first local sliding manifold. If there is a chattering along the quantized boundaries of x_2, then the first QSM emerges and the trajectory slides along it until the quantized value of x_1 changes.

Now, we show that, if condition (11.13) in Theorem 11.1 is not satisfied, it is possible that the trajectory of the QSMC system cannot arrive at the dead-zone area of the uniform quantizer. Assume the initial state is $s > 0$, $\hat{x}_2 > 0$ and $|\alpha| < \frac{(|a_1|+|a_2-c_1|)q}{2}$, then the dynamic of x_2 changes into $\dot{x}_2 = a_1e_1 + (a_2 - c_1)e_2 - c_1x_2 - \alpha$. The terms $a_1e_1 + (a_2 - c_1)e_2$ are like a kind of disturbance as we said before. So if α is not large enough, there will be two adjacent quantized cells whose quantized values are $(\hat{x}_1, \hat{x}_{2n})$ and $(\hat{x}_1, \hat{x}'_{2n})$, with the properties that $x_{2n} = x'_{2n} \pm q$, the quantized value of x_1 are the same and

$$\dot{x}_2 = a_1e_1 + a_2e_2 - c_1\hat{x}_{2n} - \alpha < 0, \dot{x}_2 = a_1e_1 + a_2e_2 - c_1\hat{x}'_{2n} - \alpha > 0$$

The switching of the control part $\hat{u}_{eq}$ is of high-frequency, hence the trajectory will slide along the line $x_2 = \frac{x_{2n}+x'_{2n}}{2}$. Thus the trajectory will slide along the common boundaries of the quantized cells $x_2 = \frac{x_{2n}+x'_{2n}}{2}$. Note in this situation, $\hat{u}_s$ is a constant and $\hat{u}_{eq}$ will change according to the location of x. The input $\hat{u}$ switches as follows

$$\hat{u} = \begin{cases} a_1\hat{x}_1 + (a_2 - c_1)\hat{x}_2 - \hat{u}_s, & \text{if } x \text{ in the cell } (\hat{x}_1, \hat{x}_{2n}), \\ a_1\hat{x}_1 + (a_2 - c_1)\hat{x}'_2 - \hat{u}_s, & \text{if } x \text{ in the cell } (\hat{x}_1, \hat{x}'_{2n}). \end{cases} \tag{11.17}$$

It is $\hat{u}_{eq}$ in the controller (11.10) that becomes a switching controller. The SMC emerges in the system even in the reaching phase due to the structure of $\hat{u}_{eq}$ varies based on $\hat{x}$.

Observe that when the states stay in region $|x| < \frac{q}{2}$, $\hat{u} = 0$ and the control is not in effect, there exists a dead-zone area. If the system is unstable without the control, the states will converge to a stable limit cycle as shown in Fig. 11.7(a). Trajectories both inside and outside the limit cycle move towards it. ■

Remark 11.1. Compared with SMC system (11.1)-(11.3), the sign of the QSM $\hat{s}$ is no longer a constant and may change in the reaching phase. As shown in Fig. 11.6, e.g., if the initial point is located in region I and the trajectory reaches the QSM through region IV, in which $\text{sgn}(\hat{s})$ turns into -1 or 0 although the QSM is not reached.

Remark 11.2. In the sliding phase, the trajectory of the conventional SMC system (11.1)-(11.3) slides along the linear switching line. For the QSMC system (11.8), the QSM only happens when the trajectory slides along the boundaries of x_2 in each quantized cells. There is no switching in $\hat{u}$ when the state is located inside the quantized cells.

The uniform quantizer is simple and widely used, but it has a dead-zone area which is proportionally amplified following with the value of quantizing level q. The closed-loop system state can only converge to region $|x| = \frac{q}{2}$. To avoid this, we will propose an alternative state quantizer – the logarithmic quantizer in the next part.

11.3.2 The Equivalent-Control Based SMC System with Logarithmic Quantized State Feedback

As introduced in Section 11.3.1, the uniform quantizer has a dead-zone area, which makes the system converge and chatter around the area $\hat{x} = 0$. We now propose the QSMC system with logarithmic quantizer for improvement.

The dynamic behavior of the logarithmic QSMC system is similar to that of uniform QSMC system, as shown in Fig. 11.8(a). The length of the local quantizer equals to the quantizing level of x_1 due to the term $\hat{x}_1$ included in the QSM $\hat{s}$. Once the trajectory reaches the first local QSM as the point G in Fig. 11.8(a), it will slide along the common boundary of the adjacent quantized cells Q_1 and Q_1' as the rectangles in the figure. Note the quantized cells are no longer squares and have different lengths and widths. In the sliding phase, each local uniform QSM is equal whereas for the logarithmic QSM, the length of $\hat{s}^k$, $k = 1, 2, ..., m$, equals to $\frac{2\delta}{1-\delta^2}\rho^i\mu$ in the quantizing band $\hat{x}_1 = \rho^i\mu$, $i = 0, \pm1, \pm2, ...$ Assume $i \le r$, $r \in N^+$, then the smallest quantization level next to $\hat{x} = 0$ equals $|\rho^r\mu|$. Hence, the dead-zone area of the logarithmic QSMC system can be controlled by r and turn into as small as possible by choosing a suitable integer r.

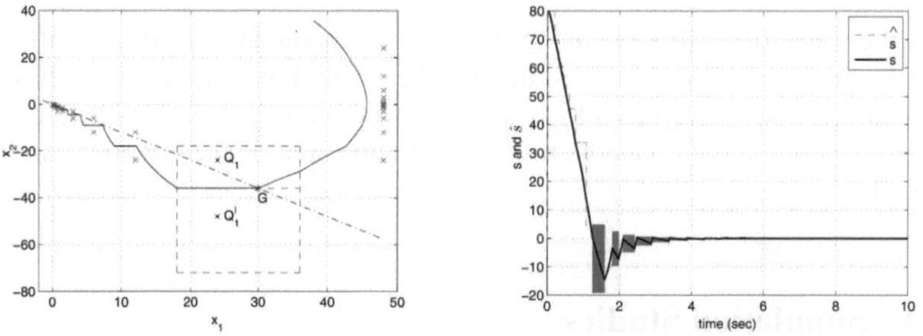

(a) The phase portrait. (b) Time responses of s and $\hat{s}$.

Fig. 11.8 The effect of the logarithmic quantizer to the SMC system

The sufficient condition for the stability of the logarithmic QSMC system is different from that of the uniform QSMC system in Theorem 11.1 and can be derived in the following theorem.

Theorem 11.2. *Consider the second-order equivalent-control based SMC system (11.8)-(11.10), with the application of logarithmic quantizer between the system state feedback and control input.* e_{1max} *and* e_{2max} *are the largest quantizing error of the system states. If the condition*

$$\alpha \ge |a_1 e_{1max}| + |(a_2 - c_1)e_{2max}| \tag{11.18}$$

is satisfied, then the following statements hold:

1. *The sliding mode will converge to the region $|s| \leq (1+c_1)e$ monotonically, where e is the quantization error when the first local QSM happens.*
2. *The close-loop system is globally bounded.*
3. *In steady state, the trajectory is chattering around the region bounded by $|x| \leq \rho^r \mu_0$, and the quantized state is bounded by $|\hat{x}| = \rho^r \mu_0$, where $i \leq r$ in (11.6).*

Proof. Similar to Theorem 11.1, we need $|\alpha| \geq a_1 e_1 + (a_2 - c_1)e_2$. For the logarithmic quantizer the error $|e| \leq \delta|x|$, e decrease as the value of x decrease as introduced in Section 11.2. When condition (11.18) is satisfied, s will converge monotonically in the reaching phase. Fig. 11.8(b) shows the time responses of s and $\hat{s}$ and it can be observed that in the sliding phase, the error between s and $\hat{s}$ decreases following the decrease of the quantization error.

In the region the sign of $\hat{s}$ is not changed, $\hat{u}_s$ is a constant and equals to $\pm\alpha$, but switching still exists due to $\hat{u}_{eq}$. If (11.18) is not satisfied, according to the dynamic of x_2

$$\dot{x}_2 = a_1 e_1 + (a_2 - c_1)e_2 - c_1 x_2 - \alpha sgn(\hat{s}) \tag{11.19}$$

It may emerge that two adjacent quantized cells of x with the same value of $\hat{x}_1 = \hat{x}_{1k}$, viz., $(\hat{x}_{1k}, \hat{x}_{2k})$ and $(\hat{x}_{1k}, \hat{x}'_{2k})$, have the following properties

$$\dot{x}_2 = a_1 e_1 + a_2 e_2 - c_1 \hat{x}_{2k} - \alpha < 0,\ \dot{x}_2 = a_1 e_1 + a_2 e_2 - c_1 \hat{x}'_{2k} - \alpha > 0$$

Thus the trajectory will slide along the common boundaries of the quantized cells $\hat{x}_2 = \hat{x}_{2k}$ and $\hat{x}_2 = \hat{x}'_{2k}$. In this situation, the switching caused by $\hat{u}_{eq}$ has a high frequency and $\hat{u}_{eq}$ becomes a kind of SMC whenever $\hat{u}_s$ does not change. The resulting system cannot converge to the dead-zone area of the quantizer around the original. ■

11.4 Simulation Studies

We now present some simulation results to verify the theoretical results presented in Section 11.3. In the simulations, the discontinuous system equations are digitally integrated using a ZOH at discrete moments, with the sampling period $h = 0.001$ second. In all phase plane diagrams, the solid lines are system trajectories whereas the cross signals are the quantized states. The dash-dot lines represent the QSM band in uniform QSMC systems and the designed sliding mode in logarithmic QSMC systems.

First, we show the trajectories of the second-order uniform QSMC system with the parameters selected as: $a_1 = -1$, $a_2 = -3$, $x(0) = (16, 8)$, $\alpha = 20$ and $c_1 = 1.2$. The quantization level q can be tailored to provide an anticipated response. Fig. 11.9 demonstrates trajectories with quantizing level $q = 0.5$, $q = 1$, $q = 2$ and $q = 3$, respectively. As indicated in theoretical discussions, the trajectories approach and slide along the QSM. Moreover, the trajectories converge to the dead-zone area whose magnitude increases with q. Fig. 11.10 demonstrates the quantization error

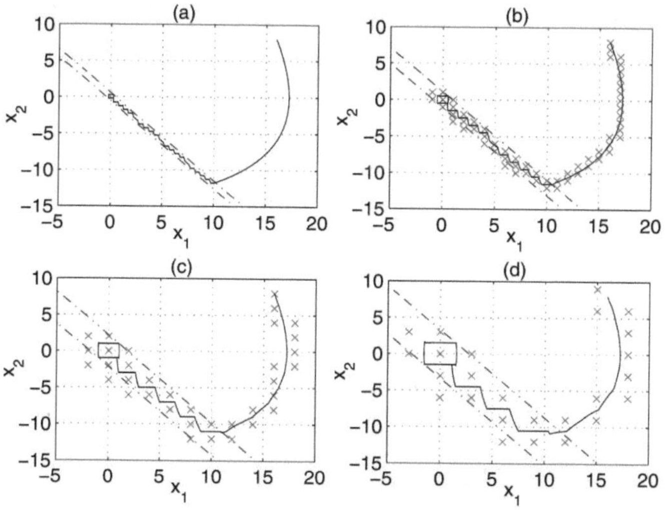

Fig. 11.9 $x_1 - x_2$ phase plane of the uniform QSMC system with different q. (a) $q = 0.5$, (b) $q = 1$, (c) $q = 2$, (d) $q = 3$.

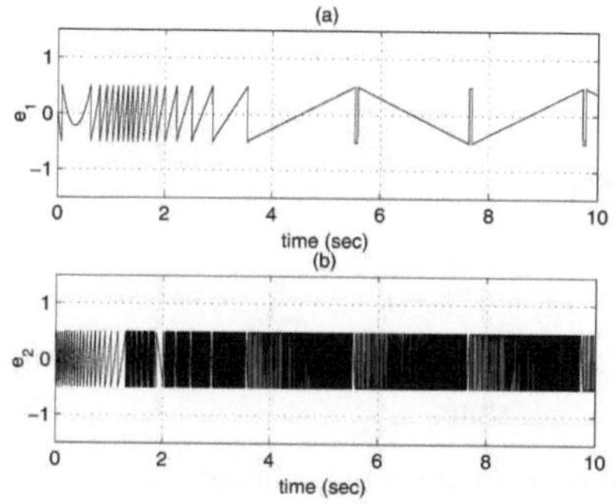

Fig. 11.10 The quantization error of the uniform QSMC system. (a) e_1, (b) e_2.

of the system in Fig. 11.9(b). Observe that the boundaries of e_1 and e_2 keep the same value of $\frac{q}{2}$ in the control process due to the characteristic of the uniform quantizer.

As the second example to illustrate Theorem 11.1, let us compare the impact of control parameter α on the system as shown in Fig. 11.11 where the designed system has parameter setting as $a_1 = -4, a_2 = -2, x(0) = (6,9), q = 1$ and $c_1 = 2.1$. The control parameter decreases from $\alpha = 20$ to $\alpha = 2$. It can be obtained that the smaller α, the worse the control effect and the trajectory in Fig. 11.11(d) cannot converge to zero. In Fig. 11.12 the state response and control input of the system in Fig. 11.11(d) are depicted. Fig. 11.12(a) shows $\hat{u}_s$ never changes whereas $\hat{u}_{eq}$ holds on switching according to the state located in different cells. The small value of α results in x_2 never converges to zero and x_1 cannot converge due to $\dot{x}_1 = x_2$.

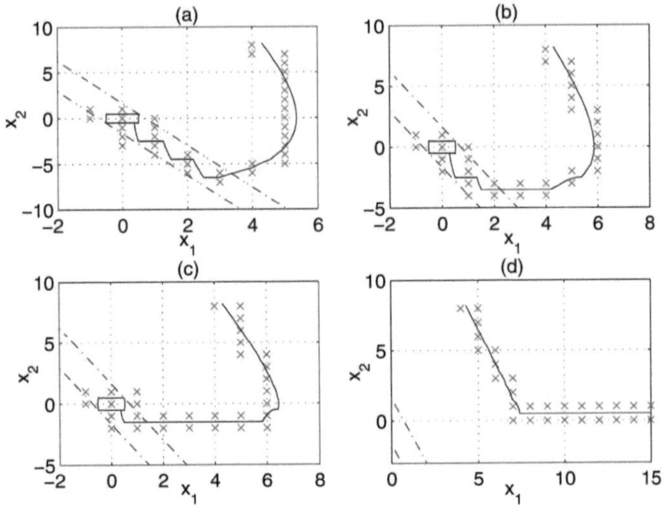

Fig. 11.11 $x_1 - x_2$ phase plane of the uniform QSMC system with different α. (a) $\alpha = 20$, (b) $\alpha = 10$, (c) $\alpha = 5$, (d) $\alpha = 2$.

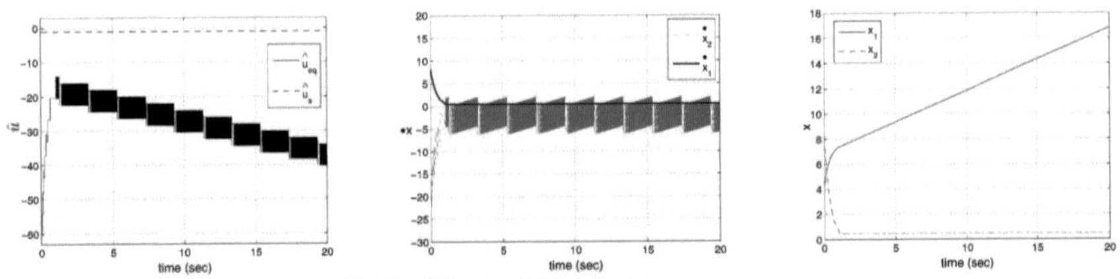

(a) The system input $\hat{u}_{eq}$ and $\hat{u}_s$.

(b) The derivative of system state.

(c) The system state x_1 and x_2.

Fig. 11.12 System response with uniform quantizer

Next examples depict the QSMC system performances with the logarithmic quantization.

The effect of the quantization density ρ on the response of the QSMC system is shown in Fig. 11.13. The parameters are set as $a_1 = -1$, $a_2 = -1$, $x(0) = (14.3, 18.2)$, $\mu = 4$ and $c_1 = 1.3$. The quantization density is changed from $\rho = 0.2$ to $\rho = 0.8$. Comparing the plots in Fig. 11.13, one can conclude that the quantization density has a tremendous effect on the system performance. Furthermore, similar to the effect of q in uniform QSMC systems, ρ can be chosen to provide a desirable QSM. The bigger the ρ chosen, the denser the QSM and the trajectory is closer to the case without a quantizer.

Fig. 11.14 shows the quantization error of the system in Fig. 11.13(d). We select the first 4 seconds which includes the reaching and sliding phases. Observe that the quantization error is reduced when the state is close to the original point.

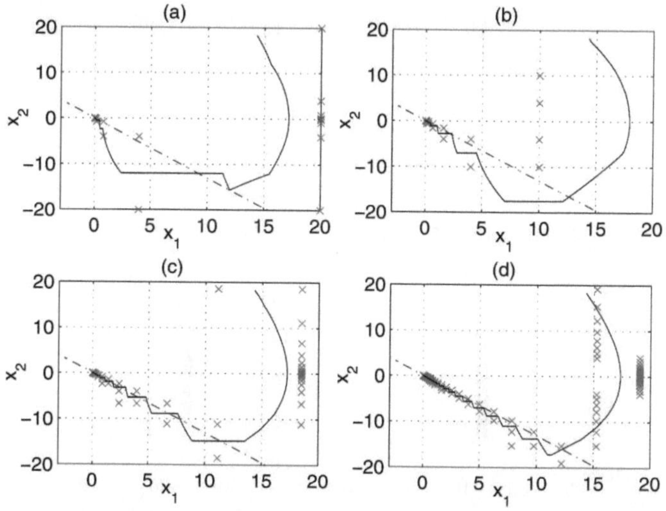

Fig. 11.13 $x_1 - x_2$ phase plane of the logarithmic QSMC system with different μ. (a) $\mu = 0.2$, (b) $\mu = 0.4$, (c) $\mu = 0.6$, (d) $\mu = 0.8$.

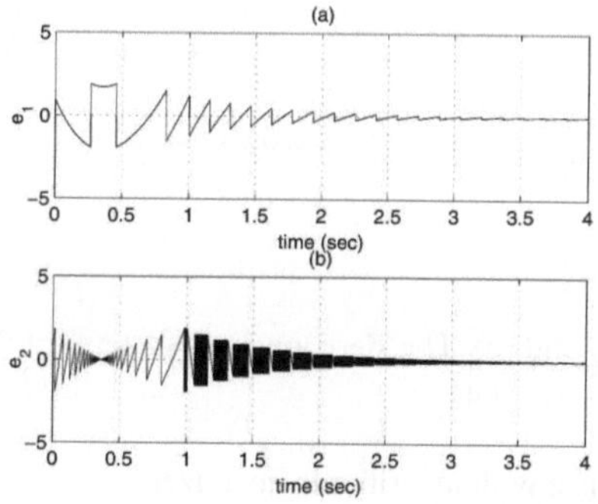

Fig. 11.14 The quantization error of the logarithmic QSMC system. (a) e_1, (b) e_2.

The plots in Fig. 11.15 represent the cases with the following parameters $a_1 = -10$, $a_2 = 2$, $x(0) = (-20.3, 12.2)$, $\mu = 4$, $\rho = 0.5$ and $c_1 = 0.8$. Here different values of α are chosen. The trajectories in Fig. 11.15(a)-(b) can converge to the original points. The smaller values of α in Fig. 11.15(c)-(d) result in the trajectories not converging to the original points but chattering around $(-12, 0)$ and $(-24, 0)$, respectively, as indicated in Theorem. 11.2. Let us illustrate Fig. 11.15(c) using the time response plots in Fig. 11.16. It can be obtained from Fig. 11.16(a) that the system control $\hat{u}_s$ dose not change from the moment approximately at 4.4 seconds whereas $\hat{u}_{eq}$ switches during the time and causes the chattering of state. Hence, the trajectory chatters around the cell boundary $x_1 = 12$. The situation in Fig. 11.15(d) is similar to that in Fig. 11.15(c).

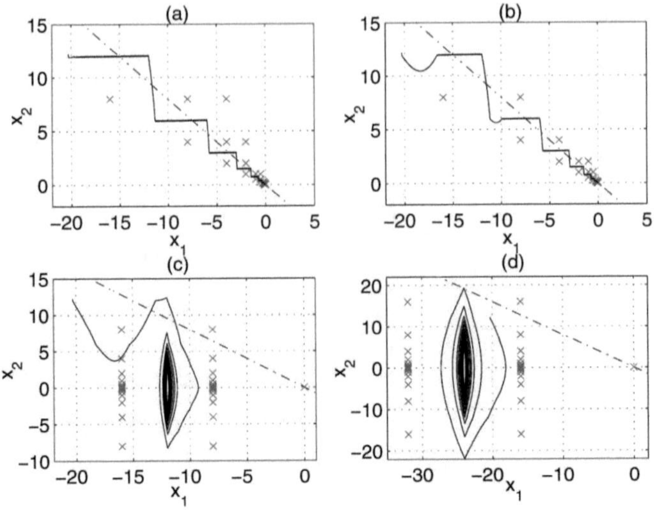

Fig. 11.15 $x_1 - x_2$ phase plane of the logarithmic QSMC system with different α. (a) $\alpha = 40$, (b) $\alpha = 25$, (c) $\alpha = 10$, (d) $\alpha = 2$.

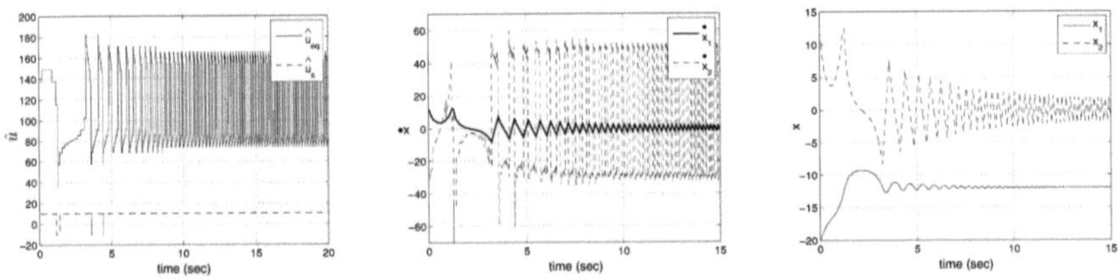

(a) The system input $\hat{u}_{eq}$ and $\hat{u}_s$. (b) The derivative of system state. (c) The system state x_1 and x_2.

Fig. 11.16 System response with logarithmic quantizer

Finally, Fig. 11.17 shows the cases with different values of quantizer parameter r. The $r \in N^+$ is the biggest value of i in (11.5) as we introduced before. The dead-zone area for the logarithmic quantizer is $x_{1,2} \in (-\frac{1}{1+\delta}\rho^r, \frac{1}{1+\delta}\rho^r)$ as shown in (11.6). Here we choose various values of r and it can be obtained that if r is small, the value of $\frac{1}{1+\delta}\rho^r$ is large, thus the dead-zone area of the logarithmic quantizer is similar to that of the uniform quantizer as shown in Fig. 11.17(a)-(b). The parameters of the system in the figure are set as $a_1 = -2$, $a_2 = -6$, $x(0) = (-10.3, -18.2)$, $\mu = 5$, $\rho = 0.6$, $\alpha = 40$ and $c_1 = 1.1$. In Fig. 11.17(a), $r = 3$ and the dead-zone area is $x_{1,2} \in [-\frac{1}{1+\delta}\rho^3, \frac{1}{1+\delta}\rho^3]$, in which $\hat{x}$ and $\hat{u}$ equals zero. When $r = 15$ in Fig. 11.17(c), the dead-zone area is smaller as the zoom in area in Fig. 11.17(d). Hence the parameter r can be designed large enough to improve the control accuracy.

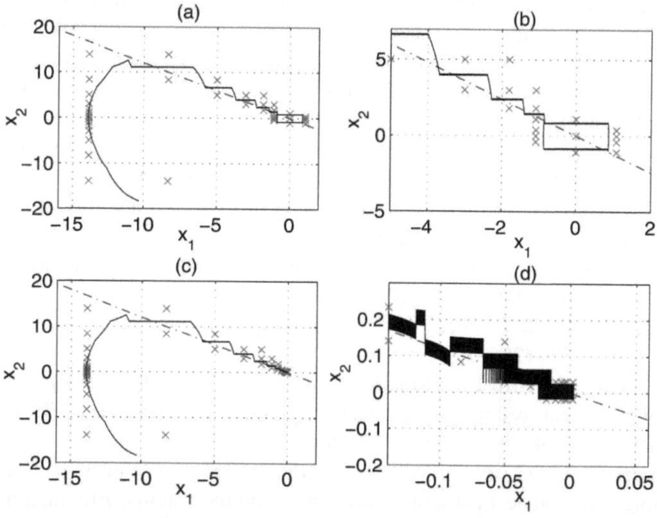

Fig. 11.17 Phase plane of the QSMC system with logarithmic quantizer. (a) $r = 3$, (b) Zoom in of plot (a) around the original, (c) $r = 15$, (d) Zoom in area of plot (c) around the original.

11.5 Conclusion

In this chapter, following the equivalent-control based SMC approach, we have explored the dynamic characteristic of the second order equivalent-control based SMC systems with quantized state feedback. We have also proposed a novel sliding mode – QSM and the uniform and logarithmic quantizers have been adopted between the system state output and control input. The QSM design provides a systematic way to guarantee a stable piecewise constant sliding manifold. Future work will be focused on multi-input and multi-output SMC systems.

References

1. Chakrabarty, S., Bandyopadhyay, B.: Quasi sliding mode control with quantization in state measurement. In: Proceedings of 37th Annual Conference on IEEE Industrial Electronics Society, Australia, pp. 3971–3976 (2011)
2. Corradini, M., Ippoliti, G., Orlando, G., et al.: Robust control of hybrid plants in the presence of quantization errors. In: Proceedings of IEEE International Conference on Control and Automation, New Zealand, pp. 234–239 (2009)
3. Delchamps, D.F.: Stabilizing a linear system with quantized state feedback. IEEE Transactions on Automatic Control 35(8), 916–924 (1990)
4. Edwards, C., Spurgeon, S.K.: Sliding mode control, theory and applications. CRC Press, Taylor & Francis Group (1998)
5. Elia, N., Mitter, S.K.: Stabilization of linear systems with limited information. IEEE Transactions on Automatic Control 4, 1384–1400 (2001)

6. Furuta, K., Pan, Y.: Discrete-time VSS control for continuous-time systems. In: Proceedings of the First Asian Control Conference, Tolyo, Japan, pp. 337–380 (1994)
7. Galias, Z., Yu, X.: Analysis of zero-order holder discretization of two dimensional sliding-mode control systems. IEEE Transactions on Circuits and Systems-II: Express Briefs 55(12), 1269–1273 (2008)
8. Gao, W., Wang, Y., Homaifa, A.: Discrete-time variable structure control systems. IEEE Transactions on Industrial Electronics 42(2), 117–122 (1995)
9. Hung, J.Y., Gao, W., Hung, J.C.: Variable structure control: a survey. IEEE Transactions on Industrial Electronics 40(1), 2–22 (1993)
10. Levant, A.: Sliding order and sliding accuracy in sliding mode control. International Journal of Control 58(6), 1247–1263 (1993)
11. Liberzon, D.: Hybrid feedback stabilization of systems with quantized signals. Automatica 39, 1543–1554 (2003)
12. Liu, T., Jiang, Z., Hill, D.J.: Small-gain based output-feedback controller design for a class of nonlinear systems with actuator dynamic quantization. IEEE Transactions on Automatic Control 57(5), 1326–1332 (2012)
13. Milosavljevic, C.: General conditions for the existence of a quasi-sliding mode on the switching hyperplane in discrete variable structure systems. Automation and Remote Control 3, 36–44 (1985)
14. Rasool, F., Nguang, S.K.: Quantised robutst H_∞ output feedback control of discrete-time systems with random communication delays. Control Theory and Applications 4, 2252–2262 (2010)
15. Shkolnikov, I.A., Shtessel, Y.B., Adhami, R.: Digital sliding modes and quasi-exact tracking discrete-valued signals. In: Proceedings of IEEE 9th International Workshop on Variable Structure Systems, Italy, pp. 17–22 (2006)
16. Su, W.C., Drakunov, S., Özgüner, Ü.: An $O(T^2)$ boundary layer in sliding mode for sampled-data systems. IEEE Transactions on Automatic Control 45, 482–485 (2000)
17. Utkin, V.I.: Variable structure systems with sliding mode. IEEE Transactions on Automatic Control 22(2), 212–222 (1977)
18. Utkin, V.I., Drakunov, S.: On discrete-time sliding mode control. In: Proceedings of IFAC Symposium on Nonlinear Control Systems, Capri, Italy, pp. 384–389 (1989)
19. Wang, B., Yu, X., Chen, G.: ZOH discretization effect on single-input sliding mode control systems with matched uncertainties. Automatica 45, 119–125 (2009)
20. Xia, X., Zinober, A.S.I.: Periodic orbits from delta-modulation of stable linear systems. IEEE Transactions on Automatic Control 49(6), 1376–1380 (2004)
21. Yan, Y., Yu, X., Yu, S.: Quantization effect on sliding-mode control of a second-order dynamical system. In: Proceedings of IEEE 12th International Workshop on Variable Structure Systems, India, pp. 243–247 (2012)
22. Yan, Y., Yu, X., Yu, S.: Quantization effect on a second-order dynamical system under sliding-mode control. International Journal of Bifurcation and Chaos (2012) (accepted)
23. Xu, J.X., Zheng, F., Lee, T.: On Sampled Data Variable Structure Control Systems. In: Young, K.D., Özgüner, Ü. (eds.) Variable Structure Systems, Sliding Mode and Nonlinear Control. LNCIS, vol. 247, pp. 69–92. Springer, Heidelberg (1999)
24. Yu, S., Yu, X., Qian, W.: Time delayed discrete variable structure control with quasi-sliding modes. In: Yu, X., Xu, J.-X. (eds.) Advances in Variable Structure Systems: Analysis, Integration and Applications, pp. 84–92. World Scientific, Singapore (2000)
25. Yu, S.H., Tsai, Y.H.: Sliding-mode quantized control of a class-D audio power amplifier. In: Proceedings of the 2010 International Power Electronics Conference, Japan, pp. 553–556 (2010)

26. Yu, X., Chen, G.: Discretization behaviors of equivalent control based variable structure systems. IEEE Transactions on Automatic Control 48(9), 1641–1646 (2003)
27. Yu, X., Wang, B., Galias, Z., et al.: Discretization effect on equivalent control based multi-input sliding mode control systems. IEEE Transactions on Automatic Control 53(6), 1563–1569 (2008)
28. Yu, X., Wang, B., Li, X.: Computer-controlled variable structure systems: The state of the art. IEEE Transactions on Industrial Informatics 8(2), 197–205 (2012)
29. Yun, S.W., Choi, Y.J., Park, P.: Dynamic output-feedback guaranteed cost control for linear systems with uniform input quantization. Nonlinear Dynamics 62, 95–104 (2010)
30. Zheng, B.C., Yang, G.H.: Quantised feedback stabilisation of planar systems via switching-based sliding-mode control. IET Control Theory and Applications 6, 149–156 (2012)

Chapter 12
On Discontinuous Observers for Second Order Systems: Properties, Analysis and Design

Jaime A. Moreno

Abstract. Smooth observers are able to converge asymptotically to the actual value of the state, in the case where no measurement noise and no persistently acting perturbations are present. Under the same conditions continuous observers can converge in finite time. However, they are unable to converge if a perturbation/uncertainty is present. In order to achieve finite time and exact convergence in the presence of perturbations, it is necessary to use discontinuous injection terms. In this chapter, some recent developments in this direction for second order systems will be presented and the results will be illustrated by means of simple examples. It will be also shown that by including non globally Lipschitz injection terms the convergence time of the observers can be made independent of the initial condition. The restriction to the two dimensional case is due to the fact that all proofs are done by means of Lyapunov functions, that are only available for planar systems. However, this has as advantage that the treatment is mainly tutorial, and provides on the one side an easy introduction to the topic, and on the other side it presents in the simplest case the main results that are (probably) valid for the general case. We hope to be able to provide a similar treatment of the general case in the near future.

Key words: Sliding Modes, Variable Structure Control, Lyapunov Methods, Discontinuous Observers, Second Order Systems.

12.1 Introduction and Problem Statement

We will consider the class of (second order) systems that are described by the (possibly multivalued or discontinuous) differential equation

---

Jaime A. Moreno
Eléctrica y Computación, Instituto de Ingeniería,
Universidad Nacional Autónoma de México, 04510 Mexico D.F., Mexico
e-mail: `JMorenoP@ii.unam.mx`

B. Bandyopadhyay et al. (Eds.): *Advances in Sliding Mode Control*, LNCIS 440, pp. 243–265.
DOI: 10.1007/978-3-642-36986-5_12

$$\begin{aligned}\dot{x}_1 &= f_1(x_1,u) + x_2 + \delta_1(t,x,u)\,,\\ \dot{x}_2 &= f_2(x_1,x_2,u) + \delta_2(t,x,u,w)\,,\\ y &= x_1\end{aligned} \tag{12.1}$$

where $x_1 \in \mathbb{R}$, $x_2 \in \mathbb{R}$ are the states, $u \in \mathbb{R}^m$ is a known input, $w \in \mathbb{R}^r$ represents an unknown input and $y \in \mathbb{R}$ is the measured output. f_1 is a known continuous function and f_2 corresponds to a known possibly discontinuous or multivalued function. δ_1 and δ_2 represent uncertain terms. The measured variables are x_1 and the known input u. It is assumed that system (12.1) has solutions in the sense of Filippov [8].

When the uncertainty $\delta_1(t,x,u) \equiv 0$ in (12.1) the observability map is

$$\mathcal{O}(x,u,w) = \begin{bmatrix} y \\ \dot{y} \end{bmatrix} = \begin{bmatrix} x_1 \\ f_1(x_1,u) + x_2 \end{bmatrix},$$

which is clearly globally invertible for every known and unknown input u and w. In the absence of unknown input w system (12.1) (with $\delta_1(t,x,u) \equiv 0$ and $\delta_2(t,x,u,w=0) \equiv 0$) is *uniformly observable for every input [10–12]*. When there is an unknown input w the system (with $\delta_1(t,x,u) \equiv 0$) is said to be *strongly observable [13, 25]*. In both cases it is theoretically possible to determine the unmeasured state x_2 from the measurement of x_1. Note that if the uncertain term $\delta_1(t,x,u) \neq 0$ observability is lost, and it is imposible to determine exactly the state x_2.

Many second order systems are described by equations (12.1). For example (12.1) can represent a mechanical system when $\delta_1(t,x,u) \equiv 0$ and $f_1(x_1,u) \equiv 0$, where x_1 corresponds to the (measured) position and x_2 is the velocity. u can represent a control force (or torque) and w can correspond to uncertain parameters or forces. If there exist Coulomb friction forces or in the presence of back-slash or hysteretic phenomena the functions $f_2(x,u)$ and/or $\delta_2(t,x,u,w)$ are discontinuous or multivalued.

Many other systems, although not represented by (12.1) in original coordinates, can be brought to (12.1) by a (local or global) state diffeomorphism. In particular, it is well known that smooth systems (without uncertainties) and that are uniformly observable for every input [10–12] can be transformed to the form (12.1).

12.1.1 Objectives

Our aim in this chapter is to propose an observer that is able to estimate the unmeasured state x_2 from the measurement of x_1. It is clear from the observability analysis of the previous paragraph that this will be possible in an exact manner only if the perturbation term $\delta_1(t,x,u) \equiv 0$ (we are only considering the case without measurement noise).

Since many existing observer algorithms can be used for this purpose, we will list the distinguishing properties of the proposed observer:

1. It is able to estimate exactly the state x_2 after a *finite time* and *robustly with respect to uncertainties/perturbations*, represented by $\delta_2(t, x, u, w)$ in (12.1), that are persistent. In order to achieve this feature it is necessary to introduce discontinuous functions in the injection terms of the observer. It is important to note that finite time convergence can be achieved without discontinuous injection terms (just with continuous but not locally Lipschitz continuous ones at zero), but only in the absence of uncertainties/perturbations. See subsections 12.2.2, 12.4.1, 12.4.2.
2. The proposed observer is able to converge in a finite time that is *independent of the initial condition* of the plant and of the observer. In order to achieve this property it is required to introduce not globally Lipschitz injection terms. See subsections 12.2.2, 12.4.4.
3. The observer is able to deal with a known function f_1 that is continuous but not necessarily Lipschitz (globally or locally). The function f_2 can be discontinuous, it does not have to be locally or globally Lipschitz in x_1 and it can grow linearly in x_2. See subsections 12.2.2, 12.2.3, 12.4.3.
4. When a bounded uncertainty/perturbation δ_1 is present, the estimation error will be bounded. The same will be true in presence of measurement noise. See subsections 12.2.2, 12.2.3, 12.4.5.
5. The design of the observer proposed in the chapter is in the spirit of the High-Gain (HG) observer: the observer constants are parametrized in terms of a single gain, that has to be set large enough to meet the convergence, robustness and convergence time required. See subsections 12.2.1, 12.2.2, 12.4.
6. All proofs are based on Lyapunov's method. The Lyapunov functions used here are of quadratic type, so that the mathematical machinery required is very similar to what is needed for linear systems. See Section 12.4.
7. The proposed method can be considered as a generalization and improvement of other observer design methods in the literature. See subsections 12.2.3, 12.3.

In order to put in perspective the first two properties, we will in the next subsection illustrate in a simple simulation example the behavior with respect to finite time convergence, robustness to uncertainties/perturbations and the convergence time with increasing initial estimation error for two typical observer design methods: High-Gain Observer [7, 11, 15] and (First Order) Sliding Mode Observer [27].

12.1.2 Simulation Example

Consider a simple (mechanical) system described by

$$\dot{x}_1 = x_2\,, \quad \dot{x}_2 = w(t) \tag{12.2}$$

where x_1 is the measured position, x_2 is the (unmeasured) velocity and w is the unknown applied force. Note that w can represent unmodeled nonlinear and discontinuous phenomena as hysteresis, back-slash or Coulomb friction. The trajectories of the plant, with initial conditions $x_1(0) = 2$ and $x_2(0) = 1$, are shown in Figure

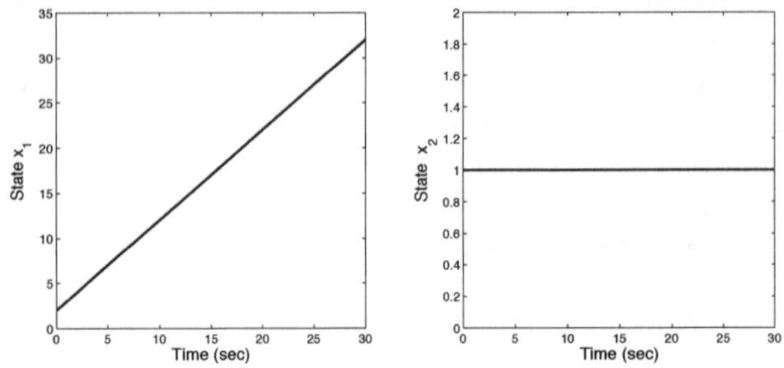

Fig. 12.1 Plant's trajectories with vanishing unknown input $w(t) = 0$

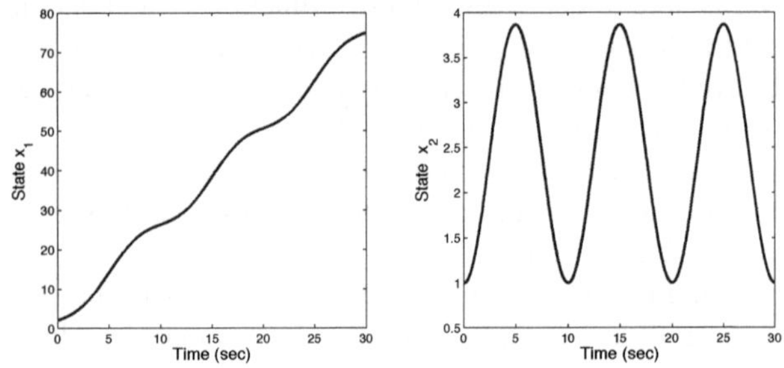

Fig. 12.2 Plant's trajectories with a periodic unknown input $w(t)$

12.1 in the case $w = 0$ and in Figure 12.2 when $w(t) = 0.9\sin(0.2\pi t)$. We will use this conditions for all the following simulations and those in Section 12.3.

12.1.2.1 A Linear Observer

The linear observer

$$\dot{\hat{x}}_1 = -l_1\gamma(\hat{x}_1 - x_1) + \hat{x}_2, \quad \dot{\hat{x}}_2 = -l_2\gamma^2(\hat{x}_1 - x_1)$$

with appropriately designed gains $l_1 > 0$, $l_2 > 0$ and $\gamma > 0$ is known to provide an exponentially convergent estimate of the velocity in the absence of unknown input. This can be seen in the simulation in Figure 12.3, where the gains have been selected as $l_1 = l_2 = \gamma = 1$. The initial conditions of the observer have been selected as $\hat{x}(0) = [-2, -1]$.

This is clear from the analysis of the dynamical behavior of the estimation errors $e_1 = \hat{x}_1 - x_1$, $e_2 = \hat{x}_2 - x_2$, given by

$$\dot{e}_1 = -l_1 e_1 + e_2, \quad \dot{e}_2 = -l_2 e_1 - w(t).$$

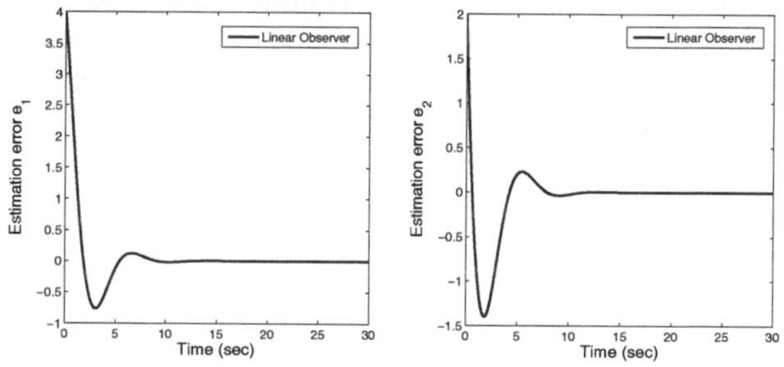

Fig. 12.3 Estimation errors of the Linear Observer without unknown input

However, it is also clear from the last equation that in the presence of a non vanishing unknown input w the estimation error will be unable to converge to zero. This is also illustrated in the simulation in Figure 12.4, with the same gains and periodic unknown input.

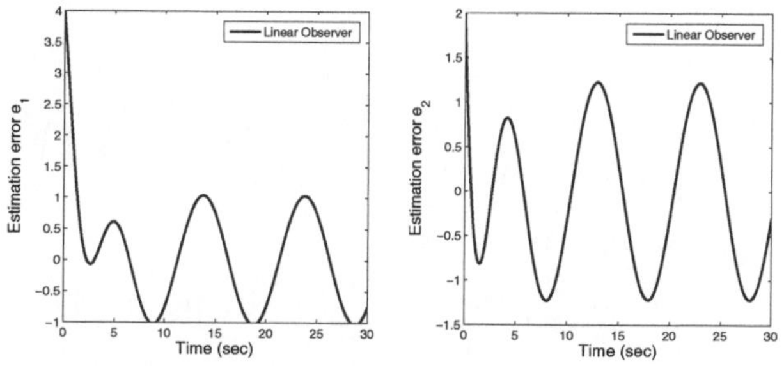

Fig. 12.4 Estimation errors of the Linear Observer with unknown input

In synthesis, the linear observer converges asymptotically (not in finite time) and is not able to converge to the true value of the unmeasured state in the presence of an unknown input. In fact, finite time convergence is impossible for any observer having locally Lipschitz continuous injection terms, and the convergence in the presence of persistent unknown inputs is also impossible for any continuous observer.

12.1.2.2 A Discontinuous First Order Sliding Mode Observer

In order to alleviate the problem, we consider a (First Order) Sliding Modes (SM) Observer [27], that has discontinuous injection terms, and has the form

$$\dot{\hat{x}}_1 = -l_1 \text{sign}(\hat{x}_1 - x_1) + \hat{x}_2, \quad \dot{\hat{x}}_2 = -l_2 \text{sign}(\hat{x}_1 - x_1).$$

However, this observer is also unable to either converge in finite time or to estimate the velocity correctly in the presence of an unknown input. This is illustrated in Figures 12.5 and 12.6, where $l_1 = l_2 = 1$, and the same initial conditions as for the linear observer, i.e. $\hat{x}(0) = [-2, -1]$, have been used.

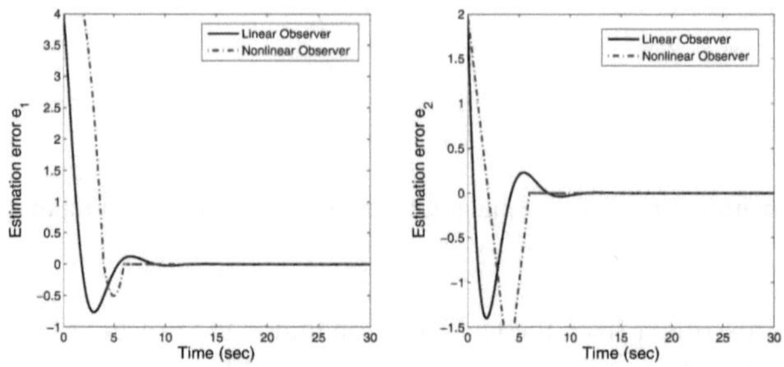

Fig. 12.5 Estimation errors for the Linear and the SM Observers without unknown input

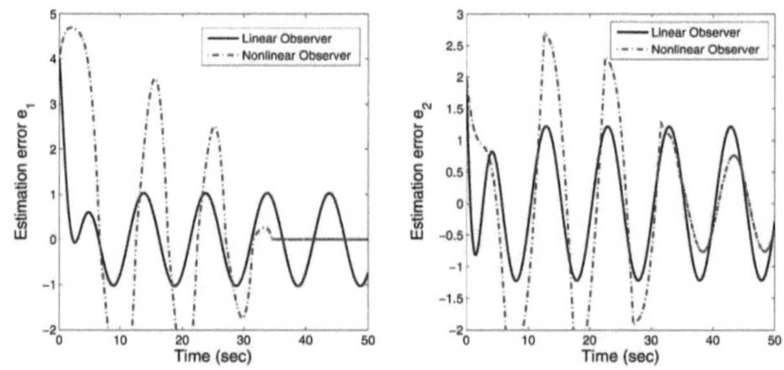

Fig. 12.6 Estimation errors of the linear and the SM Observers with unknown input

Finally, we can observe that for the linear observer (and also for the sliding mode observer) the larger the initial estimation error, the larger the convergence time (see Figure 12.7, where the initial state for the observer has been set to $\hat{x}(0) = 500[-2, -1]$, and compare with Figure 12.3). This means that it is difficult to estimate a priori the time required by the observer to provide a good estimation of the velocity.

12.2 The Proposed Observer: Design Method and Properties

In order to achieve the features for the observer, that have been listed in Subsection 12.1.1, in this section we propose a (discontinuous) observer, named *Generalized Super-Twisting Observer (GSTO)*, for the plant (12.1). We also describe how

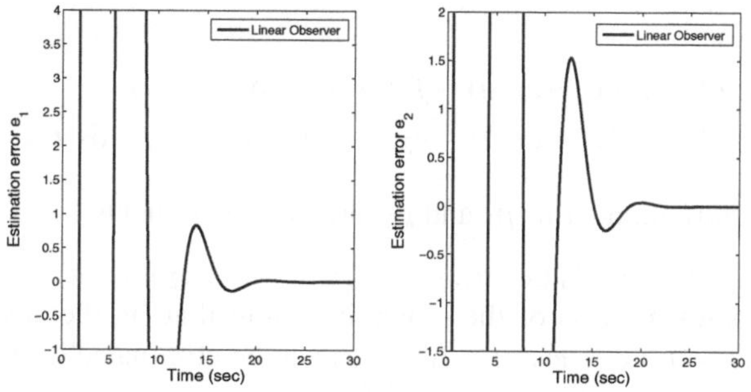

Fig. 12.7 Estimation error of the Linear Observer without UI with very large initial conditions

it is designed and discuss its properties. The proofs of the results will be given in Section 12.4.

12.2.1 The Generalized Super-Twisting Observer (GSTO)

When the plant is given in the form (12.1), the proposed GSTO has the form

$$\begin{aligned}\dot{\hat{x}}_1 &= -l_1\gamma\phi_1(e_1) + f_1(\hat{x}_1, u) + \hat{x}_2\,,\\ \dot{\hat{x}}_2 &= -l_2\gamma^2\phi_2(e_1) + f_2(\hat{x}_1, \hat{x}_2, u)\,,\end{aligned} \tag{12.3}$$

where $e_1 = \hat{x}_1 - x_1$, and $e_2 = \hat{x}_2 - x_2$ are the state estimation errors. $l_1 > 0$ and $l_2 > 0$ are positive, $\gamma > 0$ is an observer gain that has to be selected large enough to assure the convergence of the observer. The injection nonlinearities ϕ_1 and ϕ_2 are of the form

$$\phi_1(e_1) = \mu_1 |e_1|^{\frac{1}{2}} \operatorname{sign}(e_1) + \mu_2 |e_1|^q \operatorname{sign}(e_1)\ ,\ \mu_1\ ,\ \mu_2 \geq 0\,, \tag{12.4}$$

$$\phi_2(e_1) = \frac{\mu_1^2}{2}\operatorname{sign}(e_1) + \mu_1\mu_2\left(q + \frac{1}{2}\right)|e_1|^{q-\frac{1}{2}}\operatorname{sign}(e_1) + \mu_2^2|e_1|^{2q-1}\operatorname{sign}(e_1)\ , \tag{12.5}$$

where μ_1 and μ_2 are non negative constants, not both zero, and $q \geq \frac{1}{2}$ is a real number. Note that ϕ_1 and ϕ_2 are related, since $\phi_2(e_1) = \phi_1'(e_1)\phi_1(e_1)$, that they are both monotonically increasing functions of e_1 and ϕ_1 is continuous while ϕ_2 is discontinuous at $e_1 = 0$. Solutions of the observer (12.3) are understood in the sense of Filippov [8]. The state estimation errors (i.e. the estimation error vector $e = [e_1, e_2]^T$) satisfy the differential equation

$$\begin{aligned}\dot{e}_1 &= -l_1\gamma\phi_1(e_1) + e_2 + \rho_1(t, e, x, u)\\ \dot{e}_2 &= -l_2\gamma^2\phi_2(e_1) + \rho_2(t, e, x, u, w)\,,\end{aligned} \tag{12.6}$$

where

$$\rho_1(t,e_1,x,u) = f_1(x_1+e_1,u) - f_1(x_1,u) - \delta_1(t,x,u) \tag{12.7}$$
$$\rho_2(t,e,x,u,w) = f_2(x_1+e_1,x_2+e_2,u) - f_2(x_1,x_2,u) - \delta_2(t,x,u,w)\,. \tag{12.8}$$

Each of the perturbation terms ρ_1 and ρ_2 has two components:

- $\rho_{1f} = f_1(x_1+e_1,u) - f_1(x_1,u)$, $\rho_{2f} = f_2(x_1+e_1,x_2+e_2,u) - f_2(x_1,x_2,u)$ are due to the known terms of the dynamics. Note that (in the absence of noise) the term $\rho_{1f} = f_1(x_1+e_1,u) - f_1(x_1,u)$ can be eliminated if one uses $f_1(y,u)$ instead of $f_1(\hat{x}_1,u)$ in the observer (12.3).
- $\rho_{1\delta} = -\delta_1$, $\rho_{2\delta} = -\delta_2$ due to the uncertain/perturbation terms δ_1 and δ_2.

Each term has a different influence on the behavior of the observer, and this will be discussed below.

If the dynamics of the plant is given by

$$\begin{aligned}\dot{z}_1 &= F_1(t,z_1,z_2,u,w)\,,\\ \dot{z}_2 &= F_2(t,z_1,z_2,u,w)\,,\\ y &= H(t,z_1,z_2,u)\end{aligned} \tag{12.9}$$

where $z_1 \in \mathbb{R}$, $z_2 \in \mathbb{R}$ are the states, and it can be transformed into the form (12.1) by a global diffeomorphism $x = \Phi(z)$, an observer in original coordinates can be obtained from (12.3) as

$$\begin{aligned}\frac{d}{dt}\hat{z} &= F(t,\hat{z}_1,\hat{z}_2,u,0) - \left(\frac{\partial\Phi(\hat{z})}{\partial z}\right)^{-1}\begin{bmatrix} l_1\gamma\phi_1(\hat{y}-y) \\ l_2\gamma^2\phi_2(\hat{y}-y)\end{bmatrix},\\ \hat{y} &= H(t,\hat{z}_1,\hat{z}_2,u)\,.\end{aligned} \tag{12.10}$$

12.2.2 Observer Design

In this subsection we will discuss how to design the gains l_1, l_2, γ and q of the observer, so that, in the absence of perturbation $\delta_1 = 0$ (and measurement noise), the estimation error e converges in finite time to the origin, and robustly with respect to a perturbation δ_2, when δ_2 and the terms ρ_{1f} and ρ_{2f} satisfy some growth conditions (to be specified later). Moreover, the effect of the gains in the convergence time will be discussed. When a perturbation δ_1 is present, we know from the observability properties that it is impossible to obtain convergence to zero of the estimation error. In this case we show that "practical" stability is achieved.

We impose the following growth conditions on the perturbation terms (when $\delta_1 = 0$):

Property 12.1. We assume that there exist a real number $0 \le r$, two real numbers $\frac{1}{2} \le s_1 \le s_2$ and non negative (real) constants α_0, α_1, α_2, β_1 and β_2 such that

$$\begin{aligned} |\rho_1| &\leq \beta_1 |e_1|^{s_1} + \beta_2 |e_1|^{s_2}, \\ |\rho_2| &\leq \alpha_0 + \alpha_1 |e_1|^r + \alpha_2 |e_2|. \end{aligned} \quad (12.11)$$

The next Theorem provides a procedure to design the observer (See the proof in Section 12.4):

Theorem 12.1. *Assume that $\delta_1 = 0$. Suppose further that the perturbation terms satisfy Property 12.1. Select the parameter q such that*

$$q \geq \max\left\{1, s_2, \frac{r+1}{2}\right\}.$$

Select $l_1 > 0$ and $l_2 > 0$ arbitrarily, what implies that the matrix

$$A_l = \begin{bmatrix} -l_1 & 1 \\ -l_2 & 0 \end{bmatrix}$$

is Hurwitz. Set $\mu_1 > 0$ and $\mu_2 > 0$. Under these conditions there exists a value $\gamma_0 > 0$ such that for every $\gamma > \gamma_0$ the state estimation error $e(t)$ converges to zero in finite time, for every initial condition and robustly with respect to the perturbations satisfying (12.11). *Moreover, if $q > 1$ the convergence time is upper bounded by a constant, independent of the initial estimation error. Furthermore, if the perturbation δ_1 is a signal uniformly bounded for all the time, the estimation error $e(t)$ will be ultimately and uniformly bounded ⟦15 . Sect. 9.2⟧, i.e. there exists a positive constant b and a finite time T such that $\|e(t)\| \leq b$ for all $t > T$.*

12.2.3 Discussion of the Observer and Its Properties

It is important to note that from the inequalities (12.11), due to the parameter α_0, the perturbation ρ_2 does not have to vanish at the equilibrium point, i.e. when $e = 0$, and despite of this the estimation error can converge to zero. This situation appears for example when a persistent perturbation, due to an external unknown input w, is acting on the system (see (12.8)). Convergence under non vanishing unknown perturbations is imposible for continuous systems. The GSTO is able to achieve this property due to the discontinuous term in the injection function ϕ_2 (see (12.5)). This is a distinguishing feature of the GSTO, since a continuous observer, as the well-known High-Gain Observer (HGO) ⟦3, 7, 11, 15⟧ cannot achieve this property.

For different values of the parameters (μ_1, μ_2, q) some important particular cases are recovered:

(HG) The linear (or High Gain) Observer is recovered when $(\mu_1, \mu_2, q) = (0, 1, 1)$, so that $\phi_1(x_1) = x_1$, $\phi_2(x_1) = x_1$, and its properties can be derived in the same form as for Theorem 12.1. However, to avoid confusion in the redaction they are not included in the Theorem (see the related results in ⟦23⟧). The GST observer has much stronger properties, as described in the listing in paragraph 12.1.1.

(STA) The classical Super-Twisting Algorithm (STA), originally proposed in [[13]], is obtained by setting $(\mu_1, \mu_2, q) = (1, 0, q)$, so that $\phi_1(x_1) = |x_1|^{\frac{1}{2}} \operatorname{sign}(x_1)$, $\phi_2(x_1) = \frac{1}{2}\operatorname{sign}(x_1)$. In this case $\phi_2(x_1)$ is a discontinuous function. The algorithm has been used for observation in mechanical systems by [[6]]. A comparison of (some of) the properties of the ST and the GST Observers is done in Section 12.3.

(H) A Homogeneous Algorithm is obtained if $\phi_1(x_1) = |x_1|^q \operatorname{sign}(x_1)$, $\phi_2(x_1) = q|x_1|^{2q-1} \operatorname{sign}(x_1)$, for $q \geq \frac{1}{2}$. In this case system (12.6) without perturbations is homogeneous [[4, 18]]. When $q = \frac{1}{2}$ the previous ST algorithm is recovered. For $\frac{1}{2} < q < 1$ the algorithm is continuous but not locally Lipschitz, and it is able to converge in finite time. However, it is not able to converge to zero when a non vanishing perturbation δ_2 is present. When $q > 1$ the algorithm is smooth, but not globally Lipschitz, and although it converges only asymptotically its convergence time is uniform in the initial conditions. An algorithm combining both terms (with $q < 1$ and $q > 1$) can be obtained in the same framework as the GSTO (see [[23]]), and it combines both convergence properties. This is in the spirit of the observers designed in a recursive manner by [[1]]. The GSTO is not recursive, and it provides the whole set of all possible gains. The structure of the injection terms is different, and so is also the Lyapunov function used for the proof. Finally, the insensitivity properties of the GSTO when there is a persistent perturbation δ_2 cannot be achieved by these continuous algorithms.

(UD) The Uniform differentiator introduced in [[5]] is recovered when $q = \frac{3}{2}$.

(GSTA) For $q = 1$ the Generalized Super-Twisting Algorithm (GSTA) proposed in [[21]] is obtained.

The design method in Theorem 12.1 resembles the standard procedure for High-Gain observers (HGO) [[3, 11, 15]] in which a gain has to be designed high enough to assure the convergence. The design method presented here differs from the one that can be derived from [[23]] (and originally proposed in [[21]] for the case $q = 1$) since in [[21, 23]] the design of the gains l_1 and l_2 requires the solution of a Riccati Algebraic Equation when there are perturbations. Here this is much simpler, since only the gain γ has to be set large enough.

The value of the gain γ required to assure the convergence depends on the growth conditions of the perturbation terms (12.11) and on the selected gains l_1 and l_2 (see (12.17) for an expression of the gain γ_0 and subsection 12.4.3). Instead of calculating this gain explicitly, what can be a difficult task, it is possible to tune the observer by increasing γ until its performance is acceptable. Note that increasing γ results also in a smaller convergence time, as can be seen from the convergence time estimations provided by (12.19) and (12.20). Moreover, the larger γ is selected the smaller will be the effect of the perturbation δ_1 on the estimation error (see subsection 12.4.5), but the estimation error cannot be better than a certain minimal bound, depending on the size of the perturbation δ_1 (see equation 12.25). This is coherent with the observability analysis for the system (see Section 12.1), that indicates that x_2 can be estimated at the best within an error of the size of the pertubation δ_1. However, as for HGO, a large gain γ can produce a large peaking in the initial transient of the observer, what is an undesirable effect (see subsection 12.4.2). Furthermore, in the presence of measurement noise, a large gain γ will amplify the effect of noise in the estimation error.

12.3 Simulation Example (Continued).

In order to illustrate some of the properties of the observer proposed in the previous section, we will perform a simulation study using the example presented in subsection 12.1.2. In particular we want to show the effect of the two terms in the injection nonlinearity ϕ_1 (12.4) (and the corresponding ones in ϕ_2 (12.5)), that are obtained setting $\mu_1 = 0$ or $\mu_2 = 0$.

12.3.1 Super-Twisting Observer

In this subsection we design an observer, derived from the GSTO (12.3) by setting $\mu_2 = 0$. One obtains the well-known Super-Twisting Algorithm (STA), that has been proposed by Levant [[14]] as a differentiator and also for control in [[9, 13]]. In [[6]] this algorithm has been used as an observer for mechanical systems, that correspond to (12.1) without the known nonlinearities f_1 and f_2. The observer is given by (with $\gamma = 1$)

$$\dot{\hat{x}}_1 = -l_1 \left|e_1\right|^{\frac{1}{2}} \operatorname{sign}(e_1) + \hat{x}_2\,,$$
$$\dot{\hat{x}}_2 = -l_2 \operatorname{sign}(e_1)\,.$$

The effect of the discontinuous term in ϕ_2 is twofold:

- Convergence in finite time to zero of the estimation error. This can be seen in the simulation in Figure 12.8, where the unknown input is $w = 0$, and the same initial conditions for plant and observer as in the linear case were used.
- More importantly: The convergence in finite time to zero error is kept despite of a non vanishing unknown input $w(t) = 0.9 \sin(0.2\pi t)$. This can be appreciated in Figure 12.9. This is a distinguishing feature of this observer, and it is clearly due to the discontinuity in ϕ_2.

The Super-Twisting Observer (STO) has, however, some disadvantages:

1. The convergence time grows very fast (and unboundedly) with the size of the initial estimation error. This can be observed in Figure 12.10, where the initial condition of linear and ST Observers is $\hat{x}(0) = 500[-2, -1]$, i.e. 500 times its value for Figure 12.8. One notes here that the convergence time of the STO grows faster than that for the Linear Observer.
2. When the bound of the perturbation δ_2 is larger than the gain l_2, then the STO can diverge. It is not possible to assure the desirable property that a bounded perturbation produces a bounded estimation error (see [[20, 23]] for more details).

3. The STO is not able to assure global convergence in the presence of known terms f_1 and/or f_2 in the plant's model (12.1).

The two last features are not illustrated in the simulations.

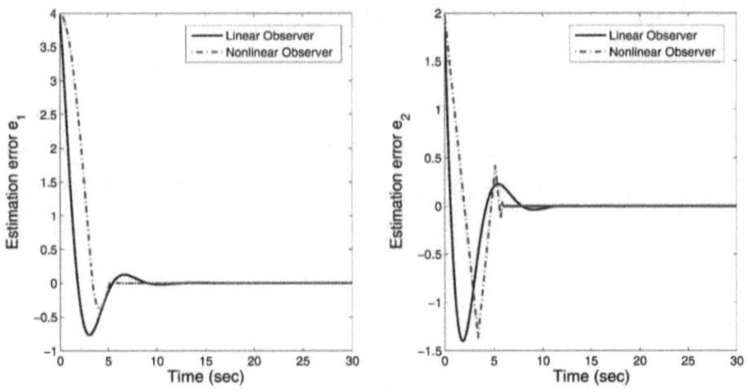

Fig. 12.8 Estimation error for the Linear and the Super-Twisting Observers without unknown input

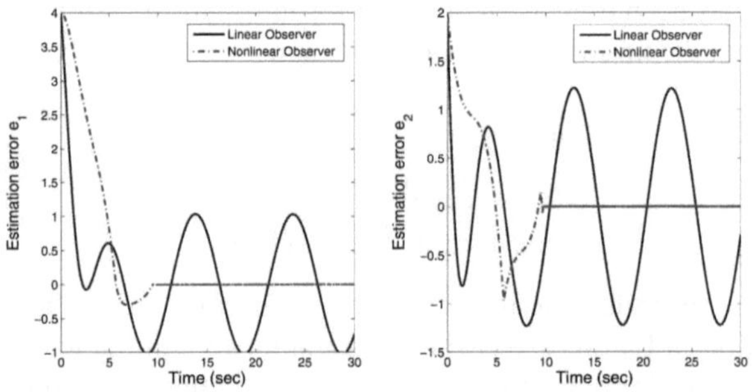

Fig. 12.9 Estimation error for the Linear and the Super-Twisting Observers with unknown input

12.3.2 Generalized Super-Twisting Observers

All these drawbacks of the STO can be solved by the GSTO (12.3) proposed here. For the simulation we have used the Generalized Super-Twisting Observer (GSTO) (12.3) with $q = \frac{3}{2}$, $\mu_1 = \mu_2 = 1$ and $\gamma = 1$. Figures 12.11 and 12.12 illustrate two features: i) The GSTO converges to zero in finite time, with or without unknown input. ii) The convergence time is basically the same for very large initial estimation error conditions. This nice feature of the GSTO is due to the introduction of a nonlinear term with a power q larger than one.

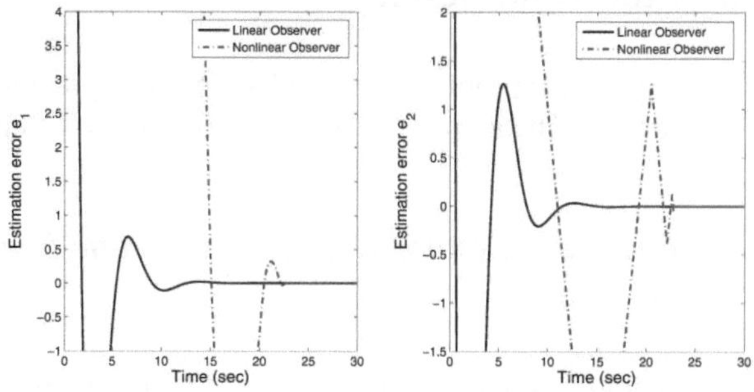

Fig. 12.10 Estimation error for the linear and the Super-Twisting Observer without UI with large initial conditions

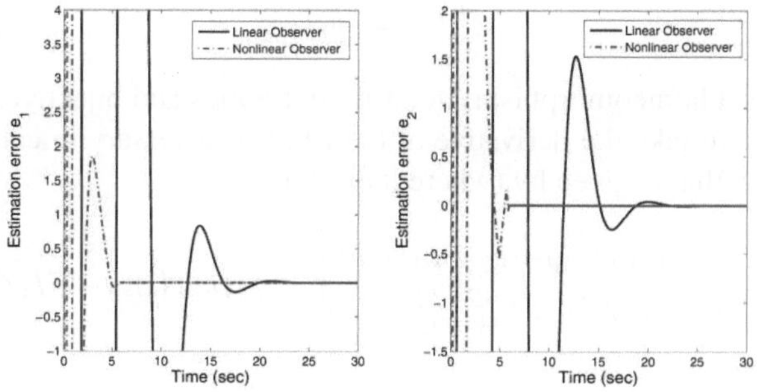

Fig. 12.11 Estimation error for the Linear and the Generalized Super-Twisting Observers without unknown input and large initial conditions, i.e. $\hat{x}(0) = 10[-2, -1]$

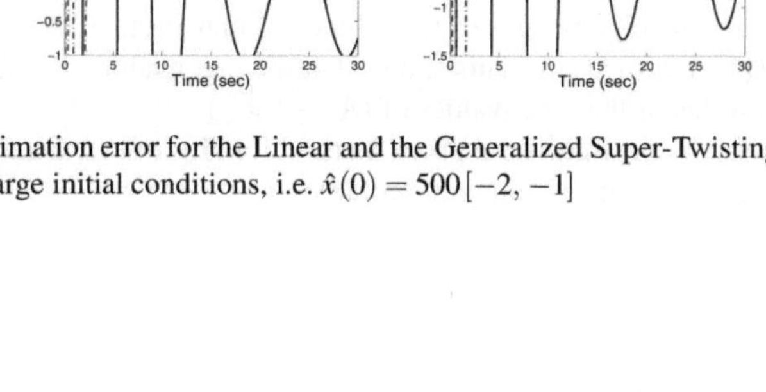

Fig. 12.12 Estimation error for the Linear and the Generalized Super-Twisting Observer with UI with very large initial conditions, i.e. $\hat{x}(0) = 500[-2, -1]$

12.4 Proofs of the Main Results

In this section we provide the proofs of the results presented previously. In particular, we provide a proof for Theorem 12.1. We proceed in several steps.

12.4.1 *The Convergence Proof Using a Quadratic Lyapunov Function*

In [[19–21,23]] a quadratic Lyapunov function (LF), that is continuous but not Lipschitz continuous, has been introduced for the analysis of the convergence and robustness properties of Super-Twisting-like algorithms. This Lyapunov function is quadratic not in the state vector, but in a vector

$$\varepsilon^T = \varphi^T(e) = \left[\phi_1(e_1)\ ,\ e_2\right]\ , \tag{12.12}$$

where $\varphi(e)$ is a homeomorphism (i.e. it is continuous and bijective, with a continuous inverse). To take the derivative of the LF it is necessary to calculate the time derivative of ε, that is given by (where it exists)

$$\dot{\varepsilon} = \phi_1'(e_1)\begin{bmatrix} -l_1\gamma\phi_1(e_1) + e_2 + \rho_1(t,e) \\ -l_2\gamma^2\phi_1(e_1) + \frac{\rho_2(t,e)}{\phi_1'(e_1)} \end{bmatrix} = \phi_1'(e_1)\left\{(A_0 - \Gamma L_0 C_0)\,\varepsilon + \tilde{\rho}\right\}\ ,$$

with

$$A_0 = \begin{bmatrix} 0 & 1 \\ 0 & 0 \end{bmatrix},\ L_0 = \begin{bmatrix} l_1 \\ l_2 \end{bmatrix},\ C_0 = \left[1,0\right],\ \Gamma = \begin{bmatrix} \gamma & 0 \\ 0 & \gamma^2 \end{bmatrix}$$

and

$$\tilde{\rho}(t,\varepsilon,\cdot) = \begin{bmatrix} \rho_1(t,e,\cdot) \\ \left(\frac{2|e_1|^{\frac{1}{2}}}{\mu_1 + 2q\mu_2|e_1|^{q-\frac{1}{2}}}\right)\rho_2(t,e,\cdot) \end{bmatrix}_{e=\varphi^{-1}(\varepsilon)}\ ,$$

where we have used the error equation (12.6). Note that the characteristic polynomial of the matrix $(A_0 - \Gamma L_0 C_0)$ is

$$p(s) = \det(s\mathbb{I} - (A_0 - \Gamma L_0 C_0)) = s^2 + \gamma l_1 s + \gamma^2 l_2 = (s - \gamma\lambda_1)(s - \gamma\lambda_2)$$

where λ_1, λ_2 are the eigenvalues of the (Hurwitz) matrix $A_l = (A_0 - L_0 C_0)$, i.e matrix $(A_0 - \Gamma L_0 C_0)$ with $\gamma = 1$. This shows that the eigenvalues of $(A_0 - \Gamma L_0 C_0)$ are $\gamma\lambda_1$, $\gamma\lambda_2$, multiples of the eigenvalues of $(A_0 - L_0 C_0)$.

Similar to the by now classical proof method for High-Gain Observer [[3,11,15]] we introduce here a further change of variables

$$\xi = \theta\Gamma^{-1}\varepsilon = \begin{bmatrix} \frac{\theta}{\gamma}\varepsilon_1 \\ \frac{\theta}{\gamma^2}\varepsilon_2 \end{bmatrix},$$

where $\theta > 0$ is an arbitrary positive constant, and we obtain (since $\Gamma^{-1}A_0\Gamma = \gamma A_0$ and $C_0\Gamma = \gamma C_0$)

$$\dot{\xi} = \theta\Gamma^{-1}\phi_1'(e_1)\left\{(A_0 - \Gamma L_0 C_0)\frac{1}{\theta}\Gamma\xi + \tilde{\rho}\right\} = \phi_1'(e_1)\left\{\gamma(A_0 - L_0C_0)\xi + \theta\Gamma^{-1}\tilde{\rho}\right\}.$$

Using a quadratic Lyapunov function (see [[23]])

$$V(\xi) = \xi^T P \xi$$

where $P = P^T > 0$ is the unique, symmetric and positive definite solution of the Algebraic Lyapunov Equation

$$(A_0 - L_0C_0)^T P + P(A_0 - L_0C_0) = -Q,$$

for $Q = Q^T > 0$, an arbitrary positive definite and symmetric matrix. The derivative of V along the solutions of the error equation (almost everywhere) is given by

$$\begin{aligned}\dot{V} &= \phi_1'(e_1)\left\{\gamma\xi^T\left[(A_0 - L_0C_0)^T P + P(A_0 - L_0C_0)\right]\xi + 2\xi^T P\theta\Gamma^{-1}\tilde{\rho}\right\}\\ &= \phi_1'(e_1)\left\{-\gamma\xi^T Q\xi + 2\xi^T P\theta\Gamma^{-1}\tilde{\rho}\right\} \qquad (12.13)\\ &\leq \phi_1'(e_1)\left\{-\gamma\lambda_{\min}\{Q\}\|\xi\|^2 + 2\|\xi\|\|P\|\left\|\theta\Gamma^{-1}\tilde{\rho}\right\|\right\}\end{aligned}$$

where $\lambda_{\min}\{Q\}$ is the minimal eigenvalue of Q, $\|\xi\|$ is the Euclidean norm of ξ and $\|P\| = \lambda_{\max}\{P\}$ is the induced (Euclidean) norm of matrix P. Recall that $\phi_1'(e_1) \geq 0$ since $\phi_1(e_1)$ is monotone increasing.

Here we will consider the case that $\delta_1 = 0$, and we will assume here that there exist some constants k_1, k_2 such that the perturbation terms satisfy the following restrictions

$$|\tilde{\rho}_1(t,e_1)| = |\rho_1(t,e_1)| \leq k_1|\phi_1(e_1)| = k_1\left(\mu_1 + \mu_2|e_1|^{q-\frac{1}{2}}\right)|e_1|^{\frac{1}{2}} \qquad (12.14)$$

and

$$|\tilde{\rho}_2(t,e)| = \left(\frac{2|e_1|^{\frac{1}{2}}}{\mu_1 + 2q\mu_2|e_1|^{q-\frac{1}{2}}}\right)|\rho_2(t,e)| \leq k_2\left(\phi_1^2(e_1) + e_2^2\right)^{\frac{1}{2}}. \qquad (12.15)$$

Below, in subsection 12.4.3, it will be proved that (12.14-12.15) follow from the Property 12.1. Using the relations

$$\frac{\gamma}{\theta}\xi_1 = \varepsilon_1 = \phi_1(e_1), \quad \frac{\gamma^2}{\theta}\xi_2 = \varepsilon_2 = e_2. \qquad (12.16)$$

we obtain the inequalities

$$\left\|\theta\Gamma^{-1}\tilde{\rho}\right\|^2 = \frac{\theta^2}{\gamma^2}\tilde{\rho}_1^2 + \frac{\theta^2}{\gamma^4}\tilde{\rho}_2^2 \le \left(\frac{\theta^2}{\gamma^2}k_1^2 + \frac{\theta^2}{\gamma^4}k_2^2\right)\phi_1^2(e_1) + \frac{\theta^2}{\gamma^4}k_2^2 e_2^2$$
$$= \left(k_1^2 + \frac{1}{\gamma^2}k_2^2\right)\xi_1^2 + k_2^2\xi_2^2 \le k^2\|\xi\|^2$$

for

$$k^2 \ge \max\left\{k_1^2 + \frac{1}{\gamma^2}k_2^2, k_2^2\right\}.$$

This implies that

$$\dot{V} \le -\phi_1'(e_1)\left(\gamma\lambda_{\min}\{Q\} - 2k\lambda_{\max}\{P\}\right)\|\xi\|^2$$

so that $\dot{V}$ is negative definite for a sufficiently large gain γ, i.e. for

$$\gamma > \gamma_0 \triangleq 2k\frac{\lambda_{\max}\{P\}}{\lambda_{\min}\{Q\}}. \tag{12.17}$$

This can always be achieved, since P and Q are independent of γ and k decreases with γ.

Recall the standard inequality for quadratic forms

$$\lambda_{\min}\{P\}\|\xi\|_2^2 \le \xi^T P\xi \le \lambda_{\max}\{P\}\|\xi\|_2^2 ,$$

where

$$\|\xi\|_2^2 = \xi_1^2 + \xi_2^2 = \frac{\theta^2}{\gamma^2}\phi_1^2(e_1) + \frac{\theta^2}{\gamma^4}e_2^2$$
$$= \frac{\theta^2}{\gamma^2}\left(\mu_1^2|e_1| + 2\mu_1\mu_2|e_1|^{q+\frac{1}{2}} + \mu_2^2|e_1|^{2q}\right) + \frac{\theta^2}{\gamma^4}e_2^2$$

is the Euclidean norm of ξ. Note that the inequality

$$|e_1|^{\frac{1}{2}} \le \frac{1}{\mu_1}|\phi_1(e_1)| \le \frac{\gamma}{\mu_1\theta}\|\xi\| \le \frac{\gamma}{\mu_1\theta\lambda_{\min}^{\frac{1}{2}}\{P\}}V^{\frac{1}{2}}(\xi) \tag{12.18}$$

is satisfied for $\mu_1 > 0$, and therefore

$$-\frac{1}{|e_1|^{\frac{1}{2}}} \le -\frac{\mu_1\theta}{\gamma\|\xi\|} \le -\frac{\mu_1\theta\lambda_{\min}^{\frac{1}{2}}\{P\}}{\gamma}V^{-\frac{1}{2}}(\xi) .$$

Since

$$\phi_1'(e_1) = \frac{1}{2}\mu_1\frac{1}{|e_1|^{\frac{1}{2}}} + q\mu_2|e_1|^{q-1}$$

it follows that

$$\begin{aligned}\dot{V} &\leq -\left(\gamma\lambda_{\min}\{Q\} - 2k\|P\|\right)\left(\frac{1}{2}\mu_1\frac{1}{|e_1|^{\frac{1}{2}}} + q\mu_2|e_1|^{q-1}\right)\|\xi\|^2 \\ &\leq -\left(\gamma - \frac{2k\|P\|}{\lambda_{\min}\{Q\}}\right)\lambda_{\min}\{Q\}\left(\frac{1}{2}\mu_1^2\frac{\theta}{\gamma}\|\xi\| + q\mu_2|e_1|^{q-1}\|\xi\|^2\right) \\ &\leq -\mu_1^2\frac{\theta(\gamma-\gamma_0)\lambda_{\min}\{Q\}}{2\gamma\lambda_{\max}^{\frac{1}{2}}\{P\}}V^{\frac{1}{2}}(\xi) - \mu_2\frac{q(\gamma-\gamma_0)\lambda_{\min}\{Q\}}{\lambda_{\max}\{P\}}|e_1|^{q-1}V(\xi),\end{aligned}$$

where we have used the definition of γ_0 in (12.17), and therefore $V(\xi(t))$ is monotonically decreasing, and the origin is asymptotically stable.

12.4.2 *About the Convergence Velocity of the Error*

From the differential inequality satisfied by the LF, it is possible to estimate the convergence velocity of the state estimation errors. We will do this explicitly for two (simple) cases.

12.4.2.1 The Case When $\mu_1 \neq 0$ and q Is Arbitrary

From the differential inequality satisfied by the Lyapunov function it follows that

$$\dot{V} \leq -\mu_1^2\frac{\theta(\gamma-\gamma_0)\lambda_{\min}\{Q\}}{2\gamma\lambda_{\max}^{\frac{1}{2}}\{P\}}V^{\frac{1}{2}}(\xi).$$

Since the solution of the differential equation

$$\dot{v} = -\gamma_1 v^{\frac{1}{2}}, \quad v(0) = v_0 \geq 0$$

is given by

$$v(t) = \left(v_0^{\frac{1}{2}} - \frac{1}{2}\gamma_1 t\right)^2 \text{ if } \gamma_1 > 0,$$

it follows from the comparison principle that

$$V(t) \leq \left(V^{\frac{1}{2}}(\xi_0) - \frac{1}{2}\mu_1^2\frac{\theta(\gamma-\gamma_0)\lambda_{\min}\{Q\}}{2\gamma\lambda_{\max}^{\frac{1}{2}}\{P\}}t\right)^2,$$

before the finite convergence time. This implies that

$$\lambda_{\min}\{P\}\|\xi(t)\|^2 \leq \xi^T(t)P\xi(t) \leq \left(\left(\xi_0^T P\xi_0\right)^{\frac{1}{2}} - \frac{1}{2}\mu_1^2\frac{\theta(\gamma-\gamma_0)\lambda_{\min}\{Q\}}{2\gamma\lambda_{\max}^{\frac{1}{2}}\{P\}}t\right)^2,$$

and therefore

$$\|\xi(t)\| \le \frac{1}{\lambda_{\min}^{\frac{1}{2}}\{P\}}\left(\xi_0^T P \xi_0\right)^{\frac{1}{2}} - \frac{1}{2}\mu_1^2 \frac{\theta(\gamma-\gamma_0)\lambda_{\min}\{Q\}}{2\gamma\lambda_{\min}^{\frac{1}{2}}\{P\}\lambda_{\max}^{\frac{1}{2}}\{P\}} t\,.$$

In original coordinates (see (12.16)), and noting that (for $\gamma \ge 1$)

$$\frac{\theta}{\gamma^2}\|\varepsilon(t)\| \le \left\| \begin{bmatrix} \frac{\theta}{\gamma}\phi_1(e_1(t)) \\ \frac{\theta}{\gamma^2}e_2(t) \end{bmatrix} \right\| = \|\xi(t)\| \le \frac{\theta}{\gamma}\|\varepsilon(t)\|,$$

one obtains that

$$\|\varepsilon(t)\| \le c_P\gamma\|\varepsilon_0\| - \mu_1^2 \frac{\gamma(\gamma-\gamma_0)\lambda_{\min}\{Q\}}{4\lambda_{\min}^{\frac{1}{2}}\{P\}\lambda_{\max}^{\frac{1}{2}}\{P\}} t\,,\; c_P = \sqrt{\frac{\lambda_{\max}\{P\}}{\lambda_{\min}\{P\}}},$$

where c_P is the condition number of matrix P. The finite convergence time can be estimated by

$$T(\varepsilon_0) \le \frac{4\lambda_{\max}\{P\}}{\mu_1^2(\gamma-\gamma_0)\lambda_{\min}\{Q\}}\|\varepsilon_0\|\,. \tag{12.19}$$

We notice that the convergence time can be made as small as desired by increasing the gain γ. However, the initial deviation term, given by $c_P\gamma$ grows also with the gain γ. This corresponds to the peaking phenomenon, well-known for High-Gain Observers [[15]].

12.4.2.2 The Case When $\mu_1 \ne 0$ and $q = 1$

The Lyapunov function satisfies the differential inequality

$$\dot V \le -\mu_1^2 \frac{\theta(\gamma-\gamma_0)\lambda_{\min}\{Q\}}{2\gamma\lambda_{\max}^{\frac{1}{2}}\{P\}} V^{\frac{1}{2}}(\xi) - \mu_2 \frac{(\gamma-\gamma_0)\lambda_{\min}\{Q\}}{\lambda_{\max}\{P\}} V(\xi)\,.$$

From the solution of the Differential Equation (See e.g. [[23]])

$$\dot v = -\gamma_1 v^{\frac{1}{2}} - \gamma_2 v\,,\quad v(0) = v_0 \ge 0,$$

given by

$$v^{\frac{1}{2}}(t) = \exp\left(-\frac{1}{2}\gamma_2 t\right) v_0^{\frac{1}{2}} - \frac{\gamma_1}{\gamma_2}\exp\left(-\frac{1}{2}\gamma_2 t\right)\left[\exp\left(\frac{1}{2}\gamma_2 t\right) - 1\right],$$

and the comparison principle [[15]] it follows that

$$V^{\frac{1}{2}}(\xi(t)) \le \exp\left(-\frac{1}{2}\gamma_2 t\right)\left[V^{\frac{1}{2}}(\xi_0) + \frac{\gamma_1}{\gamma_2}\right] - \frac{\gamma_1}{\gamma_2},$$

or

$$V^{\frac{1}{2}}(\xi(t)) \leq \exp\left(-\frac{1}{2}\mu_2 \frac{(\gamma-\gamma_0)\lambda_{\min}\{Q\}}{\lambda_{\max}\{P\}} t\right)\left[V^{\frac{1}{2}}(\xi_0) + \frac{\mu_1^2 \theta \lambda_{\max}^{\frac{1}{2}}\{P\}}{2\mu_2\gamma}\right] - \frac{\mu_1^2 \theta \lambda_{\max}^{\frac{1}{2}}\{P\}}{2\mu_2\gamma}.$$

This implies that

$$\|\xi(t)\| \leq c_P \left\{\exp\left(-\frac{1}{2}\mu_2 \frac{(\gamma-\gamma_0)\lambda_{\min}\{Q\}}{\lambda_{\max}\{P\}} t\right)\left[\|\xi_0\| + \frac{\mu_1^2\theta}{2\mu_2\gamma}\right] - \frac{\mu_1^2\theta}{2\mu_2\gamma}\right\}.$$

In original coordinates (see (12.16)) results (for $\gamma \geq 1$)

$$\|\varepsilon(t)\| \leq \gamma c_P \left\{\exp\left(-\frac{1}{2}\mu_2 \frac{(\gamma-\gamma_0)\lambda_{\min}\{Q\}}{\lambda_{\max}\{P\}} t\right)\left[\|\varepsilon_0\| + \frac{\mu_1^2}{2\mu_2}\right] - \frac{\mu_1^2}{2\mu_2}\right\}.$$

When $\mu_1 > 0$ the Finite Convergence time can be estimated as

$$T(\varepsilon_0) \leq \frac{2\lambda_{\max}\{P\}}{\mu_2(\gamma-\gamma_0)\lambda_{\min}\{Q\}} \ln\left(\frac{2\mu_2}{\mu_1^2}\|\varepsilon_0\| + 1\right). \tag{12.20}$$

It is clear that this time can be made arbitrarily small by selecting a gain γ sufficiently large. However, the initial bound (for $t = 0$), given by

$$\|\varepsilon(0)\| \leq \gamma \frac{\lambda_{\max}^{\frac{1}{2}}\{P\}}{\lambda_{\min}^{\frac{1}{2}}\{P\}} \|\varepsilon_0\|,$$

also grows with the gain γ, which corresponds to the Peaking Phenomenon.

12.4.3 About the Restrictions on the Perturbations

Here we show that (12.14-12.15) follow from the Property 12.1. To show (12.14) it suffices to consider the case

$$|\rho_1| \leq \beta_0 |e_1|^s, \quad \frac{1}{2} \leq s \leq q.$$

It is clear that there exists a constant k_1 such that

$$|\tilde{\rho}_1(t,e_1)| = |\rho_1(t,e_1)| \leq k_1 |\phi_1(e_1)| = k_1\left(\mu_1 + \mu_2 |e_1|^{q-\frac{1}{2}}\right)|e_1|^{\frac{1}{2}}.$$

To show (12.15) suppose that

$$|\rho_2| \leq \alpha_0 + \alpha_1 |e_1|^r + \alpha_2 |e_2|, \quad 0 \leq r \leq 2q-1, q \geq 1.$$

We will show that there exists a constant $k_2 > 0$ such that

$$\left|\tilde{\rho}_2\right| \le \left(\frac{2\left|e_1\right|^{\frac{1}{2}}}{\mu_1 + 2q\mu_2\left|e_1\right|^{q-\frac{1}{2}}}\right)\left(\alpha_0 + \alpha_1\left|e_1\right|^r + \alpha_2\left|e_2\right|\right) \le k_2\left(\phi_1^2\left(e_1\right) + e_2^2\right)^{\frac{1}{2}}. \tag{12.21}$$

It is clear that the previous inequality follows if the following three are satisfied:

$$\left(\frac{2\left|e_1\right|^{\frac{1}{2}}}{\mu_1 + 2q\mu_2\left|e_1\right|^{q-\frac{1}{2}}}\right)\alpha_0 \le k_{21}\left(\phi_1^2\left(e_1\right) + e_2^2\right)^{\frac{1}{2}} \tag{12.22}$$

$$\left(\frac{2\left|e_1\right|^{\frac{1}{2}}}{\mu_1 + 2q\mu_2\left|e_1\right|^{q-\frac{1}{2}}}\right)\alpha_1\left|e_1\right|^r \le k_{22}\left(\phi_1^2\left(e_1\right) + e_2^2\right)^{\frac{1}{2}} \tag{12.23}$$

$$\left(\frac{2\left|e_1\right|^{\frac{1}{2}}}{\mu_1 + 2q\mu_2\left|e_1\right|^{q-\frac{1}{2}}}\right)\alpha_2\left|e_2\right| \le k_{23}\left(\phi_1^2\left(e_1\right) + e_2^2\right)^{\frac{1}{2}}. \tag{12.24}$$

The inequality (12.22) is equivalent to inequality (12.23) for $r = 0$. So we prove (12.23), which is equivalent to

$$4\alpha_1^2\left|e_1\right|^{2r+1} \le k_{22}^2\left(\mu_1 + 2q\mu_2\left|e_1\right|^{q-\frac{1}{2}}\right)^2\left(\phi_1^2\left(e_1\right) + e_2^2\right).$$

Extracting the two terms with the highest and the lowest power of e_1 in the right hand side of the previous inequality one obtains that

$$k_{22}^2\left(\mu_1^4\left|e_1\right| + 4q^2\mu_2^4\left|e_1\right|^{4q-1}\right) \le k_{22}^2\left(\mu_1 + 2q\mu_2\left|e_1\right|^{q-\frac{1}{2}}\right)^2\left(\phi_1^2\left(e_1\right) + e_2^2\right),$$

and therefore (12.23) follows if

$$4\alpha_1^2\left|e_1\right|^{2r+1} \le k_{22}^2\left(\mu_1^4\left|e_1\right| + 4q^2\mu_2^4\left|e_1\right|^{4q-1}\right).$$

Clearly there exists a constant k_{22} if $1 \le 2r+1 \le 4q-1$, or equivalently if $0 \le r$ and $r \le 2q-1$. So both (12.23) and (12.22) are satisfied.

Now we show that (12.24) is fulfilled. It follows from the simple observation that for $q \ge 1$ the function $1/\phi_1'(e_1)$ is bounded by a constant, i.e.

$$\frac{2\left|e_1\right|^{\frac{1}{2}}}{\mu_1 + 2q\mu_2\left|e_1\right|^{q-\frac{1}{2}}} \le M.$$

We can conclude that (12.21) is satisfied.

12.4.4 On the Convergence Uniform in the Initial Conditions

When $q > 1$ it is affirmed in the Theorem 12.1 that there is a constant value $T > 0$ so that all the trajectories will converge to zero within a time lesser than T, i.e. for every initial condition. The previous paragraphs show that this property does not follow from the quadratic Lyapunov function, which is a well-known fact, as it is discussed in detail in [[23]]. In that reference a non-quadratic LF has been proposed to show the uniformity in the initial conditions property. A similar procedure can be used in our case, but the details are too long to be presented here. We refer the reader to references [[5, 23]] for those details.

12.4.5 *The Effect of a Non Vanishing Perturbation* δ_1

So far we have considered only the case when the perturbation $\delta_1 = 0$. If we take into account this term in the derivative of the LF (see (12.13)) we obtain

$$\begin{aligned}
\dot{V} &= \phi_1'(e_1)\left\{-\gamma\xi^T Q\xi + 2\xi^T P\theta\Gamma^{-1}\left(\tilde{\rho}_0 + \begin{bmatrix}-1\\0\end{bmatrix}\delta_1\right)\right\} \\
&\leq \phi_1'(e_1)\left\{-\gamma\lambda_{\min}\{Q\}\|\xi\|^2 + 2\|\xi\|\|P\|\left(\left\|\theta\Gamma^{-1}\tilde{\rho}_0\right\| + \frac{\theta}{\gamma}|\delta_1|\right)\right\} \\
&\leq -\phi_1'(e_1)\left\{(\gamma-\gamma_0)\lambda_{\min}\{Q\}\|\xi\| - 2\lambda_{\max}\{P\}\frac{\theta}{\gamma}|\delta_1|\right\}\|\xi\|
\end{aligned}$$

where $\tilde{\rho}_0$ represents $\tilde{\rho}$ without the term δ_1, and we have assumed that $\tilde{\rho}_0$ satisfies Property 12.1. If the gain γ is set larger than the corresponding γ_0 in (12.17) it is clear that $\dot{V} < 0$ outside a ball containing the origin, i.e. for

$$\|\xi\| > \frac{2\lambda_{\max}\{P\}\theta|\delta_1|}{\gamma(\gamma-\gamma_0)\lambda_{\min}\{Q\}}.$$

Using standard arguments [[15]] it follows that the trajectories are ultimately uniformly bounded, if δ_1 is bounded. Moreover, in original coordinates

$$\|\varepsilon(t)\| > \frac{2\lambda_{\max}\{P\}}{\lambda_{\min}\{Q\}}\frac{\gamma}{(\gamma-\gamma_0)}|\delta_1|, \tag{12.25}$$

which implies that the final bound has an infimum value that can be approached the larger the gain γ is selected. A similar proof (see also [[23]]) can be used to show the boundedness of the estimation error when the perturbation δ_2 is bounded, but its bound is larger than the one used to set the gain γ, or when there is measurement noise.

12.5 Conclusions

We have presented in this chapter a unified method to design a class of discontinuous observers for second order systems. It generalizes and improves several other known methods, as for example the High-Gain Observer, the Super-Twisting Observer and the Uniform Differentiator, enhancing their properties. We have restricted the treatment to the two dimensional case for two reasons: i) We present all proofs in a unified Lyapunov framework, which is at the moment only available for planar systems. ii) We provide a tutorial presentation that allows an easy introduction to the topic and also presents the main results in the simplest case.

Much work is still necessary to complete the program. In particular a discussion of the effect of measurement noise is crucial for estimation, that has not been included here. It is clear that increasing the gain γ will improve the performance of the observer with respect to convergence velocity and reduction of the effect of the perturbations (unknown input), but it will also increase the effect of noise, and viceversa. So a clear trade-off between estimation error due to noise and to perturbations/unknown inputs is to be considered. For High-Gain Observers (used as differentiators) this has been done recently in ⟦26⟧, where a method to optimize the gain γ has been presented. For the GSTO there are some preliminary results ⟦2⟧.

It is also clear that the extension of the results for higher order systems is an important step, that is part of ongoing research. Applications of the observers are manifold. In ⟦22⟧ they are applied for a class of chemical reactors, output feedback control is presented in a Lyapunov framework in ⟦24⟧. We hope to be able to provide a similar treatment of the general case in the near future.

Acknowledgements. The author gratefully acknowledges the financial support from PAPIIT, UNAM, grant IN111012, and Fondo de Colaboracin del II-FI, UNAM, IISGBAS-165-2011.

References

1. Andrieu, V., Praly, L., Astolfi, A.: Homogeneous approximation, recursive observer design and output feedback. SIAM Journal of Control and Optimization 47(4), 1814–1850 (2008)
2. Marco Tulio, A.-B., Moreno, J.A., Fridman, L.M.: Optimal Gain for the Super-Twisting Differentiator in the Presence of Measurement Noise. In: The 2012 American Control Conference (ACC 2012), Montréal, Canada, June 27-29, pp. 6154–6159 (2012)
3. Besanccon, G.: An Overview on Observer Tools for Nonlinear Systems. In: Besanccon, G. (ed.) Nonlinear Observers and Applications. LNCIS, vol. 363, pp. 1–33. Springer, Heidelberg (2007)
4. Baccioti, A., Rosier, L.: Lyapunov functions and stability in control theory, 2nd edn. Springer, New York (2005)
5. Cruz-Zavala, E., Moreno, J.A., Fridman, L.: Uniform Robust Exact Differentiator. IEEE Trans. on Automatic Control 56(11), 2727–2733 (2011), doi:10.1109/TAC.2011.2160030
6. Davila, J., Fridman, L., Levant, A.: Second-Order Sliding- Modes Observer for Mechanical Systems. IEEE Transactions on Automatic Control 50(11), 1785–1789 (2005)
7. Esfandiari, F., Khalil, H.K.: Output feedback stabilization of fully linearizable systems. Int. J. Control 56, 1007–1037 (1992)
8. Filippov, A.F.: Differential equations with discontinuous right-hand side, 304 p. Kluwer, Dordrecht (1988)

9. Fridman, L., Levant, A.: Higher order sliding modes. In: Barbot, J.P., Perruquetti, W. (eds.) Sliding Mode Control in Engineering, pp. 53–101. Marcel Dekker, New York (2002)
10. Gauthier, J.-P., Bornard, G.: Observability for any $u(t)$ of a class of nonlinear systems. IEEE Trans. Aut. Cont. 26, 922–926 (1981)
11. Gauthier, J.-P., Hammouri, H., Othman, S.: A simple observer for nonlinear systems. Applications to bioreactors. IEEE Trans. Automatic Control 37, 875–880 (1992)
12. Gauthier, J.-P., Kupka, I.: Deterministic Observation Theory and Applications. Cambridge University Press, Cambridge (2001)
13. Hautus, M.L.J.: Strong detectability and observers. Linear Algebra and its Applications 50, 353–368 (1983)
14. Khalil, H.: High-Gain Observers in Nonlinear Feedback Control. In: Nijmeijer, H., Fossen, T. (eds.) New Directions in Nonlinear Observer Design. LNCIS, vol. 244, pp. 249–268. Springer, Heidelberg (1999)
15. Khalil, H.K.: Nonlinear Systems, 3rd edn., 750 p. Prentice–Hall, Upsaddle River (2002)
16. Levant, A.: Sliding order and sliding accuracy in sliding mode control. International Journal of Control 58(6), 1247–1263 (1993)
17. Levant, A.: Robust Exact Differentiation via Sliding Mode Technique. Automatica 34(3), 379–384 (1998)
18. Levant, A.: Homogeneity approach to high-order sliding mode design. Automatica (41), 823–830 (2005)
19. Moreno, J.A., Osorio, M.: A Lyapunov approach to second-order sliding mode controllers and observers. In: 47th IEEE Conference on Decision and Control, CDC 2008, pp. 2856–2861 (2008)
20. Moreno, J.A., Osorio, M.: Strict Lyapunov functions for the Super-Twisting Algorithm. IEEE Trans. on Automatic Control 57(4), 1035–1040 (2012), doi:10.1109/TAC.2012.2186179
21. Moreno, J.A.: A Linear Framework for the Robust Stability Analysis of a Generalized Supertwisting Algorithm. In: Proc. 6th Int. Conf. Elect. Eng., Comp. Sci. and Aut. Cont (CCE 2009), Mexico, November 10-13, pp. 12–17 (2009)
22. Moreno, J.A., Alvarez, J., Rocha-Cozatl, E., Diaz-Salgado, J.: Super-Twisting Observer-Based Output Feedback Control of a Class of Continuous Exothermic Chemical Reactors. In: 2010 IFAC 9th International Symposium on Dynamics and Control of Process Systems, DYCOPS 2010. Leuven, Belgium, July 5-7 (2010)
23. Moreno, J.A.: Lyapunov Approach for Analysis and Design of Second Order Sliding Mode Algorithms. In: Fridman, L., Moreno, J., Iriarte, R. (eds.) Sliding Modes. LNCIS, vol. 412, pp. 113–149. Springer, Heidelberg (2011)
24. Moreno, J.A.: A Lyapunov Approach to Output Feedback Control using Second Order Sliding Modes. IMA Journal of Mathematical Control and Information (2012), doi:10.1093/imamci/dnr036 (published on line January 2, 2012)
25. Moreno, J.A., Dochain, D.: Global observability and detectability analysis of uncertain reaction systems and observer design. International Journal of Control 81, 1062–1070 (2008)
26. Vasiljevic, L.K., Khalil, H.K.: Error bounds in differentiation of noisy signals by high-gain observers. Systems & Control Letters 57, 856–862 (2008)
27. Utkin, V., Guldner, J., Shi, J.: Sliding Mode Control in Electro-Mechanical Systems, 2nd edn. CRC Press, Taylor & Francis (2009)

Chapter 13
Multirate Functional Observer Based Discrete-Time Sliding Mode Control

S. Janardhanan and Neeli Satyanarayana

Abstract. In this chapter, a brief introduction to the functional observer, multirate output sampling and discrete-time sliding mode control will be given. Using the concept of multirate sampling of plant output, a functional observer based sliding model controller has been designed for systems with uncertainty. The necessary and sufficient conditions are given for the design of the proposed controller using the minimal number of output samples. A numerical example is considered to demonstrate the procedure and efficacy of the approach.

13.1 Introduction

A control system is an implemented strategy used to cause a system to behave in a desired manner. In the literature, different methods of controller implementation have been discussed which utilize the states of the system, to control its dynamics. A sliding mode controller alters the state of the system in order to match the desired performance criteria [21]. Control engineering is one of many areas where digital computer technology has made a great impact. Hence, most practical systems which are continuous in nature are studied based on their sampled-data form. Much progress has been made on sampled feedback stabilization [22], [23]. In a sampled-data system, the continuous-time system is driven by a control that is piecewise constant over the sampling interval τ sec. Therefore, it is quite natural to extend the technique of continuous sliding mode control to discrete-time control systems. Considerable efforts have been made in the study of the concept of Discrete Sliding Mode (DSM) controller design [28]- [34].

S. Janardhanan · Neeli Satyanarayana
Department of Electrical Engineering, Indian Institute of Technology Delhi, India
e-mail: `janas@ee.iitd.ac.in`, `neeli.satya@gmail.com`

B. Bandyopadhyay et al. (Eds.): *Advances in Sliding Mode Control*, LNCIS 440, pp. 267–281.
DOI: 10.1007/978-3-642-36986-5_13

Most of the sliding mode control methods require full state feedback. But, measurement of the full state vector is practically not possible. Therefore, one has to either abandon the concept of state feedback based controller design or use an approximation of the states instead of actual states. Normally, the latter option is simple. Hence, one has to substitute the original states with the approximated states in the control law. Hence, a method must be devised to estimate the unmeasurable state variables. The problem of observing the state vector for deterministic linear time-invariant multivariable systems has been the object of numerous studies ever since the original work of Luenberger [1]. The observer will utilize the available inputs and output of the original system to construct the state vector. In the literature, a wide variety of observers are designed with respect to procedure and dimension [1]- [7]. This led to state-observer based sliding mode control [26].

In general, the sliding mode controller for the LTI system with state vector x has the structure

$$u(k) = Fx(k) + h(c^T x(k)) \tag{13.1}$$

where, $Fx(k)$ is the linear component of the control which forms the equivalent control and $h(c^T x(k))$ is the control responsible for the reaching phase. The techniques proposed in [26], [27] compute the state-vector $x(k)$ and then derive the control signal using the expression described above. However, it is worthy to note that it is not absolutely necessary to compute the entire state vector $x(k)$, the knowledge of the linear state combinations, $Fx(k)$ and $c^T x(k)$ would be sufficient to compute the desired control input. Therefore, only an implicit functional observer needs to be constructed. The concept of linear functional observers has been around for more than four decades [2]. Using the concept of multirate sampling of the plant output, the design of functional observers [37] and/or sliding mode control [38] can also be developed. The degree of simplification achievable by [38] is more pronounced than achievable by [26], [27].

No physical system can be accurately modeled for control design purposes. The sliding mode control technique is well known for its robustness against model uncertainties, parameter variations and external disturbances as indicated in [29]. The equivalent structure for the sliding mode controller for systems with uncertainty can be represented structurally as

$$u(k) = Fx(k) + F_1 d(k) + h(c^T x(k)) \tag{13.2}$$

where $d(k)$ is the uncertainty in the system. Janardhanan and Bandyopadhyay [33], proposed a multirate output feedback based quasi-sliding mode control strategy for LTI systems with bounded unmatched uncertainty in its dynamics. In [34], a new control technique is developed for systems with matched uncertainty by combining the Bartoszewiczs [35] reaching law approach and the fast output sampling technique, which avoids chattering. In [36], G.D.Reddy et al. proposed a similar controller to the one in [34]. However, the algorithms proposed in [33], [34], and [36] are still based on full state estimation and an assumption on the disturbance bounds. As mentioned earlier, it may not always be necessary to estimate the entire state vector for implementation of (13.2).

This chapter addresses the problem of sliding mode control design based on the concept of functional observer theory and multirate sampling of the plant output for systems with uncertainty. The proposed method of design requires fewer output samples as compared to [33] or [34]. Further, the proposed method uses only the system output and past inputs to implement the control, making the algorithm more practical in comparison to state feedback based methods.

13.2 Functional Observers

Linear functional observers estimate linear functions of the state vector of a system without estimating all the individual states. Such functional estimates are useful in feedback control system design because the control signal is often a linear combination of the states, and it is possible to utilize a linear functional observer to directly estimate the feedback control signal. Such an observer effectively reduces the complexity of the control system. There has been a considerable amount of research carried out on the subject of functional state estimation for well over four decades ever since the concept was introduced by D.G. Luenberger [2]. Many design algorithms to estimate desired linear functions of the state vector have been proposed in [8]- [15], and [16]- [19] for LTI systems with and without uncertainty.

13.2.1 Linear Time-Invariant Systems without Uncertainty

Consider the linear time-invariant system described

$$\dot{x}(t) = Ax(t) + Bu(t) \tag{13.3a}$$
$$y(t) = Cx(t) \tag{13.3b}$$
$$w_d(t) = F_d x(t) \tag{13.3c}$$

where the state vector $x(t) \in \mathbf{R}^n$, the input vector $u(t) \in \mathbf{R}^r$, and output vector $y(t) \in \mathbf{R}^m$. The vector $w_d(t) \in \mathbf{R}^v$ is a function of state vector that is to be estimated. $A \in \mathbf{R}^{n\times n}$, $B \in \mathbf{R}^{n\times r}$, $C \in \mathbf{R}^{m\times n}$, and $F_d \in \mathbf{R}^{v\times n}$ are constant matrices. It is assumed that the matrix pair (C,A) is observable and $rank(C) = m$.

The problem being considered here is the design of a functional observer with the smallest possible effective order which has arbitrarily assignable eigenvalues. The observer state $z \in \mathbf{R}^l$ is governed by a finite-dimensional system

$$\dot{z}(t) = Dz(t) + Ly(t) + Gu(t) \tag{13.4}$$

and the reconstructed functional $\hat{w}_d(t)$ is governed by an algebraic relation

$$\hat{w}_d(t) = My(t) + Nz(t) \tag{13.5}$$

The constant matrices D, L, G, M, and N are to be determined, such that they satisfy the well-known observer equations [[10]],

$$TA = DT + LC \tag{13.6a}$$
$$F_d = NT + MC \tag{13.6b}$$
$$G = TB \tag{13.6c}$$

such that $\hat{w}_d(t) - F_d x(t) \to 0$ as $t \to \infty$ for all $u(t)$, where $z(t) \to Tx(t)$.

13.2.2 Linear Time-Invariant Systems with Uncertainty

In most realistic control systems, there is frequently some degree of uncertainty surrounding the plant. For example, some of the plant parameters may not be known, or the plant may be subject to unknown disturbances, or uncertainties as in applications like fault detection and identification and unknown external excitation. These uncertainties can often be incorporated into the system model by treating them as unknown inputs.

Consider the following LTI system with perturbations

$$\dot{x}(t) = Ax(t) + Bu(t) + Dd(t) \tag{13.7a}$$
$$y(t) = Cx(t) \tag{13.7b}$$
$$w_d(t) = F_d x(t) \tag{13.7c}$$

where the the disturbance vector $d(t) \in \mathbf{R}^p$, and $D \in \mathbf{R}^{n \times p}$ is a constant vector. For the ensuing discussion, the matrix pair (C, A) is assumed to be observable, and $rank(D) = p$. The term $Dd(t)$ in (13.7a) has been used to represent additive disturbance as well as different types of modelling uncertainties [[19]]. Designing of disturbance decoupled linear functional observers of state have been reported in [[16]]- [[19]].

The functional observer that can estimate (13.7c) has the same structure as (13.4) and (13.5). The constant matrices D, L, G, M, and N are to be determined if matrix equations [[19]] such that $TD = 0$ is satisfied in addition to (13.6) and D being a Hurwitz matrix. Namely, all the eigenvalues of D are inside the unit-circle. A constructive procedure is presented in [[17]] to estimate a linear function of state vector with an order $l \geq \frac{v(n-m)}{(m-p)}$ with a constraint that the number of functionals are equal to the number of inputs $(v = r)$. In [[18]], existence conditions are given for observing a linear function of state vector with an l-th order $(m \leq l \leq n-m)$. Hou and Muller [[16]], presented an easy and systematic procedure to design a decentralized linear functional observer.

13.2.3 Problems with Observers

Even with best efforts to reduce the order of the observers [20], two clear disadvantages can be pointed out. Namely, increase in the order of the system and possibility of producing an unstable controller. Further, the accuracy of the observer will be improve after long time. Hence, search for non-dynamic observer designs has been an ongoing process. This problems of increase in observer dimension and accuracy of estimation are effectively addressed by design of controllers based on multirate sampling of plant output.

13.3 Multirate Output Sampling

In this section, we introduce essential tools and results for the development of a multirate model for sampled-data systems as required for subsequent developments. Multirate systems are those systems which use more than one sampling rate. In multirate output systems, the plant output is observed at a faster rate than the input is updated. The i-th plant output is detected N_i times for every update of plant input, which happens once in sampling period of τsec [24]. For computational simplicity, all the outputs are assumed to be sampled at a uniform rate .

To illustrate the approach in this work, we consider the system to be sampled at the two different rates $\frac{1}{\tau}$ and $\frac{1}{\Delta}$, with $\Delta = \frac{\tau}{N}$, where N is a positive integer. In the sampled-data system, the continuous-time system (13.7a), (13.7b)) is driven by a control that is piecewise constant over a sampling interval τ. The equivalent discrete-time system approximation for a sampling period τ will be

$$x(k+1) = \Phi_\tau x(k) + \Gamma_\tau u(k) + D_\tau d(k) \tag{13.8a}$$
$$y(k) = Cx(k) \tag{13.8b}$$

where $x(k) = x(k\tau)$, $u(k)$ =value of $u(t)$ over $[k\tau,(k+1)\tau)$, and the system matrices Φ_τ, Γ_τ, and D_τ are given by

$$\Phi_\tau = e^{A\tau}, \Gamma_\tau = \int_0^\tau e^{A\lambda} B \mathrm{d}\lambda, D_\tau = \int_0^\tau e^{A\lambda} D \mathrm{d}\lambda$$

Assume that the system (13.7a), (13.7b) is sampled with a sampling time of $\Delta = \frac{\tau}{N}$sec. Then we have

$$x(k+1) = \Phi_\Delta x(k) + \Gamma_\Delta u(k) + D_\Delta d(k) \tag{13.9a}$$
$$y(k) = Cx(k) \tag{13.9b}$$

where

$$\Phi_\Delta = e^{A\Delta}, \Gamma_\Delta = \int_0^\Delta e^{A\lambda} B \mathrm{d}\lambda, D_\tau = \int_0^\Delta e^{A\lambda} D \mathrm{d}\lambda$$

Note that the system matrices in (13.8a) are related as

$$\Phi_\tau = \Phi_\Delta^N, \Gamma_\tau = \sum_{i=0}^{N-1} \Phi_\Delta^i \Gamma_\Delta, D_\tau = \sum_{i=0}^{N-1} \Phi_\Delta^i D_\Delta$$

Consider the output of the system (13.7b) being sampled at every $\Delta = \frac{\tau}{N}$ sec and input being updated once in every sampling period τ sec. Then we have

$$y(k\tau + i\Delta) = Cx(k\tau + i\Delta) \tag{13.10}$$

where $i = 0, 1, 2, \cdots, (N-1)$. Then, these output samples can be represented as stacked output as

$$y_{k+1} = C_0 x(k) + D_0 u(k) + C_d d(k) \tag{13.11}$$

where constant matrices $C_0 \in \mathbf{R}^{Nm\times n}$, $D_0 \in \mathbf{R}^{Nm\times r}$, $C_d \in \mathbf{R}^{Nm\times p}$ and output vector $y_{k+1} \in \mathbf{R}^{Nm\times 1}$ are given as

$$C_0 = \begin{bmatrix} C \\ C\Phi_\Delta \\ \vdots \\ C\Phi_\Delta^{N-1} \end{bmatrix}, D_0 = \begin{bmatrix} 0 \\ C\Gamma_\Delta \\ \vdots \\ C\sum_{i=0}^{N-2} \Phi_\Delta^i \Gamma_\Delta \end{bmatrix}, C_d = \begin{bmatrix} 0 \\ CD_\Delta \\ \vdots \\ C\sum_{i=0}^{N-2} \Phi_\Delta^i D_\Delta \end{bmatrix};$$

where

$$y_{k+1} = \begin{bmatrix} y(k\tau) \\ y(k\tau + \Delta) \\ \vdots \\ y((k+1)\tau - \Delta) \end{bmatrix}$$

13.3.1 Relationship between System State and Fast Output

From (13.8a) and (13.11), the relationship between the system state $x(k)$ and the fast output y_k can be calculated in the following manner [[33]].

Consider the output equation of (13.11) being multiplied by C_0^T

$$C_0^T y_{k+1} = C_0^T (C_0 x(k) + D_0 u(k) + C_d d(k)) \tag{13.12}$$

From this, the value of $x(k)$ can be obtained in terms of y_{k+1} and $u(k)$ as

$$\begin{aligned} x(k) = &(C_0^T C_0)^{-1} y_{k+1} - (C_0^T C_0)^{-1} C_0^T D_0 u(k) \\ &+ (C_0^T C_0)^{-1} C_0^T C_d d(k) \end{aligned} \tag{13.13}$$

Substituting the value of $x(k)$ from (13.13) in the state equation (13.8a), the relationship between $x(k+1)$ and y_{k+1}, or equivalently, $x(k)$ and y_k can be derived to be

$$x(k) = L_y y_k + L_u u(k-1) + L_d d(k-1) \tag{13.14}$$

where

$$L_y = \Phi_\tau (C_0^T C_0)^{-1} C_0^T \tag{13.15}$$
$$L_u = \Gamma_\tau - \Phi_\tau (C_0^T C_0)^{-1} C_0^T D_0 \tag{13.16}$$
$$L_d = D_\tau - \Phi_\tau (C_0^T C_0)^{-1} C_0^T C_d \tag{13.17}$$

13.3.2 Advantages of Multirate Output Sampling

The main advantage of this multirate output feedback approach, as it is evident from the above explanation, is that all the objectives that could be achieved through full state feedback (i.e., using full state information) could now be achieved with the use of past output and input samples, and without the use of a dynamical observer. Further, as the state estimation error dynamics do not exist (i.e., the error is dependent only on the disturbance alone), the accuracy is much greater than any other dynamical observer.

13.4 Motivation of Multirate Output Sampling Based Functional Estimation

Most of the design techniques for sliding mode control assume that the entire state vector is available for measurement, which is not true in many practical systems. Hence, an observer based sliding mode controller is to be implemented. But, the problem associated with observer based design is that the order of the overall system will be increased. With increase in the order, the effective complexity of the control algorithm increases, as a result computational efficiency decreases due to the higher computational burden. With conventional observers, the accuracy between state and its estimate will be improved after a number of samples as compared to just one sampling period in the case of multirate output sampling based state estimation.

The problem associated with the existing method [33] is that the ratio of the sampling frequencies, viz., the integer N, is to be chosen to be greater than or equal to the observability index of the system [39]. In [37], and [38] a multirate output sampling based functional observer and/or sliding mode controller is proposed for systems without uncertainty. The problem of designing a functional observer based sliding mode controller with minimal number of samples for systems with uncertainty is still open. This chapter addresses the problem of sliding mode control design based on the concept of functional observer theory and multirate sampling of the plant output for systems with uncertainty. The proposed method of design requires fewer output samples. Further, the proposed method is based on multirate output feedback which makes the algorithm more practical in comparison to state feedback based methods.

13.5 Discrete-Time Sliding Mode Control

Most realistic control systems operate in an environment where unknown disturbances of one type or another are present. Disturbance are, by definition, plant inputs which cannot be manipulated and are not completely known a priori. Further, due to the limited availability of outputs, controllers based on state variables cannot be implemented. Hence, controller design for systems having restricted output and unknown disturbance is a challenging area for researchers.

Variable structure control with sliding modes is widely recognized in the control research community. The sliding mode control technique is well known for its robustness against model uncertainties, parameter variations and external disturbances as indicated in [21]. Sliding mode control design generally involves two main steps: firstly, the selection of sliding surface which has stable reduced-order dynamics assigned by the designer, and secondly the synthesis of a switching control law to force the closed-loop system trajectories onto and subsequently to remain on the sliding surface. The use of digital computers and samplers in the control circuitry, has made the use of a discrete-time system representation more justifiable for controller design than continuous-time representation.

Therefore, it is quite natural to extend the technique of continuous sliding mode control to discrete-time control systems. Considerable efforts have been put in the study of the concept of discrete-time sliding mode (DSM) controller design [28], [29]. In the case of DSM design, the control input is applicable only at certain sampling instants, and the control effort is constant over the entire sampling period. Hence, the system states move about the sliding mainfold but are unable to stay on it. Hence, the terminology, quasi-sliding mode (QSM).

Consider the representation of a linear discrete-time system with sampling period τ sec

$$\begin{aligned} x(k+1) &= \Phi_\tau x(k) + \Gamma_\tau u(k) + D_\tau d(k) \\ y(k) &= Cx(k) \end{aligned} \tag{13.18}$$

For the ensuing discussion, the following is assumed to be valid.

1. It is assumed that the nominal system pairs (Φ_τ, C) and (Φ_τ, Γ_τ) are detectable and controllable respectively with the matrices Γ_τ, and C being full rank.
2. The disturbance part in (13.18) has a known upper bound $d_0(k)$.

$$\| d(k) \| < d_0(k)$$

where $\|\cdot\|$ denotes standard Euclidean norm and the disturbance also satisfies the matching condition

$$Rank(\Gamma_\tau) = Rank(\begin{bmatrix}\Gamma_\tau & D_\tau\end{bmatrix}) \tag{13.19}$$

For the VSC of a discrete-time system, the reaching law proposed in [29] has the following form

$$s(k+1) - s(k) = -q\tau s(k) - \varepsilon\tau sgn(s(k)) \tag{13.20}$$

where $\tau > 0$ is the sampling period, $\varepsilon > 0$, $q > 0$, and $1 - q\tau > 0$ and $s(k)$ is the switching function which is actually linear combination of the states and defined as

$$s(k) = c^T x(k) \tag{13.21}$$

Consider an incremental change in $s(k)$

$$\begin{aligned} s(k+1) - s(k) = c^T \{\Phi_\tau x(k) + \Gamma_\tau u(k) + D_\tau d(k)\} \\ -c^T x(k) \end{aligned} \tag{13.22}$$

Solving for the control expression from (13.22) and (13.20), gives

$$u(k) = F_1 x(k) + F_2 d(k) + \gamma sgn(s(k)) \tag{13.23}$$

where F_1, F_2, and γ are given by

$$F_1 = -(c^T\Gamma_\tau)^{-1}[c^T\Phi_\tau + q\tau c^T - c^T] \tag{13.24a}$$

$$F_2 = -(c^T\Gamma_\tau)^{-1}c^T D_\tau \tag{13.24b}$$

$$\gamma = -(c^T\Gamma_\tau)^{-1}\varepsilon\tau \tag{13.24c}$$

It may be noted that the control in (13.23) is not implementable using state feedback alone as there is no information about the disturbance $d(k)$.

13.6 Multirate Output Feedback Based Discrete-Time Sliding Mode Controller for LTI Systems with Uncertainty

In this section, we apply the results developed in the previous sections for implementation of the control law (13.23). Consider the continuous time system (13.3a) and its discrete-time representation (13.18). The implementation of the sliding mode control law (13.23) using fast output sampling is as follows:

If the functions dependent on state and uncertainty $F_1x(k) + F_2d(k)$ and $c^Tx(k)$ are known, it is possible to compute the control law (13.23) in terms of the fast sampled output and previous input.

Theorem 13.1. *For the system (13.18), the implementation of controller of the form (13.23) is possible by an estimate*

$$u(k) = M_1 y_k + G_1 u(k-1) + \gamma sgn\{M_2 y_k + G_2 u(k-1)\} \tag{13.25}$$

where G_1, G_2 given by

$$G_1 = F_1 \Gamma_\tau - M_1 D_0 \tag{13.26}$$

$$G_2 = c^T \Gamma_\tau - M_2 D_0 \tag{13.27}$$

provided the following conditions hold:

$$\left[F_1 \Phi_\tau \; F_1 D_\tau + F_2 \right] = M_1 \left[C_0 \; C_d \right] \tag{13.28}$$

$$\left[c^T \Phi_\tau \; c^T D_\tau \right] = M_2 \left[C_0 \; C_d \right] \tag{13.29}$$

Proof. The control law (13.23) can be equivalently expressed as

$$u(k+1) = F_1 x(k+1) + F_2 d(k+1) + \gamma sgn(s(k+1)) \tag{13.30}$$

where F_1, F_2 and γ are given by (13.24a), (13.24b) and (13.24c). Using (13.18), the control law (13.30) can be represented as

$$u(k+1) = F_1 \Phi_\tau x(k) + F_1 D_\tau d(k) + F_2 d(k+1) + F_1 \Gamma_\tau u(k) + \gamma sgn(s(k+1)) \tag{13.31}$$

With an assumption of slowly varying disturbance, (13.31) can be written as

$$u(k+1) = F_1 \Phi_\tau x(k) + \{F_1 D_\tau + F_2\} d(k) + F_1 \Gamma_\tau u(k) + \gamma sgn(s(k+1)) \tag{13.32}$$

Now, consider the functional

$$g(k) = \begin{bmatrix} F_1 \Phi_\tau & F_1 D_\tau + F_2 \\ c^T \Phi_\tau & c^T D_\tau \end{bmatrix} z(k) \tag{13.33}$$

where $z(k) = \left[x^T(k) \; d^T(k) \right]^T$. If $g(k)$ is known, then the control law $u(k)$ can be computed in terms of fast sampling output and previous control input given by (13.25). For there to be a solution to (13.28) and (13.29),

$$Rank \begin{bmatrix} F_1 \Phi_\tau & F_1 D_\tau + F_2 \\ c^T \Phi_\tau & c^T D_\tau \\ C_0 & C_d \end{bmatrix} = Rank \left[C_0 \; C_d \right]$$

or equivalently

$$\mathscr{R}\begin{bmatrix} F_1\Phi_\tau & F_1D_\tau + F_2 \\ c^T\Phi_\tau & c^TD_\tau \end{bmatrix} \subseteq \mathscr{R}\begin{bmatrix} C_0 & C_d \end{bmatrix} \tag{13.34}$$

where $\mathscr{R}[\cdot]$ refers to the row range space of $[\cdot]$. This condition (13.34) can be equivalently written as (13.28) and (13.29).

Remark 13.1. The feasibility of (13.34) does not require the condition that the number of output samples N be greater than the observability index of the plant.

Therefore, the condition of complete observability of the system is relaxed for estimation of the functionals. This is one of the important features of the proposed technique.

From the definition of $g(k)$ in (13.33), one can compute the values of the equivalent control and sliding function as

$$\begin{aligned} F_1\Phi_\tau x(k) + \{F_1D_\tau + F_2\}d(k) &= \begin{bmatrix} I_r & 0 \end{bmatrix} g(k) \\ c^T\Phi_\tau x(k) + c^TD_\tau d(k) &= \begin{bmatrix} 0 & I_r \end{bmatrix} g(k) \end{aligned}$$

where I_r is an identity matrix of order r. Now, substituting these values in (13.25), one can implement the sliding mode control.

From (13.34), it is clear that for the computation of (13.25), it is not necessary to estimate the full-state of the system. The proposed technique uses fewer samples than any other method based on multirate sampling of the plant output and the observability condition is not a necessary condition.

13.7 Numerical Example and Simulation Results

To illustrate the proposed method of controller design and its advantages, let us consider a numerical example of the form (13.3a) with $n = 10, m = 3, r = 1$. The system matrices are

$$A = \begin{bmatrix} -2 & 0.5 & 1.5 & 2 & 1.6 & -1 & 0.75 & 2 & -1.25 & 0 \\ 0.4 & 1 & -2 & 1.2 & 3 & 0 & -2 & 1 & -1 & -3 \\ 0.8 & 1.5 & 1 & -1 & 2 & 0.75 & -4 & 3.5 & 0.5 & -2 \\ 0 & -1 & 2 & 1.5 & 0.75 & -2 & 0.5 & 0 & 4 & -3 \\ 1 & -1 & 2 & -0.5 & -2 & 0.5 & 1.5 & -1.5 & -5 & 1 \\ -2 & 1 & 0.5 & 1.5 & 0.75 & 2 & 2.5 & -3 & 3.5 & -1 \\ -3 & 1 & 1.5 & 0 & 2 & -1 & 0.75 & 0.5 & 2.5 & 3 \\ -1 & 0.4 & 1.5 & 2 & -2 & 5 & 4.5 & 0 & 1 & 2.5 \\ 2 & 0.5 & -1 & -0.5 & -2 & 1.5 & 2.5 & 4 & -3 & -1.5 \\ -1 & 0 & 2.5 & 1.5 & 1 & 3 & 0.75 & 2 & 3 & 4 \end{bmatrix}$$

$$B = \begin{bmatrix} 0.5 & 1 & -2 & 0.75 & -1 & 3 & 2.5 & 1.5 & -1 & 1 \end{bmatrix}^T$$

$$D = \begin{bmatrix} 0.25 & 0.5 & -1 & 0.375 & -0.5 & 1.5 & 1.25 & 0.75 & -0.5 & 0.5 \end{bmatrix}^T$$

and

$$C = \begin{bmatrix} 1 & 0.5 & -1 & 0 & 2 & 2.5 & 1.5 & -1 & -2 & 0.75 \\ -2 & 0.5 & 1 & -2.5 & 0.75 & 1.5 & 3 & 0 & -1 & -0.5 \\ -3 & 2.5 & 1 & -1 & 0.5 & 0.75 & 1.5 & -2 & -0.5 & 0 \end{bmatrix}$$

The plant is discretized with $\tau = 0.45$ sec. The sliding function gain is

$$c = \left(-0.0147\ -0.019\ -0.007\ 0.005\ -0.03\ -0.0103\ 0.0283\ 0.0455\ 0.0297\ 0.0961 \right)$$

This system has an observability index equal to 4. However, the condition (13.34) is satisfied with $N = 3$. For the simulation purpose, a slowly varying disturbance signal of $d(k) = 5\sin(k/20)\exp(-k/500)$ is used.

Since no information about the initial state is available for the multirate controller, an estimated initial state of $X_e = \begin{bmatrix} -5 & -1 & 2.5 & -2.25 & 0.5 & -1.5 & 0 & 1 & 3 & 0.75 \end{bmatrix}^T$ is used to generate the multirate output feedback control signal for $k = 0$. Fig. 13.1, Fig. 13.2, and Fig. 13.3 show the simulation results of proposed technique. Fig. 13.1 shows response of control input, Fig. 13.2, and Fig. 13.3 show the time responses sliding function and output of the system. It can be seen that the sliding mode control is emulated faithfully for the uncertain system using proposed multirate output feedback control.

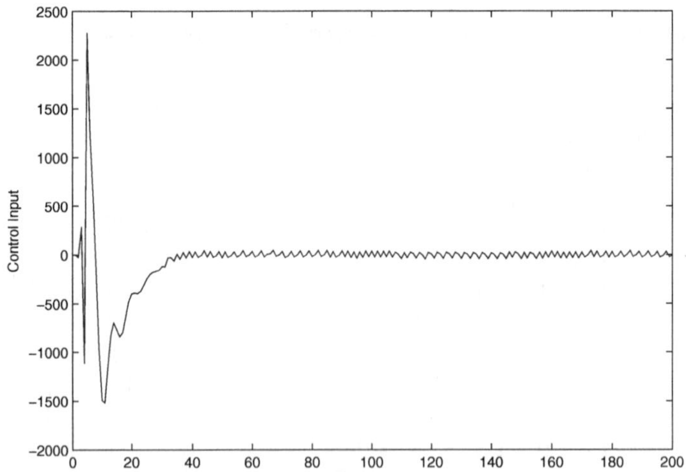

Fig. 13.1 Response of control input

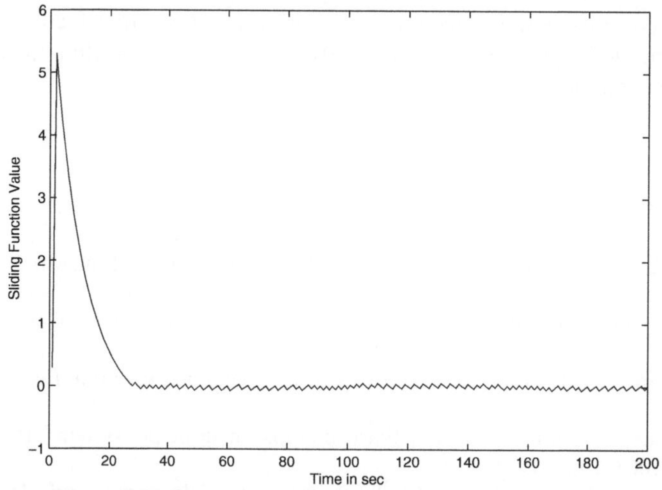

Fig. 13.2 Response of sliding function value

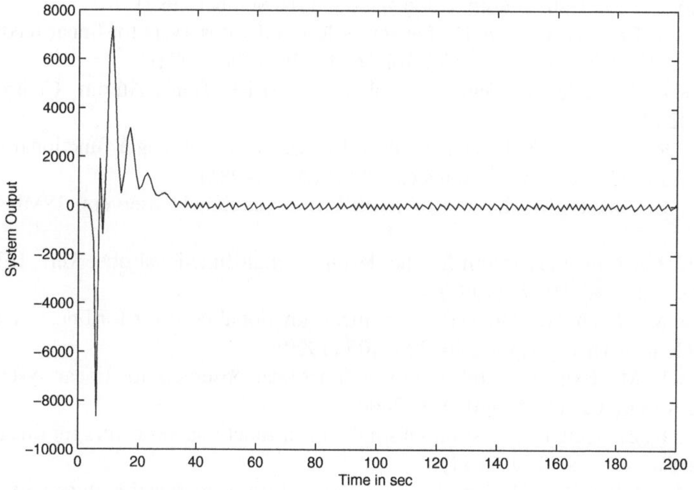

Fig. 13.3 Response of system output

13.7.1 Conclusions

A method for computation of discrete-time sliding mode control for a class of uncertain systems, using the concepts of multirate sampling and functional observation, is proposed in this chapter. It has been demonstrated that the number of output samples required is less than the observability index, which in turn is the minimum value required for the computation of the full state vector. The use of output feedback as opposed to state feedback makes the design more practical. It is also shown that full state observability is not a necessary condition for the output feedback controller design. The design procedure and its advantages have been illustrated through a numerical example.

References

1. Luenberger, D.G.: Observing the state of a linear system. IEEE Tran. Mil. Elecrron. MIL-8, 74–80 (1964)
2. Luenberger, D.G.: An introduction to observer. IEEE Tran. Autom. Contr. 16, 596–602 (1971)
3. Cumming, S.D.G.: Design of observers of reduced.order. Elecrron. Letters 5, 213–214 (1969)
4. Gopinath, B.: On the control of linear multiple input-output systems. Bell Syst. Tech. J. 50, 1063–1081 (1971)
5. Wonham, W.M., Morse, A.S.: Feedback invariants of linear multivariable systems. Automatica 8, 93–100 (1972)
6. Emre, E.: Pole assignment by dynamic feedback. Inr. J. Contr. 33(2), 381–394 (1981)
7. Syrmos, V.L., Syrmos, G.L.: Computational design techniques for reduced-order compensators. IEEE Tran. Autom. Control 37(11) (November 1992)
8. Fortmann, T.E., Williamson, D.: Design of low-order observers for linear feedback control laws. IEEE Trans. Autom.Control AC-17, 301–308 (1972)
9. Murdoch, P.: Design of degenerate observers. IEEE Trans. Autom. Control AC-19, 441–442 (1974)
10. Gupta., R.M., et al.: A direct procedure for the design of single functional observers. IEEE Trans. Circuits and Systems cas-28(4) (April 1981)
11. Kondo, E., Takata, M.: Design of multi-linear functional observers. JSME 20 (April 1977)
12. Tsui, C.-C.: A new algorithm for the design of multifunctional observers. IEEE Trans. Autom. Contr. 30, 89–93 (1985)
13. Aldeen, M., Trinh, H.: Reduced-order linear functional observer for linear systems. IEE Proc. Control Theory Appl. 146, 399–405 (1999)
14. Darouach, M.: Existence and design of functional observers for linear systems. IEEE Trans. Autom. Contr. 45, 940–943 (2000)
15. Rotella, F., Zambettakis, I.: Minimal single linear functional observers for linear systems. Automatica 47, 164–169 (2011)
16. Hou, M., Muller, P.C.: Design of decentralized linear state function observers. Automatica 11, 1801–1805 (1994)
17. Trinh, H., Ha, Q.: Design of linear functional observers for linear systems with unknown inputs. Int. J. Syst. Sci. 31, 741–749 (2000)
18. Trinh, H., Fernando, T., Nahavandi, S.: Design of reduced-order functional observers for linear systems with unknown inputs. Asian J. Control 6, 514–520 (2004)
19. Trinh, H., Tran, T.D., Fernando, T.: Disturbance decoupled observers for systems with unknown inputs. IEEE. Trans. Autom. Control 53, 2397–2402 (2008)
20. Er, M.J., Anderson, B.D.O.: Design of reduced-order multirate output linear functional observer-based compensator. Automatica 31, 237–242 (1995)
21. Utkin, V.I.: Variable structure systems with sliding modes. IEEE Tran. Autom. Control 22(2), 212–222 (1977)

22. Astrom, K.J., Wittenmark, B.: Computer Control System–Theory and Design. Prentie-Hall Inc., Englewood Cliffs (1984)
23. Chen, T.W., Francis, B.: Optimal Sampled-Data Control Systems. Springer, London (1995)
24. Hagiwara, T., Araki, M.: Design of a stable state feedback controller based on the multi-rate sampling of the plant output. IEEE Trans. Automat. Contr. AC-33, 812–819 (1988)
25. Werner, H.: Robust control of a laboratory flight simulator by nondynamic multirate output feedback. In: Proceeding of 35th International Conference on Control, pp. 1575–1580 (December 1996)
26. Saaj, C.M., Bandyopadhyay, B., Unbehauen, H.: A new algorithm for discrete-time sliding mode control using fast output sampling feedback. IEEE Trans. Ind. Electr 49(3), 518–523 (2002)
27. Sarpturk, S.Z., Istefanopulos, Y., Kaynak, O.: On the stability of discrete-time sliding mode systems. IEEE Tran. Autom. Control AC-32, 930–932 (1987)
28. Furta, K.: Sliding mode control of a discrete system. Syst. Control. Lett. 14(2), 144–152 (1990)
29. Gao, W., Wang, Y., Homaifa, A.: Discrete-time variable structure control systems. IEEE Trans. Ind. Electron. 42(2), 117–122 (1995)
30. Tang, C.Y., Misawa, E.A.: Discrete variable structure control for linear multivariable systems: the state feedeback case. In: Proceeding of Amer. Control Conf., pp. 14–118 (June 1998)
31. Bandyopadhyay, B., Janardhanan, S.: Discrete-time sliding mode control: A multirate-output feedback approach. LNCIS, vol. 323, 147 p. Springer, Heidelberg (2005)
32. Bandyopadhyay, B., Janardhanan, S.: Discrete-time Sliding Mode Control using Multi-rate Output Feedback. LNCIS, vol. 334, pp. 351–371. Springer, Berlin (2006)
33. Janardhanan, S., Bandyopadhyay, B.: Discrete sliding mode control of systems with unmatched uncertainty using multirate output feedback. IEEE Trans. Autom. Control AC-51(6), 1030–1035 (2006)
34. Janardhanan, S., Bandyopadhyay, B.: Output feedback sliding-mode control for uncertain systems using fast output sampling technique. IEEE Trans. Ind. Electron. AC-53(5), 1677–1682 (2006)
35. Bartoszewicz, A.: Discrete-time quasi-sliding-mode control strategies. IEEE Trans. Ind. Electron. AC-45(4), 633–637 (1998)
36. Reddy, G.D., et al.: Discrete-time output feedback sliding mode control for spatial control of a large PHWR. Automatica 45, 2159–2163 (2009)
37. Janardhanan, S., Satyanarayana, N.: Design of Multirate Output Feedback Sampling based Functional Observer and State Feedback. In: Proceeding of 11th International Conference on Control, Automation, Robotics and Vision (ICARCV 2010), Singapore, pp. 1297–1301 (December 2010)
38. Janardhanan, S., Inamdar, S.: Multirate Output Feedback Sliding Mode Control Design Using Reduced Order Functional Observer. In: Proceeding of 11th International Workshop on Variable Structure Systems (VSS 2010), Mexico City, Mexico, pp. 395–398 (June 2010)
39. Chen, C.T.: Linear System Theory and Design. Oxford University Press, New York (1999)

Chapter 14
Observers with Discrete-Time Measurements in the Sliding Mode Output-Feedback Stabilization of Nonlinear Systems

Elisabetta Punta

Abstract. The chapter investigates the problem of designing an observer for nonlinear nonaffine systems with discrete-time measurements (continuous-discrete-time systems). The chapter considers the variable-structure control of nonlinear systems when the state vector is not completely available and the output measurements are discrete-time; the use of suitably designed observers is required. The strategy of introducing integrators in the input channel is exploited to enlarge the class of tractable control systems. An observer is proposed and conditions are found under which the convergence to the unique ideal solution is proven for both system and observer. The control problem is solved by forcing a sliding regime for the observer, while satisfying an exponential stability criterion for the observation error state equation.

14.1 Introduction

This chapter deals with nonlinear systems nonaffine in the control law when, due to incomplete state availability, the design of sliding mode control calls for suitable observation procedure.

It is proven in [[1]] that the control problem has a solution for perfectly known nonlinear nonaffine systems, provided some uniqueness conditions, [[2]], are satisfied by the coupled state-observer system, and a nonlinear matrix inequality involving the Jacobian matrices of the observer has a solution. This method is "differentiator free", nevertheless in some cases the posed convergence conditions result to be too restrictive.

In [[3]] nonlinear nonaffine systems are considered and novelties with respect to [[1]] are presented. Integrators are introduced in the input channel, [[4]], [[5]], [[6]],

Elisabetta Punta
National Research Council of Italy, Institute of Intelligent Systems for Automation (CNR-ISSIA), Via De Marini, 6 - 16149 Genoa, Italy
e-mail: elisabetta.punta@cnr.it

B. Bandyopadhyay et al. (Eds.): *Advances in Sliding Mode Control*, LNCIS 440, pp. 283–297.
DOI: 10.1007/978-3-642-36986-5_14

with the aim of strongly simplifying the convexity constraints required to ensure the global convergence of the coupled state-observer system to the unique ideal one.

In the present chapter we consider nonlinear nonaffine control systems. Integrators are introduced in the input channel in order to deal with a larger class of nonlinear nonaffine control systems. A full-order observer is designed and the relevant convergence conditions are found. The sliding motion of the state-observer coupled system on a sliding manifold in the state space of the observer is guaranteed. The analysis of the closed-loop robustness of the proposed scheme is performed with respect to the discrete-time availability of the measurements of the system. Conditions are posed about the considered system and the accessible measurements.

This chapter investigates the problem of designing an observer for nonlinear nonaffine systems with discrete-time measurements (continuous-discrete-time systems). The contribution of the chapter is in the context of output feedback under perfect plant knowledge and with discrete-time measurements.

The use of continuous-discrete observers to estimate the state of nonlinear systems has already been investigated in the literature, [[7]], [[8]], [[9]]. In particular sliding mode observers have been developed in presence of sampled output information, [[10]], [[11]], [[12]].

The practical issues relevant to differentiators, [[13]], [[14]], are not addressed in the chapter.

The chapter is organized as follows. Section 14.2 proposes the statement of the considered variable-structure control problem. Integrators are introduced in the input channel in order to deal with a larger class of nonlinear nonaffine control systems. Conditions are posed about the considered system. In Section 14.3 an observer with continuous time measurements is designed and the relevant convergence conditions are found. In the following Section 14.4 it is considered the case when the state vector is not completely available and the output is accessible via discrete-time measurements: the use of a suitably designed observer is required. Finally a detailed example and simulation results are presented in Section 14.5.

Throughout the chapter a prime denotes transpose and $|\cdot|$ is the Euclidean norm or the induced matrix norm.

14.2 Problem Statement

We consider the nonlinear nonaffine control system

$$\dot{\eta} = \varphi(t, \eta, u) \qquad t \geq 0, \tag{14.1}$$

where $\varphi : [0, +\infty) \times \Omega \times R^m \to R^n$ is a Carathéodory mapping, $\eta \in R^n$ is the state vector, Ω is an open set of R^n, $u \in R^m$ is the available control vector.

The state vector is not completely available and the output vector $\zeta \in R^k$ is expressed by the following equation

$$\zeta = \rho(\eta), \tag{14.2}$$

where ρ is of class C^2.

The output is accessible via discrete-time measurements

$$\zeta_i = \rho\left(\eta\left(t_i\right)\right), \tag{14.3}$$

where t_i, $i = 0, 1, \ldots$, is the sequence of positive real numbers, the sampling instants, defined as $t_{i+1} = t_i + \delta$, $t_0 = 0$, $i = 0, 1, \ldots$, and the constant $\delta > 0$ is the measurement sampling interval.

The sliding manifold is

$$\xi\left(\eta\right) = 0, \tag{14.4}$$

with $\xi\left(\eta\right) \in R^m$.

We assume that $n \geq m$,

$$\xi = \xi\left(\eta\right) : \Omega \to R^m,$$

ξ is $C^2\left(\Omega\right)$, and the $m \times n$ Jacobian matrix

$$\xi_\eta = \frac{\partial \xi}{\partial \eta}\left(\eta\right) \qquad \text{has maximum rank } m \tag{14.5}$$

for $\eta \in \Omega$.

The objective is to control the state variables $\eta\left(t\right)$, $t \geq 0$, of the control system (14.1) in order to guarantee that the sliding output

$$\xi\left[\eta\left(t\right)\right] \to 0 \quad \text{as} \quad t \to +\infty.$$

In the following section a solution is proposed, which introduces integrators in the input channel. This procedure, traditionally implemented in order to reduce the chattering phenomenon, allows to consider a larger class of nonlinear nonaffine control systems and results in a strongly simplified convexity condition, [[3]].

14.2.1 *The Introduction of Integrators in the Input Channel*

Consider the control system (14.1) and sliding manifold (14.4).

Let us define the following augmented control system

$$\dot{\eta} = \varphi\left(t, \eta, u\right) \quad \dot{u} = v, \quad t \geq 0, \tag{14.6}$$

with control vector $v \in R^m$. We measure $\zeta_1 = \rho_1\left(u\right)$, where $\rho_1 : R^m \to R^m$ and the Jacobian matrix $\rho_{1u} = \dfrac{\partial \rho_1}{\partial u}\left(u\right)$ has maximum rank m. The output ζ_1 is accessible via discrete-time measurements

$$\zeta_{1i} = \rho_1\left(u\left(t_i\right)\right),$$

where t_i, $i=0,1,\ldots$, is the sequence of positive real numbers, the sampling instants, defined as $t_{i+1}=t_i+\delta$, $t_0=0$, $i=0,1,\ldots$, and the constant $\delta>0$ is the measurement sampling interval.

Assume that φ, ξ are both of class C^2 everywhere. For almost every t, the first time derivative of ξ is given by

$$\dot{\xi}=\xi_\eta(\eta)\,\varphi(t,\eta,u)$$

We introduce a new sliding output

$$s=\dot{\xi}+\Lambda\xi, \tag{14.7}$$

where $\Lambda=\operatorname{diag}(\lambda_j)$, $\lambda_j>0$, $j=1,\ldots,m$, is a constant $m\times m$ diagonal matrix.

Let the augmented state vector $x=(\eta',u')'\in R^{n+m}$ and the measured vector $y=(\zeta',\zeta_1')'\in R^{k+m}$, we can write

$$\begin{aligned}\dot{x}=\begin{bmatrix}\dot{\eta}\\ \dot{u}\end{bmatrix}=\begin{bmatrix}\varphi(t,\eta,u)\\ 0\end{bmatrix}+\begin{bmatrix}0\\ I\end{bmatrix}v=\\ =A(t,x)+Bv=f(t,x,v)\end{aligned} \tag{14.8}$$

$$y=\begin{bmatrix}\zeta\\ \zeta_1\end{bmatrix}=\begin{bmatrix}\rho(\eta)\\ \rho_1(u)\end{bmatrix}=h(x),$$

where $A(t,x)=\begin{bmatrix}\varphi(t,\eta,u)\\ 0\end{bmatrix}$, $B=\begin{bmatrix}0\\ I\end{bmatrix}$, $f(t,x,v)=A(t,x)+Bv$, and $h(x)=\begin{bmatrix}\rho(\eta)\\ \phi(u)\end{bmatrix}$. The control vector is v and the sliding output s is defined by (14.7).

Consider (14.6), (14.2) and (14.7). The following new variable-structure control problem can be defined

$$\dot{x}=f(t,x,v),\quad t\geq 0,\quad \text{state equation}, \tag{14.9}$$

$$\dot{u}=v,\quad \text{control equation}, \tag{14.10}$$

$$y=h(x),\quad \text{output equation}, \tag{14.11}$$

$$y_i=h(x(t_i)),\quad \text{discrete-time measurement equation}, \tag{14.12}$$

$$s(t,x)=0,\quad \text{sliding manifold}, \tag{14.13}$$

where t_i, $i=0,1,\ldots$, is the sequence of positive real numbers, the sampling instants, defined as $t_{i+1}=t_i+\delta$, $t_0=0$, $i=0,1,\ldots$, and the constant $\delta>0$ is the measurement sampling interval. The vector field $f(t,x,v)$ is defined by (14.8).

The output function h is such that

$$|h(x)|\leq\psi_1|x|,\quad \psi_1>0,\quad \forall t,x,\quad t\geq 0. \tag{14.14}$$

Assume that the Jacobian matrix

$$\xi_\eta(\eta)\,\varphi_u(t,\eta,u) \quad \text{is everywhere nonsingular.} \tag{14.15}$$

The objective is to control, by the vector v, the state variables $(\eta'(t),u'(t))'$, $t \geq 0$, of the augmented system in order to guarantee the sliding property

$$s(t) \to 0 \quad \text{as} \quad t \to +\infty.$$

14.3 Nonlinear Observer with Continuous Time Measurement

In this section it is considered the case when the output is accessible via continuous-time measurement.

The nonlinear observer with continuous-time measurements, [3], for system (14.9) is defined as

$$\dot{\hat{x}} = f(t,\hat{x},v) + N_1\,[y(t) - h(\hat{x})]. \tag{14.16}$$

We have x, $\hat{x} \in R^{(n+m)}$, $y \in R^{(k+m)}$, u, v, $s \in R^m$ and $m \leq n$. $N_1 \in R^{(n+m)\times(k+m)}$ is a constant matrix. The function f is defined by (14.8). The functions f, h, s are continuously differentiable in x, with f measurable in t and continuous in (x, v).

If (14.15) holds for system (14.9) and (14.16), then for every $t \geq 0$, y, $\hat{x}$ there exists a unique solution

$$v_{1*}(t,y,\hat{x}) \tag{14.17}$$

of the equation

$$s_t(t,\hat{x}) + s_x(t,\hat{x})\{f(t,\hat{x},\cdot) + N_1\,[y - h(\hat{x})]\} = 0,$$

where N_1 is as in (14.16). The mapping v_{1*} is by definition *the observer's equivalent control* corresponding to the output y.

We consider solutions in $[0,+\infty)$ (either in the Filippov or a.e. sense) to (14.9), (14.11), (14.16) corresponding to the observer's equivalent control, i.e. solutions to

$$\dot{x} = f(t,x,v_{1*}(t,h(x),\hat{x})), \tag{14.18}$$

$$\dot{\hat{x}} = f(t,\hat{x},v_{1*}(t,h(x),\hat{x})) + N_1\,[h(x) - h(\hat{x})], \tag{14.19}$$

the existence of which is guaranteed by previous conditions (14.15) and (14.17).

By specializing the results in [1] and [3] to the coupled state-observer system (14.9) and (14.16), we obtain the following. If there exist matrices $N_1 \in R^{(n+m)\times(k+m)}$, $M_1 \in R^{(n+m)\times(n+m)}$, positive numbers α_1, $\omega_1 \in R^+$ such that the eigenvalues of M_1 are between α_1 and ω_1, and positive number $\varepsilon_1 \in R^+$ such that the following matrix inequality holds

$$M_1(f_x - N_1 h_x) + (f_x - N_1 h_x)' M_1 \leq -\varepsilon_1 I, \tag{14.20}$$

and if $|s_x(t,x)| \leq L$ everywhere, for some positive constant L, then for every $t \geq 0$

$$|s(t,x(t)) - s(t,\hat{x}(t))| \leq L\left(\frac{\omega_1}{\alpha_1}\right)^{\frac{1}{2}} |x(0) - \hat{x}(0)| \exp(c_1 t),$$

where $\omega_1 c_1 = -\varepsilon_1$, for every a.e. solution $(x', \hat{x}')'$ in $[0, +\infty)$ to (14.18) and (14.19).

The observer (14.16) is a nonlinear system with available state vector $\hat{x}$. We assume that the control vector v can be designed to reach in finite time the observer sliding manifold $s(t,\hat{x}) = 0$. Then, from previous inequality, once $s(t,\hat{x}) = 0$, we have that

$$|s(t,x(t))| \leq L\left(\frac{\omega_1}{\alpha_1}\right)^{\frac{1}{2}} |x(0) - \hat{x}(0)| \exp(c_1 t).$$

The previous conditions (14.15) and (14.17) guarantee that the sliding motion of the state-observer coupled system on $s(t,\hat{x}) = 0$ is described by (14.18) and (14.19), independently of the nature (continuous or discontinuous) of the control v suitably designed to enforce $s(t,\hat{x}) = 0$ and actually applied.

14.4 Nonlinear Observer with Discrete-Time Measurement

Consider the system (14.9) and the sliding manifold (14.13). The control objective is to steer to zero the sliding output $s(t,x)$ by the vector v.

The state variables x are not available. The vector $y = h(x)$ is the accessible output, the measurements y_i of which are received at discrete-times t_i, with a fixed sampling period δ, according to (14.12).

The observer with discrete-time measurements for system (14.9) is designed as the following continuous-discrete-time observer defined by the following hybrid system

$$\begin{cases} \dot{\bar{x}}(t) = f(t,\bar{x},v) - Ph(\bar{x}), \quad t \in [t_i, t_{i+1}), \\ \bar{x}(t_i) = \bar{x}(t_i^-) + P\left[y_i - h\left(\bar{x}\left(t_i^-\right)\right)\right], \end{cases} \tag{14.21}$$

where $y_i = h(x_i)$ and $x_i = x(t_i)$, the sampling instants are defined as $t_{i+1} = t_i + \delta$, $t_0 = 0$, $i = 0, 1, \ldots$, the constant $\delta > 0$ is the measurement sampling interval, and $P \in R^{(n+m)\times(k+m)}$ is a constant matrix, which will be specified in the sequel. The functions f, h and s are defined by (14.8), (14.11) and (14.13). The functions f, h, s are continuously differentiable in x, with f measurable in t and continuous in (x, v).

Since (14.15) holds for system (14.9) and observer (14.21), then for every $t \geq 0$, y_i, $\bar{x}$ there exists a unique solution

$$v_*(t, y_i, \bar{x}) \tag{14.22}$$

of the equation

$$s_t(t,\bar{x}) + s_x(t,\bar{x})\left[f(t,\bar{x},\cdot) - Ph(\bar{x})\right] = 0,$$

where P is as in (14.21). The mapping v_* is by definition *the observer's equivalent control* corresponding to the measurements y_i of the output y.

We consider solutions in $[0,+\infty)$ (either in the Filippov or a.e. sense) to (14.9), (14.11), (14.21) corresponding to the observer's equivalent control, i.e. solutions to

$$\dot{x} = f\left(t,x,v_*\left(t,y_i,\overline{x}\right)\right), \tag{14.23}$$

$$\dot{\overline{x}} = f\left(t,\overline{x},v_*\left(t,y_i,\overline{x}\right)\right) - Ph\left(\overline{x}\right), \tag{14.24}$$

the existence of which is guaranteed by previous conditions (14.15) and (14.22).

Assumption 14.1. *The observer (14.21) is a perfectly known nonlinear system with available state vector $\overline{x}$. The control vector v can be designed to reach in finite time the observer sliding manifold $s(t,\overline{x}) = 0$.*

We assume that the sliding output is designed such that the state of the observer on the sliding manifold of the observer is exponentially stable.

Assumption 14.2. *For every a.e. solution $\overline{x}$ in $[0,+\infty)$ to (14.24), on $s(t,\overline{x}) = 0$, we assume that there exist a matrix $M \in R^{(n+m)\times(n+m)}$, two positive numbers α, $\varepsilon \in R^+$ such that the eigenvalues of M are between α and ε, and a positive number $\beta \in R^+$ such that the first time derivative of the Lyapunov function $V(\overline{x}) = \overline{x}' M \overline{x}$ satisfies*

$$\dot{V} \leq -\beta V \leq -\beta\alpha\left|\overline{x}\right|^2.$$

Consider the coupled state-observer system (14.9) and (14.21), the following theorem can be stated.

Theorem 14.1. *Consider the system (14.9) and the observer (14.21), for which the previously posed conditions, particularly Assumption 14.1 and 14.2, hold.*
Assume that it is possible to find a symmetric matrix $Q \in R^{(n+m)\times(n+m)}$ such that the eigenvalues of Q are between two positive numbers μ and κ, and such that the eigenvalues of the symmetric part of $Q(f_x - Ph_x)$ are less or equal $-\nu$ everywhere, being ν a positive number.
Assume moreover that $|s_x(t,x)| \leq D$ everywhere, for some constant D.

Provided μ, κ, and ν are such that $c_1 = \left(\frac{\nu}{\kappa} - 2\frac{\kappa^2}{\nu\mu}\psi^2\right) > 0$ and $c_2 = \left(\beta - 2\frac{\kappa^2}{\nu\alpha}\psi^2\right) > 0$ with $\psi = |P|\,\psi_1$, ψ_1, P, α, and β defined by (14.14), (14.21), and Assumption 14.2, then for every $t \geq 0$

$$\left|s\left(t,x(t)\right)\right| \leq D\left[\frac{\kappa}{\mu}\left|x(0) - \overline{x}(0)\right|^2 + \frac{\varepsilon}{\mu}\left|\overline{x}(0)\right|^2\right]^{\frac{1}{2}} \exp(-ct), \tag{14.25}$$

where $2c = \min(c_1,c_2)$, for every a.e. solution $(x',\overline{x}')'$ in $[0,+\infty)$ to (14.23) and (14.24), such that $s(0,\overline{x}(0)) = 0$.

Proof.
Set

$$W(t) = V_1(t) + V_2(t)$$
$$= [x(t) - \overline{x}(t)]' Q [x(t) - \overline{x}(t)] + \overline{x}'(t) M \overline{x}(t), \quad t \geq 0.$$

Then a.e. for $t \geq 0$,

$$\dot{W}(t) = 2[x(t) - \overline{x}(t)]' Q$$
$$\{f(t,x,v_*) - f(t,\overline{x},v_*) + Ph(x) - P[h(x) - h(\overline{x})]\} + \dot{V}_2(t),$$

where

$$v_*(t) = v_*(t, y_i, \overline{x}).$$

Therefore $\dot{W}(t) =$

$$= 2[x(t) - \overline{x}(t)]' Q \left[\int_0^1 (f_x - Ph_x)\, da\right] [x(t) - \overline{x}(t)]$$
$$+2[x(t) - \overline{x}(t)]' QPh(x) + \dot{V}_2(t),$$

where f_x and h_x are evaluated at $(t, \overline{\alpha}(a,t), v_*)$ and $\overline{\alpha}(a,t) = ax(t) + (1-a)\overline{x}(t)$.
We have $\dot{W}(t) =$

$$= \int_0^1 (x - \overline{x})' \left[Q(f_x - Ph_x) + (f_x - Ph_x)' Q'\right] (x - \overline{x})\, da$$
$$+2(x - \overline{x})' QPh(x) + \dot{V}_2(t),$$

from which, recalling that the symmetric part of a square matrix A is by definition $\frac{A+A'}{2}$,

$$\dot{W}(t) \leq -2\int_0^1 \nu (x - \overline{x})' (x - \overline{x})\, da + 2(x - \overline{x})' QPh(x) + \dot{V}_2(t)$$

$$\leq -2\tfrac{\nu}{\kappa} V_1(t) + 2(x - \overline{x})' QPh(x) + \dot{V}_2(t);$$

then

$$\dot{W}(t) \leq -2bV_1(t) + 2(x - \overline{x})' QPh(x) + \dot{V}_2(t),$$

with $b = \frac{\nu}{\kappa}$.
From previous inequality, by standard computations, we obtain

$$\dot{W}(t) \leq -bV_1(t) + \frac{1}{b} h'(x) P' QPh(x) + \dot{V}_2(t),$$

then

$$\dot{W}(t) \leq -bV_1(t) + \frac{\kappa}{b} |Ph(x)|^2 + \dot{V}_2(t)$$

and by (14.14)

$$\dot{W}(t) \leq -bV_1(t) + \frac{\kappa}{b} \psi^2 |(x - \overline{x}) + \overline{x}|^2 + \dot{V}_2(t),$$

where $\psi = |P|\, \psi_1$.

This can be rewritten as

$$\dot{W}(t) \leq -bV_1(t) + 2\frac{\kappa}{b\mu}\psi^2 V_1(t) + 2\frac{\kappa}{b}\psi^2 |\overline{x}|^2 + \dot{V}_2(t)$$

and since Assumption 14.2 holds

$$\dot{W}(t) \leq -\left(b - 2\frac{\kappa}{b\mu}\psi^2\right) V_1(t) - \left(\beta - 2\frac{\kappa}{b\alpha}\psi^2\right) V_2(t).$$

If μ, κ, and ν are such that

$$c_1 = \left(\frac{\nu}{\kappa} - 2\frac{\kappa^2}{\nu\mu}\psi^2\right) > 0$$

and

$$c_2 = \left(\beta - 2\frac{\kappa^2}{\nu\alpha}\psi^2\right) > 0$$

with $\psi = |P|\,\psi_1$, ψ_1, P, α, and β defined by (14.14), (14.21), and Assumption 14.2, we obtain that

$$\dot{W}(t) \leq -2cW(t),$$

where $2c = \min\left[\left(\frac{\nu}{\kappa} - 2\frac{\kappa^2}{\nu\mu}\psi^2\right), \left(\beta - 2\frac{\kappa^2}{\nu\alpha}\psi^2\right)\right]$.

If we set

$$W_1(t) = W(t)\exp(2ct),$$

we have that $\dot{W}_1(t) \leq 0$, thus giving

$$W(t) = W(0)\exp(-2ct).$$

Then

$$\begin{aligned}\mu|x-\overline{x}|^2 + \alpha|\overline{x}|^2 &\leq W(t)\\ &\leq \left[\kappa|x(0)-\overline{x}(0)|^2 + \varepsilon|\overline{x}(0)|^2\right]\exp(-2ct).\end{aligned}$$

We can conclude

$$|x-\overline{x}| \leq \left[\frac{\kappa}{\mu}|x(0)-\overline{x}(0)|^2 + \frac{\varepsilon}{\mu}|\overline{x}(0)|^2\right]^{\frac{1}{2}} \exp(-ct),$$

$$|s(t,x) - s(t,\overline{x})| \leq D|x-\overline{x}|$$

and therefore (14.25) since it holds $s(t,\overline{x}) = 0$.

□

14.5 Example

We consider the following variable-structure control system

$$\begin{aligned}\dot{\eta} &= \varphi(t,\eta,u) = \\ &= \begin{bmatrix} (3+\sin^2(\omega t))(\eta_2-\eta_1)-4(\eta_1+\eta_1^3) \\ (2-\sin^2(\omega t))\eta_1-4(\eta_2+\eta_2^3)+\rho_1(u) \end{bmatrix}, \\ \dot{u} &= v,\end{aligned}$$

where $\eta \in R^2, u \in R$ and $\rho_1(u) = \sqrt{3}(u-1)+\sqrt{u^2+3}$; $\omega = 2\pi$; the output equation has the form $\zeta = \rho(\eta) = \eta_1$, $\zeta \in R$; the sliding manifold is designed as $\xi(\eta) = \eta_1 - \eta_2 = 0$, $\xi \in R$; it is trivial to verify that the corresponding zero-dynamics is asymptotically stable.

Let the augmented state vector $x = (\eta', u')'$; we consider the variable-structure control system

$$\begin{aligned}\dot{x} &= f(t,x,v) = A(t,x)+Bv = \\ &= \begin{bmatrix} (3+\sin^2(\omega t))(x_2-x_1)-4(x_1+x_1^3) \\ (2-\sin^2(\omega t))x_1-4(x_2+x_2^3)+\rho_1(x_3) \\ 0 \end{bmatrix} \\ &+ \begin{bmatrix} 0 \\ 0 \\ 1 \end{bmatrix} v,\end{aligned} \tag{14.26}$$

where $x \in R^3$ and $\rho_1(x_3) = \sqrt{3}(x_3-1)+\sqrt{x_3^2+3}$.

The sliding manifold is designed as

$$s(x) = \dot{\xi} + \Lambda\xi = A_1(t,x) - A_2(t,x) + \Lambda(x_1-x_2) = 0, \tag{14.27}$$

where $s \in R$, $A_1(t,x)$ and $A_2(t,x)$ are respectively the first and second element of the vector field $A(t,x)$ in (14.26), and $\Lambda \in R$ is chosen $\Lambda = 10$. On $s(t,x) = 0$ the system (14.26) is stable.

The output vector $y \in R^2$ is

$$y = \begin{bmatrix} \zeta \\ \zeta_1 \end{bmatrix} = h(x) = \begin{bmatrix} x_1 \\ \rho_1(x_3) \end{bmatrix}. \tag{14.28}$$

The state vector x is not completely available. The vector $y = h(x)$ is the accessible output, the measurements y_i of which are received at discrete-times t_i, with a fixed sampling period δ, according to (14.12).

The observer with discrete-time measurements for system (14.26) is designed as the following continuous-discrete-time observer defined by the following hybrid system

$$\begin{cases} \dot{\overline{x}}(t) = f(t,\overline{x},v) - Ph(\overline{x}), \quad t \in [t_i, t_{i+1}), \\ \overline{x}(t_i) = \overline{x}(t_i^-) + P[y_i - h(\overline{x}(t_i^-))], \end{cases} \tag{14.29}$$

where $y_i = h(x_i)$ and $x_i = x(t_i)$, the sampling instants are defined as $t_{i+1} = t_i + \delta$, $t_0 = 0$, $i = 0, 1, \ldots$, the constant $\delta = 0.2\,sec$ is the measurement sampling interval, and $P \in R^{3\times 2}$ is a constant matrix, which will be specified in the sequel.

In particular the first equation of the observer (14.29) for system (14.26)–(14.27) takes the form

$$\dot{\overline{x}} = f(t,\overline{x},v) - Ph(\overline{x}) =$$

$$= \begin{bmatrix} (3+\sin^2(\omega t))(\overline{x}_2 - \overline{x}_1) - 4(\overline{x}_1 + \overline{x}_1^3) \\ (2-\sin^2(\omega t))\overline{x}_1 - 4(\overline{x}_2 + \overline{x}_2^3) + \rho_1(\overline{x}_3) \\ v \end{bmatrix} \tag{14.30}$$

$$-P\begin{bmatrix} \overline{x}_1 \\ \rho_1(\overline{x}_3) \end{bmatrix},$$

where $P \in R^{3\times 2}$; $x_1(0) = 1.5$, $x_2(0) = -1.5$ and $x_3(0) = -1.5$.

We have the two jacobian matrices $f_x(t,x) =$

$$\begin{bmatrix} -(3+\sin^2(\omega t)) - 4(1+3x_1^2) & (3+\sin^2(\omega t)) & 0 \\ (2-\sin^2(\omega t)) & -4(1+3x_2^2) & \left(\sqrt{3}+\frac{x_3}{\sqrt{x_3^2+3}}\right) \\ 0 & 0 & 0 \end{bmatrix}$$

and

$$h_x(t,x) = \begin{bmatrix} 1\;0 & 0 \\ 0\;0 & \left(\sqrt{3}+\frac{x_3}{\sqrt{x_3^2+3}}\right) \end{bmatrix}.$$

Let us choose $Q = \begin{bmatrix} 1\;0\;0 \\ 0\;1\;0 \\ 0\;0\;1 \end{bmatrix}$ and $P = \begin{bmatrix} 8\;0 \\ 5\;1 \\ 0\;1 \end{bmatrix}$. It is easy to verify that the symmetric part of $\theta = Q(f_x - Ph_x)$, that is the matrix $\dfrac{\theta+\theta'}{2} =$

$$\begin{bmatrix} -(15+\sin^2(\omega t)+3x_1^2) & 0 & 0 \\ 0 & -4(1+3x_2^2) & 0 \\ 0 & 0 & -\left(\sqrt{3}+\frac{x_3}{\sqrt{x_3^2+3}}\right) \end{bmatrix},$$

is globally negative definite, independently of u, v and on the chosen sliding manifold. The conditions of Theorem 1 holds.

Let us consider the sliding output $s(t,\overline{x}) = A_1(t,\overline{x}) - A_2(t,\overline{x}) + \Lambda(\overline{x}_1 - \overline{x}_2)$, $\Lambda = 10$, and its first time derivative, which can be expressed as

$$\dot{s}(t) = \Phi(t,\overline{x},y) - \left(\sqrt{3}+\frac{\overline{x}_3}{\sqrt{\overline{x}_3^2+3}}\right)v, \tag{14.31}$$

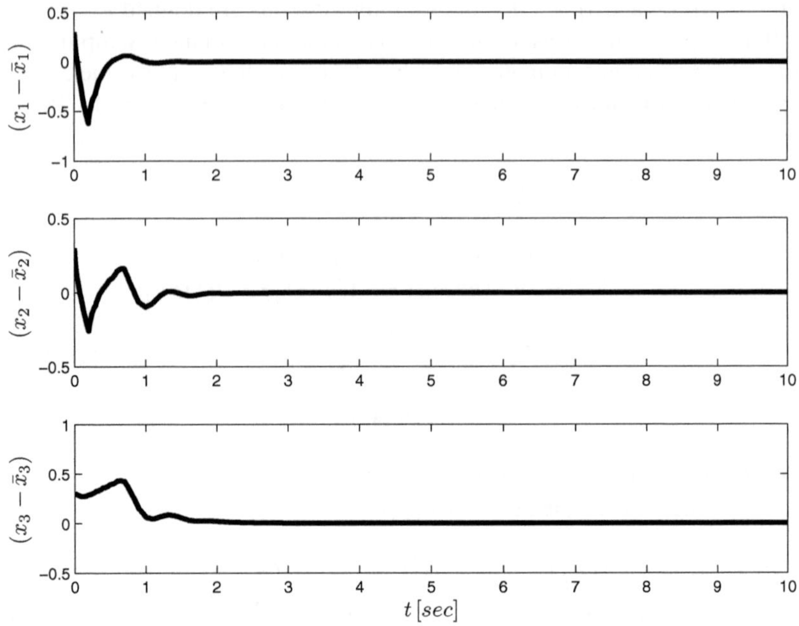

Fig. 14.1 The observation error vector $(x - \overline{x})$ converges to zero exponentially

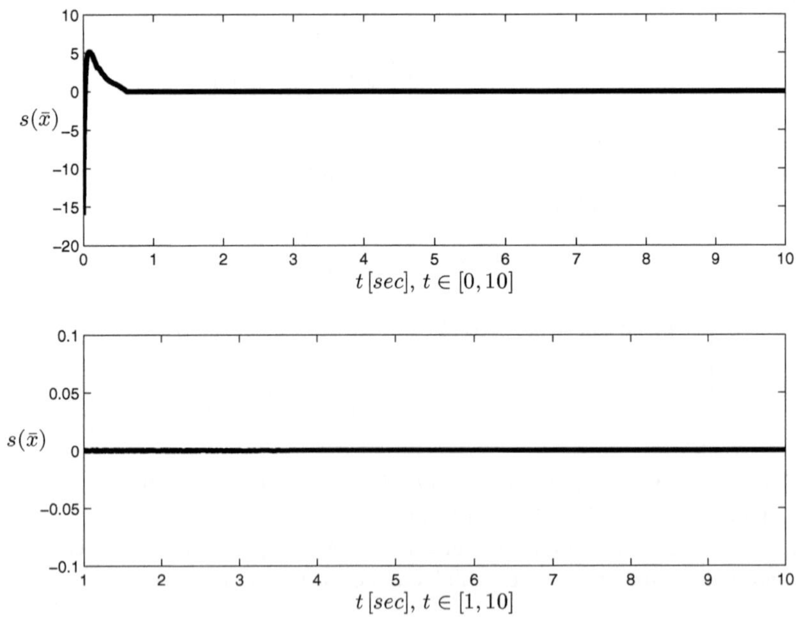

Fig. 14.2 The sliding output $s(\overline{x})$ converges to zero in finite time

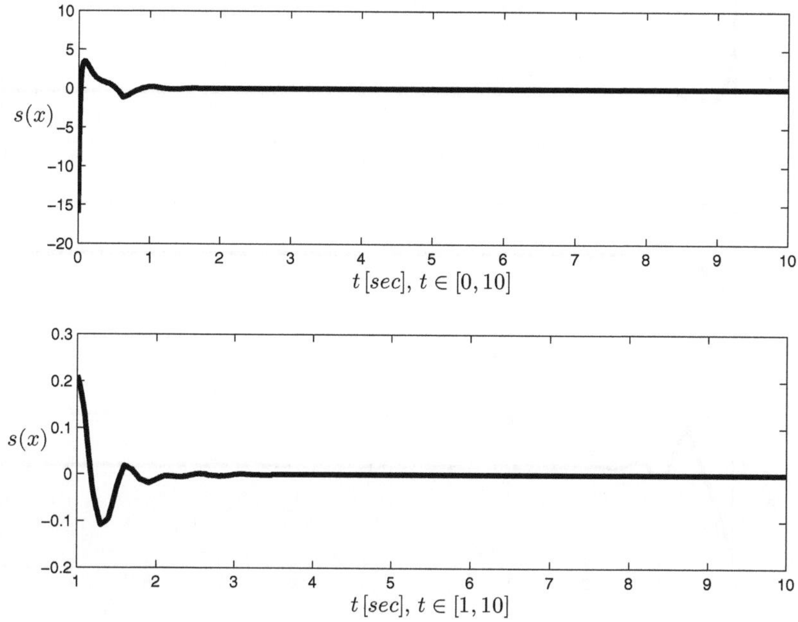

Fig. 14.3 The sliding output $s(x)$ converges to zero exponentially

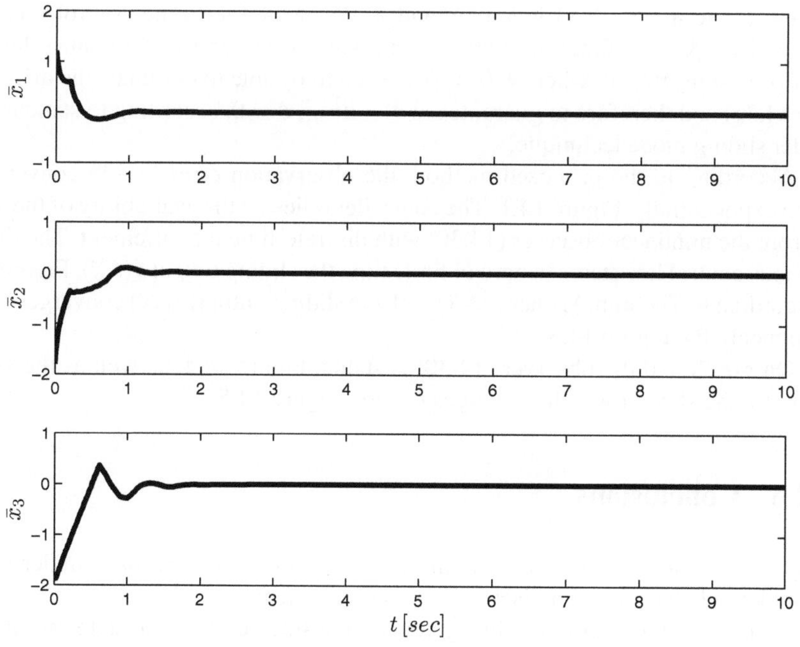

Fig. 14.4 On $s(\bar{x}) = 0$ the observer's state vector $\bar{x}$ converges to zero

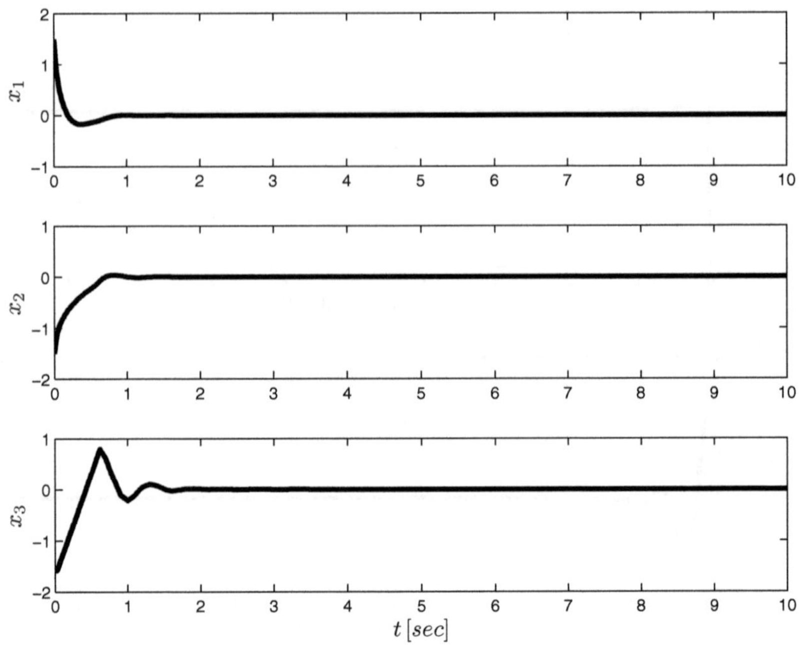

Fig. 14.5 The system's state vector x converges to zero

where the term $\Phi(t,\bar{x},y)$ is known and in the second term the control v is modulated by a known function with constant sign. It is applied the control law $v = -\tilde{K}(t,\bar{x})\,\mathrm{sign}[s(t,\bar{x})]$, where $\tilde{K}(t,\bar{x})$ is chosen to be able to dominate the drift terms in (14.31) and therefore to guarantee $s\dot{s} \leq -\varepsilon^2 |s|$, $\varepsilon \neq 0$, according to standard first order sliding mode technique.

According to the proposed method, the observation error $(x-\bar{x})$ converges to zero exponentially, Figure 14.1. The controller relies on the availability of the vector $\bar{x}$ from the nonlinear observer (14.30) with discrete-time measurement. The discontinuous control law steers to zero in finite time the sliding output $s(t,\bar{x})$, Figure 14.2. According to Theorem 1, once $s(t,\bar{x}) = 0$, the sliding output $s(t,x)$ converges to zero exponentially, Figure 14.3.

On $s(t,\bar{x}) = 0$ the observer (14.30) is stable, Figure 14.4, as well as the system (14.26), the state of which converges to zero, Figure 14.5.

14.6 Conclusions

The contribution of the chapter is in the context of output feedback under perfect plant knowledge and with discrete-time measurements.

A class of nonlinear nonaffine systems is considered when the state vector is not completely available and the output is accessible via discrete-time measurements; the use of suitably designed observers is required.

The proposed methodology introduces integrators in the input channel and combines sliding mode and Luenberger-like observers.

The procedure considers an augmented state and a new control, which is the first time derivative of the original one. The strategy attains chattering reduction, while ruling out possible ambiguous behaviors.

A full-order observer is proposed and conditions are found under which the convergence to the unique ideal solution is proven for both system and observer despite the discrete-time measurements.

References

1. Bartolini, G., Zolezzi, T.: Dynamic output feedback for observed variable-structure control systems. Systems & Control Letters 7(3), 189–193 (1986)
2. Bartolini, G., Zolezzi, T.: Control of nonlinear variable structure systems. J. Math. Anal. Appl. 118, 42–62 (1986)
3. Bartolini, G., Punta, E.: Reduced-order observer in the sliding-mode control of nonlinear nonaffine systems. IEEE Trans. Automatic Control 55(10), 2368–2373 (2010)
4. Bartolini, G., Ferrara, A., Usai, E.: Chattering avoidance by second order sliding modes control. IEEE Trans. Automatic Control 43(2), 241–247 (1998)
5. Bartolini, G., Ferrara, A., Usai, E., Utkin, V.I.: On multi-input chattering-free second order sliding mode control. IEEE Trans. Automatic Control 45(9), 1711–1717 (2000)
6. Boiko, I., Fridman, L., Pisano, A., Usai, E.: Analysis of chattering in systems with second-order sliding-modes. IEEE Trans. Automatic Control 52(11), 2085–2102 (2007)
7. Deza, F., Busvelle, E., Gauthier, J., Rakotopara, D.: High gain estimation for nonlinear systems. Systems & Control Letters 18(4), 295–299 (1992)
8. Ali, T.A., Postoyan, R., Lamnabhi-Lagarrigue, F.: Continuous discrete adaptive observers for state affine systems. Automatica 45(12) (2009)
9. Andrieu, V., Nadri, M.: Observer design for lipschitz systems with discrete-time measurements. In: Proc. 49th IEEE Conference on Decision and Control, Atlanta, GA, USA (2010)
10. Salgado, I., Moreno, J., Chairez, I.: Sampled output based continuous second order sliding mode observer. In: Proc. 11th International Workshop on Variable Structure Systems, Mexico City, Mexico (2010)
11. Salgado, I., Moreno, J., Chairez, I., Fridman, L.: Design of mixed luenberger and sliding continuous mode observer using sampled output information. In: Proc. 49th IEEE Conference on Decision and Control, Atlanta, GA, USA (2010)
12. Han, X., Fridman, E., Spurgeon, S.: A sliding mode observer for fault reconstruction under output sampling: A time-delay approach. In: Proc. 50th IEEE Conference on Decision and Control and European Control Conference (CDC-ECC) Orlando, FL, USA (2011)
13. Levant, A.: Sliding order and sliding accuracy in sliding mode control. Int. J. of Control 58(6), 1247–1263 (1993)
14. Levant, A.: Robust exact differentiation via sliding mode technique. Automatica 34(3), 379–384 (1998)

Chapter 15
Discrete-Time Sliding-Mode-Based Differentiation

Arie Levant and Miki Livne

Abstract. Homogeneous sliding-mode-based differentiators provide for the high-accuracy robust finite-time-exact estimation of derivatives. It is shown that their discrete-time implementation misses the homogeneity, and respectively features worse accuracy with respect to the sampling time interval. Detailed analysis of the asymptotic accuracy is provided in both cases of constant and variable sampling intervals.

15.1 Introduction

Sliding-mode (SM) control is one of the most popular approaches to the problem of control under heavy uncertainty conditions. The idea is to keep some properly chosen function (sliding variable) at zero by permanent control switching. Sliding mode is accurate and insensitive to disturbances [8,22]. Yet standard sliding modes feature the so-called chattering effect [3,8,17,24], and require the control to appear already in the first derivative of the sliding variable. In other words the relative degree of the sliding variable is to be 1.

High order sliding modes (HOSMs) [4,13–22] were created to remove the above restrictions by hiding the switching in the higher derivatives of the sliding variable. One actually only needs to know the relative degree of the sliding variable in order to apply HOSM controllers. By artificially increasing the relative degree, one can remove the dangerous types of chattering [4, 14, 18, 22]. Such controllers directly solve the control problem, if the sliding variable is a tracking error. Their ultimate accuracy is the result of their homogeneity features [16].

One of the main applications of sliding-mode control is the robust practical differentiation and observation [5, 6, 11, 14–16, 22, 24]. HOSM methods allow

Arie Levant · Miki Livne
Tel-Aviv University, Ramat-Aviv, 69978 Tel-Aviv, Israel
e-mail: levant@post.tau.ac.il, miki.livne@gmail.com

B. Bandyopadhyay et al. (Eds.): *Advances in Sliding Mode Control*, LNCIS 440, pp. 299–312.
DOI: 10.1007/978-3-642-36986-5_15

estimation of n derivatives, provided the $(n+1)$th-derivative magnitude has a known bound. Contrary to popular high-gain observers, ⟦1⟧ exact derivative estimations are obtained in the absence of noises. The HOSM differentiators are built as a recursive chain of second-order sliding modes. Their main features are based on the differentiator homogeneity in the space of differentiator errors. The corresponding differential inclusion is shown to be finite-time stable with a negative homogeneity degree. The respective accuracy in the presence of infinitesimal input noises is shown to be asymptotically optimal. In the absence of noises, the accuracy of the ith derivative is of the order of τ^{n-i+1}, where τ is the sampling interval.

The recently published differentiator modification ⟦7⟧ features faster convergence and even uniformly bounded transient time. Higher order terms are added to the differentiator to this end. Nevertheless, the differentiator is still described by the same homogeneous finite-time stable inclusion in a small vicinity of zero in the error space. Respectively it has the same asymptotic accuracy.

HOSM-based differentiators have already found a lot of theoretical and practical applications ⟦4, 20, 22⟧. The above listed differentiator features are widely considered as giving sufficient theoretical basis for practical implementation. Yet recently the situation was found to be not so simple. Indeed, the above features were proved under the assumption that though the sampling takes place at discrete time instants, the whole system still evolves in continuous time. In reality the differentiator is a computer-based dynamic system. Since the system is not smooth, the integration is to be performed by the Euler method. To emulate a continuous-time system the computer integration step should be much less than the sampling period. It not only complicates the implementation, but also turns the choice of the integration step into the state of the art. Note that in practice the implementation is usually based on only one Euler step performed between the successive samplings.

It is proved in this paper that the homogeneity is still preserved for the first-order discrete-time differentiator, which means that its one-Euler-step discrete-time implementation is fully justified. Nevertheless, the higher-order discrete-time differentiator is shown to loss its homogeneity. As a result, the above-presented asymptotic accuracy with respect to the sampling step deteriorates. At the same time the discrete-time differentiator remains asymptotically exact. Moreover, its optimal asymptotic accuracy with respect to the noises contaminating the input is preserved, which explains its successful implementation in numerous practical projects.

15.2 Discrete-Time Differentiation

15.2.1 Discrete-Time Differentiation and Homogeneity

It is well-known that exact practical differentiation of general smooth functions is impossible due to always present small noises. Indeed, one can easily approximate any smooth function by another smooth function having very large derivatives. These two functions are indistinguishable due to the possible presence of noises, which means that any *exact* differentiator would inevitably differentiate the noises.

The usual condition under which the practical exact differentiation is possible, is the condition that the $(n+1)$th order derivative of the function has a known maximal magnitude. In that case one can estimate the first n function derivatives. The explanation is that one cannot approximate a function of this class by another function of the same class without approximating its derivatives at the same time [[12]]. Respectively, the main idea of differentiation based on control methods is to construct a dynamic system tracking the input function without the knowledge of its derivatives.

Let the input be $f(t) = f_0(t) + \eta(t) \in \mathbf{R}$, where $\eta(t)$ is a Lebesgue-measurable bounded noise, $|\eta| \leq \varepsilon$. Neither ε nor the properties of the noise are known. The function $f_0(t)$ is an unknown, n-times differentiable smooth component of f, which is to be restored together with its n derivatives. The last derivative $f_0^{(n)}$ is known to be Lipschitzian with the constant L, which means that $f_0^{(n+1)}$ exists almost everywhere and satisfies $f_0^{(n+1)}(t) \in [-L, L]$. The general differentiator [[1, 15]] usually has the form

$$\begin{aligned} \dot{z}_i &= \varphi_i(z_0 - f) + z_{i+1}, \quad i = 0, ..., n-1 \\ \dot{z}_n &= \varphi_n(z_0 - f), \end{aligned} \tag{15.1}$$

where φ_i is a scalar function of scalar argument. The system is understood in the Filippov sense [[9]] to allow discontinuities of φ_i and f. Subtracting $f_0^{(i+1)}$ from both sides, denoting $\sigma_i = z_i - f_0^{(i)}$ and using $f_0^{(n+1)}(t) \in [-L, L]$ with $\eta = 0$ obtain

$$\begin{aligned} \dot{\sigma}_i &\in \varphi_i(\sigma_0) + \sigma_{i+1}, \qquad i = 0, ..., n-1 \\ \dot{\sigma}_n &\in \varphi_i(\sigma_0) + [-L, L], \end{aligned} \tag{15.2}$$

which is a differential inclusion in the error space $\sigma_0, \sigma_1, ..., \sigma_n$.

With properly chosen functions φ_i, inclusion (15.2) becomes homogeneous and finite-time stable. The homogeneity means that some positive number m_i (called *the weight* or *the homogeneity degree*, [[2]]) is assigned to each coordinate σ_i, $\deg \sigma_i = m_i > 0$. Also the time t gets its weight $\deg t = p > 0$ (called *minus* homogeneity degree of the inclusion [[16]]) so that the transformation

$$(t, \sigma_0, \sigma_1, ..., \sigma_n) \mapsto (\kappa^p t, \kappa^{m_0}\sigma_0, \kappa^{m_1}\sigma_1, ..., \kappa^{m_n}\sigma_n) \tag{15.3}$$

preserves the trajectories of (15.2) with any positive κ. Recall also that a function Ξ is called homogeneous of the degree m, $\deg \Xi = m$, if

$$\Xi(\kappa^{m_0}\sigma_0, \kappa^{m_1}\sigma_1, ..., \kappa^{m_n}\sigma_n) = \kappa^m \Xi(\sigma_0, \sigma_1, ..., \sigma_n).$$

It is easy to see that all weights can be proportionally changed. Thus, let $\deg t = 1$. Due to the segment present in the last nth equation of (15.2) the only possible weight of σ_n is 1, which implies $\deg \sigma_i = n - i + 1$, $i = 0, ..., n$ [[16]]. In order to keep the homogeneity, the function $\varphi_i(\sigma_0)$ is chosen proportional to $|\sigma_0|^{(n-i)/(n+1)}$ (see further).

Let the input be sampled at the time instants t_k with the time interval $\tau_k = t_{k+1} - t_k$. Meantime suppose that $\tau_k = \tau = \text{const}$. The natural Euler discrete-time implementation of (15.1) produces

$$\begin{aligned} z_i(t_{k+1}) &= z_i(t_k) + \tau\varphi_i(z_0(t_k) - f(t_k)) + \tau z_{i+1}(t_k), \;\; i = 0, ..., n-1 \\ z_n(t_{k+1}) &= z_n(t_k) + \tau\varphi_n(z_0(t_k) - f(t_k)). \end{aligned} \tag{15.4}$$

Subtract $f_0^{(i)}(t_{k+1})$ from both sides of (15.4), using $f_0^{(i)}(t_{k+1}) = f_0^{(i)}(t_k) + \tau f_0^{(i+1)}(t_k) + (\tau^2/2) f_0^{(i+2)}(\xi_{ik})$, $\xi_{ik} \in [t_k, t_{k+1}]$, $i \leq n-2$. Taking into account $\eta \in [-\varepsilon, \varepsilon]$, with $n \geq 2$ obtain

$$\begin{aligned} \sigma_i(t_{k+1}) &\in \sigma_i(t_k) + \tau\varphi_i(\sigma_0(t_k) + [-\varepsilon, \varepsilon]) + \tau\sigma_{i+1}(t_k) - (\tau^2/2) f_0^{(i+2)}(\xi_{ik}), \\ & \qquad i = 0, 1, ..., n-2; \\ \sigma_{n-1}(t_{k+1}) &\in \sigma_{n-1}(t_k) + \tau\varphi_{n-1}(\sigma_0(t_k) + [-\varepsilon, \varepsilon]) + \tau\sigma_n(t_k) + (\tau^2/2)[-L, L], \\ \sigma_n(t_{k+1}) &\in \sigma_n(t_k) + \tau\varphi_n(\sigma_0(t_k) + [-\varepsilon, \varepsilon]) + \tau[-L, L]. \end{aligned} \tag{15.5}$$

If the term $f_0^{(i+2)}(\xi_{ik})$ were absent in (15.5), deg $\tau = 1$, deg $\varepsilon = n+1$ could be set, and the inclusion would be homogeneous with respect to the homogeneity transformation

$$G_\kappa : (t, \tau, \varepsilon, \sigma_0, \sigma_1, ..., \sigma_n) \mapsto (\kappa t, \kappa\tau, \kappa^{n+1}\varepsilon, \kappa^{n+1}\sigma_0, \kappa^n\sigma_1, ..., \kappa\sigma_n) \tag{15.6}$$

in the sense that trajectories of the system with parameters τ, ε were bijectively transferred onto trajectories with parameters $\kappa\tau$, $\kappa^{n+1}\varepsilon$. In such a case, the discrete-time differentiator (15.4) would preserve the asymptotic properties of the original homogeneous continuous-time differentiator. Unfortunately, in general, the term $f_0^{(i+2)}(\xi_{ik})$ with $n \geq 2$ is not zero.

Obviously with $n = 1$ the first line of (15.5) does not appear, implying the homogeneity of system (15.5) with respect to (15.6). We will see that, as a result, the one-Euler-step discrete-time implementation preserves all standard differentiator features with $n = 1$.

15.2.2 Asymptotic Accuracy of Discrete-Time HOSM Differentiator with Constant Sampling Intervals

The recursive form of the nth-order homogeneous HOSM differentiator [15] is

$$\begin{aligned} \dot{z}_0 &= v_0, \;\; v_0 = -\lambda_n L^{1/(n+1)} |z_0 - f|^{n/(n+1)} \operatorname{sign}(z_0 - f_0) + z_1, \\ \dot{z}_i &= v_i, \;\; v_i = -\lambda_{n-i} L^{1/(n-i+1)} |z_i - v_{i-1}|^{(n-i)/(n-i+1)} \operatorname{sign}(z_i - v_{i-1}) + z_{i+1}, \\ \dot{z}_n &= -\lambda_0 L \operatorname{sign}(z_n - v_{n-1}), \end{aligned} \tag{15.7}$$

where $i = 1,2,...,n-1$. As previously $f(t) = f_0(t) + \eta(t)$, $\eta \leq \varepsilon$. Here z_i is the estimation of $f_0^{(i)}$, and parameters λ_i of differentiator (15.7) are chosen in advance for each n. An infinite sequence of parameters λ_i can be built, valid for all n [[15]]. In particular, one can choose $\lambda_0 = 1.1$, $\lambda_1 = 1.5$, $\lambda_2 = 2$, $\lambda_3 = 3$, $\lambda_4 = 5$, $\lambda_5 = 8$ [[17]], which corresponds to differentiators of the order n, $1 \leq n \leq 5$.

In the absence of noises the equalities $z_i = f_0^{(i)}$ are established in finite time. In the presence of a sampling noise with the maximal magnitude ε, the accuracy $|z_i - f_0^{(i)}| = O(\varepsilon^{i/(k+1)})$ is obtained, and these asymptotics cannot be improved [[15]].

Excluding v_i from (15.7) obtain the standard form (15.1) with

$$\varphi_i(\sigma_0) = -\tilde{\lambda}_{n-i}L^{(i+1)/(n+1)}|\sigma_0|^{(n-i)/(n+1)}\text{sign}(\sigma_0). \tag{15.8}$$

It is easy to see that the new coefficients $\tilde{\lambda}_0, \tilde{\lambda}_1, ..., \tilde{\lambda}_k > 0$ are calculated from (15.7) according to the recursive formula $\tilde{\lambda}_n = \lambda_n$, $\tilde{\lambda}_j = \lambda_j \tilde{\lambda}_{j+1}^{j/(j+1)}$, $j = n-1, n-2, \ldots, 0$. Note that also $\tilde{\lambda}_0 = \lambda_0$. It is easy to check that (15.2), (15.8) is homogeneous with $\deg t = 1$, $\deg \sigma_i = n-i+1$, $i = 0,...,n$. It is always assumed in the following that the parameters λ_i are properly chosen, so that (15.2), (15.8) is finite time stable. The following Theorem calculates the asymptotic accuracy of the Euler discretization of the differentiator (15.7) with a constant sampling interval.

Theorem 15.1. *Let the input $f(t)$ of the discrete-time differentiator* (15.4) *consist of an n-smooth function $f_0(t)$, $n \geq 1$, and a Lebesgue-measurable additive noise not exceeding ε in its absolute value. Suppose that the input derivatives $\ddot{f}_0(t), ..., f_0^{(n)}(t)$ are bounded in absolute value, $|f_0^{(j)}| \leq D_j$, $j = 2,3,...,n$, and $f_0^{(n)}(t)$ is a Lipschitzian function with the Lipschitz constant L, $D_{n+1} = L$. Let $\tau > 0$ be the sampling interval of the differentiator* (15.4). *Define*

$$\delta_i = \max\{\varepsilon^{(n-i+1/(n+1)}, \tau^{(n-i+1)}, \tau D_{i+1}\}, \quad i = 1,...,n. \tag{15.9}$$

Then there exist such constants $\mu_i > 0$ that the accuracy

$$|z_0 - f_0| \leq \mu_0 \max\{\varepsilon, \tau^{n+1}\}, \quad |z_i - f_0^{(i)}| \leq \mu_i \delta_i, \quad i = 1,...,n \tag{15.10}$$

is established in finite time and is kept forever. The constants μ_i only depend on the differentiator parameters $\lambda_0, \lambda_1, \ldots, \lambda_n$ and L.

Remark 15.1. Notice that for small enough input noise the accuracy $|z_0 - f_0| \leq \mu_0 \tau^{n+1}$ is provided.

The following Lemma is repeatedly used in the sequel.

Lemma 15.1. *Let a differential inclusion be finite-time stable and homogeneous with a negative homogeneity degree. Then its Euler discrete-time approximation is also homogeneous with the weight of the sampling periods equal to the weight of time. Its asymptotic accuracy is the same as of the continuous-time system with discrete sampling.*

Proof. The Euler system is an approximation of the continuous-time finite-time stable system. Thus, similarly to [16] it has an invariant finite-time stable attractor. The homogeneity implies the needed asymptotics [16]. ■

Now prove the Theorem.

Proof. Introduce the variables $(\gamma_0, ..., \gamma_{n+1})$ defined at each sampling instant t_k as the divided differences: $\gamma_0(t_k) = f_0(t_k)$, $\gamma_1(t_k) = (\gamma_0(t_{k+1}) - \gamma_0(t_k))/\tau$, ..., $\gamma_{n+1}(t_k) = (\gamma_n(t_{k+1}) - \gamma_n(t_k))/\tau$. It is well-known that $\gamma_i(t_k) = f_0^{(i)}(\xi_k)$ for some $\xi_k \in [t_k, t_{k+i}]$, $i = 0, ..., n$, and $\gamma_{n+1}(t_k) \in [-L, L]$. Letting $s_i = z_i - \gamma_i$ obtain from (15.4) that,

$$
\begin{aligned}
s_i(t_{k+1}) &\in s_i(t_k) + \tau\varphi_i(s_0(t_k) + [-\varepsilon, \varepsilon]) + \tau s_{i+1}(t_k), \quad i = 0, ..., n-1 \\
s_n(t_{k+1}) &\in s_n(t_k) + \tau\varphi_n(s_0(t_k) + [-\varepsilon, \varepsilon]) + \tau[-L, L].
\end{aligned}
\tag{15.11}
$$

Note that this coordinate transformation does not affect the differentiator dynamics. Inclusion (15.11) is the homogeneous discretization of the finite-time stable homogeneous differential inclusion (15.2) with the sampled noise η. Thus, according to Lemma 15.1 the accuracy $|s_i| \leq \mu_i \max\{\varepsilon^{(n-i+1)/(n+1)}, \tau^{(n-i+1)}\}$ is provided in finite time with some coefficients μ_i only depending on the differentiator parameters. Now, $|z_i(t_k) - f_0^{(i)}(t_k)| \leq |z_i(t_k) - \gamma_i(t_k)| + |\gamma_i(t_k) - f_0^{(i)}(t_k)| \leq \nu_i \max\{\varepsilon^{(n-i+1)/(n+1)}, \tau^{(n-i+1)}\} + (t_{k+i} - t_k)\sup|f_0^{(i+1)}|$, and the needed asymptotics are obtained (note that with $i = 0$, we get $t_{k+i} - t_k = 0$). ■

Remark 15.2. Notice that for small enough input noise the accuracy $|z_0 - f_0| \leq \mu_0\tau^{n+1}$ is provided.

15.2.3 Variable Sampling Intervals

The discrete-time implementation of (15.1) with variable sampling step takes the form

$$
\begin{aligned}
z_i(t_{k+1}) &= z_i(t_k) + \tau_k\varphi_i(z_0(t_k) - f(t_k)) + \tau_k z_{i+1}(t_k), \quad i = 0, ..., n-1 \\
z_n(t_{k+1}) &= z_n(t_k) + \tau_k\varphi_n(z_0(t_k) - f(t_k)).
\end{aligned}
\tag{15.12}
$$

where $\tau_k = t_{k+1} - t_k$.

Similarly to the procedure described in Subsection 15.2.1, subtracting $f_0^{(i)}(t_{k+1})$ from both sides of (15.12), and using $\eta \in [-\varepsilon, \varepsilon]$, obtain the structure of inclusion (15.5) with variable sampling step

$$\sigma_i(t_{k+1}) \in \sigma_i(t_k) + \tau_k \varphi_i(\sigma_0(t_k) + [-\varepsilon,\varepsilon]) + \tau_k \sigma_{i+1}(t_k) - (\tau_k^2/2) f_0^{(i+2)}(\xi_{ik}),$$

$$\sigma_{n-1}(t_{k+1}) \in \sigma_{n-1}(t_k) + \tau_k \varphi_{n-1}(\sigma_0(t_k) + [-\varepsilon,\varepsilon]) + \tau_k \sigma_n(t_k) + (\tau_k^2/2)[-L,L],$$

$$\sigma_n(t_{k+1}) \in \sigma_n(t_k) + \tau_k \varphi_n(\sigma_0(t_k) + [-\varepsilon,\varepsilon]) + \tau_k[-L,L], \tag{15.13}$$

where $i = 0,1,...,n-2$, with $n \geq 2$ and $\xi_{ik} \in [t_k, t_{k+1}]$. As it was mentioned above, the presence of the high-order derivatives of f_0 on the right-hand side of (15.13) destroys its homogeneity features.

Theorem 15.2. *Let the input f(t) of the discrete-time differentiator* (15.12) *satisfy the conditions of Theorem* 15.1, *and the sampling intervals be bounded from above by* $\tau > 0$. *Then the accuracy*

$$|\sigma_i| \leq \mu_i \delta^{n-i+1} \tag{15.14}$$

is established in finite-time for some constants $\mu_i > 0$ *and kept forever;*

$$\delta = \max\left\{ \left(\frac{\tau}{2}D_2\right)^{1/n}, \left(\frac{\tau}{2}D_3\right)^{1/(n-1)}, ..., \left(\frac{\tau}{2}D_n\right)^{1/2}, \tau, \varepsilon^{1/(n+1)} \right\}. \tag{15.15}$$

The constants μ_i *depend only on the differentiator parameters* $\lambda_0, \lambda_1, \ldots, \lambda_n$ *and L.*

Remark 15.3. Note that with $n = 1$ the standard accuracy $|\sigma_0| \leq \mu_0 \max(\tau^2, \varepsilon)$, $|\sigma_1| \leq \mu_1 \max(\tau, \varepsilon^{1/2})$ is obtained.

Proof. Consider a piecewise-linear continuous function $s(t) = (s_0(t), ..., s_n(t))$ defined by the equations

$$s(t) = s(t_k) + (t - t_k)v_k, \quad t_k \leq t \leq t_{k+1}. \tag{15.16}$$

The vector $v_k \in F(s(t_k))$ is arbitrarily taken from the set

$$F(s(t_k)) = \begin{cases} \varphi_i(s_0(t_k) + [-\varepsilon,\varepsilon]) + s_{i+1}(t_k) + \bar{\delta}^{n-i}[-1,1], \\ \varphi_{n-1}(s_0(t_k) + [-\varepsilon,\varepsilon]) + s_n(t_k) + \bar{\delta}[-1,1], \\ \varphi_n(s_0(t_k) + [-\varepsilon,\varepsilon]) + [-L,L], \end{cases} \tag{15.17}$$

where $i = 0,1,...,n-2, n \geq 2$. With $\bar{\delta}$ large enough, each solution of (15.13) satisfies (15.16), (15.17) in the sense that for each interval $[t_k, t_{k+1}]$, there exist $v_k \in F(s(t_k))$ such that the same values are got at the same sampling times.

Now, the needed asymptotics is obtained immediately from the following lemma with properly chosen $\bar{\delta}$.

Lemma 15.2. *There exist constants* γ_i *such that* $|s_i(t_k)| \leq \gamma_i \left[\max\{\bar{\delta}, \tau, \varepsilon^{1/n+1}\}\right]^{n-i+1}$ *holds after finite transient time.*

Proof. Constructed successively on the intervals $[t_k, t_{k+1}]$, the function $s(t)$ is absolutely continuous and $|\dot{s}_i(t)|$ is bounded ($t \neq t_k$). Therefore, every solution of the discrete system (15.16), (15.17) satisfies the following differential inclusion [9]

$$\begin{aligned} \dot{s}_i(t) &\in \varphi_i(s_0(t_k) + [-\varepsilon, \varepsilon]) + s_{i+1}(t_k) + \bar{\delta}^{n-i}[-1,1], \\ \dot{s}_n(t) &\in \varphi_n(s_0(t_k) + [-\varepsilon, \varepsilon]) + [-L, L], \end{aligned} \tag{15.18}$$

where $i = 0, 1, ..., n-1$ and $t \in (t_k, t_{k+1})$. Consider now the disturbed continuous-time inclusion,

$$\begin{aligned} \dot{s}_i &\in \varphi_i(s_0(t - \tau[0,1]) + [-\varepsilon, \varepsilon]) + s_{i+1}(t - \tau[0,1]) + \bar{\delta}^{n-i}[-1,1], \\ \dot{s}_n &\in \varphi_n(s_0(t - \tau[0,1]) + [-\varepsilon, \varepsilon]) + [-L, L] \end{aligned} \tag{15.19}$$

where $i = 0, 1, ..., n-1$. Obviously, each solution of inclusion (15.18) satisfies inclusion (15.19) almost everywhere. Indeed, for each $t \in [t_k, t_{k+1}]$ there exists $\varpi \in [0,1]$ such that $t - \tau\varpi = t_k$.

Define $\delta = \max\left\{\bar{\delta}, \tau, \varepsilon^{1/(n+1)}\right\}$ and consider also the continuous disturbed inclusion

$$\begin{aligned} \dot{s}_i &\in \varphi_i(s_0(t - \delta[0,1]) + \delta^{n+1}[-1,1]) + s_{i+1}(t - \delta[0,1]) + \delta^{n-i}[-1,1], \\ \dot{s}_n &\in \varphi_n(s_0(t - \delta[0,1]) + \delta^{n+1}[-1,1])) + [-L, L], \end{aligned} \tag{15.20}$$

for $i = 0, 1, ..., n-1$, $n \geq 1$. Since the parameters τ, ε and $\bar{\delta}$ are enlarged, every solution of inclusion (15.19) also satisfies inclusion (15.20) almost everywhere [9]. Similarly to [16], the disturbed differential inclusion (15.20) has a finite time attracting set under the homogeneous transformation

$$G_\kappa : (\delta, t, s_0, ..., s_n) \mapsto (\kappa\delta, \kappa t, \kappa^{n+1} s_0, ..., \kappa s_n). \tag{15.21}$$

Therefore, with some fixed value δ_0 of the parameter δ, there are such constants $c_i > 0$ that the inequalities $|s_i| \leq c_i$ are kept after the finite-time transient. Take now arbitrary value of δ. After the transformation (15.21) with $\kappa = \delta/\delta_0$, $|s_i| \leq \gamma_i \delta^{n-i+1}$, where $\gamma_i = c_i/\delta_0^{n-i+1}$. ■

Taking

$$\bar{\delta} = \max\left\{\left(\frac{\tau}{2}D_2\right)^{1/n}, \left(\frac{\tau}{2}D_3\right)^{1/(n-1)}, ..., \left(\frac{\tau}{2}D_n\right)^{1/2}, \frac{\tau}{2}L\right\}$$

the needed asymptotics are obtained from Lemma 15.2. ■

Remark 15.4. Let the noise be absent, $\varepsilon = 0$. It is clear that when τ tends to zero the formulas (15.14), (15.15) imply the accuracy $\sigma_i = O(\tau^{(n-i+1)/n})$. Let now τ take on some fixed (possibly small) value. After the discrete-time differentiator enters its steady state, the derivatives of the input function keep on changing continuously. As a result, the *local* accuracy (15.14), (15.15) is in fact determined by the *local* bounds of the input derivatives taken instead of the larger global bounds D_j in (15.15). Thus, in the special case of an input function with large derivatives, the asymptotic accuracy of the form $\sigma_i = O(\tau^{\alpha(n-i+1)})$ can be observed with some $\alpha \in [1/n, 1]$ depending on the comparative sizes of the terms from (15.15).

15.3 Simulation Results

Both cases of constant and variable sampling intervals are considered. The goal is to demonstrate the convergence of the discretized differentiator and to show the compliance of the observed asymptotic accuracies with the upper estimations provided by the theorems.

Consider the discrete-time differentiator (15.4) of the order $n = 3$ with the noise-free input function $f(t) = f_0(t) = \sin(2t) + 5\cos(t)$. The differentiator with $\lambda_0 = 1.1$, $\lambda_1 = 1.5$, $\lambda_2 = 2$, $\lambda_3 = 3$, $L = 21$ gets the non-recursive form

$$\begin{aligned}
z_0(t_{k+1}) &= z_0(t_k) + [-3L^{1/4}|z_0(t_k) - f(t_k)|^{3/4}\operatorname{sign}(z_0(t_k) - f(t_k)) + z_1(t_k)]\tau_k,\\
z_1(t_{k+1}) &= z_1(t_k) + [-4.16L^{2/4}|z_0(t_k) - f(t_k)|^{2/4}\operatorname{sign}(z_0(t_k) - f(t_k)) + z_2(t_k)]\tau_k,\\
z_2(t_{k+1}) &= z_2(t_k) + [-3.06L^{3/4}|z_0(t_k) - f(t_k)|^{2/4}\operatorname{sign}(z_0(t_k) - f(t_k)) + z_3(t_k)]\tau_k,\\
z_3(t_{k+1}) &= z_3(t_k) + [-1.1L\operatorname{sign}(z_0(t_k) - f(t_k))]\tau_k, \ L = 21.
\end{aligned} \tag{15.22}$$

The considered sampling intervals τ_k vary in the range from $\tau_{min} = 0.001$ to the upper bound τ with the increments proportional to $\Delta\tau = 10^{-4}$. While in the constant sampling interval case the interval $\tau_k = \tau =$ constant is taken in each run, in the case of variable sampling intervals, τ_k is generated by a random value generator with the uniform probability density function in the range $[\tau_{min}, \tau]$. Various values of τ were taken.

The discrete-time differentiator errors are evaluated as the maximal absolute value $\|z_j - f_0^{(j)}\|_{\infty,\tau,[a,b]}$, $j = 0,...,3$, of the considered sampled derivative estimation error obtained on a fixed finite time interval $[a,b]$ entirely lying in the steady state time region.

Convergence of the discrete-time differentiator with the constant sampling interval $\tau = 10^{-3}$ is demonstrated in Figure 15.1. In order to check the asymptotics with constant sampling intervals, a number of runs were performed with different values of sampling intervals τ and the errors $\|z_j - f_0^{(j)}\|_{\infty,\tau,[10,14]}$, $j = 0,...,3$ were calculated using the specified norm.

Denote $y_j = \ln\|z_j - f_0^{(j)}\|_{\infty,\tau,[10,14]}$, $j = 0,...,3$ and $x = \ln\tau$. The logarithmic scale plot generated for the derivative estimations of the orders 0-3 is demonstrated in Figure 15.2. There are four straight lines in Figure 15.2, whose slopes represent the asymptotic accuracy orders of the discrete-time differentiator. Three parallel lines of the graphs of y_1, y_2, y_3 have slopes equal to 1, and a single line with slope 4

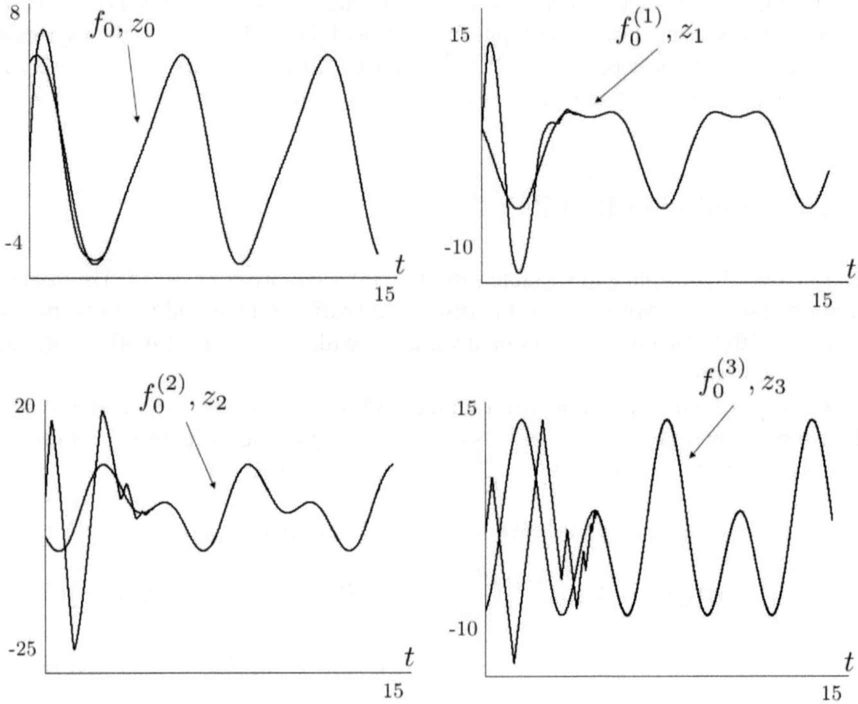

Fig. 15.1 Convergence of the discrete-time differentiator using constant sampling intervals ($\tau = 10^{-3}$)

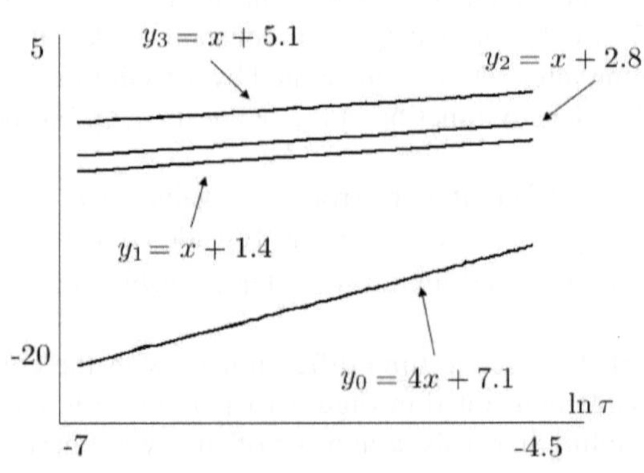

Fig. 15.2 Asymptotics of the discrete-time differentiator with constant sampling intervals

corresponds to y_0. This exactly corresponds to the estimations of Theorem 15.1: $z_0 - f_0 = O(\tau^{n+1})$ with $n = 3$, whereas for any $j > 0$ the asymptotic accuracy $z_j - f_0^{(j)} = O(\tau)$ is kept. The intersections of the lines with the ordinate axis correspond to the logarithms of the proportionality coefficients μ_j from Theorem 15.1.

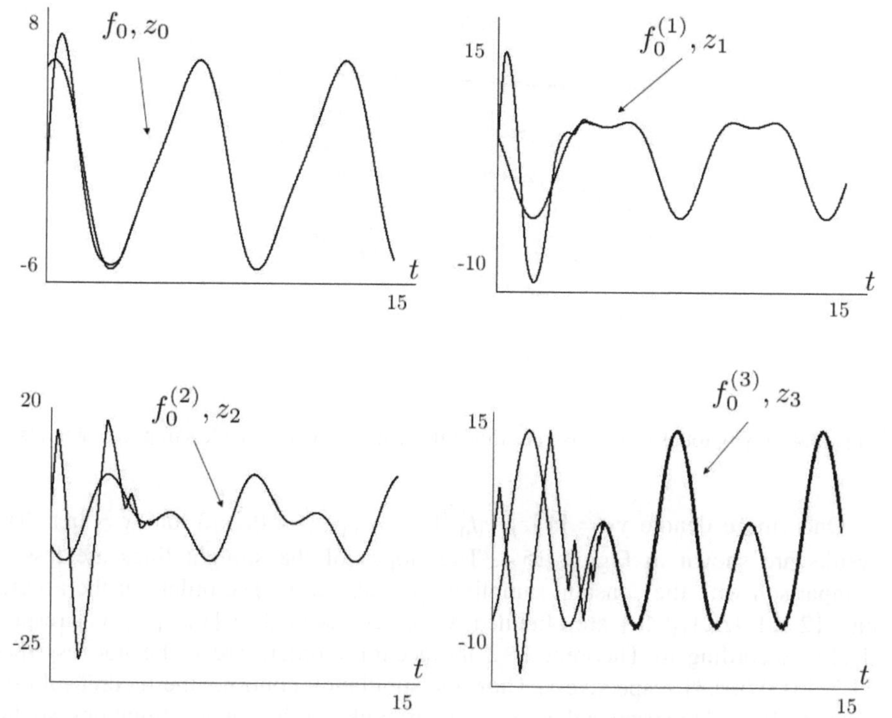

Fig. 15.3 Convergence of the discrete-time differentiator using variable sampling intervals $0 < \tau_k \leq \tau = 10^{-3}$

Now consider the case of variable sampling intervals. The same differentiator (15.22) and the same input are taken. For each value of τ, a set of randomly generated sampling intervals $\{\tau_k\}$ is produced with $0 < \tau_k \leq \tau$. Then the set $\{\tau_k\}$ is exploited in the run with the specified value of τ.

The convergence of the discrete-time differentiator with variable sampling intervals, $0 < \tau_k \leq \tau = 10^{-3}$, is demonstrated in Figure 15.3. Note that in the case when the sampling interval changes gradually, i.e. "smoothly", the asymptotics for the constant sampling interval are practically preserved (Theorem 15.1). The accuracy deteriorates, when sampling periods τ_k are taken randomly, though the convergence is still observed. As it is expected, the accuracy degrades in comparison with the constant-sampling-intervals' case. For example, it is indicated by the thickness of the plot for $f_0^{(3)}, z_0$ in Figure 15.3. Consider the corresponding asymptotic accuracies.

Similarly to the previous case the estimation errors are calculated as $\|z_j - f_0^{(j)}\|_{\infty,\tau,[12,13]}$, $j = 0,...,3$. Note that the interval $[10,14]$ is replaced by the interval $[12,13]$, since in the variable sampling intervals' case the transient time is longer. Also a slightly shorter interval $[12,13]$ is taken in order to reduce the fluctuations in the plots.

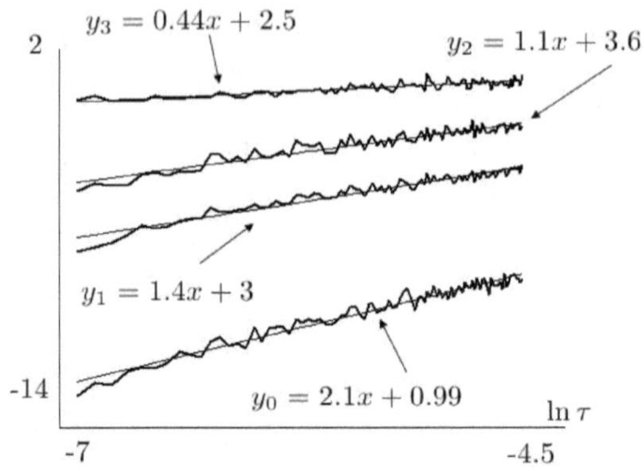

Fig. 15.4 Asymptotics of the discrete-time differentiator with variable sampling intervals

Once more denote $y_j = \ln \|z_j - f_0^{(j)}\|_{\infty,\tau,[12,13]}$, $j = 0,...,3$ and $x = \ln \tau$. The results are shown in Figure 15.4. The slopes of the straight lines are less in comparison with the constant sampling intervals' case. The orders of the accuracies $\{2.1, 1.4, 1.1, 0.44\}$ are obtained, which correspond to $\{y_0, y_1, y_2, y_3\}$ respectively. According to Theorem 15.2 the accuracy orders are to be not less than $\{1.3, 1.0, 0.6, 0.3\}$ respectively. Thus, the simulation confirms the theoretical estimations. Indeed the proven theorems only provide for the upper estimations, so that the observed asymptotic accuracies can be higher than in theory.

15.4 Conclusions

The accuracy of the discretized differentiators is shown to depend on how the corresponding dynamic system is integrated in real time. The previously known results were obtained under the assumption of ideal *continuous-time integration* of the dynamic system with *discrete-time* measurements. Robustness and exactness of standard homogeneous SM differentiators implemented by means of one-Euler-step integration method are proved, and their asymptotic accuracy is calculated.

In the case of the first-order differentiation the standard accuracies proved for the continuous-time integration are proved to be preserved in the discrete integration case, both with constant and variable sampling intervals.

The discrete case accuracy of higher order differentiators is significantly worse than with the continuous time integration due to the loss of homogeneity of the error system. In particular, if the sampling period is constant, differentiation accuracies are proportional to the sampling period and the upper bound of derivatives of the first order and higher. Fractional powers of the maximal sampling period appear in the accuracies when the sampling interval is variable.

The results are also valid for any modifications of the differentiator, which preserve its homogeneity properties, when the derivative estimation errors are close to zero. In particular it is true for [[7]].

References

1. Atassi, A., Khalil, H.: Separation results for the stabilization of nonlinear sys-tems using different high-gain observer designs. Systems & Control Letters 39(3), 183–191 (2000)
2. Bacciotti, A., Rosier, L.: Liapunov Functions and Stability in Control Theory. Springer, London (2005)
3. Bartolini, G.: Chattering phenomena in discontinuous control systems. Internat. J. Systems Sci. 20, 2471–2481 (1989)
4. Bartolini, G., Pisano, A.E.P., Usai, E.: A survey of applications of second-order sliding mode control to mechanical systems. International Journal of Control 76(9/10), 875–892 (2003)
5. Bartolini, G., Pisano, A., Usai, E.: First and second derivative estimation by sliding mode technique. Journal of Signal Processing 4(2), 167–176 (2000)
6. Bejarano, F., Fridman, L.: High order sliding mode observer for linear systems with unbounded unknown inputs. International Journal of Control 83(9), 1920–1929 (2010)
7. Cruz-Zavala, E., Moreno, J.: L., F.: Uniform second-order sliding mode observer for mechanical systems. In: Proc. of 11th International Workshop VSS, June 26-28, pp. 14–19 (2010)
8. Edwards, C., Spurgeon, S.: Sliding Mode Control: Theory and Applications. systems and control book series. Taylor & Francis. Taylor & Francis (1998)
9. Filippov, A.: Differential Equations with Discontinuous Right-Hand Sides. Mathematics and Its Applications. Kluwer Academic Publishers (1988)
10. Fridman, L.: Chattering analysis in sliding mode systems with inertial sensors. International Journal of Control 76(9/10), 906–912 (2003)
11. Kobayashi, S., Suzuki, S., Furuta, K.: Frequency charachteristics of levant's differentiator and adaptive sliding mode differentiator. International Journal of Systems Science 38(10), 825–832 (2007)
12. Kolmogoroff, A.N.: On inequalities between upper bounds of consecutive derivatives of an arbitrary function defined on an infinite interval. Amer. Math. Soc. Transl. 2, 233–242 (1962)
13. Levant, A.: Sliding order and sliding accuracy in sliding mode control. International J. Control 58, 1247–1263 (1993)
14. Levant, A.: Robust exact differentiation via sliding mode technique. Automatica 34(3), 379–384 (1998)
15. Levant, A.: Higher order sliding modes, differentiation and output-feedback control. International J. Control 76(9/10), 924–941 (2003)
16. Levant, A.: Homogeneity approach to high-order sliding mode design. Automatica 41(5), 823–830 (2005)
17. Levant, A.: Quasi-continuous high-order sliding-mode controllers. IEEE Trans. Aut. Control 50(11), 1812–1816 (2005)
18. Levant, A.: Chattering analysis. IEEE Trans. Aut. Control 55(6), 1380–1389 (2010)
19. Levant, A.: Discretization issues of high-order sliding modes. In: Proc. of 18th IFAC Congress, Milano, Italy (2011)

20. Levant, A., Pridor, A., Gitizadeh, R., Yaesh, I., Ben-Asher, J.Z.: Aircraft pitch control via second-order sliding technique. AIAA Journal of Guidance, Control 23(4), 586–594 (2000)
21. Plestan, F., Glumineau, A., Laghrouche, S.: A new algorithm for high-order sliding mode control. International Journal of Robust and Nonlinear Control 18(4/5), 441–453 (2008)
22. Shtessel, Y.B., Shkolnikov, I.A.: Aeronautical and space vehicle control in dynamic sliding manifolds. International Journal of Control 76(9/10), 1000–1017 (2003)
23. Utkin, V.I.: Sliding modes in control and optimization. Springer Verlag, Berlin, Germany (1992)
24. Yu, X., Xu, J.X.: Nonlinear derivative estimator. Electronic Letters 32(16) (1996)

Chapter 16
Sliding Mode Control in Heavy Vehicle Safety

H. Imine and L. Fridman

Abstract. In this chapter, an original approach to heavy vehicles rollover risk prediction is presented and validated experimentally. It is based on the calculation of the LTR (Load Transfer Ratio) which depends on the estimated vertical forces using high order sliding mode observers. Previously, a tractor model is developed. The validation tests were carried out on an instrumented truck rolling on the road at various speeds and lane-change manoeuvres. Many scenarios have been experienced: driving straight, curved trajectories, zigzag manoeuvre and brake tests to emphasize the rollover phenomenon and its prediction to set off an alarm for the driver. In this study, the vehicle dynamic parameters (masses, inertias, stiffness..) and the static forces infrastructure characteristics (road profile, radius of curvature, longitudinal and lateral slope, skid resistance) are measured or calculated before the tests.

Key words: Heavy vehicle modeling, Rollover, Sliding mode observer, Estimation, Prediction.

16.1 Introduction

Statistics show that accidents related to heavy goods vehicle (HGV) are more dangerous than those of passenger vehicles ([39], [13]). While they constitute only

H. Imine
LEPSIS-IFSTTAR, Laboratory for road operation, perception,simulators and simulations, The French institute of science and technology for transport, development and networks, Paris, 75732, France
e-mail: hocine.imine@ifsttar.fr

L. Fridman
UNAM, Departamento de Ingeniería, de Control y Robótica División de Ingeniería Eléctrica, Facultad de Ingeniería UNAM, Mexico
e-mail: lfridman@servidor.unam.mx

B. Bandyopadhyay et al. (Eds.): *Advances in Sliding Mode Control*, LNCIS 440, pp. 313–340.
DOI: 10.1007/978-3-642-36986-5_16

3% of vehicles in traffic, heavy vehicles are involved in 10% of accidents with fatalities. Furthermore, the fatality rate is twice as high when a HGV is involved. Rollover is one of the most frequent accidents (20%) and causes significant damage to the vehicles and injuries to its driver and passengers. Several anti-rollover systems and rollover warning systems were developed to assist or warn the driver ([[4]], [[1]], [[18]]). Most of the current prevention systems have some limitations, because they are based on real time measurements without any prediction of the vehicle dynamics. When the HGV behavior and infrastructure are well known, it is possible to be closer to the safety limit while maintaining an acceptable risk level. But with less information, the rollover risk increases, and the driver must reduce its risk by reducing the vehicle speed. Therefore, it is important to take into account the most relevant uncertainties and parameters variations of heavy vehicle system.

As mentioned in ([[7]], [[25]]), the rollover occurs when the lateral acceleration equals or exceeds the vehicle's rollover limit (which may be assisted by roadway cross-fall or camber). Lateral acceleration in a curve is highly sensitive to speed. The required speed to produce rollover reduces as the radius of curvature reduces. Roll stability is influenced by the centre of gravity height (COG), the effective track width provided by the axles and tires, and the suspensions characteristics. The COG height is affected by the chassis height and the heavy vehicle load. This performance measure is evaluated in terms of the steady-state lateral acceleration at which all wheels on the inside of curvature have lifted off the road surface. This is accomplished by increasing the steer angle of a vehicle until all axles on one side of a given vehicle lift off. The rollover can occur when one wheel of the same axle of the vehicle, lifts off the road surface. Previous work was done on rollover of heavy vehicles and several simulation results were presented ([[19]], [[5]], [[31]], [[23]], [[41]], [[25]]). Most of these works have not presented experimental results. Indeed, the instrumentation of a heavy vehicle is very expensive and not easily reached to all. In this chapter, the first experimental results of a developed predictive rollover system of heavy vehicle are presented.

The predictive system is based on the calculation of the Load Transfer Ratio (LTR) which is an indicator of rollover stability. This LTR is defined as the proportion of load on one side of a vehicle unit transferred to the other side in a transient manoeuvre. Thus, it depends on vertical forces that are estimated via a third order sliding mode observer ([[24]], [[22]], [[25]]). It estimates in the same and finite time, positions, speeds and accelerations of the heavy vehicle ([[36]], [[37]], [[16]], [[17]], [[9]], [[28]]). Many researches have been performed in order to estimate vertical forces of the vehicle and many simulation results are presented ([[26]], [[33]]). However, in these works, the dynamic parameters of the vehicle are supposed well known which is not always true. Khemoudj et al [[32]] have developed method to estimate vertical forces without knowledge of dynamic parameters. However, in these works the validation is done only with simulations using software simulator. In this work some of dynamic parameters, namely suspension stiffness and unsprung mass have been identified. This identification permits to improve the quality of the estimation ([[40]], [[14]]). One notices also that most of the published works presented only simulation results without real validation using an instrumented vehicle. This is due to the fact that the instrumentation of vehicle using the existing devices is expensive and difficult to install (more information in the site *www.kistler.com*).

In the presence of uncertainties and perturbation, the sliding mode observer is proved to be the most interesting tool ([[2]], [[43]]). The observer outputs are the estimated state variables of the vehicle (positions and especially the COG height, speeds and accelerations). Then the vertical forces acting on the wheels, which depend on the road inputs are deduced from these estimated variables. In order to show the effectiveness of the proposed system, some validation tests were carried out on an instrumented truck driving on the road at various speeds and maneuvers. Many scenarios have been experienced: driving in straight line, in curve, zigzag and brake tests to emphasize the rollover phenomenon and its prediction and send an alarm to the driver with recommended speed in order to avoid the rollover. This chapter is organized as follows: the second section is devoted to the description of the heavy vehicle model. In the third section, identification of uncertainties and perturbations of heavy vehicle is developed. In section four, the used sliding mode observer to predict rollover is developed. Instrumented heavy vehicle is described in the section four. In section five, some validation results are presented. Finally some conclusions and perspectives are given in the last section.

16.2 System Modelling

Various studies have dealt with the heavy vehicle modeling taking into account the infrastructure characteristics ([[3]], [[6]], [[1]], [[20]], [[26]]). To go in this way, the measurements carried out on the real vehicle are exploited in order to constitute a database which was used to study the various driving risks such as rollover or jack-knifing. In this part, a heavy vehicle model using the infrastructure database is developed. A tractor with 2 axles and 5 degrees of freedom is considered and represented in figure 16.1.

The tractor chassis (with the mass M) is suspended on its axles through two suspension systems. The tire of the wheel i is modeled by the springs with coefficients k_i and the suspension is modeled by both springs with coefficient K_i and damper elements B_i.

The suspension is modeled as the combination of spring and damper elements. The front view of this model is shown in figure 16.2.

The wheel masses are given by m_i $(i = 1, \cdots, 4)$. At the tire contact, the road profile, longitudinal and lateral slope, skid resistance and radius of curvature are considered as the inputs of the system. The road profile is represented by the variable u_i $(i = 1, \cdots, 4)$. The relative yaw and pitch angles of the heavy vehicle are neglected. Therefore, the dynamic model of the vehicle derived from the Lagrangian's equations is given by:

$$M(q)\ddot{q} + B(q,\dot{q})\dot{q} + K(q) = F_g \tag{16.1}$$

Fig. 16.1 Instrumented heavy vehicle

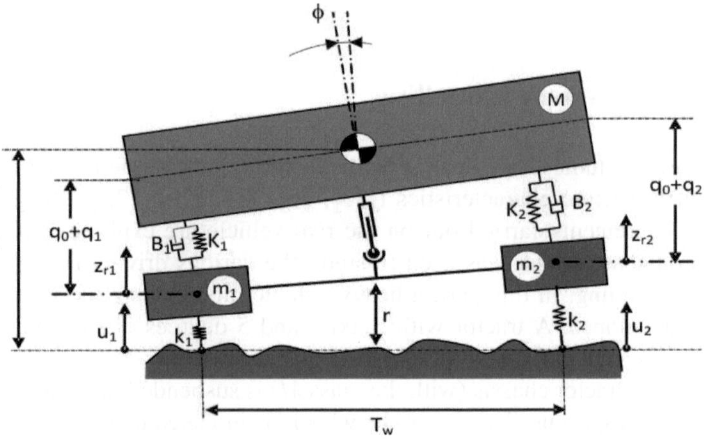

Fig. 16.2 Suspension model

where $M \in \Re^{5\times5}$ is the inertia matrix (mass matrix), $B \in \Re^{5\times5}$ is the matrix taking into account the damping effects, $K \in \Re^5$ is the springs stiffness vector and $F_g \in \Re^5$ is the generalized forces. The coordinates variable vector $q \in \Re^5$ is defined by:

$$q = [q_1, q_2, q_3, q_4, \phi]^T \tag{16.2}$$

where q_1 and q_2 are respectively the left and right front suspension deflection of the tractor, q_3 and q_4 are respectively the left and right rear suspension deflection of the tractor and ϕ is the roll angle. The vertical acceleration of the chassis (tractor's body) is function of vector q and its time derivative $\dot{q}$:

$$\ddot{z} = f(q, \dot{q}) \tag{16.3}$$

where the variable z is the vertical displacement of the tractor sprung mass which is the centre height of the gravity. The vertical displacements of the wheels with respect to the ground (road) are represented by z_{ri} $(i = 1, \cdots, 4)$ and can be calculated as follows:

$$\begin{cases} z_{r1} = z - (q_0 + q_1) + \frac{T_w}{2}\sin(\phi) - r \\ z_{r2} = z - (q_0 + q_2) - \frac{T_w}{2}\sin(\phi) - r \end{cases} \tag{16.4}$$

where q_0 is the static distance between the COG and the axles of the vehicle, T_w is the tractor track width and r is the wheel's radius. From equation (16.4), the centre height of gravity z is as follows :

$$z = \frac{1}{2}(z_{r1} + z_{r2} + q_1 + q_2) + q_0 + r \tag{16.5}$$

The vertical accelerations of the wheels are obtained using the following equations:

$$\begin{cases} \ddot{z}_{r1} = (B_1\dot{q}_1 + K_1\frac{T_w}{2}\sin(\phi) + B_1\frac{T_w}{2}\cos(\phi)\dot{\phi} \\ \qquad +K_1 q_1 - k_1 z_{r1} + k_1 u_1)/m_1 \\ \ddot{z}_{r2} = (B_2\dot{q}_2 - K_2\frac{T_w}{2}\sin(\phi) - B_2\frac{T_w}{2}\cos(\phi)\dot{\phi} \\ \qquad +K_2 q_2 - k_2 z_{r2} + k_2 u_2)/m_2 \end{cases} \tag{16.6}$$

The normal forces F_{ni}, $i = 1, \cdots, 4$ acting on the wheels are calculated using the following expression:

$$F_{ni} = F_{ci} + k_i(u_i - z_{ri}), \; i = 1, \cdots, 4 \tag{16.7}$$

where F_{ci} is the static force due to the static mass of the vehicle. In this study, the force generated by damping effects is neglected comparing to the spring forces $k_i(u_i - z_{ri})$. On the other hand, the dynamic rolling of the vehicle is described using the following differential equation:

$$I_{xx}\ddot{\phi} = ma_y h\cos(\phi + \varsigma) + mgh\sin(\phi + \varsigma) - C_R\dot{\phi} - K_R\phi \tag{16.8}$$

where I_{xx} is the inertia moment in the roll axis, C_R represents a damping coefficient of the roll motion, K_R is spring coefficient of the roll motion, $\dot{\phi}$ is the roll rate, $\ddot{\phi}$ is the roll acceleration with respect to the road, ς is the lateral slope, h is the centre height of gravity with respect to the roll axis, g represents the gravity acceleration and a_y is the lateral acceleration of the heavy vehicle. This latter variable is calculated during vehicle modelling. It depends on the lateral and longitudinal forces. It can also be measured using accelerometer sensor as described in the following chapter on the experimentation.

16.3 Perturbations and Parameters Identification

Sliding mode based observers are presented as an alternative to the problem of observation of perturbed systems. In particular, High Order Sliding Mode (HOSM) based observers can be considered as a successful technique for the state observation of perturbed systems due to their high precision and robust behavior with respect to parametric uncertainties. In this section, we show how the higher order sliding mode concept can be applied for observation of uncertainties and parameter identification of heavy vehicle ([[29]], [[21]], [[27]]). In order to develop the observer, let us rewrite the equation 16.1 in state form as follows:

$$\begin{cases} \dot{x}_1 = x_2 \\ \dot{x}_2 = f(x_1, x_2) + F_g(x_1, u) \\ y = x_1 \end{cases} \tag{16.9}$$

where $x_1 = [q_1\ q_2\ \phi]$ is the state vector representing the measured outputs vector of the system, x_2 represents its speeds, f is a vector of nonlinear analytical function and F_g is an unknown input vector computed as follows: $F_g = [-F_{g1} - F_{g2}\ 0]^T = [-\frac{F_{Z1}}{m_1}$ $-\frac{F_{Z2}}{m_2}\ 0]^T$, where F_{Z1} and F_{Z2} are respectively the right and left impact forces, m_1 and m_2 represent respectively the right and left wheels mass. Before developing the sliding mode observer, let us consider the following assumptions: 1. The state is bounded. 2. The system inputs are bounded. In order to estimate the vertical forces and identify parameters of the system, let us rewrite the system 16.9 :

$$\begin{cases} \dot{x}_{11} = x_{21} \\ \dot{x}_{21} = a_1 \varphi_1(x_{11}) - F_{g1} \\ \dot{x}_{12} = x_{22} \\ \dot{x}_{22} = a_2 \varphi_2(x_{12}) - F_{g2} \end{cases} \tag{16.10}$$

where $x_{11} = q_1, x_{21} = \dot{q}_1, x_{12} = q_2, x_{22} = \dot{q}_2$. The unknown vectors of parameters are represented by a_1 and a_2 such as: $a_1 = \left[k_1(\frac{m_1 - m}{m_1})\ k_2\right]$, $a_2 = \left[k_1\ k_2(\frac{m - m_2}{m_2})\right]$, $\varphi_1 = \frac{q_1}{m}$ and $\varphi_2 = \frac{q_2}{m}$. where m represents the unsprung mass of the vehicle. Assuming $(x_1, x_2) = (x_{11}, x_{21})$ or $(x_1, x_2) = (x_{12}, x_{22})$, $\varphi = \varphi_1$ or $\varphi = \varphi_2$ and $a = a_1$ or $a = a_2$,and in order to observe states, the following second order observer is developed: ([[12]], [[10]]) :

$$\begin{cases} \dot{\hat{x}}_1 = \hat{x}_2 + \lambda\, |\tilde{x}_1|^{1/2}\ sign(\tilde{x}_1) \\ \dot{\hat{x}}_2 = \bar{a}\varphi(\hat{x}_1, \hat{x}_2) + \alpha\ sign(\tilde{x}_1) \end{cases} \tag{16.11}$$

where $\hat{x}_1$ and $\hat{x}_2$ are respectively the estimations of x_1 and x_2, $\tilde{x}_1 = x_1 - \hat{x}_1 \in \Re$ is the estimation error, the variable $\bar{a}$ represents a vector of the nominal values of the vector parameters. In this case, the dynamic estimation errors are calculated as follows:

$$\begin{cases} \dot{\tilde{x}}_1 = \tilde{x}_2 - \lambda\, |\tilde{x}_1|^{1/2}\ sign(\tilde{x}_1) \\ \dot{\tilde{x}}_2 = \tilde{a}\ \varphi(x_1, \tilde{x}_2) + \zeta - \alpha\ sign(\tilde{x}_1) \end{cases} \tag{16.12}$$

with $\tilde{a} = a - \bar{a}$ is the estimation error of the vector a, $\zeta = -F_{g1}(x_1, u)$ or $\zeta = -F_{g2}(x_1, u)$ and $\tilde{x}_2 = x_2 - \hat{x}_2$. Since the accelerations of the system are bounded, the variable α can be minored, satisfying the inequality:

$$\alpha > 2\left|\dot{\hat{x}}_2\right| \tag{16.13}$$

On the other hand, from [11], the gains of the matrix λ satisfying the inequality, can be selected as :

$$\lambda > \sqrt[2]{\frac{2}{\alpha - 2\left|\dot{\hat{x}}_2\right|} \frac{(\alpha + 2\left|\dot{\hat{x}}_2\right|)(1+p)}{1-p}} \tag{16.14}$$

where $p \in (0,1)$ are some constants to be chosen (proof in ([10]). In order to study the observer stability, first, the convergence of $\tilde{x}_1$ and $\dot{\tilde{x}}_1$ to 0 , in finite time t_0 is proved. Then, some conditions about $\tilde{x}_2$ to ensure its convergence to 0 are deduced. Therefore, for $t \geq t_0$ the surface $\tilde{x}_2 = 0$ is attractive, leading $\hat{x}_2$ to converge towards x_2 satisfying the inequalities (16.13) and (16.14). The Super Twisting controller is insensitive to general perturbations $\rho(x_1, S) = F_g(x_1, u)$ satisfying the following conditions:

$$\left\{ \begin{array}{l} \rho_1(x_1, S) \leq k_1 \left|S\right|^{1/2} \\ \left|\frac{\partial}{\partial t}\rho_2(x_1)\right| \leq k_2 \end{array} \right| , \; S = \tilde{x}_1 \tag{16.15}$$

where $\rho(x_1, S) = \rho_1(x_1, S) + \rho_2(x_1)$, $k_1 > 0, k_2 > 0$ As described in [38], this is a strong requirement in order to ensure the complete rejection of the disturbance $\rho_1(x_1, S)$ by the Super Twisting algorithm. This allows to develop a general stability proof of this algorithm subject to the general class of disturbances in equation 16.12. The proof is based on the following quadratic Lyapunov function:

$$\left\{ \begin{array}{l} V(\zeta, u_1) = \zeta^T P \zeta \\ \zeta = [\left|S\right|^{1/2} sign(S), u_1] \\ \dot{u}_1 = -\alpha sign(S) \end{array} \right| \tag{16.16}$$

where $P = P^T > 0$ is a symmetric and positive definite solution of the following linear matrix inequality (LMI), with some constant $\varepsilon > 0$:

$$\begin{bmatrix} A^T P + PA + \varepsilon P + R & PB \\ B^T & -\Theta \end{bmatrix} \leq 0 \tag{16.17}$$

where A is the Hurwitz matrix of the system 16.12:

$$A = \begin{bmatrix} -\lambda & 1 \\ -\alpha & 0 \end{bmatrix} \tag{16.18}$$

In our case, the elements of the gains are $\lambda = 500$ and $\alpha = 1$. The matrices R and Θ take into account the perturbation bounds of the stated problem and can be considered as parameters for observer design and $B = \begin{bmatrix} 1 & 0 \end{bmatrix}^T > 0$. Then, the function 16.16 is a global strong Lyapunov function for the system 16.12. In (⟦38⟧, ⟦8⟧), it is given the proof that the system trajectories under Super Twisting control starting at $S_0 = \begin{bmatrix} S(0) & \dot{S}(0) \end{bmatrix}$ to the origin in finite time when the perturbation $\rho(x_1, S)$ of equation 16.12 is bounded by 16.15. In this case, and from 16.12, one obtains:

$$z_2 = \alpha\, sign(\tilde{x}_1) = \tilde{a}\, \varphi(x_1, \tilde{x}_2) + \zeta \tag{16.19}$$

Theoretically, the equivalent output injection is the result of an infinite switching frequency of the discontinuous term . Nevertheless, the realization of the observer produces a high switching frequency which makes the application of a filter necessary. To eliminate the high frequency component, a filter of the following form is used:

$$\tau\, \dot{\bar{z}}_2(t) = \bar{z}_2(t) + z_2(t) \tag{16.20}$$

where $\tau \in \Re$ and , $h << \tau << 1$ being a sampling step. The variable z_2 is then rewritten as follows:

$$z_2(t) = \bar{z}_2(t) + \xi(t) \tag{16.21}$$

with $\bar{z}_2(t)$ is the filtered version of $z_2(t)$ and $\xi(t)$ is the difference caused by the filtration. Nevertheless, as it is shown in ([⟦15⟧], [⟦42⟧]) that:

$$\lim_{\substack{\tau \to 0 \\ h/\tau \to 0}} \bar{z}_2(\tau, h) = z_2(t) \tag{16.22}$$

Thus, it is possible to assume that the equivalent output injection is equal to the output of the filter.

16.3.1 Perturbations Identification

In order identify the perturbation, the vector of parameters a is supposed to be known. In this case $\tilde{a} = 0$. Therefore and using the equation 16.19, the vertical force is obtained as follows:

$$\zeta = \alpha\, sign(\tilde{x}_1) \tag{16.23}$$

One recalls that this perturbation is composed of the impact force or which can be calculated as shown in the equation (16.7). One can then mention the advantages of the proposed method as following: - The measuring of the road profiles u_1 and u_2 is not necessary. - The estimation of the vertical displacements of the wheels and its derivative are also not necessary to obtain.

16.3.2 Parameters Identification

To identify the parameters of the system, we suppose that the perturbation $\zeta = 0$. That means that the road profile is supposed to be close to zero (no irregularities on the road that can affect vertically the vehicle). In this case and using the equation 16.19, we obtain:

$$z_2 = \alpha \; sign(\tilde{x}_1) = \tilde{a} \; \varphi(x_1, \tilde{x}_2) \tag{16.24}$$

Considering the unknown parameters vector a as a constant vector and in order to identify it, a linear regression algorithm, namely the least square method is applied. The time integration is given by:

$$\frac{1}{t}\int_0^t z_2(\sigma)\varphi(\sigma)^T d\sigma = \tilde{a}\frac{1}{t}\int_0^t \varphi(\sigma)\varphi(\sigma)^T d\sigma \tag{16.25}$$

The vector is then estimated by:

$$\overset{\wedge}{\tilde{a}} = \left[\int_0^t z_2(\sigma)\varphi(\sigma)^T d\sigma\right]\left[\int_0^t \varphi(\sigma)\varphi(\sigma)^T d\sigma\right]^{-1} \tag{16.26}$$

where $\overset{\wedge}{\tilde{a}}$ is the estimation of $\tilde{a}$. Let us define

$$\Gamma = \left[\int_0^t \varphi(\sigma)\varphi(\sigma)^T d\sigma\right]^{-1} \tag{16.27}$$

Its derivative is equal to

$$\dot{\Gamma} = -\Gamma\varphi(\sigma)\varphi(\sigma)^T\Gamma \tag{16.28}$$

The derivative of the vector $\overset{\wedge}{\tilde{a}}$ using the equation 16.26 gives:

$$\dot{\overset{\wedge}{\tilde{a}}} = \left[\int_0^t z_2(\sigma)\varphi(\sigma)^T d\sigma\right]\dot{\Gamma} + z_2\varphi(\sigma)^T\Gamma \tag{16.29}$$

Replacing $\dot{\Gamma}$ by its value given before and using the equation 16.26, we obtain:

$$\begin{aligned}\dot{\overset{\wedge}{\tilde{a}}} &= -\overset{\wedge}{\tilde{a}}\varphi\varphi^T\Gamma + z_2\varphi(\sigma)^T\Gamma \\ &= (-\overset{\wedge}{\tilde{a}}\varphi + z_2)\varphi^T\Gamma\end{aligned} \tag{16.30}$$

This ensures the asymptotic convergence of $\overset{\wedge}{\tilde{a}}$ to $\tilde{a}$ and consequently this allows to identify the real value of the vector . In our case, in order to obtain the unsprung masses after identification of $k_1 = a_{12}$ and $k_2 = a_{21}$, we refer to the vector a

defined previously. One notices then the identified values of $a_{11} = k_1(m_1 - m)/m_1$ and $a_{22} = k_2(m_2 - m)/m_2$. Finally, the unsprung masses are deduced as follows: $m_1 = mk_1/(k_1 - a_{11})$ and $m_2 = mk_2/(k_2 - a_{22})$.

16.4 Sliding Mode Observer for Risk Prediction

In order to evaluate the rollover risk, high order sliding mode observer is developed to estimate the state variables and the vertical forces of the vehicle ([[42]], [[35]], [[11]], [[25]], [[17]], [[30]], [[26]], [[28]]). In state space form, the system equation (16.1) can be rewritten as:

$$\begin{cases} \dot{x}_1 = x_2 \\ \dot{x}_2 = M^{-1}(F_g - B(x_1,x_2)x_2 - K(x_1)) \end{cases} \tag{16.31}$$

where $x = (x_1,x_2)^T = (q,\dot{q})^T$ is the state variables vector and $x_1 = [q_1,q_2,q_3,q_4,\phi]^T$ is the measured outputs vector of the system. The roll angle is calculated using the following formula:

$$\phi = \arcsin(\frac{q_1 - q_2}{T_w}) \tag{16.32}$$

To be able to estimate the state variables and the vertical forces, the following observer is developed and the convergence is proved ([[36]], [[17]]).

$$\begin{cases} \dot{\hat{x}}_1 = \hat{x}_2 - \lambda_0 \left|\hat{x}_1 - x_1\right|^{2/3} sign(\hat{x}_1 - x_1) \\ \dot{\hat{x}}_2 = \hat{x}_3 - \lambda_1 \left|\hat{x}_2 - \dot{\hat{x}}_1\right|^{1/2} sign(\hat{x}_2 - \dot{\hat{x}}_1) \\ \dot{\hat{x}}_3 = -\lambda_2 sign(\hat{x}_3 - \dot{\hat{x}}_2) \end{cases} \cdot \tag{16.33}$$

where $\hat{x}_1$, $\hat{x}_2$ and $\hat{x}_3$ are respectively the estimate of x_1, x_2 and $\dot{x}_2$, λ_0, λ_1 and λ_2 are the observer gains. More details about this observer can be found in [[35]]. The observer defined in (16.33) permits to estimate positions, velocities and accelerations of the system. The jerk of the system is bounded and it satisfies the inequality:

$$f^+ \geq 2|\dddot{x}_1| \tag{16.34}$$

where f^+ is some known positive scalar. The estimation errors are obtained using the equations (16.31) and (16.33) as following:

$$\begin{cases} \dot{\tilde{x}}_1 = x_2 - \hat{x}_2 + \lambda_0 \left|\hat{x}_1 - x_1\right|^{2/3} sign(\hat{x}_1 - x_1) \\ \dot{\tilde{x}}_2 = \dot{x}_2 - \hat{x}_3 + \lambda_1 \left|\hat{x}_2 - \dot{\hat{x}}_1\right|^{1/2} sign(\hat{x}_2 - \dot{\hat{x}}_1) \\ \dot{\tilde{x}}_3 = \ddot{x}_2 + \lambda_2 sign(\hat{x}_3 - \dot{\hat{x}}_2) \end{cases} \tag{16.35}$$

where $\tilde{x}_i = x_i - \hat{x}_i$ $(i = 1,\cdots,3)$ is the estimation error of the variable x_i. Chosen the ith components of λ_0, λ_1 and λ_2 as: $\lambda_0 = 3\sqrt[3]{f^+}$, $\lambda_1 = 1.5\sqrt[2]{f^+}$ and $\lambda_2 = 1.1f^+$, the estimation errors $\tilde{x}_1$, $\tilde{x}_2$ and $\tilde{x}_3$ converge in finite time t_0 toward 0 . More details about the convergence study of this observer can be found in [[16]]. In this case, by means of the equation (16.6), the vertical displacements of the wheels are estimated in finite time, since the vertical accelerations of the wheels $\ddot{z}_{ri}$ are measured using accelerometers:

$$\begin{cases} \hat{z}_{r1} = (-m_1\ddot{z}_{r1} + B_1\dot{\hat{q}}_1 + K_1\frac{T_w}{2}\sin(\hat{\phi}) \\ +B_1\frac{T_w}{2}\cos(\hat{\phi})\dot{\hat{\phi}} + K_1\hat{q}_1 + k_1u_1)/k_1 \\ \\ \hat{z}_{r2} = (-m_2\ddot{z}_{r2} + B_2\dot{\hat{q}}_2 - K_2\frac{T_w}{2}\sin(\hat{\phi}) \\ -B_2\frac{T_w}{2}\cos(\hat{\phi})\dot{\hat{\phi}} + K_2\hat{q}_2 + k_2u_2)/k_4 \end{cases} \tag{16.36}$$

From equation (16.5), the centre height of gravity $\hat{z}$ is now deduced :

$$\hat{z} = \frac{1}{2}(\hat{z}_{r1} + \hat{z}_{r2} + \hat{q}_1 + \hat{q}_2) + q_0 + r \tag{16.37}$$

Using the equation (16.7), the vertical forces F_{ni} can be estimated by:

$$\hat{F}_{ni} = F_{ci} + k_i(u_i - \hat{z}_{ri}),\ i = 1, \cdots, 4 \tag{16.38}$$

Then the Load Transfer Ratio (LTR) used to indicate the rollover risk, is calculated as follows [[1]]:

$$LTR = \frac{F_{nr} - F_{nl}}{F_{nr} + F_{nl}} \tag{16.39}$$

When $F_{nr} = 0$ ($F_{nl} = 0$) the right (left) wheels lift off the road and the rollover coefficient takes on the limit value $LTR = -1$ ($LTR = 1$). For straight driving on a horizontal road for the tire vertical forces, it holds that $F_{nr} = F_{nl}$ which means that $LTR = 0$.

16.5 Experimental Results

16.5.1 Description of the Test Bench

In order to validate theoretical study and the simulations results, an instrumented tractor of Renault Trucks company is used, as shown in figure 16.1. The vehicle is equipped with several sensors to measure the dynamics of the vehicle, such as the angular speeds, accelerations, and the suspension deflections. The figure 16.3 illustrates the added sensors which are needed for the proposed technique:

- Four sensors, LVDT (Linear Variable Differential Transformers) installed between the wheel and the chassis in order to measure the deflections of suspensions,

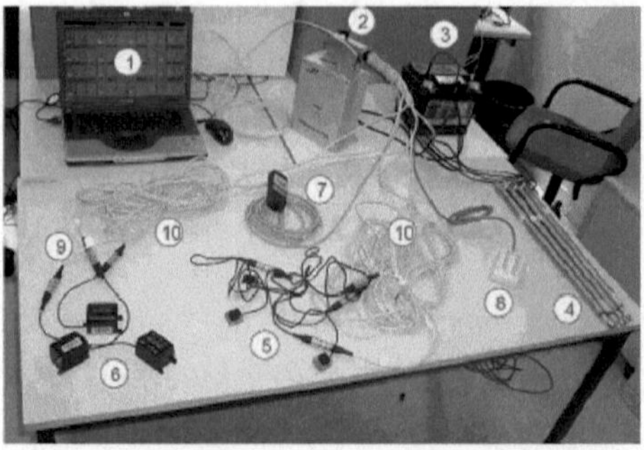

1. Control desk software
2. Micro-Autobox
3. Battery
4. LVDT sensors
5. Accelerometers
6. Gyrometers
7. Laser sensor
8. BNC connectors

Fig. 16.3 The used sensors on the Vehicle

- Four accelerometers installed on the chassis in order to measure the vertical accelerations of wheels,
- Three axial gyrometers installed on the chassis in order to measure the angular speeds (roll, pitch and yaw rate),
- Two lasers installed at the bottom of the chassis in order to measure its height.

The figure 16.4 illustrates the positions of installed sensors in the vehicle. Two LVDT sensors are installed in the front of the vehicle and two others are installed in the rear of the tractor. The two laser sensors are installed respectively in the left and in the right side in order to measure the height of the vehicle. The tri-axial gyrometer is installed in the centre of the vehicle in order to measure the three rotations of the tractor.

The LVDT are the only sensors which are necessary and needed to be added in order to product the predictive rollover system. The roll angle is deduced using LVDT sensors, as explained in the previous section. The other sensors are only used in order to test the robustness of the approach by comparing their measures to the estimated variables.

The acquisition part of the bench, consists of use of laptop computer, a dSPACE Micro AutoBox real-time hardware system, and the software: Matlab/Simulink, Real Time Workshop and the dSPACE acquisition system. This acquisition board delivers high performance and reliable data acquisition capabilities with 16 single-ended analogical inputs. It delivers both analogical and digital triggering capability, as well as two 12-bit analogical outputs, two 24-bit and 8 digital I/O lines. The sampling frequency used during the tests is 100 Hz.The algorithms were written in Matlab/Simulink, which coordinates all the data acquisition and the test measurement processes. The developed program can be easily manipulated and integrated in the vehicle.

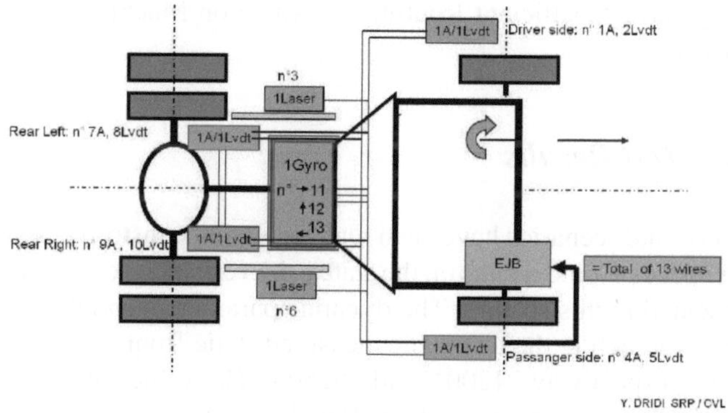

Fig. 16.4 Sensors position on the heavy vehicle

16.5.2 Infrastructure Measurements

Before the tests, the infrastructure data have been measured by different devices. The road profile is measured by Longitudinal Profile Analyzer shown in figure 16.5. The technical description and the functionalities of this device are given in [34]. The radius of curvature, longitudinal and lateral slope are measured using VANI (Véhicule d'Analyse d'Itinéraire).

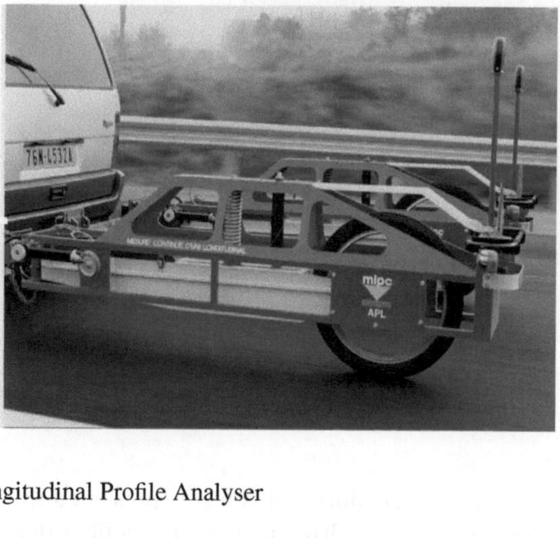

Fig. 16.5 Longitudinal Profile Analyser

This vehicle is equipped with different sensors, such as Gyrometers, GPS and lasers is realized by Regional Laboratory of Lyon, in France in 1987. The CFT (transversal friction coefficient) of the road surface is measured by SCRIM device (Sideway force Coefficient Routine Investigation Machine) which is described in *http://www.vectra.fr*.

16.5.3 Test Results

Many tests and scenarios have been realized with the instrumented vehicle driving at various speeds. Some results on the states, the vertical forces and the risk estimations are presented in this section. The dynamic parameters and the static vertical forces are measured before the tests. The measured static front left and static right vertical forces are respectively $24200N$ and $25250N$. The values of static rear left and right vertical forces are respectively $9450N$ and $12050N$.

16.5.3.1 Zigzag Test

The zigzag test is illustrated by figure 16.6. This test is very interesting for rollover study since it can cause dangerous situations. The driver changes abruptly the direction of his vehicle which implies load transfer between the left and right side of the vehicle.

Fig. 16.6 The zigzag test in practise

The steering angle of the vehicle during this test is presented in the figure 16.7. The critical times are occurred at $15s$, $30s$ and $45s$. One notices that at these times, the absolute steering angle is more than $3rad$ (180). In order to verify the condition 16.34, let us represent the jerk of the systems which corresponds to the double derivative of roll rate measured by gyrometer sensor and the jerk coming from the third derivative of suspension deflection, measured by LVDT sensor. The result is shown in the figure 16.8.

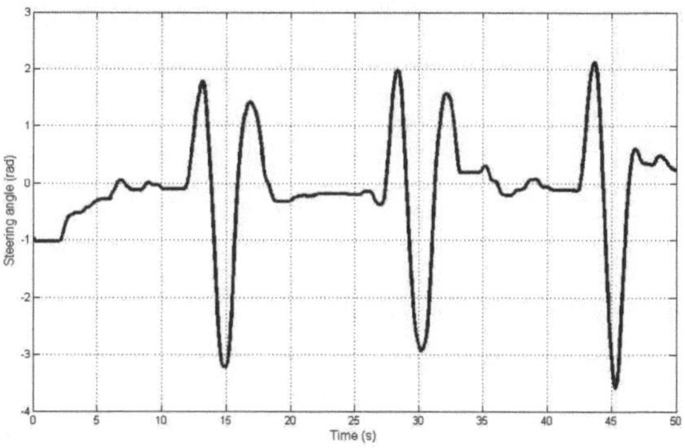

Fig. 16.7 Steering angle for zigzag test

One remarks that the maximum values of jerks of suspension deflection and roll angle are respectively $800m/s^3$ and $150rad/s^3$. In this case, the value of the gain f^+ is then deduced to be equal to 1600. The vehicle speed is shown in figure 16.9.

In the figure 16.10, suspension deflections of the front of the vehicle are estimated and compared to the measured one.

This figure shows that the observer converges quickly and the estimation error is around zero. Therefore, the two graphs are practically indistinguishable. At the critical times, the effect of the zigzag on the vehicle dynamics is clearly shown at $15s$, $30s$ and $45s$. The suspension deflection at the front right decreases from its static value $0.01m$ to $-0.03m$, whereas the suspension deflection at the front left increases from $0.01m$ to $0.025m$. From this behavior, the roll angle shown in the figure 16.11 occurred. Indeed, at the times $15s$, $30s$ and $45s$, the roll angle increased. One can notice the quality of the estimation compared to the measure. It is clearly shown that the estimated and measured roll angles are in good agreement.

The figure 16.12 shows the estimation of the centre height of gravity compared to the front left suspension deflection. Since, there is no existing sensor to measure this displacement, it's then difficult to judge the quality of the estimation.

However, one notices that at $15s$, the suspension deflection increased up to $0.025m$ and at the same time, the estimated centre height of gravity increased up to $0.71m$. The same phenomena are produced at the times $30s$ and $45s$. This implies that the estimated centre height of gravity correctly tracks the LVDT measure. This conclusion represents a good indicator to evaluate the quality of this estimation. In the figure 16.13, the vertical forces of the front wheels are presented. The force of the front left wheel is presented at the left.

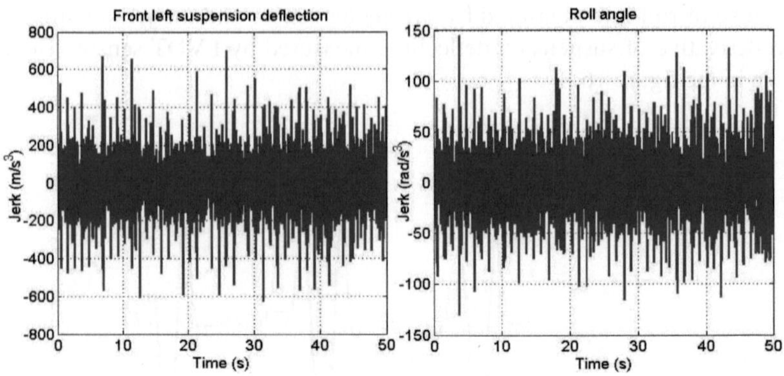

Fig. 16.8 Jerk of the system in case of zigzag test

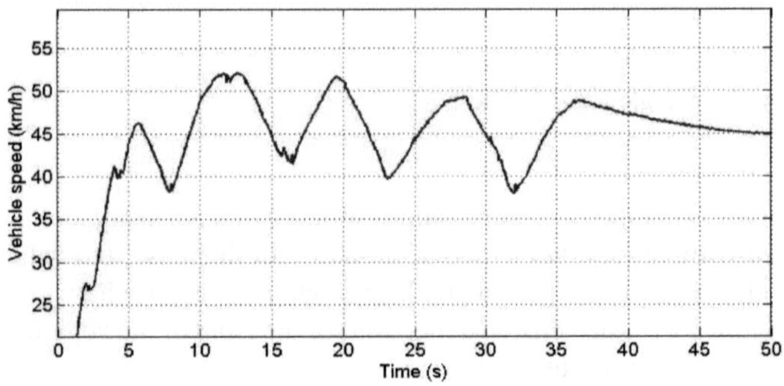

Fig. 16.9 Vehicle speed for zigzag test

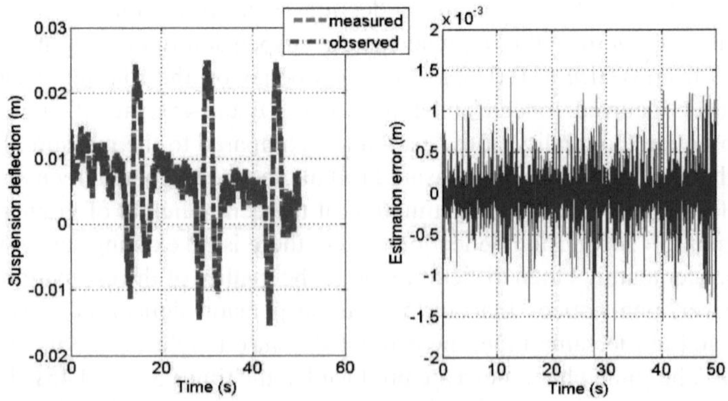

Fig. 16.10 Suspension deflection estimation

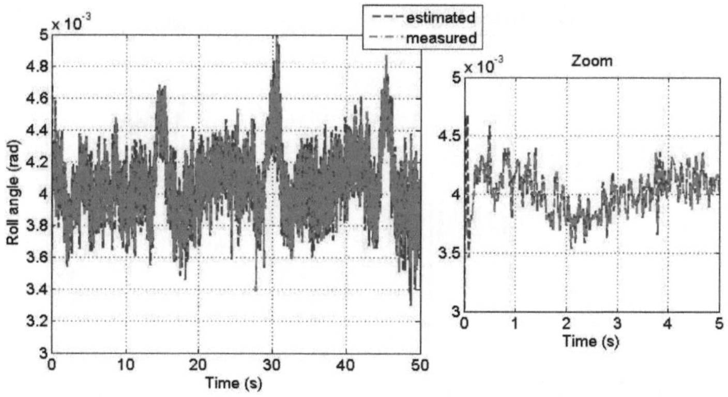

Fig. 16.11 Roll angle estimation

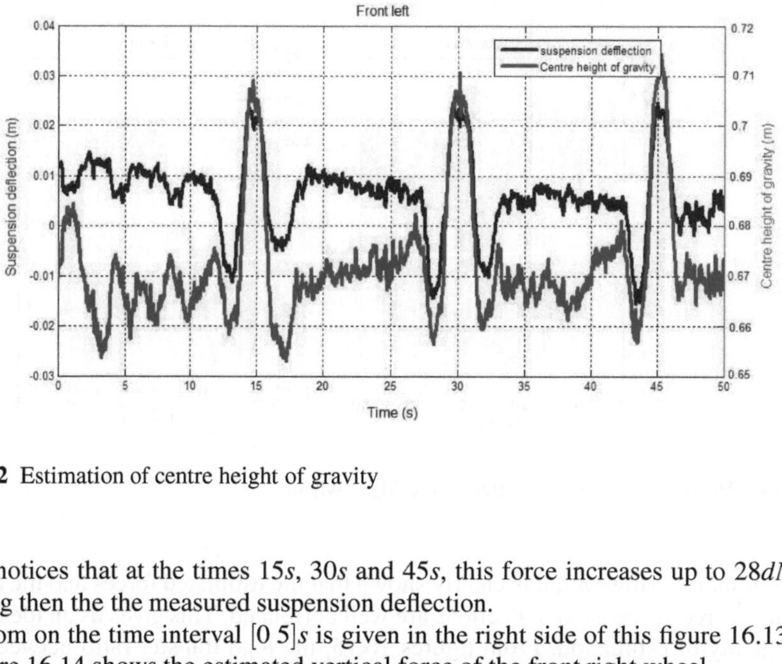

Fig. 16.12 Estimation of centre height of gravity

One notices that at the times 15*s*, 30*s* and 45*s*, this force increases up to 28*dN* following then the the measured suspension deflection.

A zoom on the time interval $[0\ 5]s$ is given in the right side of this figure 16.13. The figure 16.14 shows the estimated vertical force of the front right wheel.

As explained before, the same conclusion can be given here. indeed, at the same times, this force decreases up to 19*dN*. This phenomena can be explained by the fact, that the load transfer from the right side to the left side of the vehicle is produced. A zoom on the time interval $[0\ 5]s$ is given in the right side of th figure 16.14.

In the figure 16.15, the vertical forces are compared to suspension deflections measures. As for the centre height of gravity, there are no sensors, during this test, to measure the vertical forces. It's then difficult to conclude on the quality of the

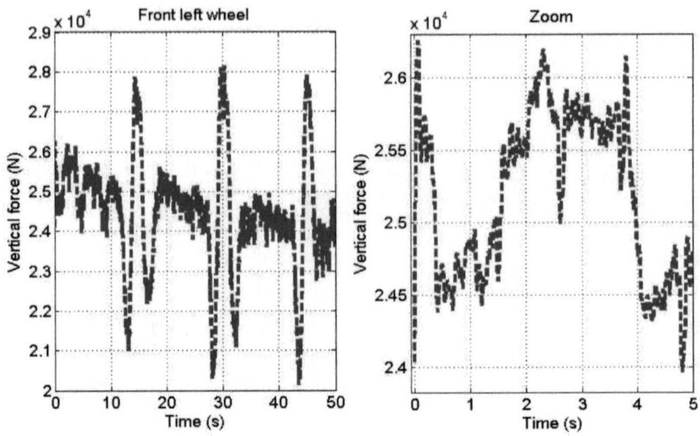

Fig. 16.13 Estimation of vertical force: left wheel

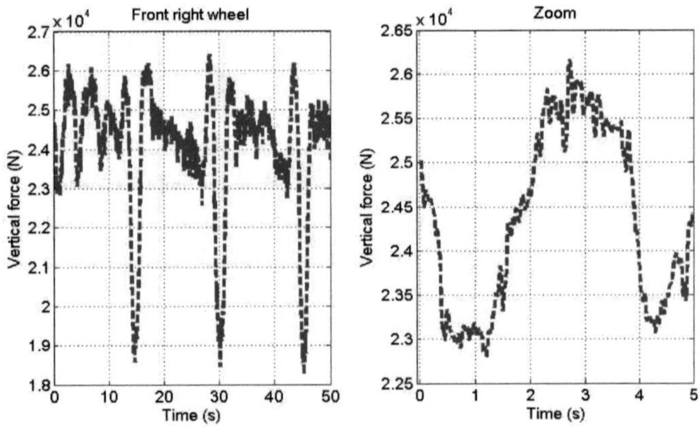

Fig. 16.14 Estimation of vertical force: right wheel

estimation. However, it is clearly shown that the estimated forces and the equivalent measured suspension deflections are well correlated. This gives us an idea about the quality of estimation. From figures 16.15, the load transfer ratio between the two wheels is calculated and shown in the figure 16.16. The values of LTR are situated between -0.15 and 0.2 .

These values are much smaller than the risk limit LTR=1, where on wheel of the same axle lifts off the road. This is due to the fact that, during the test and for safety reasons, the driver is not allowed to reach this limit. However, in order to test the approach and send an alarm to the driver, the coefficient limit of LTR is reduced to 0.2. In the figure 16.17, the identification results are shown. The suspensions stiffness k_1 and k_2 are identified with success. Compared to their nominal values

($194680N/m$ and $188540N/m$), these parameters have been identified with some variation in the time interval $[13,20]s$. This is due to the fact that at this time, the driver changes brutally the vehicle direction as shown in the steering angle of the Figure 16.7.

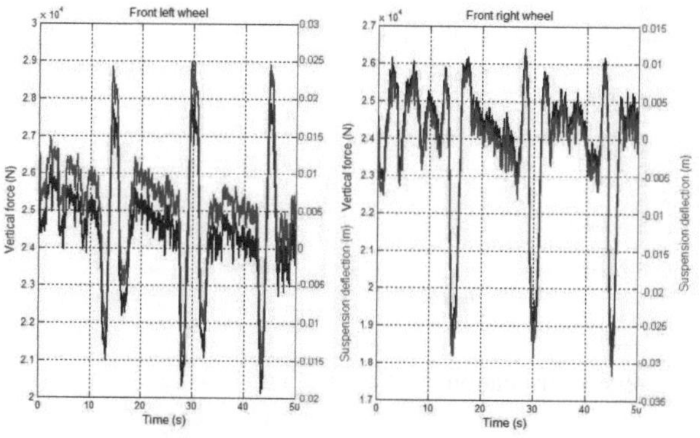

Fig. 16.15 Vertical forces estimation compared to LVDT measures

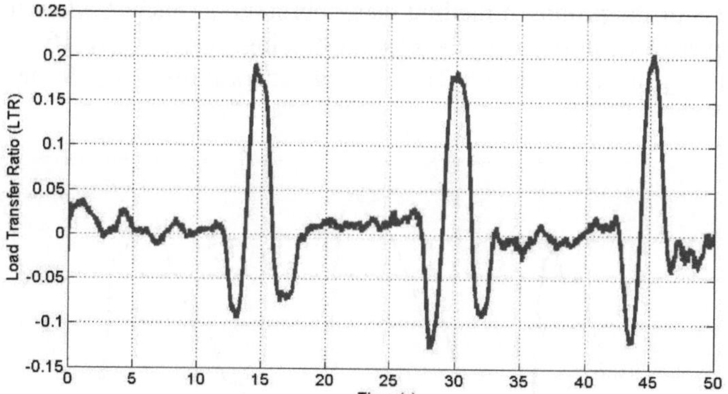

Fig. 16.16 Load Transfer Ratio (LTR)

The same remark can be given to the unsprung masses identification m_1 and m_2. The variation of these parameters occurs around their nominal values respectively of $100kg$ and $95kg$, at the same time interval $[13,20]s$.

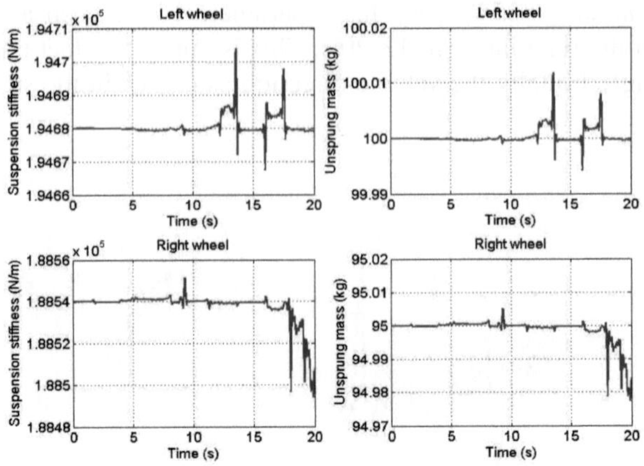

Fig. 16.17 Parameters identification

16.5.3.2 Braking test

In this section, the brake test is presented in order to show the rapidity and the robustness of the proposed method using observers. This test allow us to know if the rollover risk can occur in the case of braking. In the figure 16.18, the vehicle speed during this test is shown. The braking occurs at times 9*s*, 29*s* and 49*s*.

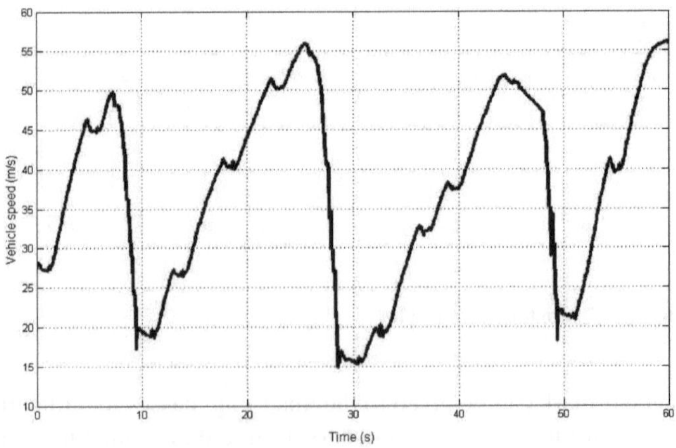

Fig. 16.18 Vehicle speed for brake test

In the following, the influence of the braking on the vehicle behavior and the rollover risk is shown.

In this case, the jerk of the system which corresponds to the double derivative of roll rate measured by gyrometer sensor and the jerk coming from the third derivative of suspension deflection measured by LVDT sensor are shown in the figure 16.19. One remarks that the jerks of suspension deflection and roll angle are respectively bounded by $800m/s^3$ and $250rad/s^3$. Also in this case, the value of the gain f^+ is chosen to be equal to 1600.

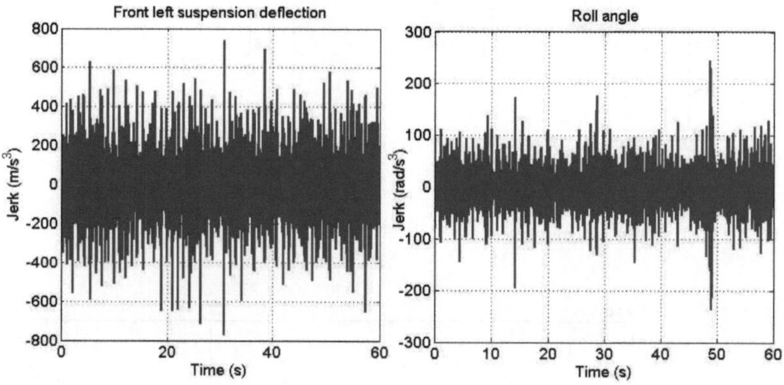

Fig. 16.19 Jerk in the brake test

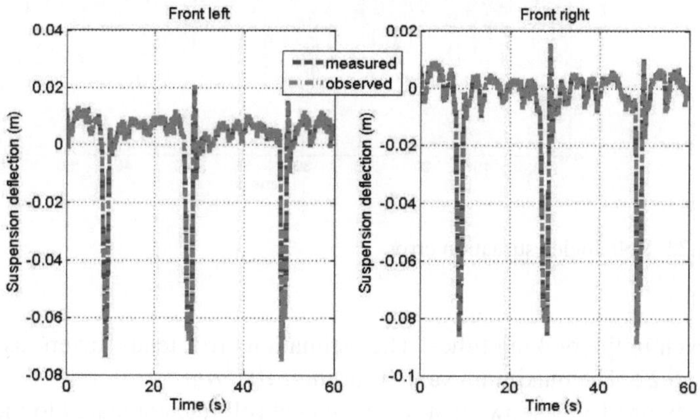

Fig. 16.20 Suspension deflection estimation

In the figure 16.20, the estimation of the suspension deflections of the front of vehicle are represented and compared to measures. At the braking times $9s$, $29s$ and $49s$, these vertical displacements decrease.

The right and the left side have almost the same value of about $-0.08m$. In this case no load transfer is occurred between the left and the right side of the vehicle.

To show the quality estimation of the roll angle, a zoom is done in the time interval $[0\ 2]s$ of the figure 16.21. One notices that in this case, the roll angle is not

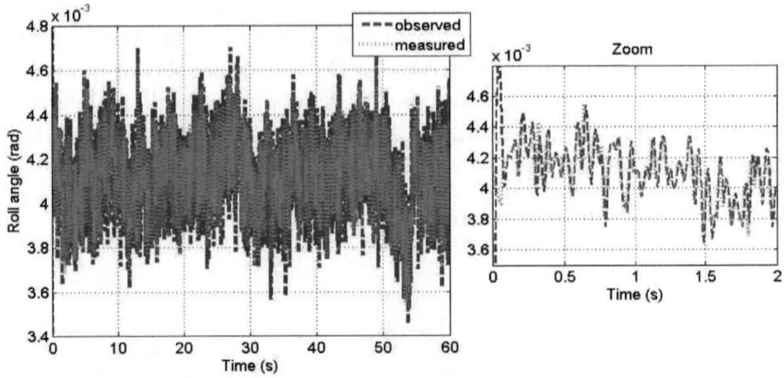

Fig. 16.21 Roll angle estimation

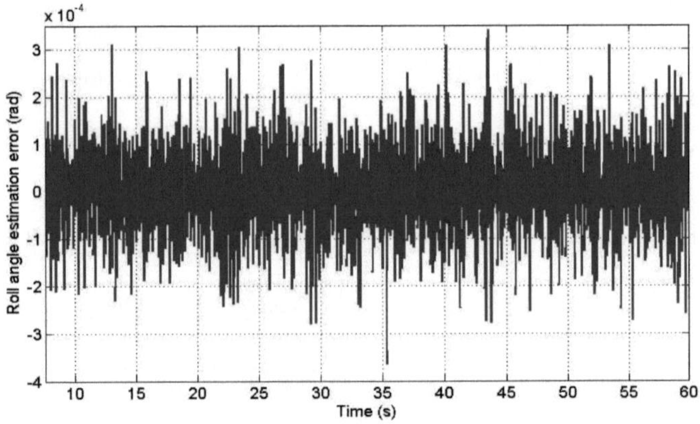

Fig. 16.22 Roll angle estimation error

high even in the braking times. The estimation error tends to zero as shown in the figure 16.22. The maximum value is around $10^{-4}rad$.

The figure 16.23 shows that the estimated roll rate compared to the gyrometer's measure. One remarks that the estimation converges quickly toward the measure.

Indeed, the minimum value for the measured roll rate is about $-0.03rad/s$ and for the estimated roll rate is about $-0.04rad/s$. The maximum value is almost the same for the two signals. It's about $0.03rad/s$. The estimation error is shown in the figure 16.24.

The figure 16.25 shows the estimation of the centre height of gravity. At the braking times $9s$, $29s$ and $49s$, the centre height of gravity increased up to $0.8m$ and between these times, the value of this displacement stays at its static value, namely $0.68m$.

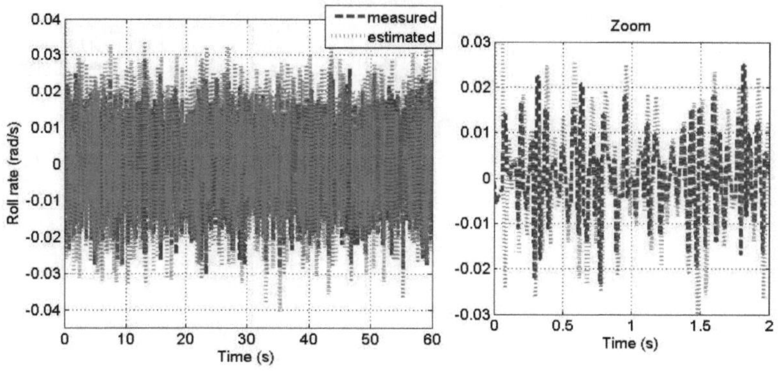

Fig. 16.23 Roll rate estimation in the case of brake test

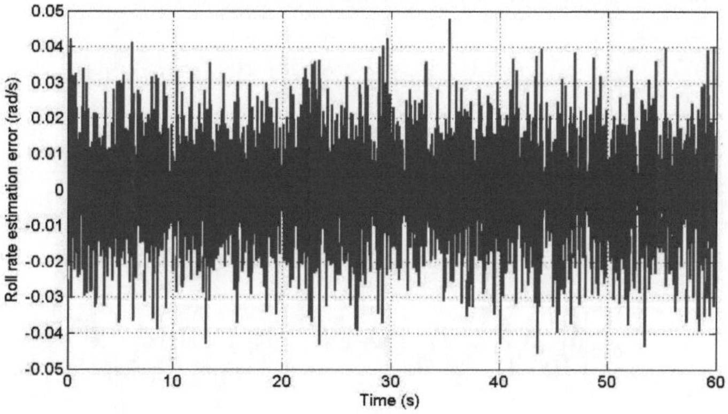

Fig. 16.24 Roll rate estimation error

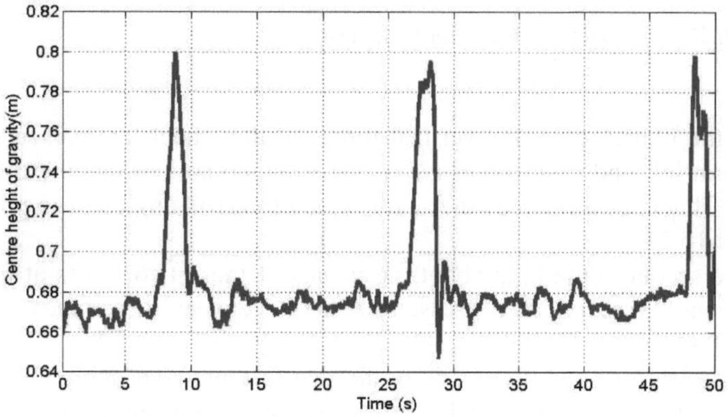

Fig. 16.25 Estimation of centre height of gravity

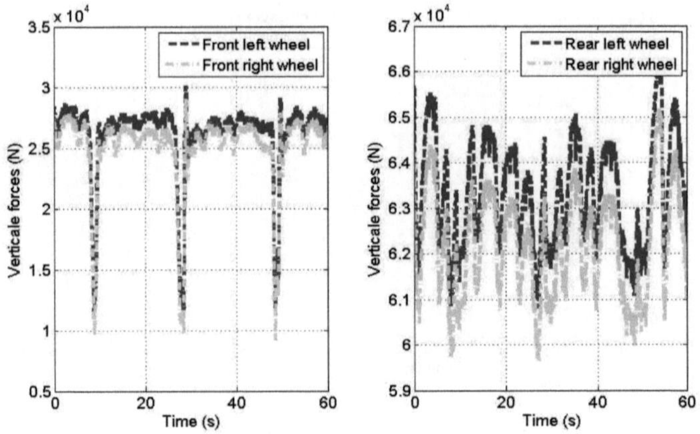

Fig. 16.26 Estimation of vertical forces in case of brake test

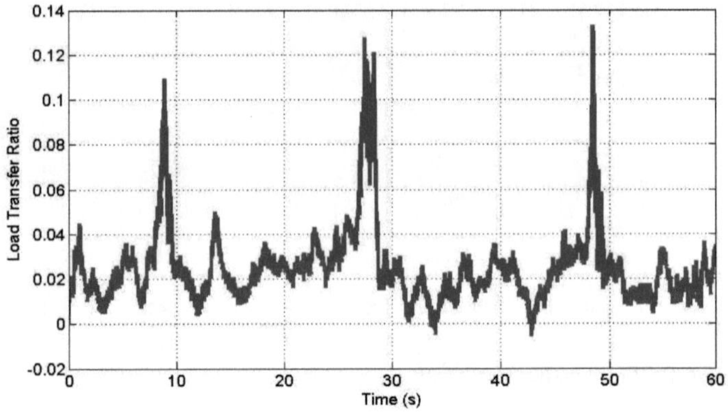

Fig. 16.27 Load Transfer Ratio for brake test

In the figure 16.26, the vertical forces of the wheels are presented. In the left, the front left and right forces are presented. One notices that these forces are quit close. That is confirmed by the small value of the roll angle shown previously in the Figure 16.21.

The second remark, is about the values of these two forces at the times 9*sec*, 29*sec* and 49*sec*, which decrease to 8000*N*. Between these times, the forces keep their static values.

In the right of the figure 16.26, the rear left and right forces are shown. These forces vary around their static values, which is conform to the braking test. The figure 16.27 shows the Load Transfer Ratio (LTR).

One notices an increase of its value from 0.02 until respectively 0.11, 0.13 and 0.135, at braking times, respectively $9sec$, $29sec$ and $49sec$. However, these values still far from the limit value of 1 and the limit fixed in this work, namely 0.2. In this condition, no rollover risk is detected and therefore, no alarm is sent to the driver.

16.6 Conclusion

In this work, an original system of heavy vehicles rollover risk prediction has been proposed. The main advantage of the method is its simplicity and it is based on vertical forces estimation using high order sliding mode observer. It has been validated experimentally on a real heavy truck rolling on the road at various speeds and lane-change manoeuvres. Good agreement has been shown between the experimental and theoretical results. In order to show the robustness of the proposed approach, two tests are presented in this study: zigzag and brake test. The results show that dynamic states are well estimated as shown in the centre height of gravity. Then, vertical forces are estimated and the rollover indicator LTR is computed. The results have been discussed. It is shown that the estimation results are quite close to experimental ones and the rollover is predicted. In this test, the LTR does not reach its limit of 1. In the real situation and for safety reason and only for this reason, we were not allowed to test this situation where one wheel lifts of the road because the tractor is not equipped with safety device. However, in order to send an alarm to the driver and to test its effect on the driver's behavior, this limit is reduced to 0.2. For this reason, the LTR limit to send the alarm is reduced to 0.2. In this case, and during the zigzag test, this limit is reached and the alarm is then sent to the driver in order to reduce his speed. The method proved in the case of brake test, that no rollover occurs (LTR$<$ 0.2). HOSMO are also employed in order to estimate vertical forces of heavy vehicle and to identify the unknown parameters. The experimental tests carried out on an instrumented tractor show the quality of this approach since the convergence of the observer is quick and is done in finite time, with errors quite close to zero. The vertical forces are also well estimated. This is noticed when the estimations are compared to the measures. The originality of this approach is the use of the equivalent control, which provides a linear regression algorithm in order to identify the unknown parameters of the system. An example of identification of the unsprung mass and the stiffness is given in this paper. In the future work, it will be interesting to test this approach in real time with an instrumented vehicle. The dynamo wheel in order to measure the impact forces will be useful in order to have a better reference to validate the impact forces estimation. The proposed method is tested on an instrumented tractor. It can be interesting to test the robustness of this approach on tractor semi-trailer.

References

1. Ackermann, J., Odenthal, D.: Damping of vehicle roll dynamics by speed-scheduled active steering. In: European Control Conference, Karlsruhe, Germany (1999)
2. Ahrens, J.H., Khalil, H.K.: High-gain observers in the presence of measurement noise: A switched-gain approach. Automatica 45(4), 936–943 (2009)
3. Cebon, D.: Interaction between heavy vehicles and roads. In: Society of Automotive Engineers, USA, pp. SP–931, 81 (1993)
4. Chen, B.C., Peng, H.: A real-time rollover threat index for sports utiliy vehicles. In: American Control Conference, San Diego, USA, pp. 1233–1237 (1999)
5. Chen, B.C., Peng, H.: Rollover warning of articulated vehicles based on a time-to-rollover metric. In: Proc. ASME Int. Congress and Exploitation (1999)
6. Chen, C., Tomizuka, M.: Dynamic modeling of articulated vehicles for automated highway systems. In: American Control Conference, pp. 653–657 (1995)
7. Dakhlallah, J., Imine, H., Sellami, Y., Bellot, D.: Heavy vehicle state estimation and rollover risk evaluation using kalman filter and sliding mode observer. In: IEEE European Control Conference, ECC 2007, Kos, Grece, pp. 3444–3449 (2007)
8. Davila, A., Moreno, J., Fridman, L.: Variable gains super-twisting algorithm: A lyapunov based design. In: IEEE American Control Conference (2011)
9. Davila, J., Fridman, L., Levant, A.: High-order sliding observation and fault detection. In: 16th Mediterranean Conference on Control and Automation, Ajaccio, Corsica, France, pp. 1699–1704 (2008)
10. Davila, J., Fridman, L., Poznyak, A.: Observation and identification of mechanical systems via second order sliding modes. In: 2006 International Workshop on Variable Structure Systems, VSS 2006, Sardinia, Italy, pp. 232–237 (2006)
11. Davila, J., Fridman, L., Poznyak, A.: Observation and identification of mechanical systems via second order sliding modes. International Journal of Control 79(10), 1251–1262 (2006)
12. Emelyanov, S.V., Korovin, S.K., Levantovsky, L.V.: Second order sliding modes in controlling uncertain systems. Soviet Journal of Computer and System Science, 63–68 (1986)
13. Evans, J., Batzer, S., Andrews, S.: Evaluation of heavy truck rollover accidents. In: 19th International Safety Conference on the Enhanced Safety of Vehicles, Washington DC, USA, pp. 05–0140–W (2005)
14. Floret-Pontet, F., Lamnabhi-Lagarrigue, F.: Parameter identification methodology using sliding mode observers. Int. Journal of Control 74(18), 1743–1753 (2001)
15. Fridman, L.: The Problem of Chattering: An Averaging Approach. In: Young, K., Ozguner, U. (eds.) Variable Structure, Sliding Mode and Nonlinear Control. LNCIS, vol. 247, pp. 363–386. Springer, Heidelberg (1999)
16. Fridman, L., Levant, A., Davila, J.: High-order sliding-mode observation and identification for linear systems with unknown inputs. In: Proc. 45th IEEE Conf. Decision Control, San Diego, CA, USA, pp. 5567–5572 (2006)
17. Fridman, L., Shtessel, Y., Edwards, C., Yan, X.: Higher-order sliding-mode observer for state estimation and input reconstruction in nonlinear systems. Int. J. Robust Nonlinear Control 18(4-5), 399–413 (2008)
18. Gaspar, P., Szaszi, I., Bokor, J.: Reconfigurable control structure to prevent the rollover of heavy vehicles. Control Engineering Practice (13), 699–711 (2005)
19. Gillespie, T.D., Karamihas, S.M.: Characterising trucks for dynamic load prediction. Int. J. Vehicle Design 1(1), 3–19 (1993)

20. Ibrahim, I.M.: Design of reducing the generated dynamic tyre loads of the articulated tanker vehicles. Int. J. of Vehicle Design Heavy Vehicle Systems 11(1), 86–99 (2004)
21. Imine, H.: Identification of heavy vehicle parameters and steering control for rollover avoidance. In: IAVSD, International Symposium on Dynamics of Vehicles on Roads and Tracks, Manchester, UK (2011)
22. Imine, H., Dolcemascolo, V.: Influence of the road profile on wheel and vehicle loads estimation. In: 9th International Symposium on Heavy Vehicle Weights and Dimensions, Pennsylvania, USA (2006)
23. Imine, H., Dolcemascolo, V.: Vertical tyre forces estimation to calculate the rollover risk of heavy vehicles. In: FISITA 2006, World Automotive Congress, Yokohama, Japan, Yokohama, Japan, p. F2006V078 (2006)
24. Imine, H., Dolcemascolo, V.: Warning system to prevent rollover of heavy vehicles. In: 9th International Symposium on Heavy Vehicle Weights and Dimensions. Pennsylvania, USA (2006)
25. Imine, H., Dolcemascolo, V.: Rollover risk prediction of heavy vehicle in interaction with infrastructure. IJHVS, International Journal of Heavy Vehicle Systems 14(3), 294–307 (2007)
26. Imine, H., Dolcemascolo, V.: Sliding mode observers to heavy vehicle vertical forces estimation. IJHVS, International Journal of Heavy Vehicle Systems 15(1), 53–64 (2008)
27. Imine, H., Fridman, L., Shraim, H.: Identification of heavy vehicle suspension parameters and estimation of vertical forces. In: VSS 2012, International Workshop on Variable Structure Systems, Bombay, India, pp. 63–69 (2012)
28. Imine, H., Fridman, L., Sraim, H., Djemai, M.: Sliding Mode Based Analysis and Identification of Vehicle Dynamics. LNCIS, vol. 414. Springer, Heidelberg (2011)
29. Imine, H., Khemoudj, O.: Dynamic parameters identification and estimation of the vertical forces of heavy vehicle. In: SAE 2011 World Congress on Automotive, Detroit, USA (2011)
30. Imine, H., Madani, T., Srairi, S.: High order sliding mode observers to estimate vertical forces: experimental results. In: ITSC, IEEE Intelligent Transportation Systems Society, Beijing, China, pp. 523–527 (2008)
31. Johansson, B., Gafvert, M.: Untripped suv rollover detection and prevention. In: 43rd IEEE Conference on Decision and Control, Atlantis, Paradise Island, Bahamas, pp. 5461–5466 (2004)
32. Khemoudj, O., Imine, H., Djema, M.: Robust observation of tractor-trailer vertical forces using inverse model and exact differentiator. SAE, International Journal of Materials and Manufacturing 3(1), 278–289 (2010)
33. Khemoudj, O., Imine, H., Djemai, M., Fridman, L.: Variable gain sliding mode observer for heavy duty vehicle tyre forces estimation. In: 11th International Workshop on Variable Structure Systems, VSS 2010, pp. 522–527. Mexico city, Mexico (2010)
34. Legeay, V., Daburon, P., Gourraud, C.: Comparaison de mesures de l'uni par l'analyseur de profil en long et par compensation dynamique. Bulletin interne, Laboratoire Central des Ponts et Chaussées, DGER/IRVAR (1996)
35. Levant, A.: Sliding order and sliding accuracy in sliding mode control. International Journal of Control 58, 1247–1263 (1993)
36. Levant, A.: Robust exact differentiation via sliding mode technique. Automatica 34(3), 379–384 (1998)
37. Levant, A.: High-order sliding modes: differentiation and output-feedback control. International Journal of Control 76(9-10), 924–941 (2003)
38. Moreno, J.: Lyapunov Approach for Analysis and Design of Second Order Sliding Mode Algorithms, ch. 4, pp. 122–150. Springer (2011)

39. ONISR: La sécurité routiére en france. bilan de l'année 2004. La documentation franccaise, sécurité routiére, Observatoire National Interministériel de sécurité routiére (2005)
40. Soderstrom, T., Stoica, P.: System Identification. Prentice Hall International, Cambridge (1989)
41. Tianjun, Z., Zong, C., Zheng, H., Tian, C., Zheng, H.: Yaw/roll stability modeling analyses and control of heavy tractor-semitrailer. In: SAE Intl. Asia Pacific Automotive Eng. Conf. (2007)
42. Utkin, V.: Sliding Modes in Control and Optimization. Springer, Berlin (1992)
43. Vasiljevic, L.K., Khalil, H.K.: Error bounds in differentiation of noisy signals by high-gain observers. Systems and Control Letters 57, 856–862 (2008)

Chapter 17
Applications of Sliding Observers for FDI in Aerospace Systems

Christopher Edwards, Halim Alwi, and Prathyush P. Menon

Abstract. This chapter presents applications of second order sliding mode observer schemes to three different aerospace problems. Two relate to ADDSAFE aircraft fault detection benchmark problems. Firstly, the detection and isolation problem associated with an actuator jam/runaway is considered and secondly an actuator oscillatory failure case is tackled. For the actuator jam/runaway scenario the actuator deflection becomes decoupled from the demand issued from the flight control computer and either remains fixed at some uncommanded point or 'runs away' to an extreme value. For the OFC problem, the reconstruction scheme requires an estimate of rod speed provided by a second order sliding mode observer. Ideally low gains in the observer are required because of the noisy environment associated with the physical system. An adaption scheme is therefore required to retain sliding in the presence of severe faults. A problem associated with fault detection in a formation flying scenario, associated with satellites is also discussed. This application to a relative degree two problem would be difficult to solve using linear observer methods.

17.1 Introduction

The study of fault detection and isolation (FDI) problems has been a popular and widely researched area. One of the techniques which has gained a good deal of

---

Christopher Edwards
Center for Systems, Dynamics and Control, CEMPS, University of Exeter
e-mail: `c.edwards@exeter.ac.uk`

Halim Alwi
Control and Instrumentation Group, Department of Engineering, University of Leicester
e-mail: `ha18@le.ac.uk`

Prathyush P Menon
Center for Systems, Dynamics and Control, Mathematical Research Institute, University of Exeter
e-mail: `P.M.Prathyush@exeter.ac.uk`

B. Bandyopadhyay et al. (Eds.): *Advances in Sliding Mode Control*, LNCIS 440, pp. 341–360.
DOI: 10.1007/978-3-642-36986-5_17

attention in recent years is sliding mode based FDI. One of the reasons is due to its robustness properties, as well as its ability to reconstruct unknown signals (faults) which may affect the system being monitored. The earliest sliding mode FDI results relied on simple residual based ideas (see for example [14, 24]). The under pinning idea in [14, 24] is to allow sliding to break in the event of a fault, and the deviation of the output estimation error away from the sliding surface indicates that a fault has occurred. In the later developments (see for example Edwards *et al* [6], Tan & Edwards [22], Jiang *et al* [15] and Kim *et al* [16]), instead of achieving detection and isolation through residuals, reconstruction of the faults has been considered. In these approaches a sliding motion is always maintained even in the presence of faults. The direct reconstruction of faults can be beneficial, especially in the case when redundancy is not available, and for sensor fault tolerant control (see for example [1]).

One of the perceived drawbacks of using sliding mode schemes in physical systems is in dealing with the discontinuities which arise from using the signum function. However there has been extensive research to obviate these difficulties ranging from simple pseudo-sliding approximations achieved through smoothing, to more advanced higher order sliding mode concepts [5, 8, 9, 17, 18, 26]. Second order sliding methods require no smoothing and allow ideal sliding motions to be achieved and therefore preserve the robustness property of sliding modes. Furthermore the inherent filtering property is beneficial for systems with noise. A recent development in second order sliding mode approaches has made a significant impression due to the inclusion of Lyapunov analysis techniques to demonstrate convergence [4, 21]. These concepts will be used in this chapter.

The Advanced Fault Diagnosis for Sustainable Flight Guidance and Control (ADDSAFE) project is a European FP7 funded consortium. The aim of ADDSAFE is to demonstrate the applicability of advanced fault detection and diagnosis (FDD) methods to support the development of sustainable aircraft. It poses challenges to improve existing FDD techniques to support new 'green' technologies allowing optimization of the aircraft structural design, improving aircraft performance and reducing the environmental footprint [19].

Two applications of 2nd order sliding mode observer schemes on the ADDSAFE aircraft benchmark problem will be presented. The first problem is associated with subtle jams and offsets in actuators which are usually automatically compensated for by the flight control system which repositions the healthy surfaces. However this can still pose a problem because drag is increased which results in excessive fuel burn. Consequently it is important to detect these incipient problems.

The second FDI problem which will be considered concerns an actuator Oscillatory Failure Case (OFC). An OFC is a type of failure in the Electrical Flight Control Systems (EFCS). When coupled with the flexible modes of the structure, OFCs can generate resonance phenomenon and cause unacceptably high vibrations and loads [10] and therefore need to be detected quickly. In the context of ADDSAFE, (as discussed in [10]) the improvement in performance of the FDI scheme allows for better optimization of the aircraft structural design, which translates to weight savings and therefore less fuel burn and a lower environmental footprint.

The final case study relates to a fault detection problem in a leader follower satellite formation situation. In the follower satellite, since the relative distance to the leader satellite is small compared to the orbit, linearizations of the Hill equations can be used to model the dynamics of the follower. These will be used as the basis of an FDI scheme to detect actuator faults in the thruster systems of the follower spacecraft. This constitutes a relative degree two problem between the measurement and the fault signals. As such traditional linear unknown input observer methods cannot be employed.

The structure of the chapter is as follows: the next section considered the applications of a sliding mode based fault detection scheme to an actuator problem; then a super twisting differentiator based scheme is used to detect the presence of oscillatory faults associated with aircraft actuators; finally a fourth order observer based on second order sliding mode principles will be developed for a specific fault detection problem in a satellite formation scenario.

17.2 Actuator Jam Problem

For this benchmark problem, a local LPV actuator model will be used for design.

17.2.1 Modeling of Hydraulic Actuator Using LPV

For the ADDSAFE benchmark problem, a LPV model representation (provided by DLR ⟦13⟧), derived from the high fidelity actuator in the ADDSAFE benchmark model will be used for design. It has the form

$$\dot{x}(t) = -K(\rho)x(t) + K(\rho)u(t) \tag{17.1}$$

$$y(t) = x(t) + f_o(t) \tag{17.2}$$

where $x(t)$ represents the deflection of the actuator and $f_o(t)$ represents the additive fault in the actuator. In ⟦13⟧ the LPV parameters ρ chosen to describe the variation of the dynamics are

$$\rho = [\rho_1, \dots \rho_4] := [m, X_{cg}, h, V_{cas}] \tag{17.3}$$

which represent mass (m), center of gravity in the x-direction (X_{cg}), altitude h, and conventional airspeed V_{cas}. As shown in ⟦13⟧, the scalar $K(\rho) > 0$ for all ρ and varies according to

$$K(\rho) = C_a(\rho) + C_b(\rho)\,\mathrm{sign}(\dot{x}(t))(x(t) + C_c(\rho)) \tag{17.4}$$

The positive scalar $C_a(\rho)$ can be interpreted as the dominant nominal gain, $C_b(\rho)$ represents the effect of deflection angle $x(t)$ and $C_c(\rho)$ represents the effect of a position offset from the trim position ⟦13⟧. These coefficients have been obtained through an affine polynomial fit to data collected on the parameter grid of $m(kg) \in [120000,\ 220000]$, $X_{cg}(\%) \in [0.21,\ 0.38]$, $h(ft) \in [0,\ 37000]$ and $V_{cas}(kt) \in [154.6,\ 176.1,\ 190.5,\ 229.6,\ 275]$. Each coefficient can be represented by

$$C_a(\rho) = C_{a,0} + C_{a,1}\rho_1 + C_{a,2}\rho_2 + C_{a,3}\rho_3 + C_{a,4}\rho_4 \tag{17.5}$$
$$C_b(\rho) = C_{b,0} + C_{b,1}\rho_1 + C_{b,2}\rho_2 + C_{b,3}\rho_3 + C_{b,4}\rho_4 \tag{17.6}$$
$$C_c(\rho) = C_{c,0} + C_{c,1}\rho_1 + C_{c,2}\rho_2 + C_{c,3}\rho_3 + C_{c,4}\rho_4 \tag{17.7}$$

17.2.2 Sliding Mode Observer

From (17.2), the fault appears at the output of the actuator model and therefore a 'sensor fault' reconstruction scheme will be employed. Consider a new state $z_f(t) \in \mathbb{R}$ which is the filtered output of $y(t)$ given by

$$\dot{z}_f(t) = -A_f z_f(t) + A_f y(t) \tag{17.8}$$

where A_f is a positive design scalar. Substituting $y(t)$ from (17.2) into (17.8) yields

$$\dot{z}_f(t) = -A_f z_f(t) + A_f x(t) + A_f f_o(t) \tag{17.9}$$

Next, augment system (17.1) and (17.9) to create a 2nd order system

$$\dot{x}_a(t) = A_a(\rho)x_a(t) + B_a(\rho)u + F_a f_o(t) \tag{17.10}$$
$$z(t) = C_a x_a(t) \tag{17.11}$$

where the augmented states $x_a(t) = \begin{bmatrix} x(t) \ z_f(t) \end{bmatrix}^{\mathrm{T}}$ and the augmented matrices

$$A_a(\rho) = \begin{bmatrix} -K(\rho) & 0 \\ A_f & -A_f \end{bmatrix}, \quad B_a(\rho) = \begin{bmatrix} K(\rho) \\ 0 \end{bmatrix}, \tag{17.12}$$
$$F_a = \begin{bmatrix} 0 \\ A_f \end{bmatrix}, \quad C_a = \begin{bmatrix} 0 \ 1 \end{bmatrix}$$

For the system in (17.10), the proposed observer has the structure

$$\dot{\hat{x}}_a(t) = A_a(\rho)\hat{x}_a(t) + B_a(\rho)u - G_l e_z(t) + G_n \nu(t) \tag{17.13}$$
$$\hat{z}_f(t) = C_a \hat{x}_a(t) \tag{17.14}$$

where the output estimation error $e_z(t) = \hat{z}_f(t) - z_f(t)$. The design parameters $G_l(\rho), G_n(\rho) \in \mathbb{R}^{2\times 1}$ are the observer gains and $\nu(t)$ is the nonlinear term used to induce the sliding motion. Consider an error $e_a = \hat{x}_a - x_a$ then subtracting (17.10) from (17.13) yields

$$\dot{e}_a(t) = A_a(\rho)e_a(t) - G_l e_z(t) + G_n \nu(t) - F_a f_o(t) \tag{17.15}$$

The objective is to force $e_z(t)$ to zero in finite time, in order to achieve a sliding mode on $\mathscr{S}_{act} = \{e_a \in \mathrm{IR}^2 : e_z = 0\}$. Here the observer gains are chosen as

$$G_n = \begin{bmatrix} 0 \\ 1 \end{bmatrix} \quad G_l = \begin{bmatrix} 0 \\ -A_f + k_2 \end{bmatrix} \tag{17.16}$$

where k_2 is a chosen positive scalar. Substituting (17.16) into (17.15) the error system can be written in expended form as

$$\begin{bmatrix} \dot{e}_x(t) \\ \dot{e}_z(t) \end{bmatrix} = \begin{bmatrix} -K(\rho) & 0 \\ A_f & -k_2 \end{bmatrix} \begin{bmatrix} e_x(t) \\ e_z(t) \end{bmatrix} + \begin{bmatrix} 0 \\ 1 \end{bmatrix} \nu(t) - \begin{bmatrix} 0 \\ A_f \end{bmatrix} f_o(t) \tag{17.17}$$

From (17.17) the reduced order sliding motion is given by

$$\dot{e}_x(t) = -K(\rho) e_x \tag{17.18}$$

Since $K(\rho) > 0$ for all ρ [13], the reduced order sliding motion is stable and $e_x \to 0$ as $t \to 0$. From the lower equation in (17.17)

$$\dot{e}_z(t) = A_f e_x - k_2 e_z(t) + \nu(t) - A_f f_o(t) \tag{17.19}$$

During sliding $e_z = \dot{e}_z = 0$ and since $e_x(t) \to 0$, equation (17.19) reduces to

$$\nu(t) = A_f f_o(t) \tag{17.20}$$

Rearranging (17.20) an estimation of the fault is obtained as

$$\hat{f}_o(t) = A_f^{-1} \nu(t) \tag{17.21}$$

which can be reconstructed online. The nonlinear injection term $\nu(t)$ has the super twisting form

$$\nu(t) = -k_1 \,\mathrm{sign}(e_z(t)) |e_z(t)|^{1/2} + z(t) \tag{17.22}$$

$$\dot{z}(t) = -k_3 \,\mathrm{sign}(e_z(t)) - k_4 e_z(t) \tag{17.23}$$

The scalars k_1, k_3, k_4 are design freedom to be chosen. Note that (17.22)-(17.23) has a similar structure to the one in [21]. For a sufficiently large scalar $\varepsilon > |\dot{f}_o(t)|$, if the gains from (17.22) and (17.23) are chosen as

$$k_1 > 2\sqrt{\varepsilon} \tag{17.24}$$

$$k_3 > \varepsilon \tag{17.25}$$

$$k_4 > \frac{k_2^2 \left(k_1^3 + \frac{5}{4} k_1^2 + \frac{5}{2}(k_3 - \varepsilon) \right)}{k_1 (k_3 - \varepsilon)} \tag{17.26}$$

then $e_z(t) = \dot{e}_z(t) = 0$ in finite time [21].

17.2.3 Simulation

The scalar A_f which defines the output filter from (17.9) has been chosen as $A_f = 0.5$. The supertwist gains from (17.24)-(17.26) have been chosen as $k_1 = 6.6408, k_2 = 0.1, k_3 = 20, k_4 = 0.3279$.

The scheme proposed above has been tested on the nonlinear high fidelity ADDSAFE benchmark model [11]. The simulations are conducted at an altitude of 37000ft, a speed of 2267kts, a weight of 185 tonnes and center of gravity of 28% MAC. Figure 17.1 shows the results from a right elevator jam during a coordinated turn manoeuvre. Figure 1(b) shows that the right elevator jam which does not respond to the command signal. Figures 1(c) shows that sliding is being maintained despite the presence of the fault. Figure 1(c) also shows good reconstruction of the fault.

17.3 OFC Problem

In this section, an adaptive second order sliding mode observer algorithm will be used to estimate an actuator oscillatory failure case. The idea is to manipulate the analytical mathematical nonlinear model of the actuator to obtain an expression for the OFC signal. Most of the parameters used in the manipulated nonlinear equation are available, except for the actuator rod speed which will be supplied by the adaptive observer.

17.3.1 Modeling of Hydraulic Actuator

The hydraulic actuator model from [10] is given by

$$\dot{x}(t) = V_c(t)\left(\frac{\Delta_p(t) - \text{sign}(i(t))\frac{F_{aero}(t)}{S}}{\Delta_{P_{ref}} + \frac{K_d(t)}{S}V_c^2(t)}\right)^{\frac{1}{2}} \tag{17.27}$$

where nominally

$$V_c(t) = K_c i(t) \tag{17.28}$$

and K_c is a conversion factor from electrical current (mA) to speed (mm/s). The current $i(t)$ is given by

$$i(t) = K(u(t) - x(t)) \tag{17.29}$$

where K is the (fixed) servo control gain. The signal $x(t)$ is the hydraulic actuator rod position and $u(t)$ is the commanded rod position (from the FCC). The fixed constants are $\Delta_{P_{ref}}$ which is the differential pressure corresponding to the maximum

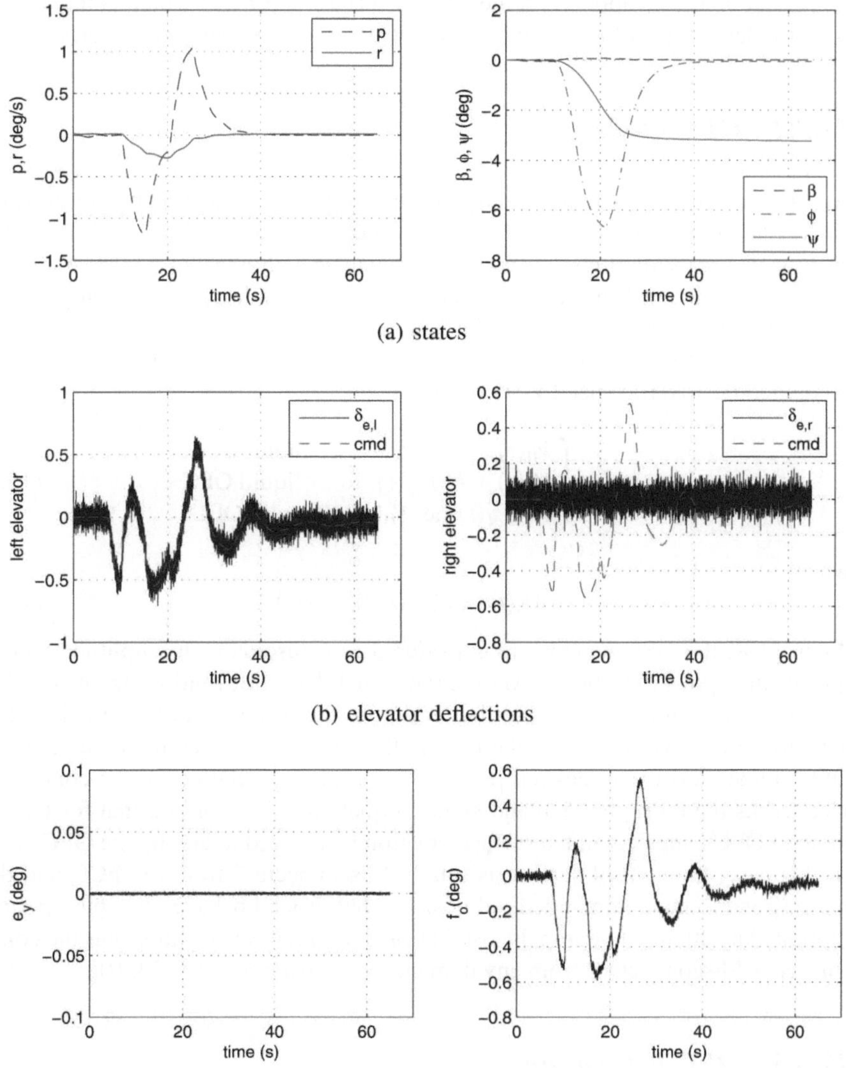

(a) states

(b) elevator deflections

(c) e_y and fault estimations

Fig. 17.1 Turn coordination: right elevator jam

rod speed[1], and S which is piston surface area. The parameters which depend on varying operational conditions (e.g. fluid temperature and/or the number of actuators used simultaneously on a given hydraulic circuit) are: $\Delta_p(t)$ which is the actual hydraulic pressure delivered to the actuator, $F_{aero}(t)$ which is the estimate of the

[1] Maximum rod speed is achieved when the servo valve is fully open and $\Delta_p(t) = \Delta_{p_{ref}}$ [10].

aerodynamic forces applied on the control surface and $K_d(t)$ which is the adjacent actuator damping coefficient (in the case of two actuators per control surface).

17.3.2 OFC Modeling

An OFC is caused by faults in any digital component (which generate unwanted sinusoidal signals) in the actuator control loop between the FCC and the control surface. These oscillations consequently propagate within the loop [10]. As in [10], only an OFC located in the servo control loop is considered. Specifically, it is assumed that the OFC source is in the analogic output signal between the FCC and the actuator (See Figure 17.2 below). In the ADDSAFE model, the OFC affects the computed/desired rod speed $V_c(t)$ so that

$$V_c(t) = \begin{cases} V_0(t) & \text{nominal} \\ V_0(t) + K_c f_{liq}(t) & \text{liquid OFC} \\ K_c f_{sol}(t) \text{ i.e. } V_0(t) = 0 & \text{solid OFC} \end{cases} \tag{17.30}$$

where

$$V_0(t) = K_c K(u(t) - x(t)) \tag{17.31}$$

As in [10], the OFC signals are considered as sinusoids with amplitude and frequency uniformly distributed over the range of 1-10Hz. Beyond 10Hz, the OFC has no effect on control surface oscillation due to the low pass characteristics of the actuator. As shown in (17.30) the liquid OFC behaves as an additive fault, and the OFC signal adds to the desired position from the FCC and hence the control surface tracks the corrupted demand signal. Equation (17.30) shows that for the case of solid OFCs, the demanded surface position is replaced totally by the OFC signal. In this case, the control surface is totally 'disconnected' from the FCC and does not respond to the commanded rod position, but instead behaves as a pure periodic motion. Any attempt to damp the oscillation does not have any impact as the control surface is 'disconnected' from any demand signal from the FCC [3, 10].

17.3.3 OFC Estimation

Consider equation (17.27) as a special case of the differential equation

$$\dot{x}(t) = g(t,x) \tag{17.32}$$

with measured output $y(t) = x(t)$. Assume that the time derivative of the function on the right hand side of (17.32) is bounded i.e.,

$$|\dot{g}(t,x)| \leq \delta \tag{17.33}$$

for some unknown constant $\delta > 0$.

Consider an observer with the following structure

$$\dot{z}_1(t) = -\kappa_1(t)|e_1(t)|^{1/2}\ \mathrm{sign}(e_1(t)) + z_2(t) \tag{17.34}$$
$$\dot{z}_2(t) = -\kappa_2(t)\ \mathrm{sign}(e_1(t)) \tag{17.35}$$

where $e_1(t) = z_1(t) - x(t)$. Subtracting (17.32) from (17.34) yields the error system

$$\dot{e}_1(t) = -\kappa_1(t)|e_1(t)|^{1/2}\ \mathrm{sign}(e_1(t)) + z_2(t) - g(t,x) \tag{17.36}$$
$$\dot{z}_2(t) = -\kappa_2(t)\ \mathrm{sign}(e_1(t)) \tag{17.37}$$

Consider $e_2(t) = z_2(t) - g(t,x)$, then (17.36)-(17.37) can be written as

$$\dot{e}_1(t) = -\kappa_1(t)|e_1(t)|^{1/2}\ \mathrm{sign}(e_1(t)) + e_2(t) \tag{17.38}$$
$$\dot{e}_2(t) = -\kappa_2(t)\ \mathrm{sign}(e_1(t)) - \dot{g}(t,x) \tag{17.39}$$

If a 2nd order sliding motion is induced, $e_1(t) = \dot{e}_1(t) = 0$, and therefore from (17.38), $e_2(t) = 0 \Rightarrow z_2(t) = g(t,x) = \dot{x}(t)$ and therefore $z_2(t)$ from (17.35) provides an estimate of rod speed $\dot{x}(t)$. Since both $x(t)$ and $\dot{x}(t)$ are known, under the three different conditions in (17.30), equation (17.27) can be rearranged to obtain an expression for the OFC. *For the liquid OFC case*

$$f_{liq}(t) = \frac{z_2(t)f(t) - V_0(t)}{K_c} \tag{17.40}$$

where

$$f(t) = \left(\frac{\Delta_{P_{ref}}}{\Delta_p(t) - \mathrm{sign}(i(t))\frac{F_{aero}(t)}{S} - z_2^2(t)\frac{K_d(t)}{S}} \right)^{\frac{1}{2}} \tag{17.41}$$

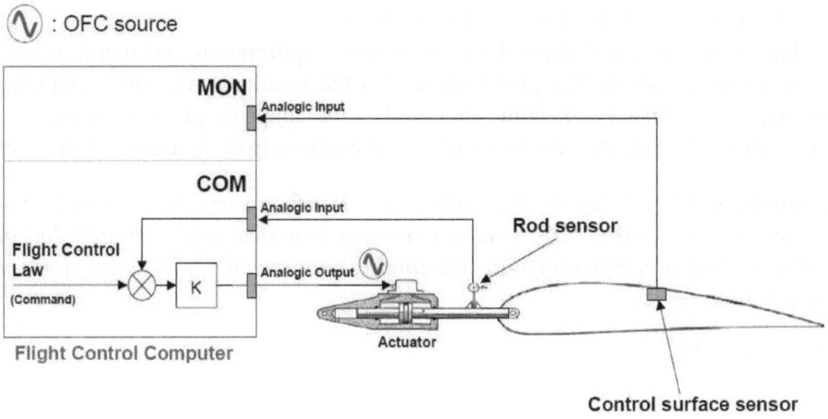

Fig. 17.2 Source of OFC in the servo control loop [10]

All the variables on the right hand side of (17.40) are available (i.e., measured or assumed to be fixed) except for the actuator rod speed. *For the solid OFC case*, since $V_0 = 0$ in (17.30), similar arguments give the estimate of the solid OFC as

$$f_{sol}(t) = \frac{z_2(t)f(t)}{K_c} \tag{17.42}$$

Here, the gains $\kappa_1(t)$ and $\kappa_2(t)$ are chosen as:

$$\kappa_1(t) = \sqrt{2\Gamma(t)} \tag{17.43}$$

$$\kappa_2(t) = 4\Gamma(t) \tag{17.44}$$

for some time varying scalar

$$\Gamma(t) = r(t) + \ell \tag{17.45}$$

where the variable ℓ is a fixed positive scalar while the varying $r(t)$ (also positive) is adapted based on the law

$$\dot{r}(t) = \begin{cases} \gamma D(|e_1(t)|^{1/2}) & \text{if } r(t) \leq r_{max} \\ 0 & \text{otherwise} \end{cases} \tag{17.46}$$

where $\gamma > 0$ is a positive design constant and the scalar $r_{max} \gg \delta$. The function $D(z) : \mathbb{R} \mapsto \mathbb{R}$ is the dead-zone function

$$D(z) = \begin{cases} 0 \text{ if } |z| < \varepsilon \\ z \text{ otherwise} \end{cases}$$

and ε is a positive scalar. The idea here is to adapt the gains when $|e_1(t)|^{1/2}$ unacceptably deviates from zero. The gain $r(t)$ will increase in magnitude according to (17.46) to force $e_1(t)$ back into a sliding regime.

The choice of $\gamma, \varepsilon, \ell$ depend on the system requirements and therefore require some design iteration. The gain ℓ represents the nominal gain when adaptation is not required, whilst γ will influence how fast the adaptive gain $r(t)$ increases. The parameter ε dictates the sensitivity of $r(t)$ to changes in $e_1(t)$ and is set to be small.

Proposition 17.1. *Using the adaptation rule (17.46) ensures the error system given in equations (17.38)-(17.39) and $r(t)$ remains bounded, and a pseudo 2nd order sliding motion is achieved in finite time forcing both e_1 and $\dot{e}_1$ to be small (depending on the choice of ε).*

Proof: See [2].

17.3.4 Simulations

The scheme above has been tested on the nonlinear high fidelity ADDSAFE benchmark model provided by AIRBUS [[12]]. The simulations are conducted at an altitude of 30000ft, Mach 0.64 (241kts), a weight of 200 tonnes and center of gravity of 30% MAC. In the benchmark model, the actuators are represented as high fidelity nonlinear models with parameters Δ_p, F_{aero} and K_d which vary based on changes in the operational conditions. For the observer design, these parameters are assumed to be fixed at constant values. Here, the control surface considered is the right inboard aileron. The design parameters from (17.45) and (17.46) used in the simulation are $\gamma = 3 \times 10^6$, $\ell = 50$ and $\varepsilon = 0.6$.

17.3.4.1 Simulation Results

Various OFC amplitudes and frequencies have been tested. For consistency and for comparison, all the tests were conducted using the same manoeuvre (a pilot longitudinal stick doublet input). For all tests, the OFC occurs at 10sec. For brevity, the results shown here represent the extreme cases of low and high amplitude and low and high frequency, to highlight the performance of the proposed scheme. The low amplitude case shows the smallest amplitude the scheme can detect, especially when masked by the noise in the system. The high frequency OFC case represents a challenge to detect the failure within the required time.

17.3.4.2 Solid OFC

Figure 17.3 shows an OFC of amplitude 0.5deg at a frequency of 0.5Hz. Figure 3(a) shows the effect of the OFC on the left elevator. Here, the OFC signal (blue solid line) totally replaces the commanded signal (red dashed line) and the elevator does not respond to the command signal from the FCC.

Figure 3(b) shows that there is no supertwist gain adaptation required for this level of OFC. Figure 3(c) shows both the estimated rod speed and the OFC. Here, again a good estimate of the rod speed is obtained as the estimate (blue solid line) overlaps the actual (red dashed line) rod speed. Subsequently, the good rod speed estimate provides a good OFC estimate.

17.3.4.3 Liquid OFC

Figure 17.4 shows the results for a liquid OFC with amplitude 1deg and a frequency of 7Hz. Due to the high frequency of the OFC, zoomed-in plots from 9-12sec are presented. Figure 4(b) shows that when the OFC occurs, the quality of sliding degrades and the gain $\Gamma(t)$ increases to regain sliding. Figure 4(c) shows the estimate of both the rod speed and the OFC. Again, during the period of sliding degradation, the rod speed and the OFC estimate slightly degrade, but quickly recover to provide

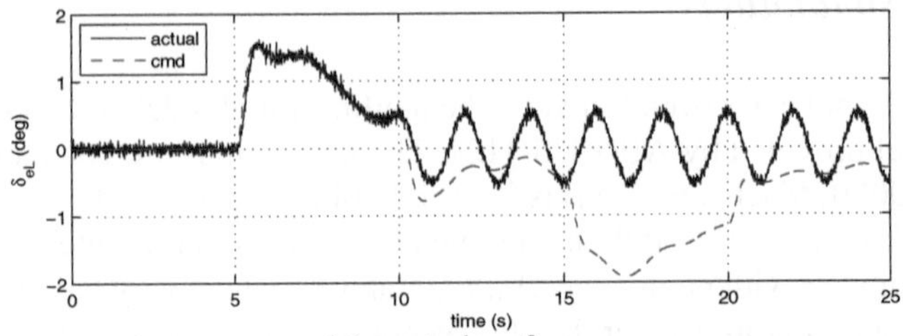

(a) control surface

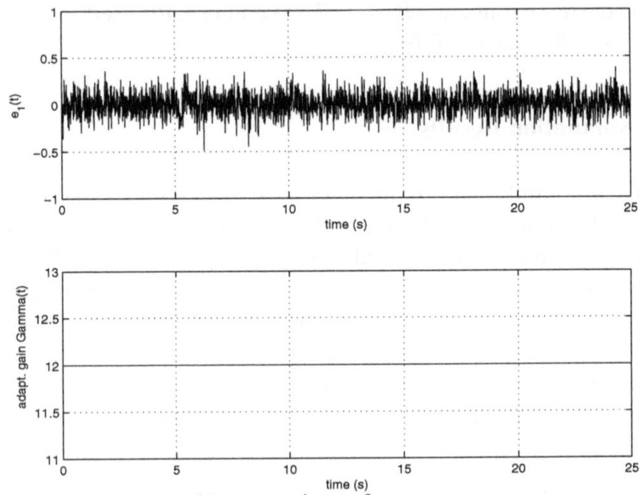

(b) supertwist performance

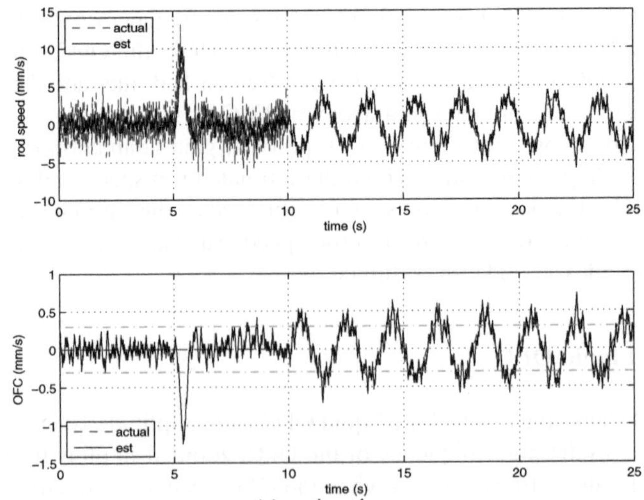

(c) estimations

Fig. 17.3 Solid OFC (amplitude 0.5, frequency 0.5)

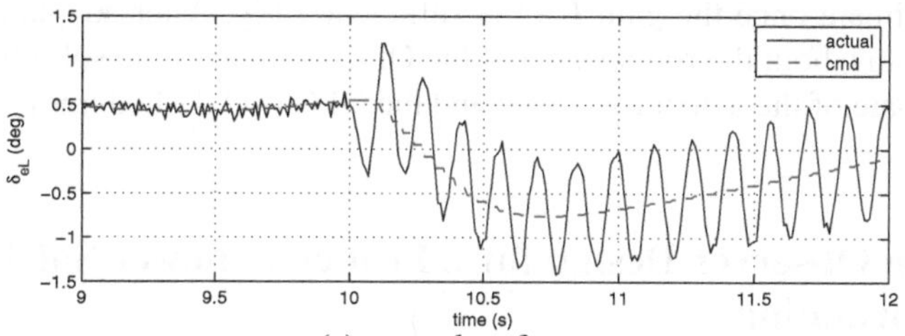

(a) control surface

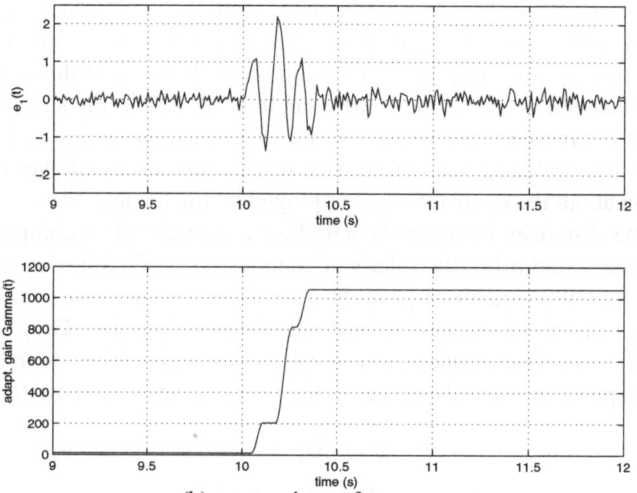

(b) supertwist performance

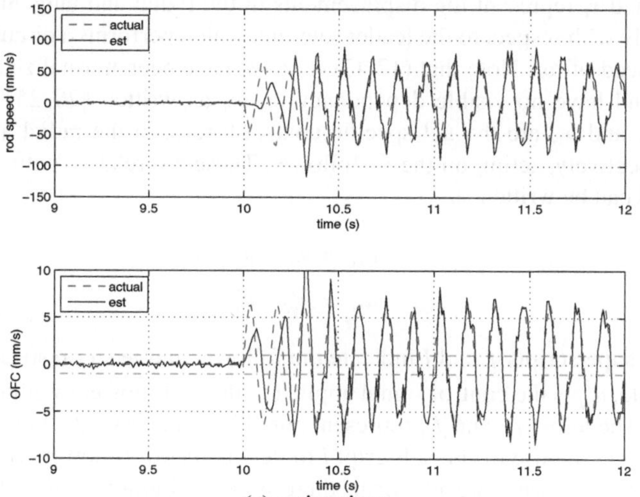

(c) estimations

Fig. 17.4 Liquid OFC (amplitude 1.0, frequency 7.0)

a good estimate once the gain $\Gamma(t)$ is sufficiently big. After adaptation of the supertwist gains, both the rod speed and the OFC estimate (blue solid line) provide a good estimate of the actual rod speed and the OFC (red dashed line).

17.4 An Observer Design for a Leader/Follower Satellite Formation

In this section a cluster of $N+1$ identical satellites, consisting of a leader satellite and N follower satellites, which are in nearby orbits, is considered. The leader satellite is on a circular Keplerian orbit and the follower satellites can measure the relative distance between all the nearby satellites as well as the leader satellite. The coupling effect between the attitude and translational dynamics of the satellites is assumed to be weak and is ignored. Also it is assumed the follower satellites have information about the control forces employed by the leader.

Since the distances between the satellites are small when compared to the diameter of the actual orbit, the relative dynamics of the ith follower satellite can be studied using Hill's equations [25]. In general, Hill's equations consist of relative dynamics in the radial, tangential and out-of-plane direction. Only the radial and tangential $(x-y)$ plane dynamics, which are coupled, is addressed in here. The Hill equations representing the dynamics in the $(x-y)$ plane can be written as:

$$\ddot{x}_i - 2\dot{y}_i - 3x_i = u_{xi} + f_{xi} \tag{17.47}$$

$$\ddot{y}_i + 2\dot{x}_i = u_{yi} + f_{yi} \tag{17.48}$$

where x_i and y_i represent the displacements in the radial and tangential directions respectively with respect to the leader satellite, which performs a circular orbit at an angular speed of ω_n. Note that (17.47) -(17.48) have been normalized with respect to time, and have no visible dependency on ω_n as written [20, 25]. The control signals u_{xi} and u_{yi} are the net specific control forces, in the radial and tangential plane respectively, acting on the ith follower. These are relative with respect to the leader and can be written as

$$u_{xi} = u^f_{xi} - u^l_{xi} \tag{17.49}$$

$$u_{yi} = u^f_{yi} - u^l_{yi} \tag{17.50}$$

where the superscripts f and l indicate the follower and leader respectively, and so for example, u^f_{xi} is the control signal applied to the ith follower satellite in the radial direction. The terms f_{xi} and f_{yi} represent possible actuator faults. For the remainder of this section the use of the subscript i to denote the ith follower will be dropped. Since all the followers are identical, there is no ambiguity in the absence of the subcript i.

To this end, for a typical follower satellite let

$$X = (x_1, x_2, x_3, x_4) = (x, \dot{x}, y, \dot{y}) \tag{17.51}$$

The nonlinear observer which is proposed here has its roots in the second order super twisting observer proposed in [7, 8, 21]. It will be designed to simultaneously robustly estimate the states and the unknown faults, $f = \mathrm{col}(f_x, f_y)$, from the measured relative position outputs (x_1, x_3) in each follower satellite.

Let the state estimate of the satellite be $\tilde{X} := \mathrm{col}(\tilde{x}_1, \tilde{x}_2, \tilde{x}_3, \tilde{x}_4)$. Consider the nonlinear observer dynamical system described by

$$\dot{\tilde{x}}_1 = \tilde{x}_2 - \tilde{k}_1 |e_1|^{\frac{1}{2}} \,\mathrm{sign}(e_1) \tag{17.52}$$

$$\dot{\tilde{x}}_2 = 3\tilde{x}_1 + 2\tilde{x}_4 - \tilde{k}_3 \,\mathrm{sign}(e_1) - \tilde{k}_2 |e_3|^{\frac{1}{2}} \,\mathrm{sign}(e_3) + u_x \tag{17.53}$$

$$\dot{\tilde{x}}_3 = \tilde{x}_4 - \tilde{k}_2 |e_3|^{\frac{1}{2}} \,\mathrm{sign}(e_3) \tag{17.54}$$

$$\dot{\tilde{x}}_4 = -2\tilde{x}_2 - \tilde{k}_4 \,\mathrm{sign}(e_3) + \tilde{k}_1 |e_1|^{\frac{1}{2}} \,\mathrm{sign}(e_1) + u_y \tag{17.55}$$

where: $e = \tilde{X} - X$, such that $e = \mathrm{col}(e_1, e_2, e_3, e_4)$. The $\tilde{k}_i \in \mathrm{I\!R}^+$, $i = 1, \ldots, 4$ represent the positive design scalar gains to be determined. This will be discussed in the sequel. When compared to the classical super-twisting observer proposed in [8], additional significant cross coupling terms $-\tilde{k}_2 |e_3|^{\frac{1}{2}} \,\mathrm{sign}(e_3)$ and $+\tilde{k}_1 |e_1|^{\frac{1}{2}} \,\mathrm{sign}(e_1)$ are present in (17.53) and (17.55). Furthermore the skew symmetry in the coupling of the states in the satellite dynamics is exploited in proposing the new nonlinear observer. The proposed nonlinear observer will be analyzed making use of the class of Lyapunov function originally proposed in [21].

The error in the state estimate of the satellite is

$$\dot{e}_1 = -\tilde{k}_1 |e_1|^{\frac{1}{2}} \,\mathrm{sign}(e_1) + e_2 \tag{17.56}$$

$$\dot{e}_2 = 3e_1 + 2e_4 - \tilde{k}_3 \,\mathrm{sign}(e_1) - \tilde{k}_2 |e_3|^{\frac{1}{2}} \,\mathrm{sign}(e_3) - f_x \tag{17.57}$$

$$\dot{e}_3 = -\tilde{k}_2 |e_3|^{\frac{1}{2}} \,\mathrm{sign}(e_3) + e_4 \tag{17.58}$$

$$\dot{e}_4 = -2e_2 - \tilde{k}_4 \,\mathrm{sign}(e_3) + \tilde{k}_1 |e_1|^{\frac{1}{2}} \,\mathrm{sign}(e_1) - f_y \tag{17.59}$$

The proposed design ensures the convergence of the error dynamics associated with the estimates of the states to zero *in finite time*.

It is assumed that the unknown faults f_x and f_y in the error dynamics satisfy a-priori known upper bounds. Specifically suppose $|f_x| \leq \delta_1$ and $|f_y| \leq \delta_2$ for known constants $\delta_1, \delta_2 \geq 0$. This assumption is similar to the one made in [20, 25].

Consider a candidate Lyapunov function $V(e)$ for the error dynamics system in (17.56) - (17.59), which is inspired by the one in [21] given by:

$$\begin{aligned} V(e) &= 2\tilde{k}_3 |e_1| + \frac{1}{2} e_2^2 + \frac{1}{2} (\tilde{k}_1 |e_1|^{\frac{1}{2}} \,\mathrm{sign}(e_1) - e_2)^2 \\ &+ 2\tilde{k}_4 |e_3| + \frac{1}{2} e_4^2 + \frac{1}{2} (\tilde{k}_2 |e_3|^{\frac{1}{2}} \,\mathrm{sign}(e_3) - e_4)^2 \end{aligned} \tag{17.60}$$

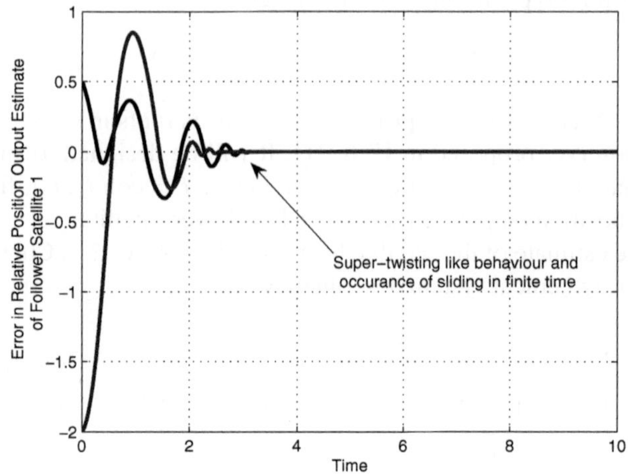

Fig. 17.5 Output estimatation errors

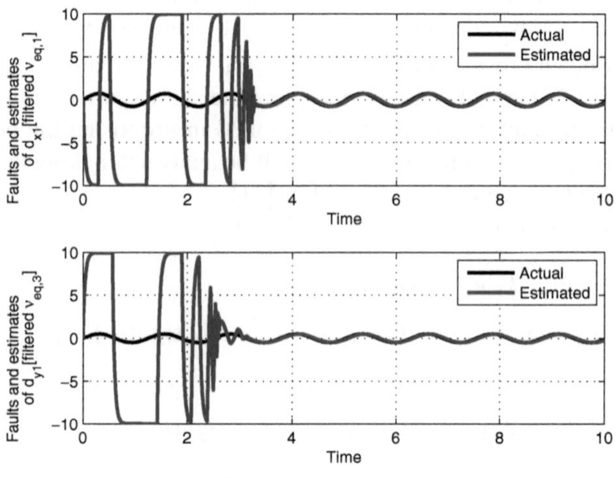

Fig. 17.6 Estimates in disturbances

For simplicity, the proposed candidate Lyapunov function can be written as a quadratic form $V(\xi) = \xi^{\mathrm{T}} P \xi$ where $\xi := \mathrm{col}(\xi_1, \xi_2)$ and $\xi_1 := \mathrm{col}(|e_1|^{\frac{1}{2}} \,\mathrm{sign}(e_1),\ e_2)$ and $\xi_2 := \mathrm{col}(|e_3|^{\frac{1}{2}} \,\mathrm{sign}(e_3),\ e_4)$. The block diagonal Lyapunov matrix

$$P = \begin{bmatrix} P_1 & \mathbf{0}_{2\times 2} \\ \mathbf{0}_{2\times 2} & P_2 \end{bmatrix} \tag{17.61}$$

where

$$P_1 = \frac{1}{2}\begin{bmatrix} 4\tilde{k}_3 + \tilde{k}_1^2 & -\tilde{k}_1 \\ -\tilde{k}_1 & 2 \end{bmatrix}, \quad P_2 = \frac{1}{2}\begin{bmatrix} 4\tilde{k}_4 + \tilde{k}_2^2 & -\tilde{k}_2 \\ -\tilde{k}_2 & 2 \end{bmatrix}$$

is radially unbounded if $\tilde{k}_3 > 0$ and $\tilde{k}_4 > 0$. It can be shown that the time derivative of $V(\xi)$ along the trajectories of the system (17.56) - (17.59) is given by

$$\dot{V}(\xi) \leq -\frac{1}{|e_1|^{\frac{1}{2}}}\xi_1^{\mathrm{T}}\tilde{Q}_1\xi_1 - \frac{1}{|e_3|^{\frac{1}{2}}}\xi_2^{\mathrm{T}}\tilde{Q}_2\xi_2 \tag{17.62}$$

where

$$\tilde{Q}_1 = \frac{\tilde{k}_1}{2}\begin{bmatrix} 2\tilde{k}_3 + \tilde{k}_1^2 - 2\delta_1 & -\tilde{k}_1 - 2\frac{\delta_1}{\tilde{k}_1} \\ -\tilde{k}_1 - 2\frac{\delta_1}{\tilde{k}_1} & 1 \end{bmatrix}$$

and

$$\tilde{Q}_2 = \frac{\tilde{k}_2}{2}\begin{bmatrix} 2\tilde{k}_4 + \tilde{k}_2^2 - 2\delta_2 & -\tilde{k}_2 - 2\frac{\delta_2}{\tilde{k}_1} \\ -\tilde{k}_2 - 2\frac{\delta_2}{\tilde{k}_1} & 1 \end{bmatrix}$$

Note that significant algebraic manipulation is necessary to achieve the structure in (17.62) because although $V(\xi)$ and $\dot{V}(\xi)$ present a decoupled block structure as given in (17.60) and (17.62), the differential equations in (17.56)-(17.59) are coupled. In achieving (17.62) the skew symmetry of the satellite plant and the additional coupling terms have been exploited.

In this situation $\dot{V}(\xi)$ is negative definite if $\tilde{Q}_1$ and $\tilde{Q}_1$ are positive definite. Provided the scalar positive gains $\tilde{k}_i$, for $i = 1,\ldots,4$, satisfy the following conditions

$$\tilde{k}_1 > 0, \quad \tilde{k}_3 > 3\delta_1 + 2\frac{\delta_1^2}{\tilde{k}_1^2} \tag{17.63}$$

$$\tilde{k}_2 > 0, \quad \tilde{k}_4 > 3\delta_2 + 2\frac{\delta_2^2}{\tilde{k}_2^2} \tag{17.64}$$

$\tilde{Q}_1$ and $\tilde{Q}_2$ are positive definite and consequently $\dot{V}(\xi)$ is negative definite for all $\xi \neq 0$ and $t > 0$. Exploiting the very specific block diagonal structure of the Lyapunov matrix in (17.61), rewrite the quadratic Lyapunov function in (17.60) as

$$V(\xi) := \underbrace{\xi_1^{\mathrm{T}}P_1\xi_1}_{V_1(\xi_1)} + \underbrace{\xi_2^{\mathrm{T}}P_2\xi_2}_{V_2(\xi_2)} \tag{17.65}$$

The functions $V_1(\xi_1)$ and $V_2(\xi_2)$ are positive definite with respect to ξ_1 and ξ_2 respectively. Then following identical arguments to those in [21], the inequality in (17.62) can be written as

$$\dot{V}(\xi) \leq -\frac{1}{|e_1|^{\frac{1}{2}}}\gamma_{min}(\tilde{Q}_1)\|\xi_1\|_2^2 - \frac{1}{|e_3|^{\frac{1}{2}}}\gamma_{min}(\tilde{Q}_2)\|\xi_2\|_2^2 \tag{17.66}$$

As argued in [21], inequality (17.66) can further be written as

$$\dot{V}(\xi) \leq -\beta_1 V_1^{\frac{1}{2}}(\xi_1) - \beta_2 V_2^{\frac{1}{2}}(\xi_2) \tag{17.67}$$

where $\beta_1 = \frac{\gamma_{min}^{\frac{1}{2}}(P_1)\gamma_{min}(\tilde{Q}_1)}{\gamma_{max}(P_1)}$ and $\beta_2 = \frac{\gamma_{min}^{\frac{1}{2}}(P_2)\gamma_{min}(\tilde{Q}_2)}{\gamma_{max}(P_2)}$ and thus it follows that

$$\dot{V}(\xi) \leq -\beta(V_1^{\frac{1}{2}}(\xi_1) + V_2^{\frac{1}{2}}(\xi_2)) \tag{17.68}$$

where $\beta = \mathtt{min}(\beta_1, \beta_2)$. Since $(V_1^{\frac{1}{2}} + V_2^{\frac{1}{2}})^2 > V_1 + V_2$, because V_1 and V_2 are positive, it can be concluded that $V_1^{\frac{1}{2}} + V_2^{\frac{1}{2}} > V^{\frac{1}{2}}$. This further implies that

$$\dot{V}(\xi) \leq -\beta V^{\frac{1}{2}} \tag{17.69}$$

and hence $V(\xi) \equiv 0$ in *finite time*. As argued above, the origin $e = 0$ is attained in *finite time*. Substituting for $e \equiv 0$ in (17.57) and (17.59) yields

$$\underbrace{\tilde{k}_3 \,\mathrm{sign}(e_1)}_{\nu_1} - f_x = 0 \tag{17.70}$$

$$\underbrace{\tilde{k}_4 \,\mathrm{sign}(e_3)}_{\nu_3} - f_y = 0 \tag{17.71}$$

Therefore $\nu_{eq,1} := f_x$ and $\nu_{eq,3} := f_y$, where $\nu_{eq,*}$ denotes the equivalent injection signals [23] necessary to maintain sliding. Thus f_x and f_y can be obtained to good accuracy by low pass filtering of ν_1 and ν_3 [23].

The following simulation shows the filtered injection signals tracking unknown sinusoidal faults/disturbances within the system. Figure 17.5 shows super-twisting-like performance. One the state estimation errors become zero after approximately 3 seconds, tracking of the unknown sinusoidal faults occurs.

17.5 Conclusions

This chapter has presented the application of second order sliding mode observer schemes to the ADDSAFE benchmark problem and a satellite formation flying problem. Two different FDD problems have been considered: firstly the detection and isolation problem associated with an actuator jam/runaway, and secondly an OFC scenario associated with the aileron actuators. Simulation results based on the full nonlinear model of the ADDSAFE aircraft, using a highly detailed model of the right inboard aileron actuator have been carried out. Both liquid and solid OFC cases have been considered. The results show good estimates of both the actuator rod speed and the OFC. A problem associated with fault detection in a formation flying scenario, associated with satellites has also been discussed. This application to a relative degree two problem would be difficult to solve using linear unknown input observer methods.

References

1. Alwi, H., Edwards, C.: Fault detection and fault-tolerant control of a civil aircraft using a sliding-mode-based scheme. IEEE Transactions on Control Systems Technology 16(3), 499–510 (2008)
2. Alwi, H., Edwards, C.: Oscillatory failure case detection for aircraft using an adaptive sliding mode differentiator scheme. In: American Control Conference, San Francisco, California, USA (2011)
3. Besch, H.M., Giesseler, H.G., Schuller, J.: Impact of electronic flight control system (EFCS) failure cases on structural design loads. Agard report 815, loads and requirements for military aircraft (1996)
4. Dávila, A., Moreno, J.A., Fridman, L.: Variable Gains Super-Twisting Algorithm: A Lyapunov Based Design. In: IEEE American Control Conference, pp. 968–973 (2010)
5. de Jager, B.: Comparison of methods to eliminate chattering and avoid steady state errors in sliding mode digital control. In: Proceedings of the IEEE VSC and Lyapunov Workshop, Sheffield, pp. 37–42 (1992)
6. Edwards, C., Spurgeon, S.K., Patton, R.J.: Sliding mode observers for fault detection. Automatica 36, 541–553 (2000)
7. Fridman, L., Davila, J., Levant, A.: Second-order sliding modes observer for mechanical systems. IEEE Trans. Autom. Control 50, 1785–1789 (2005)
8. Fridman, L., Davila, J., Levant, A.: High-order sliding-mode observation and fault detection. In: Proceedings of the Conference on Decision and Control, New Orleans, U.S.A., pp. 4317–4322 (2007)
9. Fridman, L., Levant, A.: Higher order sliding modes. In: Perruquetti, W., Barbot, J.P. (eds.) Sliding Mode Control in Engineering, pp. 53–96. Marcel Dekker, New York (2002)
10. Goupil, P.: Oscillatory failure case detection in the A380 electrical flight control system by analytical redundancy. Control Engineering Practice 18(9), 1110–1119 (2010)
11. Goupil, P., Marcos, A.: Advanced diagnosis for sustainable flight guidance and control: The european addsafe project. SAE Technical Paper 2011-01-2804 (2011)
12. Goupil, P., Puyou, G.: A high fidelity AIRBUS benchmark for system fault detection and isolation and flight control law clearance. In: European Conference for AeroSpace Sciences (EUCASS 2011) (2011)
13. Hecker, S.: Nominal and faulty LFT/LPV models. ADDSAFE report D1.3.2-3, DLR (2010)
14. Hermans, F.J.J., Zarrop, M.B.: Sliding mode observers for robust sensor monitoring. In: Proceedings of the 13th IFAC World Congress, pp. 211–216 (1996)
15. Jiang, B., Staroswiecki, M., Cocquempot, V.: Fault estimation in nonlinear uncertain systems using robust sliding–mode observers. IEE Proceedings: Control Theory & Applications 151, 29–37 (2004)
16. Kim, Y.W., Rizzoni, G., Utkin, V.: Developing a fault tolerant power train system by integrating the design of control and diagnostics. International Journal of Robust and Nonlinear Control 11, 1095–1114 (2001)
17. Levant, A.: Higher-order sliding modes, differentiation and output-feedback control. International Journal of Control 76(9-10), 924–941 (2003)

18. Luo, N.S., Feng, C.B.: A new method for suppressing chattering in variable structure feedback control systems. In: Nonlinear Control Systems Design: Science Papers of the IFAC Symposium, pp. 279–284. Pergamon, Oxford (1989)
19. Marcos, A.: Advanced fault diagnosis for sustainable flight guidance and control. In: 6th European Aeronautics Days, AERODAYS, Madrid, Spain (2011)
20. Massey, T., Shtessel, Y.: Continuous traditional and high order sliding modes for satelite formation control. AIAA Journal of Guidance Control and Dynamics 28(4), 826–831 (2005)
21. Moreno, J.A., Osorio, M.: A Lyapunov approach to second-order sliding mode controllers and observers. In: 47th IEEE Conference on Decision and Control, pp. 2856–2861 (2008)
22. Tan, C.P., Edwards, C.: Sliding mode observers for robust detection and reconstruction of actuator and sensor faults. International Journal of Robust and Nonlinear Control 13, 443–463 (2003)
23. Utkin, V.I.: Sliding Modes in Control Optimization. Springer, Berlin (1992)
24. Yang, H., Saif, M.: Fault detection in a class of nonlinear systems via adaptive sliding observer. In: Proceedings of the IEEE International Conference on Systems, Man and Cybernetics, pp. 2199–2204 (1995)
25. Yeh, H.H., Nelson, E., Sparks, A.: Nonlinear tracking control for satellite formations. AIAA Journal of Guidance Control and Dynamics 25(2), 376–386 (2002)
26. Young, K.K.D., Drakunov, S.V.: Sliding mode control with chattering reduction. In: Proceedings of the IEEE VSC and Lyapunov Workshop, pp. 188–190 (1992)

Chapter 18
Switching DSM Control of Perishable Inventory Systems with Delayed Shipments and Uncertain Demand

Przemyslaw Ignaciuk and Andrzej Bartoszewicz

Abstract. In this chapter, the concept of discrete sliding modes (DSMs) is applied to design an efficient supply policy for a class of perturbed processes with delay – goods flow control in supply chain. In the considered systems, the stock used to satisfy the unknown, time-varying demand placed at a goods distribution center; is replenished with delay from a remote supply source. The order quantity is fixed, leaving the time between the consecutive orders as a decision variable, which perfectly suits the switching nature of input signals obtained in DSM control systems. It is shown that under the proposed nonlinear policy, the stock level does not exceed the assigned storage space. Moreover, it is also demonstrated that the stock is never entirely depleted, which guarantees full demand satisfaction and maximum service level.

18.1 Introduction

Intense competition and high level of uncertainty stimulate the search for new supply policies in modern production and goods distribution systems. The improvements are particularly desired in the systems with long delivery times (companies operating on large geographical areas) and those subject to significant demand fluctuations. As recently discussed in papers [17, 19, 20], a viable approach to

Przemyslaw Ignaciuk
Institute of Information Technology, Lodz University of Technology, 215 Wólczańska St., 90-924 Łódź
e-mail: przemyslaw.ignaciuk@p.lodz.pl

Andrzej Bartoszewicz
Institute of Automatic Control, Lodz University of Technology, 18/22 Stefanowskiego St., 90-924 Łódź
e-mail: andrzej.bartoszewicz@p.lodz.pl

B. Bandyopadhyay et al. (Eds.): *Advances in Sliding Mode Control*, LNCIS 440, pp. 361–379.
DOI: 10.1007/978-3-642-36986-5_18

solving inventory control problem may be the application of formal, control-theoretic methods.

In this work, the systems in which the stock used to fulfill uncertain, variable demand imposed on a goods distribution center; deteriorates with time are considered. In the analyzed setting, it is assumed that the demand may follow any statistical distribution, and the replenishment orders are realized with non-negligible delay from a remote supply source. In order to appropriately respond to unknown demand fluctuations and to ensure stable and efficient system operation, the inherently robust control technique of sliding-mode control ⟦22⟧ is proposed. Actually, to better reflect the behavior of real inventory systems with periodic state review, the concepts of discrete sliding-mode (DSM) control ⟦1–3, 5, 14, 15⟧ are applied.

Looking at how the ordering signal evolves in time, two types of inventory systems are most commonly encountered in practice. In the first category, in addition to high efficiency and low operational costs, the primary objective is to ensure smooth ordering pattern. For this class of processes, the sliding surface may be chosen by solving dynamical optimization problem with quadratic performance index, as in ⟦7⟧. This work focuses on the second large category of inventory systems, in which the order quantity is fixed, and the instances of issuing the orders are to be determined as a result of the control action. The switching type of input signals, typically found in the traditional sliding-mode controller design is particularly well suited for this class of systems. Thus, as opposed to our earlier results in the field ⟦7–9⟧, which considered only the traditional logistic systems with non-decaying inventories, here, the more complex class of processes with deteriorating stock ⟦6, 13, 16, 18⟧ is investigated. Moreover, in contrast to ⟦7–12⟧ which refer to systems with variable order quantity, here, a nonlinear controller that determines the instances when the order of constant quantity should be placed, is proposed. This novel DSM policy with switching input signal is shown to guarantee finite, precisely determined stock level and full satisfaction of the *a priori* unknown demand for arbitrary order procurement delay. The key factor in achieving these favorable characteristics is careful selection of the sliding surface. In the system considered, the surface is chosen to incorporate the delay compensating features. The designed control strategy requires smaller warehouse capacity than the controllers proposed in ⟦7, 8, 10, 12⟧, thus offering a less costly solution if the penalty for abrupt ordering signal transitions is of little concern. Due to the careful treatment of the combined effect of delay and goods decay, it also outperforms the traditional $(r,\ Q)$ policy in terms of smaller overshoots and the resulting decreased storage space requirements.

The chapter is organized in the following way. First, in Section 18.2, the model of perishable inventory system is presented with the emphasis placed on the issues related to delay. Next, in Section 18.3, the sliding surface is designed and the DSM control policy is formulated. The properties of the proposed control system are analyzed and substantiated with formal proofs. Section 18.4 is devoted to numerical studies, whereas Section 18.5 comprises the conclusions.

18.2 Problem Statement

We consider the inventory system involving interactions among the three principal actors in supply chain: customers, distribution center, and a supply source. The system is illustrated in Fig. 18.1. The customers impose demand on the distribution center, which refills the stock from the shipments ordered at the supplier. The solid lines in Fig. 18.1 represent the flow of goods and the dashed ones reflect the flow of information. The problem addressed in this paper is to specify an ordering rule that will ensure stable flow of deteriorating goods in the presence of highly uncertain demand, when the instances of issuing the orders vary with time. It is desired to order sufficient amount of goods to satisfy the market demand, yet avoid excessive orders that might increase the purchase and holding costs, and would lead to losses due to decay. The principal obstacle in meeting balanced, cost efficient system operation is the delay between placing of an order and goods arrival at the center. The controller design should account for both the destabilizing effect of this non-negligible delay and stock deterioration while waiting for the shipment arrival.

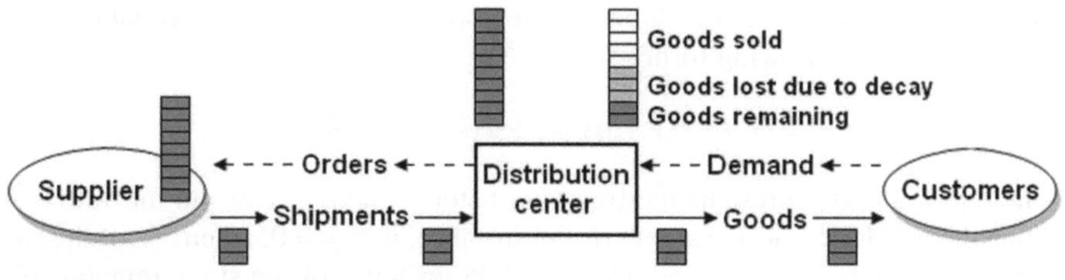

Fig. 18.1 Flow of goods and information

The model of the considered system is illustrated in Fig. 18.2. The stock level is reviewed at regular intervals kT, where T is the review period and $k = 0, 1, 2, \cdots$. The decision about the order placement is taken on the basis of the on-hand stock (the stock currently stored in the warehouse) $y(kT)$, the target stock level y_d, and the history of previous orders. The orders placed at the supplier are realized with delay L_p, assumed to be a multiple of the review period, i.e. $L_p = n_pT$, where n_p is a positive integer. The saturating integrator in the internal loop represents the operation of accumulating the stock of perishables characterized by decay factor σ, $0 \leq \sigma < 1$. In order to simplify the notation, further in the text k will be used as the independent variable in place of kT.

The demand (the number of items requested from inventory in period k) is modeled as an *a priori* unknown, bounded function of time $d(k)$,

$$0 \leq d(k) \leq d_{\max}. \tag{18.1}$$

The function $d(\cdot)$ reflects an arbitrary stochastic process with known, or uncertain parameters. If there is a sufficient number of items in the warehouse to satisfy the current demand, then the actually met demand $h(k)$ (the number of items sold to the

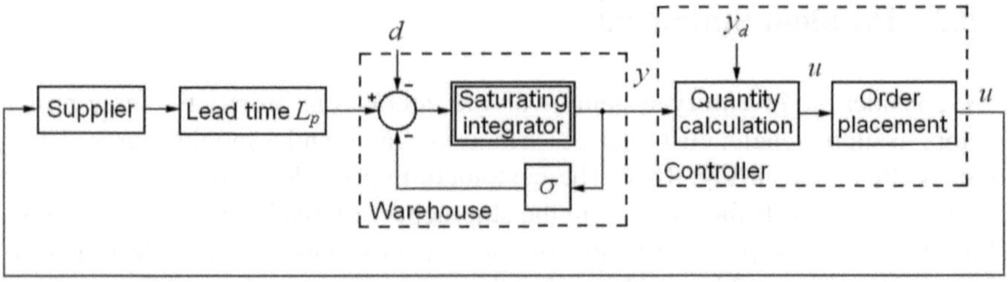

Fig. 18.2 Supply system model

customers or sent to the retailers in the distribution network) will be equal to the requested one. Otherwise, the imposed demand is satisfied only from the arriving shipments, and the surplus demand is lost (it is assumed that the sales are not back-ordered, and the excessive demand is equivalent to a missed business opportunity). Thus, one may write

$$0 \leq h(k) \leq d(k) \leq d_{\max}. \tag{18.2}$$

For the considered system with perishable goods the stock balance equation can be presented in the following form

$$y(k+1) = \rho y(k) + u(k-n_p) - h(k), \tag{18.3}$$

where $\rho = 1 - \sigma$ represents the fraction of stock which remains in the warehouse when the goods deteriorate at rate σ. For instance, if $\sigma = 0.05$, then 5% of the stock perishes in each review period and $\rho = 0.95$, or 95%, of the stock remains. Note that since $0 \leq \sigma < 1$ we have $0 < \rho \leq 1$.

It is assumed that the warehouse is initially empty, i.e. $y(0) = 0$, and the first order is placed at $k = 0$ ($u(k) = 0$ for $k < 0$). Because of the presence of delay, the first order arrives at the distribution center in period n_p, and thus $y(k) = 0$ for $k \leq n_p$. It is also assumed that the goods reach the distribution center new and deteriorate while kept in the on-hand stock. Taking into account the zero initial conditions, the stock level in arbitrary period $k \geq 0$ may be calculated from the following equation

$$\begin{aligned} y(k) &= \sum_{j=0}^{k-1} \rho^{k-1-j} u(j-n_p) - \sum_{j=0}^{k-1} \rho^{k-1-j} h(j) \\ &= \sum_{j=-n_p}^{k-n_p-1} \rho^{k-n_p-1-j} u(j) - \sum_{j=0}^{k-1} \rho^{k-1-j} h(j). \end{aligned} \tag{18.4}$$

Moreover, since $u(k) = 0$ for $k < 0$, one gets

$$y(k) = \sum_{j=0}^{k-n_p-1} \rho^{k-n_p-1-j} u(j) - \sum_{j=0}^{k-1} \rho^{k-1-j} h(j). \tag{18.5}$$

18.3 DSM Inventory Control Policy

In order to provide a fast reaction to the changes of market conditions and ensure high level of robustness, it is proposed to apply DSM control. A standard procedure of sliding-mode controller design is adopted, which consists of two phases. First, a switching function $s(k)$, which determines the overall system dynamical properties, is selected. Afterwards, in the second step, a control law is chosen to bring the system representative point onto the surface $s(k) = 0$, and to maintain the point in the vicinity of this surface afterwards in spite of the presence of parametric uncertainties and external disturbances.

18.3.1 Switching Function Design

A key point in the design of sliding-mode controllers is selection of an appropriate switching function [2, 4]. Due to the presence of non-negligible delay, in order to provide stable and fast system response, the switching function should incorporate a dead-time compensation mechanism. Here, a direct design approach is proposed, in which the dead-time compensator and the sliding surface parameters are selected in a single step. For this purpose, the delayed system dynamics is represented in an extended state space

$$\mathbf{x}(k+1) = \mathbf{A}\mathbf{x}(k) + \mathbf{b}u(k) + \mathbf{v}h(k), \tag{18.6}$$

where $\mathbf{x}(k) = [x_1(k) \quad x_2(k) \ \ldots \ x_n(k)]^T$ is the state vector with $x_1(k) = y(k)$ reflecting the on-hand stock level in period k, and the other state variables are chosen to represent the delayed input signal

$$x_j(k) = u(k-n+j-1) \tag{18.7}$$

for any $j = 2, 3, \ldots, n$. In (18.6), $\mathbf{A}$ is $n \times n$ state matrix, $\mathbf{b}$ and $\mathbf{v}$ are $n \times 1$ vectors

$$\mathbf{A} = \begin{bmatrix} \rho & 1 & 0 & \ldots & 0 \\ 0 & 0 & 1 & \ldots & 0 \\ \vdots & \vdots & \vdots & \ddots & \vdots \\ 0 & 0 & 0 & \ldots & 1 \\ 0 & 0 & 0 & \ldots & 0 \end{bmatrix}, \mathbf{b} = \begin{bmatrix} 0 \\ 0 \\ \vdots \\ 0 \\ 1 \end{bmatrix}, \mathbf{v} = \begin{bmatrix} -1 \\ 0 \\ \vdots \\ 0 \\ 0 \end{bmatrix}, \tag{18.8}$$

and the system order $n = n_p + 1 = L_p/T + 1$ depends on the process lead time.

The control objective may be formulated as the stabilization of the on-hand stock (the first state variable) at level y_d. Since the goods perish at the rate $(1-\rho)$ while kept in the warehouse, in order to maintain the on-hand stock at the desired level once y_d is reached, it needs to be refilled from the incoming shipments equal to $(1-\rho)y_d$ in the steady state. Therefore, based on (18.8), all the state variables which represent the in-bound shipments $x_2, \cdots, x_n$ should be equal to $(1-\rho)y_d$ once $y(k) = y_d$, and the desired system state is defined as

$$\mathbf{x_d} = \begin{bmatrix} x_{d,1} \\ x_{d,2} \\ \vdots \\ x_{d,n-1} \\ x_{d,n} \end{bmatrix} = \begin{bmatrix} 1 \\ 1-\rho \\ \vdots \\ 1-\rho \\ 1-\rho \end{bmatrix} y_d. \tag{18.9}$$

A DSM control strategy that meets these design objectives will be developed in a latter part of the chapter. It will also be shown how to choose a suitable reference stock level such that a number of advantageous properties in the analyzed system are achieved.

Let $\mathbf{e}(\mathrm{k}) = \mathbf{x_d} - \mathbf{x}(\mathrm{k})$ denote the closed-loop system error. One may introduce a switching function

$$s(k) = \mathbf{c}^T \mathbf{e}(k) = \mathbf{c}^T \mathbf{x_d} - \mathbf{c}^T \mathbf{x}(k), \tag{18.10}$$

where $\mathbf{c}^T = [c_1\ c_2\ \ldots\ c_n]$, $\mathbf{c}^T\mathbf{b} \neq 0$, is the vector describing the sliding hyperplane $\mathbf{c}^T\mathbf{e}(k) = 0$. Substituting (18.6) into equation $\mathbf{c}^T\mathbf{e}(k+1) = 0$ and rearranging, one obtains the equivalent control $u_{eq}(k) = (\mathbf{c}^T\mathbf{b})^{-1}\mathbf{c}^T[\mathbf{x}_d - \mathbf{A}\mathbf{x}(k)]$ and the closed-loop state matrix

$$\mathbf{A_c} = \left[\mathbf{I}_n - \mathbf{b}\left(\mathbf{c}^T\mathbf{b}\right)^{-1}\mathbf{c}^T\right]\mathbf{A}, \tag{18.11}$$

where $\mathbf{I}_\mathrm{n} = \mathrm{diag}\{1, 1, \cdots, 1\}$. In order to gain a competitive advantage, fast reaction to varying market conditions is desired. Therefore, one may intend to find such parameters of the hyperplane which will allow for the error elimination in the smallest number of steps after a change in demand – a dead-beat response. A dead-beat scheme requires all the closed-loop poles to be placed at the origin of the error state space. The characteristic polynomial of $\mathbf{A}_c$ is determined as

$$\det(z\mathbf{I}_n - \mathbf{A_c}) = z^n + \frac{c_{n-1} - \rho c_n}{c_n} z^{n-1} + \ldots + \frac{c_1 - \rho c_2}{c_n} z. \tag{18.12}$$

For all the roots of (18.12) to be placed at the origin the determinant $\det(z\mathbf{I}_\mathrm{n} - \mathbf{A_c})$ should be equal to z^n, which is satisfied when simultaneously

$$\begin{aligned} c_{n-1} &= \rho c_n, \\ c_{n-2} &= \rho c_{n-1}, \\ &\vdots \\ c_2 &= \rho c_3, \\ c_1 &= \rho c_2. \end{aligned} \tag{18.13}$$

Sequentially solving this set of equations one obtains the following vector describing the parameters of the sliding plane

$$\mathbf{c}^T = \begin{bmatrix} \rho^{n-1} & \rho^{n-2} & \rho^{n-3} & \dots & \rho & 1 \end{bmatrix} c_n. \tag{18.14}$$

Substitution of (18.14) into (18.10), yields the switching function

$$s(k) = y_d - \rho^n x_1(k) - \sum_{j=2}^{n} \rho^{n-j+1} x_j(k). \tag{18.15}$$

Without loss of generality, c_n may be selected as unity. Since $n = n_p + 1$ and $x_1(k) = y(k)$, one may rewrite (18.15) using (18.7), as

$$s(k) = y_d - \rho^{n_p+1} y(k) - \sum_{j=k-n_p}^{k-1} \rho^{k-j} u(j). \tag{18.16}$$

The value of the obtained switching function reflects the discrepancy between the current on-hand stock and its reference level ($y_d - \rho^{n_p+1} y(k)$) less the amount of open orders augmented according to the expected goods shortage due to decay. The dead-time compensating features are synthesized in the last term in (18.16).

18.3.2 DSM Controller

The DSM control law is chosen as

$$u(k) = \frac{1}{2}Q + \frac{1}{2}Q sgn[s(k)], \tag{18.17}$$

where $Q > d_{\max}$ denotes the order quantity, and $y_d \geq 0$ is the target stock level. The $sgn(s)$ function in (18.17) is defined as

$$sgn(s) = \begin{cases} -1, & \text{if } s \leq 0, \\ 1, & \text{if } s > 0. \end{cases} \tag{18.18}$$

Therefore, the controller switches between 0 and Q according to the value of $s(k)$, and

$$u(k) = \begin{cases} Q, & \text{if } y_d - \rho^{n_p+1} y(k) - \sum_{j=k-n_p}^{k-1} \rho^{k-j} u(j) > 0, \\ 0, & \text{if } y_d - \rho^{n_p+1} y(k) - \sum_{j=k-n_p}^{k-1} \rho^{k-j} u(j) \leq 0. \end{cases} \tag{18.19}$$

18.3.3 Properties of the Proposed Control System

The properties of the proposed control system will be formulated as two theorems. The first theorem shows how to select the warehouse capacity in order to always accommodate the on-hand stock and the incoming shipments. In this way, the need for costly emergency storage is eliminated. The second proposition specifies the target stock level so that full demand satisfaction can be obtained.

Theorem 18.1. *If policy (18.19) is applied to regulate the flow of goods in inventory system (18.3), the on-hand stock is always upper bounded, i.e.*

$$\underset{k\geq 0}{\forall}\; y(k) \leq y_{\max}, \tag{18.20}$$

where

$$y_{\max} = \max\left\{\rho^{-n_p-1} y_d, \rho^{-n_p}(y_d + Q)\right\}. \tag{18.21}$$

Proof. It follows from the algorithm definition and the system initial conditions that the warehouse is empty for any $k \leq n_p$. Consequently, it is sufficient to show that the proposition holds for all $k > n_p$. Let us consider some integer $l > n_p$ and the value of $s(\cdot)$ in period l. Two cases ought to be analyzed: 1) the situation when $s(l) \geq 0$, and 2) the circumstances when $s(l) < 0$.

Case 1. In the situation when $s(l) \geq 0$, directly from the definition of the switching function, (18.16), one gets

$$s(l) = y_d - \rho^{n_p+1} y(l) - \sum_{j=l-n_p}^{l-1} \rho^{l-j} u(j) \geq 0, \tag{18.22}$$

which leads to

$$y(l) \leq y_d \rho^{-n_p-1} - \rho^{-n_p-1} \sum_{j=l-n_p}^{l-1} \rho^{l-j} u(j). \tag{18.23}$$

Since $u(\cdot)$ equals either 0 or $Q > 0$, one may conclude that $y(l) \leq y_d \rho^{-n_p-1}$. This ends the first part of the proof.

Case 2. In the second part of the proof the situation when $s(l) < 0$ is considered. First, one needs to find the last period $l_1 < l$ when s was nonnegative. According to (18.16), $s(0) = y_d \geq 0$, so period l_1 indeed exists. If $s(l_1) \geq 0$, then with analogy to (18.22) and (18.23), one gets

$$s(l_1) = y_d - \rho^{n_p+1} y(l_1) - \sum_{j=l_1-n_p}^{l_1-1} \rho^{l_1-j} u(j) \geq 0, \tag{18.24}$$

and

$$y(l_1) \leq y_d \rho^{-n_p-1} - \sum_{j=l_1-n_p}^{l_1-1} \rho^{l_1-n_p-1-j} u(j). \tag{18.25}$$

On the other hand, $y(l)$ can be expressed relative to $y(l_1)$ as

$$y(l) = \rho^{l-l_1} y(l_1) + \sum_{j=l_1}^{l-1} \rho^{l-1-j} u(j-n_p) - \sum_{j=l_1}^{l-1} \rho^{l-1-j} h(j) \tag{18.26}$$

which after the sum manipulation may be represented in the following form

$$\begin{aligned}y(l) &= \rho^{l-l_1}y(l_1) + \sum_{j=l_1-n_p}^{l-n_p-1} \rho^{l-n_p-1-j}u(j) - \sum_{j=l_1}^{l-1} \rho^{l-1-j}h(j) \\ &= \rho^{l-l_1}y(l_1) + \rho^{l-l_1} \sum_{j=l_1-n_p}^{l_1-1} \rho^{l_1-n_p-1-j}u(j) \\ &\quad + \sum_{j=l_1}^{l-n_p-1} \rho^{l-n_p-1-j}u(j) - \sum_{j=l_1}^{l-1} \rho^{l-1-j}h(j).\end{aligned} \tag{18.27}$$

Consequently, after applying (18.25) to (18.27), one gets

$$\begin{aligned}y(l) &\leq \rho^{l-l_1}\left[y_d\rho^{-n_p-1} - \rho^{-n_p-1} \sum_{j=l-n_p}^{l-1} \rho^{l-j}u(j)\right] \\ &\quad + \rho^{l-l_1} \sum_{j=l_1-n_p}^{l_1-1} \rho^{l_1-n_p-1-j}u(j) + \sum_{j=l_1}^{l-n_p-1} \rho^{l-n_p-1-j}u(j) \\ &\quad - \sum_{j=l_1}^{l-1} \rho^{l-1-j}h(j) \\ &= y_d\rho^{l-n_p-1-l_1} + \sum_{j=l_1}^{l-n_p-1} \rho^{l-n_p-1-j}u(j) - \sum_{j=l_1}^{l-1} \rho^{l-1-j}h(j).\end{aligned} \tag{18.28}$$

The controller set quantity Q for the last time before l in period l_1. Consequently, the sum

$$\sum_{j=l_1}^{l-n_p-1} \rho^{l-n_p-1-j}u(j) = \rho^{l-n_p-1-l_1}u(l_1) = \rho^{l-n_p-1-l_1}Q. \tag{18.29}$$

Since $l > l_1$ and $0 < \rho \leq 1$,

$$\rho^{l-n_p-1-l_1} \leq \rho^{1-n_p-1} = \rho^{-n_p}. \tag{18.30}$$

Hence, using (18.2), the following estimate of the stock level in period l is obtained

$$y(l) \leq \rho^{l-n_p-1-l_1}y_d + \rho^{l-n_p-1-l_1}Q \leq \rho^{-n_p}(y_d + Q) \leq y_{\max}. \tag{18.31}$$

This conclusion ends the proof. □

Theorem 18.1 specifies the upper limit of the on-hand stock level ever accumulated at the distribution center. A second proposition is formulated next, which indicates how the value of y_d should be selected to ensure that the stock level is positive. This results in a scenario when; after serving the imposed demand, the stock is still greater than zero, $y(k) > 0$, and full demand satisfaction is obtained. As a consequence, in addition to maximizing the profits from the realized sales, the company also gains in the market credibility.

Theorem 18.2. *If policy (18.19) is applied to regulate the flow of goods in inventory system (18.3) with $Q > d_{\max}$, and the target stock level satisfies the following inequality*

$$y_d > \max\left\{Q\sum_{j=1}^{n_p}\rho^j, d_{\max}\sum_{j=0}^{n_p}\rho^j\right\}, \tag{18.32}$$

then there exists an instant k_L such that the stock level is strictly positive for all $k \geq k_L$.

Proof. Let us consider some integer $l \geq k_L$ and the value of signal s at instant l. Two cases may be distinguished: 1) the situation when $s(l) \leq 0$, and 2) the circumstances when $s(l) > 0$.

Case 1. When $s(l) \leq 0$, then directly from the definition of s, given by (18.16), one gets

$$s(l) = y_d - \rho^{n_p+1}y(l) - \sum_{j=l-n_p}^{l-1}\rho^{l-j}u(j) \leq 0, \tag{18.33}$$

which leads to

$$y(l) \geq \rho^{-n_p-1}\left[y_d - \sum_{j=l-n_p}^{l-1}\rho^{l-j}u(j)\right]. \tag{18.34}$$

The maximum order quantity equals Q, which implies

$$y(l) \geq \rho^{-n_p-1}\left(y_d - Q\sum_{j=1}^{n_p}\rho^j\right). \tag{18.35}$$

Using assumption (18.32), one gets $y(l) > 0$, which concludes the first part of the proof.

Case 2. In the second part of the proof the case $s(l) > 0$ is considered. If the indicated condition is satisfied for all periods $k \in [0, l]$, then the order $u(k) = Q$ is continuously placed in the indicated interval. Then since $Q > d_{\max}$, one may conclude on the basis of (18.3) that the stock level will eventually become positive. Hence, in the complementary case, one needs to investigate the value of signal s at the last moment $l_1 < l$ when it was nonnegative. Consequently, following a similar reasoning as presented in (18.33)–(18.35), one arrives at

$$s(l_1) = y_d - \rho^{n_p+1}y(l_1) - \sum_{j=l_1-n_p}^{l_1-1}\rho^{l_1-j}u(j) \leq 0, \tag{18.36}$$

and

$$y(l_1) \geq \rho^{-n_p-1}\left[y_d - \sum_{j=l_1-n_p}^{l_1-1}\rho^{l_1-j}u(j)\right] > 0. \tag{18.37}$$

The stock level in period l can be expressed relative to the one in period l_1 as in (18.26). Using (18.37) in (18.26), and performing basic algebraic manipulations, one obtains

$$\begin{aligned}y(l) &\geq \rho^{l-l_1}\rho^{-n_p-1}\left[y_d - \sum_{j=l_1-n_p}^{l_1-1}\rho^{l_1-j}u(j)\right] + \rho^{l-l_1}\sum_{j=l_1-n_p}^{l_1-1}\rho^{l_1-n_p-1-j}u(j)\\ &\quad + \sum_{j=l_1}^{l-n_p-1}\rho^{l-n_p-1-j}u(j) - \sum_{j=l_1}^{l-1}\rho^{l-1-j}h(j)\\ &= \rho^{l-l_1-n_p-1}y_d + \rho^{l-l_1-n_p-1}u(l_1) + \sum_{j=l_1+1}^{l-n_p-1}\rho^{l-n_p-1-j}u(j)\\ &\quad - \sum_{j=l_1}^{l-1}\rho^{l-1-j}h(j). \end{aligned} \tag{18.38}$$

Recall that l_1 was the last period before l when the controller did not issue an order, i.e. $u(l_1) = 0$. Afterwards, the order quantity equals Q, and the first sum in (18.38)

$$\sum_{j=l_1+1}^{l-n_p-1}\rho^{l-n_p-1-j}u(j) = Q\sum_{j=0}^{l-l_1-2-n_p}\rho^j. \tag{18.39}$$

Since for any k, $h(k) \leq d_{\max}$, the second sum in the last line in (18.38), $\sum_{j=l_1}^{l-1}\rho^{l-1-j}h(j)$, is upper-bounded by

$$d_{\max}\sum_{j=0}^{l-l_1-1}\rho^j = d_{\max}\left(\sum_{j=0}^{l-l_1-2-n_p}\rho^j + \rho^{l-l_1-1-n_p}\sum_{j=0}^{n_p}\rho^j\right). \tag{18.40}$$

Consequently, since $Q > d_{\max}$,

$$\begin{aligned}y(l) &\geq \rho^{l-l_1-n_p-1}y_d + 0 + Q\sum_{j=0}^{l-l_1-2-n_p}\rho^j - d_{\max}\sum_{j=0}^{l-l_1-2-n_p}\rho^j\\ &\quad -\rho^{l-l_1-1-n_p}d_{\max}\sum_{j=0}^{n_p}\rho^j\\ &= \rho^{l-l_1-n_p-1}\left(y_d - d_{\max}\sum_{j=0}^{n_p}\rho^j\right) + (Q-d_{\max})\sum_{j=0}^{l-l_1-2-n_p}\rho^j. \end{aligned} \tag{18.41}$$

Finally, using the theorem assumptions, $y_d > d_{\max}\sum_{j=0}^{n_p}\rho^j$ and $Q > d_{\max}$, one may conclude $y(l) > 0$. This completes the proof of Theorem 18.2. □

18.4 Simulation Examples

Performance of the developed supply policy (18.19) is verified in a series of simulation tests. The system parameters are chosen in the following way: review period $T = 1$ day, order procurement delay $L_p = n_p T = 7$ days, inventory decay factor $\sigma = 0.12$, which implies $\rho = 1 - \sigma = 0.88$, the maximum daily demand at the distribution center $d_{\max} = 100$ items, and the order quantity $Q = 110$ items.

The target stock level is selected according to the guidelines of Theorem 18.2 such that full demand satisfaction is obtained. Consequently, $y_d = 535 > 533$ items is chosen. The DSM controller performance is compared with the operation of the classical ordering rule for the systems with fixed order quantity – (r, Q) – which says to order Q items whenever the inventory position falls below r (see e.g. [21] for the description of the classical inventory policies). For fair comparison, the resupply level is set as $r = 1067$ items so that both controllers impose similar holding costs.

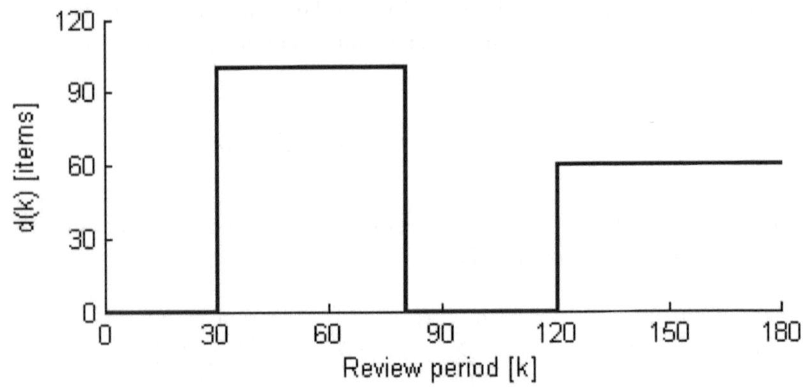

Fig. 18.3 Market demand – half-year trend

Test 1 :

In the first series of simulations, the controller performance is verified for different demand patterns. First, the response to pattern illustrated in Fig. 18.3, which reflects sudden changes in the market trend, is evaluated.

The ordering decisions taken by the DSM controller (a) and the (r, Q) policy (b) are illustrated in Fig. 18.4, and the on-hand stock level in Fig. 18.5. One can see from the graph in Fig. 18.5 that the stock level remains finite, and following the initial phase, it does not fall to zero. This implies that the imposed demand is entirely satisfied from the readily available resources, and the maximum service level is achieved. The (r, Q) policy exhibits overshoots which may lead to higher maximum stock (greater storage space is required), and occasionally may bring the stock level to zero (days 38 and 39) implying lost opportunities for selling the goods.

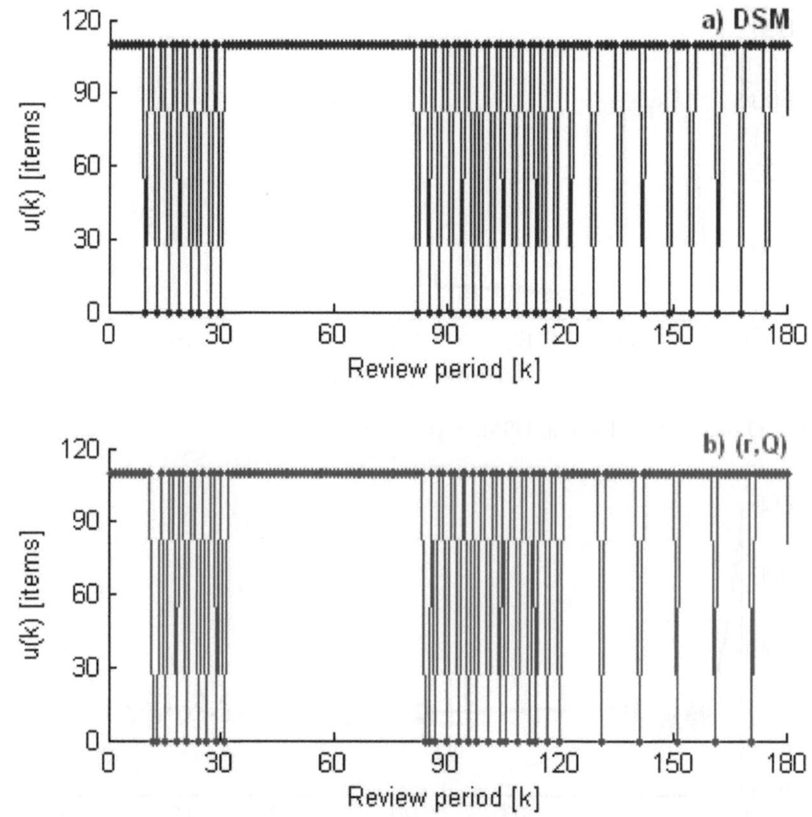

Fig. 18.4 Ordering decisions: a) DSM, b) (r, Q) policy

The evolution of the sliding variable is depicted in Fig. 18.6. The plot shows that s quickly decreases from its original value $s(0) = y_d$ to a relatively narrow band, and always remains in this band despite the presence of mismatched disturbance d, which provides a clear evidence of a properly established quasi-sliding motion in the discrete-time system.

Next, the controller performance is verified with respect to highly variable stochastic demand depicted in Fig. 18.7. The demand applied in the test follows the normal distribution with mean d_μ = 50 items and standard deviation d_δ = 30 items.

The ordering signal generated by the controllers is shown in Fig. 18.8, and the on-hand stock level in Fig. 18.9. One can notice from Fig. 18.8 that despite frequent demand variations, the ordering decisions are taken quite regularly in time. In the case of the DSM policy $u(k) = 0$ on average every 5.5 days (standard deviation 1.1 day), whereas for the (r, Q) policy $u(k) = 0$ was obtained on average every 7.2 days (standard deviation 2.8 days). The stock level remains finite and follows the trend imposed by the mean demand with small-amplitude oscillations around the trend

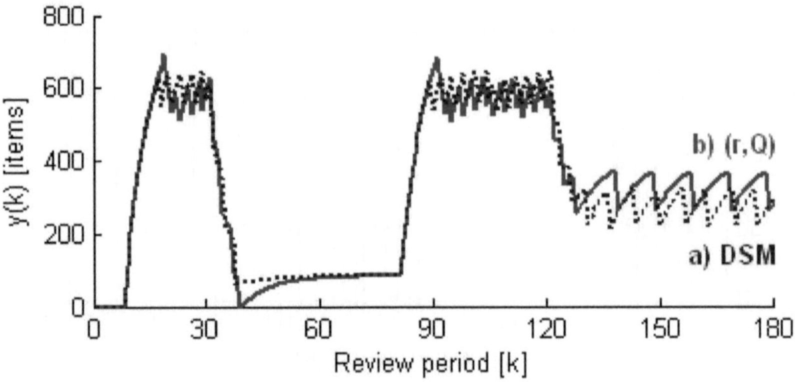

Fig. 18.5 On-hand stock level: a) DSM, b) (r, Q) policy

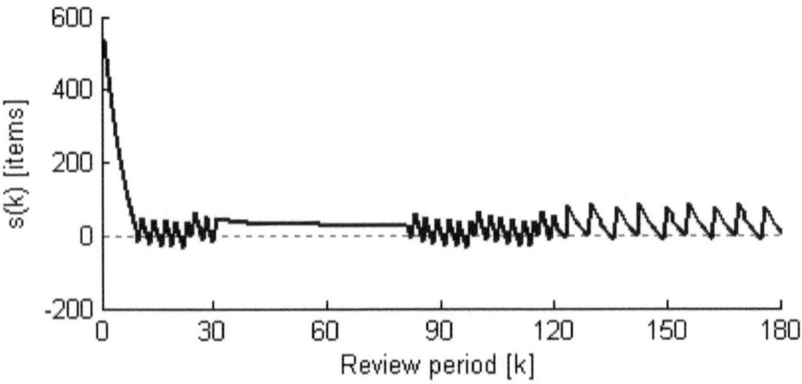

Fig. 18.6 Switching function

(with slightly increased range of variations in the case of the (r, Q) policy). Even though the demand occasionally exceeds the assumed limit of 100 items per period, the stock level remains positive implying maximum service level.

The switching function depicted in Fig. 18.10 shows that quasi-sliding motion is attained in finite time despite the presence of mismatched disturbance with high-frequency transitions.

Test 2:

In the second test, the focus is placed on the robustness issues. The controller performance is evaluated in the uncertain environment in which in addition to unpredictable demand changes (evolving as depicted in Fig. 18.3), also the decay rate and lead time exhibit unknown variations. The control action is based on the nominal parameter values specified in Test 1, whereas the true decay rate ρ fluctuates randomly

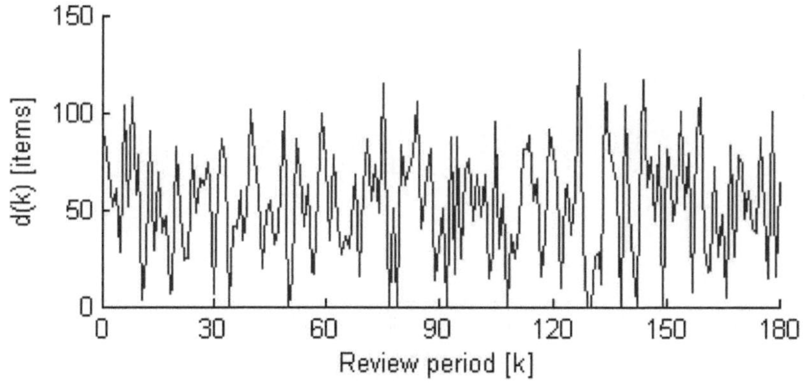

Fig. 18.7 Market demand – stochastic pattern with mean 50 items and standard deviation 30 items

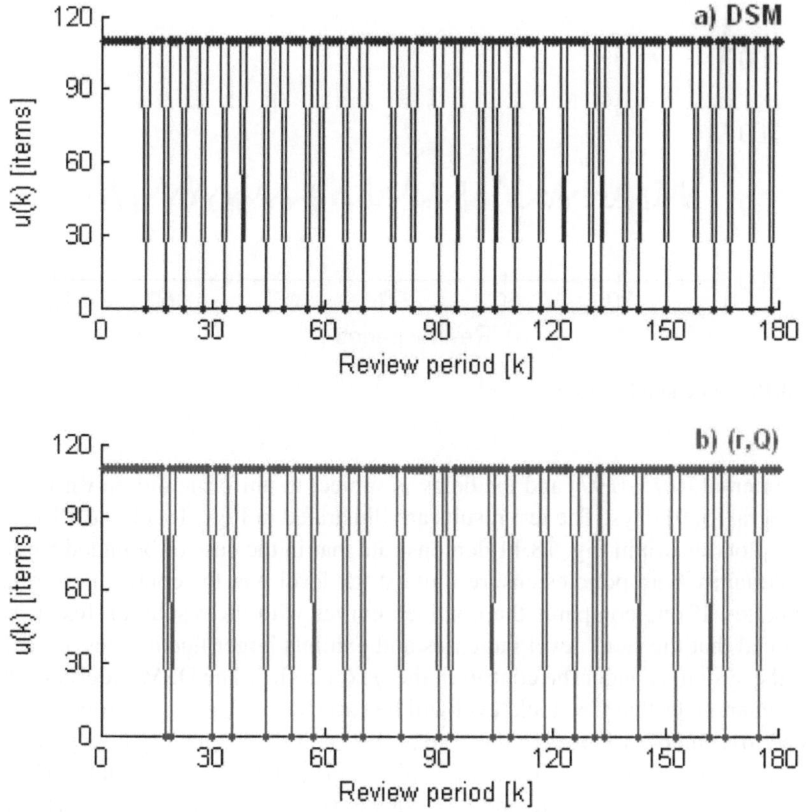

Fig. 18.8 Ordering decisions: a) DSM, b) (r, Q) policy

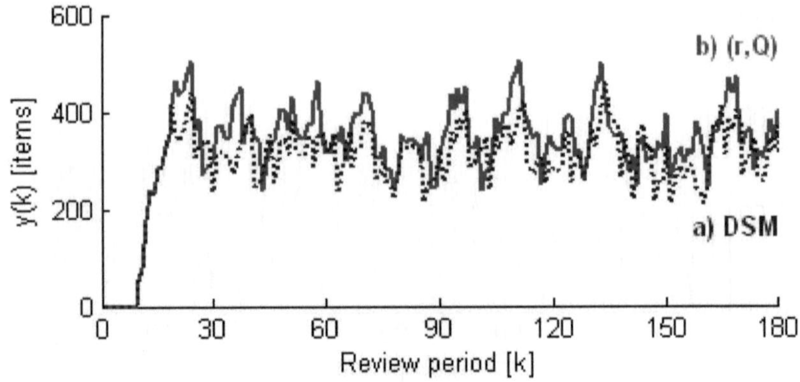

Fig. 18.9 On-hand stock level: a) DSM, b) (r, Q) policy

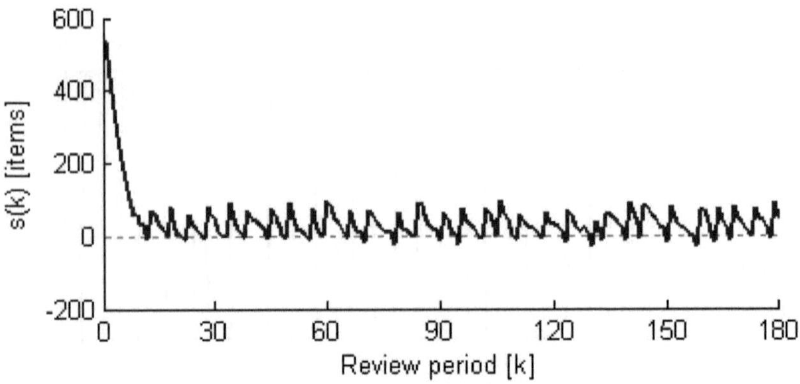

Fig. 18.10 Switching function

in the interval [0.75, 0.97] and the delay is subject to uniform random variations in the interval [5, 9] days. The test results are illustrated in Figs. 18.11 and 18.12.

The plots shown in Fig. 18.11 demonstrate that in the case of bounded parametric uncertainty both policies ensure finite stock level, yet the control performance deteriorates. If one compares the obtained curves with the results of Test 1 it can be noticed that the stock level increases and exhibits larger fluctuations, especially when the system is under the control of the (r, Q) policy. The DSM ordering rule no longer guarantees that the stock level will be strictly positive – $y(k)$ drops to zero in days 45, 61, and 67, whereas in the case of the traditional inventory policy the stock level becomes zero in days 45, 61, 62, 67, 72, 76, and 80. As a result, the (r, Q) policy imposes larger holding and lost sales costs than the DSM policy. However, both policies generate increased costs as compared to the nominal scenario.

It follows from Fig. 18.12 that the system representative point attains the sliding hyperplane $s(k) = 0$ in finite time, and remains in its proximity for all periods

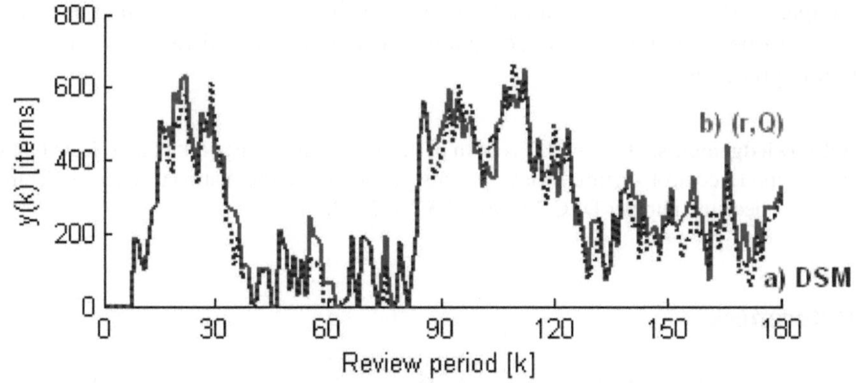

Fig. 18.11 On-hand stock level: a) DSM, b) (r, Q) policy

afterwards. However, as compared to the control of the nominal system examined in Test 1, the parametric uncertainties degrade the accuracy of the sliding-mode realization.

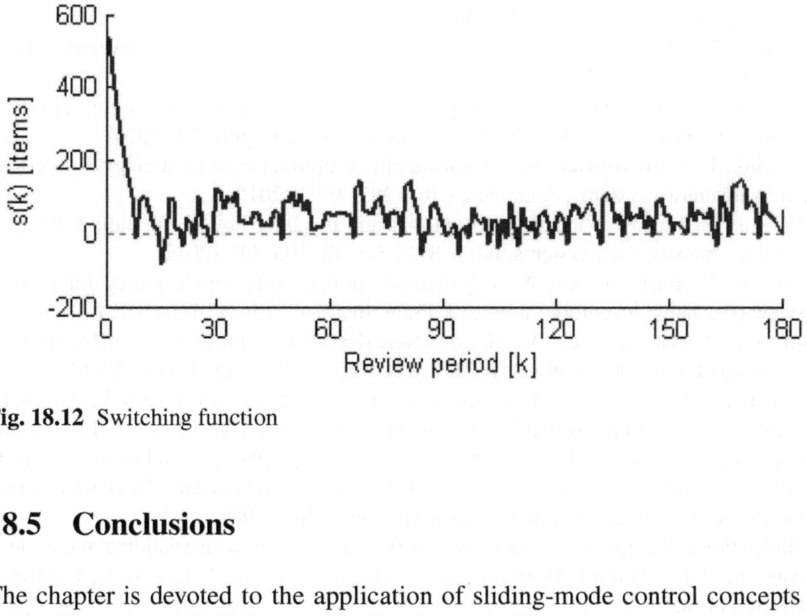

Fig. 18.12 Switching function

18.5 Conclusions

The chapter is devoted to the application of sliding-mode control concepts in the field of logistics. A DSM control policy for periodic-review inventory systems with fixed order quantity and uncertain demand was designed. The proposed policy guarantees fast system response and full demand satisfaction at the goods distribution center (maximum service level) irrespective of the pattern or statistics of demand, or the value of delay. The underlying algorithm employs only the fundamental arithmetic (additions and multiplication by a constant) and logic operations (a comparison) to decide the moment of taking the ordering decision. Thus, in addition to being intuitive, it is computationally efficient, and requires no complex tuning procedures.

Acknowledgements. This work has been performed in the framework of a project "Optimal sliding mode control of time delay systems" financed by the National Science Centre of Poland – decision number DEC 2011/01/B/ST7/02582.

References

1. Bandyopadhyay, B., Janardhanan, S.: Discrete-time sliding mode control: a multirate output feedback approach. LNCIS, vol. 323. Springer, Heidelberg (2006)
2. Bandyopadhyay, B., Fulwani, D., Kim, K.S.: Sliding mode control using novel sliding surfaces. LNCIS, vol. 392. Springer, Heidelberg (2010)
3. Bartoszewicz, A.: Discrete time quasi-sliding mode control strategies. IEEE Trans. Ind. Electron 45, 633–637 (1998)
4. Bartoszewicz, A., Nowacka-Leverton, A.: Time-varying sliding modes for second and third order systems. LNCIS, vol. 382. Springer, Heidelberg (2009)
5. Gao, W., Wang, Y., Homaifa, A.: Discrete-time variable structure control systems. IEEE Trans. Ind. Electron 42, 117–122 (1995)
6. Goyal, S.K., Giri, B.C.: Recent trends in modeling of deteriorating inventory. Eur. J. Oper. Res. 134, 1–16 (2001)
7. Ignaciuk, P., Bartoszewicz, A.: LQ optimal sliding mode supply policy for periodic review inventory systems. IEEE Trans. Autom. Control 55, 269–274 (2010)
8. Ignaciuk, P., Bartoszewicz, A.: Linear-quadratic optimal control strategy for periodic-review inventory systems. Automatica 46, 1982–1993 (2010)
9. Ignaciuk, P., Bartoszewicz, A.: LQ optimal and reaching law based sliding modes for inventory management systems. Int. J. Syst. Sci. 43, 105–116 (2012)
10. Ignaciuk, P., Bartoszewicz, A.: LQ optimal sliding-mode supply policy for periodic-review perishable inventory systems. J. Frank Inst. 349, 1561–1582 (2012)
11. Ignaciuk, P., Bartoszewicz, A.: Sliding mode dead-beat control of perishable inventory systems with multiple suppliers. IEEE Trans. Autom. Sci. Eng. 9, 418–423 (2012)
12. Ignaciuk, P., Bartoszewicz, A.: Linear-quadratic optimal control of periodic-review perishable inventory systems. IEEE Trans. Control Syst. Technol. 20, 1400–1407 (2012)
13. Karaesmen, I., Scheller-Wolf, A., Deniz, B.: Managing perishable and aging inventories: review and future research directions,". In: Kempf, K., Keskinocak, P., Uzsoy, R. (eds.) Handbook of Production Planning. Kluwer, Dordrecht (2008)
14. Milosavljević, Č.: General conditions for the existence of a quasisliding mode on the switching hyperplane in discrete variable structure systems. Autom. Rem. Control 46, 307–314 (1985)
15. Milosavljević, Č., Peruničić-Draženovic, B., Veselić, B., Mitić, D.: Sampled data quasi-sliding mode control strategies. In: Proc. IEEE Int. Conf. Ind. Technol., pp. 2640–2645 (2006)
16. Nahmias, S.: Perishable inventory theory: a review. Oper. Res. 30, 680–708 (1982)
17. Ortega, M., Lin, L.: Control theory applications to the production-inventory problem: a review. Int. J. Prod. Res. 42, 2303–2322 (2004)

18. Rafaat, F.: Survey of literature on continuously deteriorating inventory models. J. Oper. Res. Soc. 42, 27–37 (1991)
19. Sarimveis, H., Patrinos, P., Tarantilis, C.D., Kiranoudis, C.T.: Dynamic modeling and control of supply chain systems: a review. Computers Oper. Res. 35, 3530–3561 (2008)
20. Sethi, S.P., Zhang, Q.: Optimal control applications to management sciences. Special issue: Automatica 42, 1241–1428 (2006)
21. Silver, E.A., Pyke, D.F., Peterson, R.: Inventory management and production planning and scheduling. John Wiley, Chichester (1998)
22. Utkin, V.I.: Sliding modes in control and optimization. Springer, Berlin (1992)

Author Index

MIX
Papier aus verantwortungsvollen Quellen
Paper from responsible sources
FSC® C105338

If you have any concerns about our products,
you can contact us on
ProductSafety@springernature.com

In case Publisher is established outside the EU,
the EU authorized representative is:
Springer Nature Customer Service Center GmbH
Europaplatz 3, 69115 Heidelberg, Germany

Printed by Libri Plureos GmbH
in Hamburg, Germany